Advances in Intelligent Systems and Computing

Volume 874

Series editor

Janusz Kacprzyk, Polish Academy of Sciences, Warsaw, Poland
e-mail: kacprzyk@ibspan.waw.pl

The series "Advances in Intelligent Systems and Computing" contains publications on theory, applications, and design methods of Intelligent Systems and Intelligent Computing. Virtually all disciplines such as engineering, natural sciences, computer and information science, ICT, economics, business, e-commerce, environment, healthcare, life science are covered. The list of topics spans all the areas of modern intelligent systems and computing such as: computational intelligence, soft computing including neural networks, fuzzy systems, evolutionary computing and the fusion of these paradigms, social intelligence, ambient intelligence, computational neuroscience, artificial life, virtual worlds and society, cognitive science and systems, Perception and Vision, DNA and immune based systems, self-organizing and adaptive systems, e-Learning and teaching, human-centered and human-centric computing, recommender systems, intelligent control, robotics and mechatronics including human-machine teaming, knowledge-based paradigms, learning paradigms, machine ethics, intelligent data analysis, knowledge management, intelligent agents, intelligent decision making and support, intelligent network security, trust management, interactive entertainment, Web intelligence and multimedia.

The publications within "Advances in Intelligent Systems and Computing" are primarily proceedings of important conferences, symposia and congresses. They cover significant recent developments in the field, both of a foundational and applicable character. An important characteristic feature of the series is the short publication time and world-wide distribution. This permits a rapid and broad dissemination of research results.

More information about this series at http://www.springer.com/series/11156

Ajith Abraham · Sergey Kovalev
Valery Tarassov · Vaclav Snasel
Andrey Sukhanov
Editors

Proceedings of the Third International Scientific Conference "Intelligent Information Technologies for Industry" (IITI'18)

Volume 1

Editors
Ajith Abraham
Scientific Network for Innovation
and Research Excellence
Machine Intelligence Research Labs
(MIR Labs)
Auburn, WA, USA

Sergey Kovalev
Rostov State Transport University
Rostov-on-Don, Russia

Valery Tarassov
Bauman Moscow State Technical University
Moscow, Russia

Vaclav Snasel
VSB-Technical University of Ostrava
Ostrava, Czech Republic

Andrey Sukhanov
Rostov State Transport University
Rostov-on-Don, Russia

ISSN 2194-5357 ISSN 2194-5365 (electronic)
Advances in Intelligent Systems and Computing
ISBN 978-3-030-01817-7 ISBN 978-3-030-01818-4 (eBook)
https://doi.org/10.1007/978-3-030-01818-4

Library of Congress Control Number: 2018958808

This Springer imprint is published by the registered company Springer Nature Switzerland AG
The registered company address is: Gewerbestrasse 11, 6330 Cham, Switzerland

Preface

This volume of Advances in Intelligent Systems and Computing contains papers presented in the main track of IITI 2018, the Third International Scientific Conference on Intelligent Information Technologies for Industry held in September 17–21 in Sochi, Russia. The conference was jointly co-organized by Rostov State Transport University (Russia) and VŠB-Technical University of Ostrava (Czech Republic) with the participation of Russian Association for Artificial Intelligence (RAAI).

IITI 2018 is devoted to practical models and industrial applications related to intelligent information systems. It is considered as a meeting point for researchers and practitioners to enable the implementation of advanced information technologies into various industries. Nevertheless, some theoretical talks concerning the state of the art in intelligent systems and soft computing were also included into proceedings.

There were 160 paper submissions from 11 countries. Each submission was reviewed by at least three chairs or PC members. We accepted 94 regular papers (58%). Unfortunately, due to limitations of conference topics and edited volumes, the program committee was forced to reject some interesting papers, which did not satisfy these topics or publisher requirements. We would like to thank all authors and reviewers for their work and valuable contributions. The friendly and welcoming attitude of conference supporters and contributors made this event a success!

The conference was supported by Russian Fund for Basic Research (grant no. 18-07-20024 G).

September 2018

Ajith Abraham
Sergey M. Kovalev
Valery B. Tarassov
Václav Snášel
Andrey V. Sukhanov

Organization

Organizing Institutes

Rostov State Transport University, Russia
VŠB-Technical University of Ostrava, Czech Republic
Russian Association for Artificial Intelligence, Russia

Conference Chairs

Sergey M. Kovalev	Rostov State Transport University, Russia
Alexander N. Guda	Rostov State Transport University, Russia

Conference Vice-chair

Valery B. Tarassov	Bauman Moscow State Technical University, Russia

International Program Committee

Alexander I. Dolgiy	JSC "NIIAS", Rostov branch, Russia
Alexander L. Tulupyev	St. Petersburg Institute for Informatics and Automation of the Russian Academy of Sciences, Russia
Alexander N. Shabelnikov	JSC "NIIAS", Russia
Alexander N. Tselykh	Southern Federal University, Russia
Alexander P. Eremeev	Moscow Power Engineering Institute, Russia

Milan Dado	University of Žilina, Slovakia
Mohamed Mostafa	Arab Academy for Science, Technology, and Maritime Transport, Egypt
Nadezhda G. Yarushkina	Ulyanovsk State Technical University, Russia
Nashwa El-Bendary	SRGE (Scientific Research Group in Egypt), Egypt
Nour Oweis	VSB-Technical University of Ostrava, Czech Republic
Oleg P. Kuznetsov	Institute of Control Sciences of Russian Academy of Sciences
Pavol Špánik	University of Žilina, Slovakia
Petr I. Sosnin	Ulyanovsk state technical university, Russia
Petr Saloun	VSB-Technical University of Ostrava, Czech Republic
Santosh Nanda	Eastern Academy of Science and Technology, Bhubaneswar, Odisha, India
Sergey D. Makhortov	Voronezh state university, Russia
Stanislav Kocman	VŠB-Technical University of Ostrava, Czech Republic
Stanislav Rusek	VŠB-Technical University of Ostrava, Czech Republic
Svatopluk Stolfa	VSB-Technical University of Ostrava, Czech Republic
Tarek Gaber	VSB-Technical University of Ostrava, Czech Republic
Teresa Orłowska-Kowalska	Wrocław University of Technology, Poland
Vadim L. Stefanuk	Institute for Information Transmission Problems, Russia
Vadim N. Vagin	Moscow Power Engineering Institute, Russia
Vladimir V. Golenkov	Belarus State University of Informatics and Radioelectronics, Belarus
Vladimír Vašinek	VŠB-Technical University of Ostrava, Czech Republic
Yuri I. Rogozov	Southern Federal University, Russia
Zdeněk Peroutka	University of West Bohemia, Czech Republic

Organizing Committee Chair

Alexander N. Guda	Rostov State Transport University, Russia

Organizing Vice-chair

Andrey V. Sukhanov	Rostov State Transport University, Russia

Local Organizing Committee

Andrey V. Chernov	Rostov State Transport University, Russia
Anna E. Kolodenkova	Samara State Technical University, Russia
Ivan A. Yaitskov	Rostov State Transport University, Russia
Jan Platoš	VSB-Technical University of Ostrava, Czech Republic
Maria A. Butakova	Rostov State Transport University, Russia
Maya V. Sukhanova	Azov-Black Sea State Engineering Institute, Russia
Pavel Krömer	VSB-Technical University of Ostrava, Czech Republic
Vitezslav Styskala	VSB-Technical University of Ostrava, Czech Republic
Vladislav S. Kovalev	JSC "NIIAS", Russia

Contents

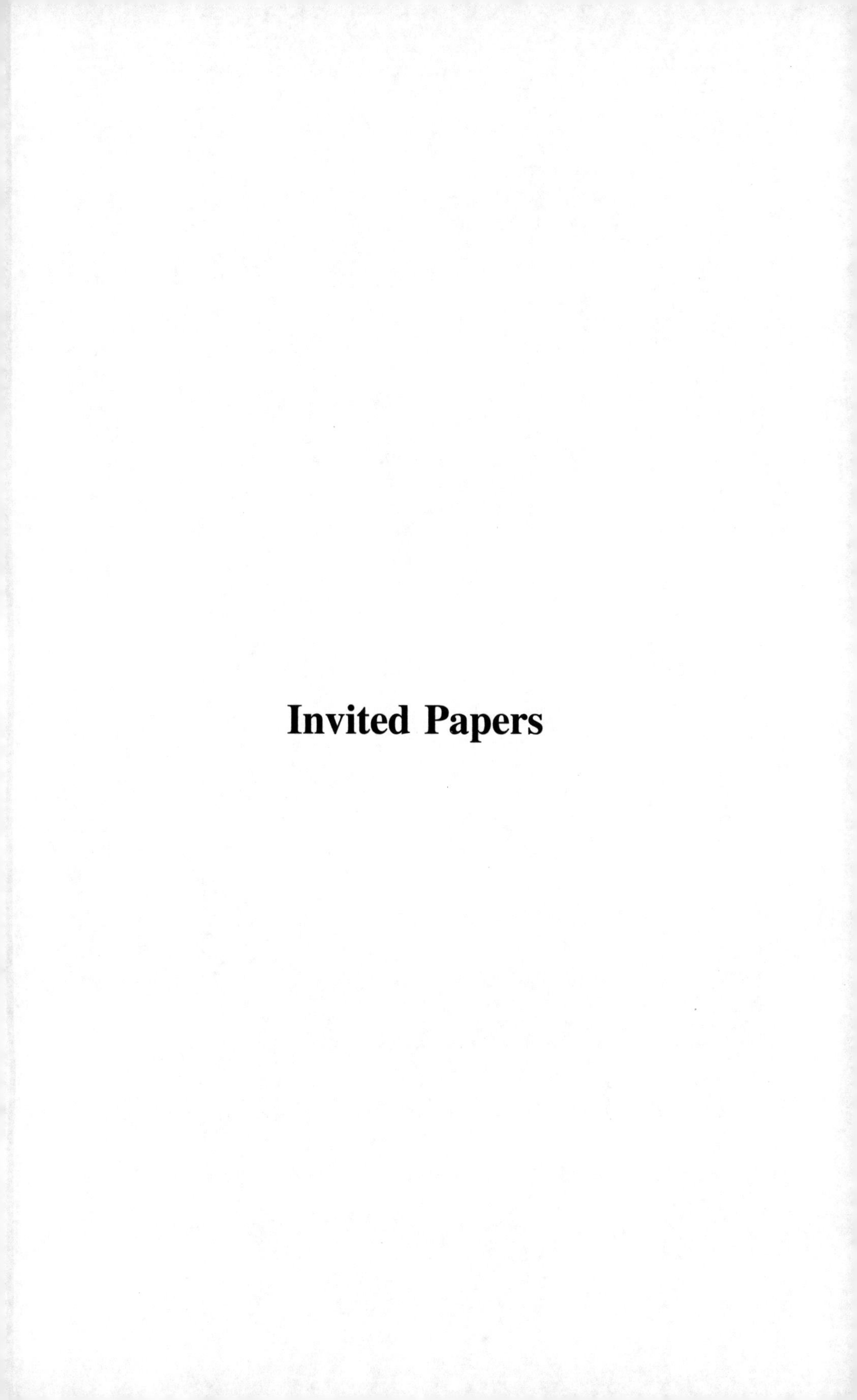

Invited Papers

Connected Vehicle Prognostics Framework for Dynamic Systems

Omar Makke and Oleg Gusikhin(✉)

Research and Advanced Engineering, Ford Motor Company,
20300 Rotunda Drive, Dearborn, MI 48124, USA
{omakke, ogusikhi}@ford.com

Abstract. Connected vehicle analytics has a promise to substantially advance vehicle prognostics and health management. However, the practical implementation of connected vehicle prognostics faces a number of challenges, such as the limitation of communication bandwidth resulting in potential loss of data that is critical for adequate prognostics models. The paper discusses a modelling framework for connected vehicle prognostics for dynamic systems that allows addressing connectivity limitations and memory constraints. The framework is based on a hybrid prognostics approach combining in-vehicle physics-based data aggregation model and cloud-based data-driven prognostics leveraging cross-vehicle and external data sources. The application of the framework is illustrated by models for brake pads wear and cabin air filter prognostics.

Keywords: Prognostics · Connected vehicles · Machine learning
Big data analytics

1 Introduction

Vehicle prognostics and health management has always been a topic of great interest for the automotive industry. It is one of the early areas of intelligent systems applications targeting on-board and service bay diagnostics and prognostics [1]. In recent years, due to increased complexity of the vehicle systems and the emergence of connected and autonomous vehicles, the interest in prognostics and health management has soared.

Connected vehicle technology brings game changing opportunities to advance vehicle prognostics and health management [2]. Connected vehicles can take advantage of cloud resources to build complex models, and they can leverage data from multiple vehicles and other external sources to infer complex relationships and offer enhanced user experience. Even existing on-board monitoring and diagnostics systems can be enhanced using connected vehicle technology. For example, modern vehicles are equipped with the oil-life monitoring system that tracks oil wear by analyzing duty-cycle and typically provide advanced notice to the driver when the oil life is at 5%. However, for scheduling purposes, it is more convenient to estimate the "time to oil change" rather than the oil wear percentage. By using connected vehicle data, it is possible to project the time to oil wear, assuming consistent duty cycle. Then, a notification containing projected time to service can be provided to the customer.

A. Abraham et al. (Eds.): IITI 2018, AISC 874, pp. 3–15, 2019.
https://doi.org/10.1007/978-3-030-01818-4_1

Extending the service need prediction to other components, such as brake pads, tires, batteries, air filter, etc. will provide a comprehensive service-scheduling package, which offers a new level of user experience and service efficiency. This capability can insure that the dealer has the right replacement parts available. It can also improve the efficiency of service parts inventory. For fleet owners, it facilitates better scheduling to minimize the impact of vehicle downtime on the fleet operations.

Connected vehicle analytics has a promise to develop such models utilizing exiting vehicle data to project the time to service without adding costly sensors or other hardware. However, practical implementation of connected vehicle prognostics faces a number of challenges such as on-board sensor data, memory availability, and limitations of the wireless communication bandwidth [3]. These limitations result in potential data loss of data which may be crucial for adequate prognostics of the given component. To overcome these challenges, the design in [4] proposed the model for connected vehicle brake prognostics, where an on-board system implements a duty-cycle aggregation model to alleviate limitations of communication channel while a cloud based model uses cross-vehicle data in lieu of the lack of direct measurement of brake wear. In this paper we generalize and formalize the approach presented in [4] by providing a theoretical framework to build such prognostics models for the components which can be modelled as dynamic systems.

This paper is organized as follows. Section 2 provides an overview of the proposed framework. Section 3 illustrates the framework using brake pad prognostics as an example. Section 4 presents the detailed mathematical foundation behind the framework. Section 5 demonstrates how the same framework can be applied to a different domain area, such as the cabin air filter prognostics. Section 6 concludes the paper and discusses future work.

2 Connected Vehicle Prognostics Framework

This section describes a prognostics framework for a class of problems commonly found in automotive vehicles. For the discussion of this paper, note that typically prognostics are classified into 3 categories [5] summarized in Fig. 1.

Fig. 1. Three different prognostics models

The first is physics based prognostics. In this approach, elaborate physics models are needed with details about many of the components material properties, such as coefficient of friction, mass, heat capacity, and other material properties. The second approach is a data-driven approach which is based solely on data without any consideration to the underlying physics. The goal here is either to use machine learning or other statistical tools in order to identify patterns in the data and make projections on when to service a component. In our framework, we will utilize a hybrid approach where an in-vehicle physics-based model generates data in a meaningful way, and this data is sent periodically to the cloud to make predictions using data-driven approaches, which feeds back a wear result to the vehicle. These in-vehicle models should be designed to be embeddable (in particular, through over the air updates) into vehicle control modules, such as modem, engine controller, brake controller, etc.

The problem can be formulated as follows. Suppose that the dynamic behavior of a component can be described with a physics-based dynamic model, requiring a vector $\mathbf{u}$ of inputs which affects a state vector $\mathbf{x}$. The wear of the component is assumed to be a function of the inputs and the states. The goal is to find a prognostics algorithm which outputs "expected time to service" by calculating the "wear" of the component, and based on the past history, predicts a time to "service component by".

To minimize the reliance on multiple parameters related to material properties and environment, machine learning is used. The algorithm must run in the vehicle within deterministic time period, must have resistance to connectivity issues, must require fixed size of storage space in the vehicle, and must consume a deterministic amount of data when transferring the data from the vehicle to the cloud. The framework which implements such algorithms with these requirements is shown in Fig. 2.

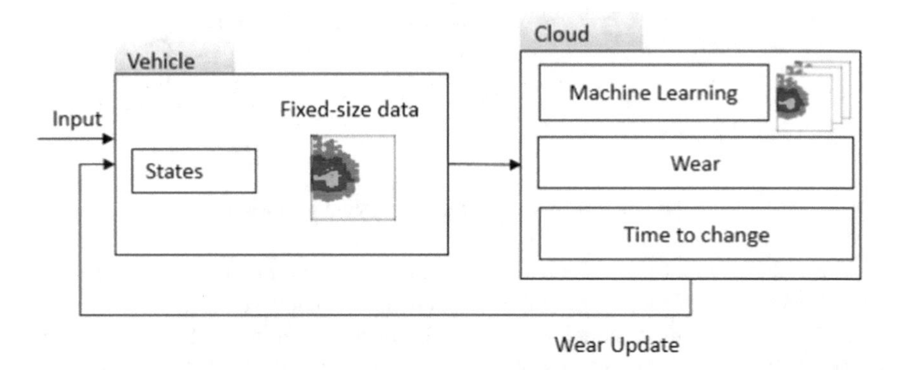

Fig. 2. Connected vehicle prognostics framework

Using this approach, a fleet manager with a known number of vehicles can purchase a data plan adequate enough to support the data transfer, or can ensure that the fleet vehicles are connected to the cloud at least once every specified period of time, such as day, a week, or similar. The vehicle generates data based on duty cycle, and the cloud utilizes this data in machine learning. As will be shown in the following sections, the method of generating data is an important factor in combining the physics approach with machine learning.

3 Brake Pad Prognostics

The main objective of the brake pad prognostics system is to give customers a close approximation of the date when they need to change their brake pads. The algorithm described here will serve as an example to help in understanding the general framework for prognostics explained in the next section. The classical approach to this problem is to create an elaborate physics model which relies heavily on material properties of the pads and execute it in the vehicle [6]. One simple form of prognostics is a distance-based prediction. A driver is asked to inspect the brake pads every 10k to 15k miles or similar, found in the vehicle's manual. Therefore, it is expected that using a physics based model, yet a simplified model compared to an elaborate 3D finite element model as shown in [7], a more accurate prediction can be made. The goal is to predict the wear so that at the next oil change pads can be replaced.

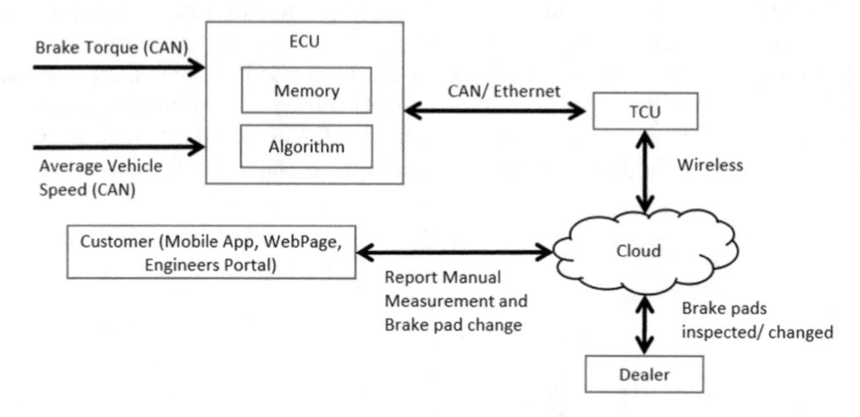

Fig. 3. Brake pad prognostics system architecture

The architecture of the brake pad prognostics system is shown in Fig. 3. In general, the characteristics of the wear on the pad depend on many variables. For the aim of prognostics, a simplified model will be used, keeping in mind that machine learning will be applied later to improve the model's accuracy. In this simplified lumped model, the wear on the brake pad is a function of the instantaneous temperature of the pad and the instantaneous energy extracted (the power) from the vehicle's kinetic energy as heat [6]. In a lab environment, one may create a setup with the average temperature of the pad being controlled so that it only fluctuates in a range ΔT (such as 50 °C fluctuations), and the pad pressure is applied to a rotating disk for measurable durations, and a table similar to Table 1 is generated. The pad mass is modelled by

$$\frac{dT}{dt} = \frac{P}{mC} - r(T - T_{\infty}) \tag{1}$$

$$\frac{dm}{dt} = -w(T, P) \tag{2}$$

Here, $w(T, P)$ is the wear from Table 1 and $T_\infty = T_{amb}$. Initial conditions are $T = T_{amb}$, $m = m_0$, and mC is the mass multiplied by specific heat capacity of the brake pad, and is assumed to be constant for a short duration of time, measured in days. It is assumed that the processor solving Eq. (1) does not have access to $w(T, P)$, and therefore it cannot calculate the change in the mass instantaneously. The temperature T in Eq. (1) is initialized to T_{amb} at ignition. This model's accuracy will deteriorate when almost all the mass of the pads is worn out, but at that time, the algorithm will already have predicted that it is time to change the brake pads. Although this approximation has errors, due to the fact that big data and machine learning will be used as will be shown later, the error is expected to be compensated for, since it is not a random error.

Table 1. Showing $w(T, P)$. Here, $a_{i,j}$ (g/s) is the average grams lost given a power at a average temperature for 1 s.

w		Temperature (Celsius)			
Power (Watts)		[-40 0]	[0 40]	[40 80]	...
	[0 100]	$a_{0,0}$	$a_{0,1}$	$a_{0,2}$	...
	[100 200]	$a_{1,0}$	$a_{1,1}$	$a_{1,2}$	...
	[200 300]	$a_{2,0}$	$a_{2,1}$	$a_{2,2}$	...
	...	...	...	...	...

The power P is derived from vehicle speed and the friction brake torque for each wheel after considering the loading effect ratios and the heat partitioning factor. Based on the calculated temperature and power at a given time t, for given (T, P), Table 1, provides the wear factor $w(T,P)$, so that the instantaneous wear is:

$$dW = w(T, P)dt \tag{3}$$

$$W = \int_0^{t_f} w(T, P)dt \tag{4}$$

Here, W is the total wear, in grams, of the brake pad until a final time t_f. Both Eqs. (3) and (4) which is dependent on material properties will be solved in the cloud. The wear in this model is a function of two variables, the state T and the input P. To solve the model, create a 2D mesh in the vehicle's memory with T and P as axes, similar to Table 1. Set a minimum and maximum for each axis:

$$-40\,^{\circ}\mathrm{C} \le T \le 400\,^{\circ}\mathrm{C}\,\mathrm{and}\ 0\mathrm{W} \le P \le 10000\mathrm{W}s \tag{5}$$

Partition each axis into non-intersecting intervals. Smaller partitions give higher accuracy but require more storage. Each axis may have different number of partitions.

This will create a 2D mesh with cells in it. Let set E_M be the set of all points M which lie in the same cell. Suppose that the algorithm runs at a rate τ ms. (e.g. 5 ms). Then, at every τ ms, calculate using Eq. (1) both $P(t)$, and $T(t)$ where t is the current time. Locate the set E_M in the 2D mesh such that E_M contains point M with coordinates ($P(t)$, $T(t)$). Then increment:

$$Value(E_M) = Value(E_M) + P(t)\tau \tag{6}$$

Then, it is evident that the set E_M is a set of points approximating infinitesimal energy values which can occur at separate times. The sum of the values of all the sets E_M for all points M in the 2D mesh is equivalent to simply integrating $P(t)$ over time. Figure 4 shows a hypothetical power curve as a function of time, and the values of the pad's temperature which fall within the same range are collected together. Since the set E_M at (T, P) is the accumulation of P(t)τ in a temperature and power range, E_M is the measure of how much energy the brake pad has dissipated at (P, T). This is also shown in Fig. 4. Since the integration formula of wear is approximated by summation, which is commutative, the final wear of Eq. (4) can be calculated by:

$$W = \sum\nolimits_{Cells} w(T, P) E_{Cell} \tag{7}$$

In order to utilize big data and machine learning, the data in the 2D mesh is sent to the cloud. The data has a known size in terms of kilobytes, and hence the requirements can be known before implementing it in the vehicle. The final architecture of the dynamic system is shown in Fig. 5.

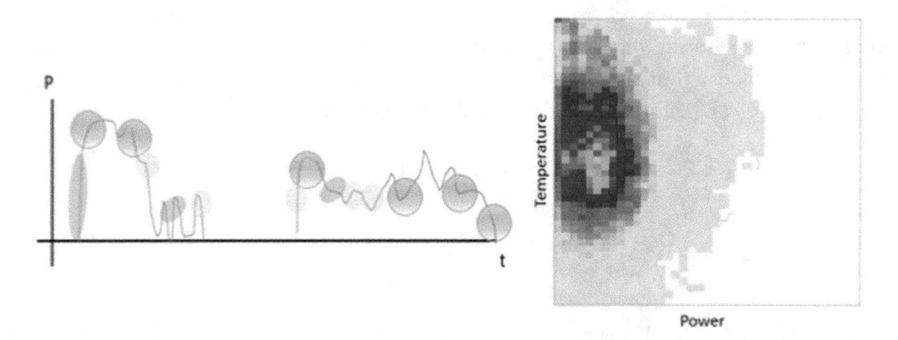

Fig. 4. Clustering a hypothetical curve based on both T and P. The intensity of each cell in the 2D mesh represents the sum of $P(t)\tau$ at different values of t, at coordinate (P,T).

The cloud can use big data and machine learning to derive the actual coefficients w (T,P) if field measurements can be provided. To achieve this, experimental vehicles can be equipped with sensors which generate data for the cloud to learn the wear factors for each cell in Fig. 4. Alternatively, measurements can be taken during a service visit at the dealer. Then Table 1 is constructed using machine learning and is made available in the cloud.

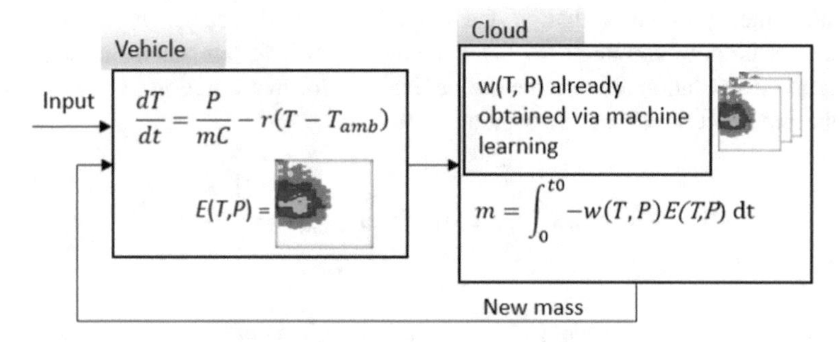

Fig. 5. Brake prognostics algorithm is separated between vehicle and cloud. Here, m is assumed constant for short durations in the vehicle.

An alternative to the aggregation method used in Eq. (6) is to only increment an integer counter in every cell, instead of aggregating a quantity. Section 4 will elaborate more about the mathematics behind this. The process is as follows. Pick any cell E_M. If the data in E_M is the accumulated energy $P(t)\tau$ at given (P, T), it is possible to divide that total energy by P to obtain the cumulative time the brake pad was used at the given temperature and power. By multiplying the cumulative time in each cell by its temperature or power coordinate, one can obtain a diagram similar to that in Fig. 4. Engineers can study temperature distributions of the usage of the pad, power distributions, or simply the time pads are used at a given temperature and power. Engineers can then make better assumptions about how the pads are used in order to improve their designs. In this approach, instead of the parameter "time" being the axis and power and temperature are the calculated values, temperature and power become the axes and time becomes the aggregated value. This can be viewed as a method to compress data generated over time, and $E(T, P)$ becomes a simple counter as will be seen in the next section.

4 Non-linear Dynamic Systems Prognostics

This section addresses prognostics of components which can be represented using non-linear models. It is desired to have a universal approach for these components. In this approach, we would like to reduce the number of experiments conducted in the lab to characterize the wear of the components, and replace them by using connected vehicles and machine learning. Without loss of generality, the initial state of the system is assumed to be known for a new part, and is 0. Suppose that the state space model for a system is

$$\dot{\mathbf{x}}^* = f(\mathbf{x}^*(t), \mathbf{u}(t)) \tag{8}$$

$$\dot{\mathbf{W}} = \mathbf{y}(t) = g(\mathbf{x}^*(t), \mathbf{u}(t)) \tag{9}$$

Under the assumption that: if the degradation of the states which are used in prognostics is slow compared to the dynamics of the system, then the states representing the degradation are assumed to be constant for that duration (a day or a week), then the input vector **x*** can be separated into:

$$\mathbf{x}^* = [x_0, x_1, \ldots, x_n | \widehat{x}_0, \widehat{x}_1, \ldots, \widehat{x}_m]^T = \begin{bmatrix} \mathbf{x} \\ - \\ \widehat{\mathbf{x}} \end{bmatrix} \tag{10}$$

$$where \left| \frac{\partial \widehat{x}_k}{\partial t} \right| \ll 1 \, for \, 0 \le k \le m \tag{11}$$

And then all $\widehat{x}_k$ can be treated as constants for short periods of time.

$\dot{\mathbf{W}}$ in Eq. (9) represents the instantaneous wear, and hence **W** is the total wear. The goal is to calculate **W** on a different processor (in the cloud) than that which calculates **x***.

W is a vector of total wear of several subcomponents of the component under monitor, and intuitively, we can select the worst value to be an answer. The correct answer will depend on the nature of the problem and its impact on customers.

The following qualitative conclusion can then be made: The state Eq. (8) captures how the component is being used, and the output Eq. (9) calculates a measure of interest, based on the usage. The vehicle calculates and aggregates the behavioral model, while the cloud calculates any **y** of interest by using a corresponding g(**x***,**u**), which itself can be learnt using big data and a true measurement of **y** using either a sensor or manual measurement of selected few components. Algorithm 1 is used to aggregate the data in the vehicle.

Algorithm 1 In-Vehicle Data Aggregation

1: Create hyperrectangle of dimension $n + r$ where n and r are the dimensions of the state and input vectors respectively
2: Identify the minimum and maximum of each axis in the hyperrectangle
4: Partition the hyperrectangle into independent partitions
5 At Key OFF -> ON: Set initial conditions and load values from memory
8: **while** KEY ON
9: Solve equation (8) numerically
10: Locate partition E in hyperrectangle containing point M identified by the values $(x_1, x_2, \ldots, x_n, u_1, u_2, \ldots, u_r)$. If any of the coordinates exceed the boundary, assume their value is the boundary
11: Increment the value in partition E identified in step 10 by 1. $E_M = E_M + 1$
12: **end while**

The equations are discretized and the errors are assumed to be negligible since a customer has few weeks to change the brake pads, oil, air filter, and not few minutes. In order to calculate **W** from Eq. 9, instead of integrating g(**x***,**u**) over time, by summing the values in time, sample by sample, we partitioned the range of g(**x***,**u**) into cells where each cell represents accumulation of values of $\dot{\mathbf{W}}$ at same pair (**x***, **u**), and summed the values in the partition (the is commutative). This is described in details as follows.

By integrating Eq. (9), **W** can be obtained by summing over all samples N as

$$\mathbf{W} = \sum_{i}^{N} g(\mathbf{x}(t_i), \mathbf{u}(t_i))\tau \tag{12}$$

Without loss of generality, assume **W** and g to be one dimensional. Since $g(\mathbf{x}(t), \mathbf{u}(t))$ is memoryless, then for the same inputs, the same output is obtained. Define c_i to be the value of g at time t_i. The value c_i represents the wear value at the conditions of (**x**, **u**). It is possible that at different values of t, (**x**, **u**) is the same, and hence g(**x, u**) is the same.

$$c_i = g(\mathbf{x}(t_i), \mathbf{u}(t_i)) \tag{13}$$

Since the summation is commutative, collect the values of g(**x**,**u**) at different times together for which the pair (**x**,**u**) are similar and call this set S_i. Therefore, S_i is a collection of values c_i. Define g_{Si} to be the sum of the values in set S_i.

$$g_{Si} = n_i c_i \tag{14}$$

Then, W can be written as

$$\mathrm{W} = \sum_{i} g_{Si}\tau = \sum_{i} c_i n_i \tau \tag{15}$$

But n_i has already been calculated in the vehicle, in step 11 of the algorithm, as E_M and therefore, the wear **W** is found to be

$$\mathrm{W} = \sum_{i} c_i E_i \tau \tag{16}$$

Note that the slow changing states, which are used for prognostics, are assumed to be constants over short durations. But, whenever the cloud updates the wear prediction, it calculates new values for $\widehat{x_k}$ and sends them to the vehicle. Hence the model approximation remains dynamic, running at two different rates, yielding a better approximation than assuming that these states are always constant. The wear calculation then feeds into a time prediction algorithm which completes the prognostics for the component. The time prediction algorithm can vary in complexity as needed, and this not an issue because it runs in the cloud where resources are not as constraint as the vehicle. Figure 6 shows the machine learning system in the cloud.

The last remaining part is to derive the values of c_i in the cloud. By either providing manual wear measurements W, or inserting sensors in a subset of the vehicles to measure this wear, we can use machine learning and optimization to derive the outputs c_i. Moreover, Eq. (16) has the form of a linear regression which also resembles a simple neural network where the activation function is just a pass-through node and in this case, the weights of the network are the coefficient c_i. and the inputs are $E_i\tau$.

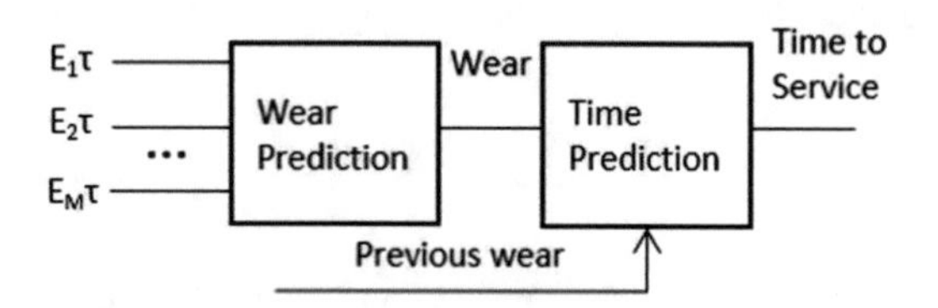

Fig. 6. Machine learning in cloud.

However, if **W** is a non-autonomous dynamic system and not a function of only **x** and **u**, as previously assumed, recurrent neural networks can be used to approximate the wear [8]. Care should be taken in this case to take enough data points throughout the wear of the component to have enough data to model the dynamics of the system. Figure 7 shows the final architecture of the system.

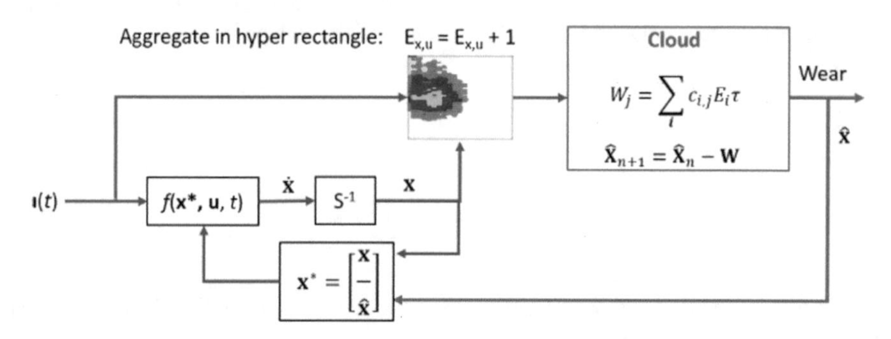

Fig. 7. The proposed prognostics system. The algorithms are executed on two compute entities, in the vehicle and in the cloud.

If needed, extra inputs can be added to the system in Fig. 6, such as geographical location, season, vehicle type, etc. when a neural network is used to add more information to the system to reduce errors unaccounted for in the models, such as effects of salt, dust, etc.

5 Cabin Air Filter Prognostics

In this section, we will illustrate the application of the framework to cabin air filter prognostics. Given air velocity U_{in} of the air feeding into the air filter, and concentration of particulates before the air filter N_{in}, it can be shown [9] that the rate of change of accumulation of particles, A, in the air filter to be:

$$\frac{dA}{dt} = U_{in}N_{in}(1 - P(A)) \tag{17}$$

In Eq. (17), *P*(*A*) is the filter penetration defined as the ratio of the particle concentration after the air filter over the particle concentration before the inlet.

$$P(A) = \frac{N_{out}}{N_{in}} \tag{18}$$

The wear here is *A*, the accumulation of particles in the air filter, and the larger A, the more wear. By integrating Eq. (17) and assuming initial wear is 0, the total wear can be found to be:

$$A(t) = \int_0^t U_{in}(\tau)N_{in}(\tau)[1 - P(A(\tau))]d\tau \tag{19}$$

To transform this model to fit in the connected prognostics framework, start by noting that the air velocity U_{in} is related to blower speed *B* in percentage, recirculation air door opening *R* in percentage, and vehicle speed *V*. Assume that there exists a sensor in the vehicle which can be used to approximate particle concentration N_{out}. Then

$$N_{in} = \frac{N_{out}}{P(A)} \tag{20}$$

The system's inputs are blower speed *B*, recirculation door position *R*, the concentration N_{in}, and the vehicle speed *V*, and the states are air velocity into the air filter U_{in}, the concentration after the filter N_{out}, and the wear *A*. *A* is assumed to be slow changing for the duration of few hours or days, depending on the region. Moreover, here, U_{in} is assumed to be approximated as a function with arguments *A*, *V*, *B* and *R*, and that it is not a dynamic system. This assumption is equivalent to saying that the steady state response of U_{in} after changing any of the inputs is reached quickly. Under this assumption, only *A*, *V*, *B* and *R* are needed to be known. Therefore,

$$U_{in} = f(A, B, R, V) \tag{21}$$

$$\mathbf{x}^* = [\mathbf{x}|\hat{\mathbf{x}}]^T = [U_{in}, N_{out}|A]^T \; and \, \mathbf{u} = [B, R, N_{in}, V]^T \tag{22}$$

Apply the framework as follows: Create a 4D hyperrectangle with axes *B*, *R*, V, and N_{out}. U_{in} is ignored because it can be calculated directly from *V*, *B*, *R*, and *A*. The value of *A* is known in the vehicle and the cloud. Knowing *P*(*A*) requires either knowing N_{in} or N_{out}, but N_{out} is chosen since it is directly measured. Partition each axis into sets where each set represents a cell. Assume air blower has 7 levels, and air recirculation door has 5 levels. N_{out} axis is divided into 20 levels, from 0 to 1000 μg/m^3. Vehicle speed is partitioned into 4 levels from 0 to 80 mph. The hyperrectangle E is sent to the cloud on timely basis based on the geographical location. The initial value of *P*(*A*) is known by the OEM. The OEM also has derived the mapping between A and $P(A)$ based on experimental data. The cloud calculates the wear for each cell E_i of value n_i in the hyperrectangle

$$\dot{\mathbf{W}}_{cell} = \dot{\mathbf{A}}_{cell} = f(A, B, R, V)\frac{N_{out}}{P(A)}(1 - P(A))n_i \tag{23}$$

And then, the total wear A is found by summing the individual wears in each cell, which happens in the cloud as shown in Eq. (24). The parameters except n_i are known from the axes coordinates, and n_i is accumulated in the vehicle.

$$A = \sum_{i}^{Cells} f(A, B, R, V)\frac{N_{out}}{P(A)}(1 - P(A))n_i \tag{24}$$

After executing Eq. (24), a new value of $P(A)$ is obtained from the already existing lookup table. The problem reduces to knowing the function f. This function can be obtained by using neural networks as global approximators, and data can be generated before production by measuring in a vehicle the air velocity under different conditions. The system is shown in Fig. 8. The function f is chosen to be in the cloud since it can be improved over time. The values of A and $P(A)$ are sent to the vehicle for use in other features.

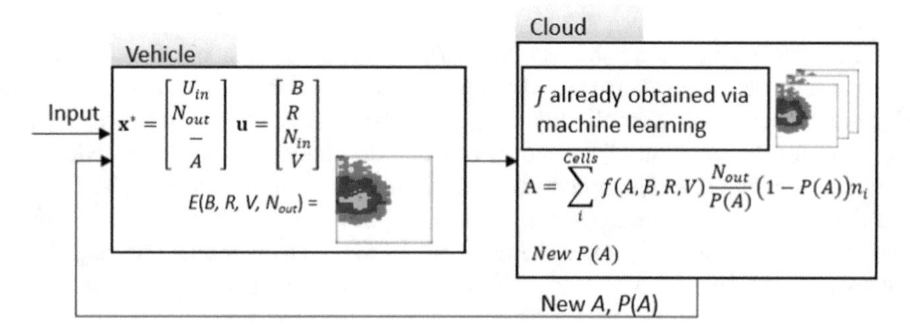

Fig. 8. In cabin air filter prognostics

6 Conclusion

In this paper, a connected vehicle prognostics framework for dynamic systems is proposed and discussed. The approach presented here generalizes and formalizes the model which is presented in [4]. In this framework, the in-vehicle physics based model generates fixed-sized data in a form suitable for machine learning and big data analysis where neural networks can be applied as universal approximators for wear functions. Since the states under monitor are assumed to change slowly over time compared to the rest of the states of the dynamic system, it is shown that it is possible to combine big data analysis and dynamic systems theory to create a hybrid solution. Furthermore, the theoretical insight is provided to show what needs to be aggregated in the vehicle, what assumptions are made, and how to map this data to machine learning inputs. The data generation is simplified to aggregating the time in a hyperrectangle which has the states and inputs of the dynamic system as axes. The application of the framework to two dynamic models: brake pad wear prediction and in cabin air filter life, is discussed. If more accuracy is desired, more elaborate models can be applied under same framework.

The cloud system can use machine learning to predict the time to service component, which can combine both machine learning and input from customers' calendar information to optimize the service scheduling for the vehicle.

This approach provides resistance to connectivity issues since the aggregated data in the vehicle has a long time window, in terms of days, to connect once to the cloud to transfer the aggregated data and obtain a new wear value.

The verification of the accuracy of the models still requires large set of field data. Upon receiving sufficient amount of data, we will analyze the prediction and if necessary, we will apply higher fidelity models. The future work includes transforming other models of components and subsystems such as fuel pumps, tires, and batteries to this framework.

References

1. Gusikhin, O., Rychtyckyj, N., Filev, D.: Intelligent systems in the automotive industry: applications and trends. Knowl. Inf. Syst. **12**, 147–168 (2007)
2. Siegel, J., Erb, D., Sarma, S.: A survey of the connected vehicle landscape–architectures, enabling technologies, applications, and development areas. IEEE Trans. Intell. Transp. Syst. **19**, 2391–2406 (2017)
3. Zhang, Y., Gantt Jr., G., Rychlinski, M., Edwards, R., Correia, J., Wolf, C.: Connected vehicle diagnostics and prognostics, concept and initial practice. IEEE: Trans. Reliab. **58**, 286–294 (2009)
4. Zagajac, J., Chopra, A., Krivtsov, V., Gusikhin, O.: Method and apparatus for connected vehicle system wear estimation and maintenance scheduling. US Patent Application US2016/0163130 A1 (2016)
5. Pecht, M.: Prognostics and Health Management of Electronics. Wiley-Interscience, New York (2008)
6. Zhang, S., Chen, W., Li, Y.: Wear of friction material during vehicle braking. In: SAE Technical Paper 2009-01-1032 (2009)
7. Bakar, A.R.A., Ouyang, H., Khai, L.C., Abdullah, M.S.: Thermal analysis of a disc brake model considering a real brake pad surface and wear. Int. J. Veh. Struct. Syst. **2**(1), 20–27 (2010)
8. Kambhampati, C., Graces, F., Warwick, K.: Approximation of non-autonomous dynamic systems by continuous time recurrent neural networks. In: Proceedings of the IEEE-INNS-ENNS International Joint Conference on Neural Networks (2000)
9. Billings, E.C.: Effects of particle accumulation in aerosol filtration. Ph.D. dissertation, California Institute of Technology. http://resolver.caltech.edu/CaltechETD:etd-09172002-113217. Accessed 14 May 2018

Human-Computer Cloud for Decision Support: Main Ontological Models and Dynamic Resource Network Configuration

Alexander Smirnov, Tatiana Levashova, Nikolay Shilov, and Andrew Ponomarev(✉)

SPIIRAS, 14th Liniya, 39, St. Petersburg 199178, Russia
{smir, tatiana.levashova, nick, ponomarev}@iias.spb.su

Abstract. Information processing systems utilizing the input received from human contributors are currently gaining popularity. One of the problems relevant to most of these systems is that they need a large number of contributors to function properly, while collecting this number of contributors may require significant effort and time. In the ongoing research, this problem is addressed by adaptation of cloud computing resource management principles to human-computer systems. The proposed human-computer cloud environment relies heavily on the use of ontologies for both resource discovery and automatic decision support workflow composition. This paper describes the set of main ontological models of the proposed human-computer cloud. Namely, the ontological model of the cloud environment, ontological model of the decision support system based on this environment and the ontology-based mechanism for workflow construction. The paper also illustrates the principles of dynamic workflow construction by an example from e-Tourism domain.

Keywords: Cloud computing · Crowdsourcing · Crowd computing Human-in-the-Loop · Human factors · Ontologies · Decision support

1 Introduction

Currently, information processing systems that rely on the input received from large and loosely connected communities of Internet users are gaining popularity and are leveraged in a growing range of applications: audio and video annotation, citizen science (e.g., [1, 2]), community sense and response (e.g., [3]), general collaborative mapping (e.g., OpenStreetMap, Google Map Maker, WikiMapia), crisis mapping (e.g., [4, 5]) to name a few.

Common problem with the systems that require human attention and human input is that each of these systems usually needs a large number of contributors to function, while collecting this number of contributors may require significant effort and time. This problem is partially alleviated by crowdsourcing platforms (like Amazon Mechanical Turk[1], Clickworker[2], Yandex.Toloka[3] etc.), that provide tools for

[1] https://www.mturk.com.
[2] https://www.clickworker.com.
[3] https://toloka.yandex.com.

A. Abraham et al. (Eds.): IITI 2018, AISC 874, pp. 16–25, 2019.
https://doi.org/10.1007/978-3-030-01818-4_2

requesters to post tasks and an interface for workers to accomplish these tasks. However, existing platforms bear two main disadvantages: (a) most of them implement only 'pull' mode in distributing tasks, therefore not providing any guarantees to the requester that his/her tasks will be accomplished, (b) they are designed for mostly simple activities (like image/audio annotation). The ongoing project is aimed on the development of a unified resource management environment, that could serve as a basis on which any human-based application could be deployed much like the way cloud computing is used nowadays to decouple computing resource management issues from application software. The proposed cloud environment addresses all three cloud models: infrastructure, platform and software. Infrastructure layer is responsible for resource provisioning, platform layer provides a set of tools for development and deployment of human-based applications, and on top this environment there are several software services leveraging human problem-solving abilities. Notably, a service allowing to build complex decision support applications. The overall architecture and main concepts of the proposed human-computer cloud (HCC) are described in detail in previous publications (e.g., [6, 7]).

The proposed concept of HCC relies heavily on the use of ontologies. Specifically, there are two main problems solved with a help of ontologies: resource discovery and automatic decision support workflow composition. To provide resource discovery, ontological representation is used to describe human skills and capabilities. There are multiple different skills and capabilities that humans can have. To describe skills and capabilities of human contributors the environment defines a very generic model that can be filled with whatever application specific skills a human is wishing to advertise. The advantage of using ontologies here is that due to the formal semantics inherent to them different ontologies can be matched (either manually via special mapping statements or even automatically [8]). As for automatic decision support workflow composition, it is done by decomposing original task into subtasks made with a help of domain knowledge represented in a form of task ontology.

This paper describes the set of main ontological models of the proposed human-computer cloud. Namely, the ontological model of the environment, ontological model of the decision-support system (DSS) based on this environment and the ontology-based mechanism for workflow construction.

Section 2 describes the proposed ontological model of the cloud environment in general. Section 3 describes the ontological model of the decision support system. Finally, in Sect. 4 describes the principles and mechanisms for on-the-fly construction of a workflow for *ad hoc* tasks with a help of task ontology.

2 Ontological Model of the Cloud Environment

Several core ontologies (of cloud computing) were analysed and evaluated as candidates to serve as a basis for the ontology of the proposed cloud environment (e.g., [9–11]). The mOSAIC model [13, 14] was chosen mainly due to two reasons: (1) it is based on the standards on the cloud computing and service-based systems, (2) it is aimed on solving the problem of interoperability of cloud services. The mOSAIC ontology is quite extensive, it contains classes for describing languages, technologies, protocols, requirements, etc. This section doesn't mention many of them to focus on the most important ones. A detailed description of all classes can be found in [13] and [14]. Main classes that are reused from the mOSAIC ontology are Resource, Service and Actors (User in the adapted ontology). In the ontology of the cloud environment, the mOSAIC model is extended by classes that allow people to be represented as a component of the cloud environment (Fig. 1).

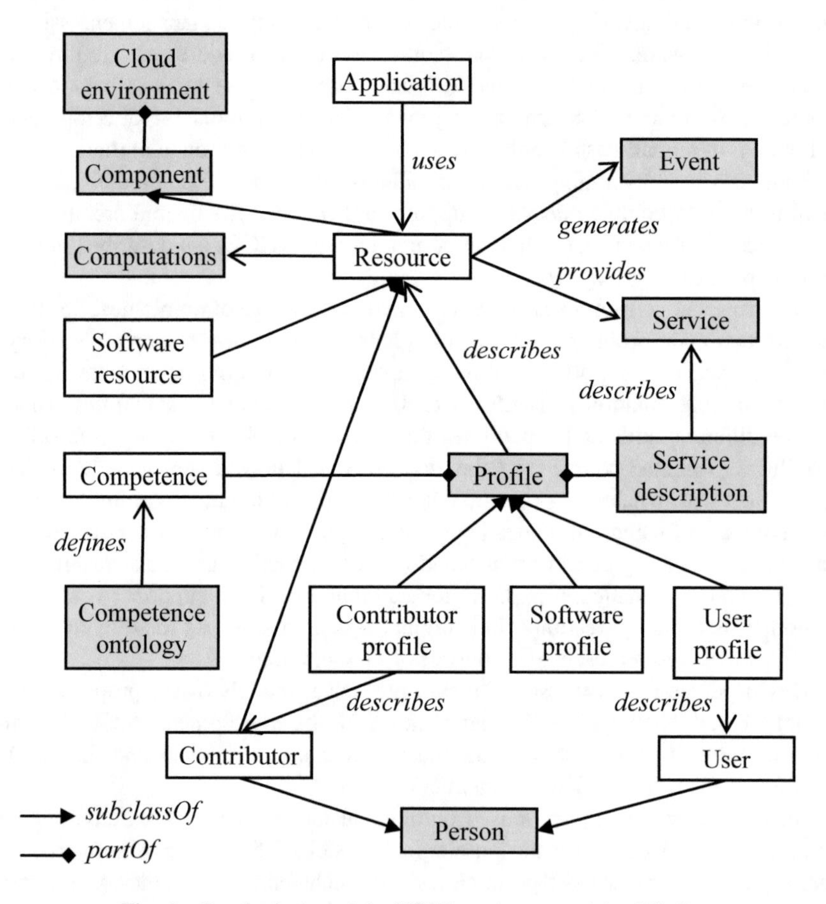

Fig. 1. Ontological model of HCC environment (simplified)

The main class of the ontology is the *Resource*, since most of the cloud environments are built around this concept, and in these environments everything (virtual machines, networks, services, etc.) is a resource [15]. Resources are presented to the user in the form of *services*, that is, the user is a consumer of cloud services. In the proposed ontology, the *user* is defined as a *person* who uses the cloud environment in accordance with their needs.

In the HCC a *person* can act two roles: a *user* and a *resource* offering services to solve certain classes of tasks. A person as a resource is represented by a class *contributor*. Other non-human resources that perform programmed actions for processing information and performing calculations are represented by a *software resource* class. Users are divided into two main categories (not shown in the figure) – *decision-maker* and the *administrator* of the environment.

All resources are described by profiles. Three types of profiles are defined: the *software profile*, the *contributor profile* and the *user profile*. The user profile contains personal information about the user, his settings, the environment, etc. In the software profile and the contributor profile, in addition to personal information, there is a description the *services* they provide and *competencies* they possess.

When registering in the cloud environment, disconnecting from it, or when performing various activities resources generate the corresponding *events*. Events change the properties of resources, including the relationship between them.

A service is an activity done by one resource for another resource or user. Information about the services provided and the terms of their provision is contained in the description of the service. The service description is a structured representation of the information mentioned.

Competence contains a qualitative assessment of the qualification of the resource for the performance of a certain type of activity. Classification of competences and possible relationships between competences are represented in the *ontology of competencies*.

Third-party applications, based on operations performed by participants in the HCC, or using tools and mechanisms provided by the cloud to develop and deploy such an application, are represented by the *application* class.

In Fig. 1, shading shows classes directly derived from to the most general class in the class taxonomy of the HCC ontology.

3 Ontological Model of Decision Support System Based on HCC

The starting point for decision support is a task formulated by the decision maker. A task describes a situation with an explicitly defined goal that needs to be achieved. The tasks that may require a solution are presented in the task ontology, which, in turn, is a part of the problem domain ontology (Fig. 2).

The conditions for the appearance of the task are located in the current situation, described by the context of the user. In accordance with the approach to context-driven decision support in a distributed information environment, the context is described by an ontological model that represents knowledge relevant to the current situation.

A complex task can be decomposed into subtasks. Each of the subtasks that need to be solved is assigned to a certain resource. The "subtask-resource" pair defines the assignment. The resource network is a description of the workflow, implemented by resources, during which the task solution is formed. As a result of the fulfillment of all assignments, that is, the solution of all the subtasks, a solution of the original task of the decision maker is formed. The result can be several alternative solutions. In this case, the decision is left for the decision maker.

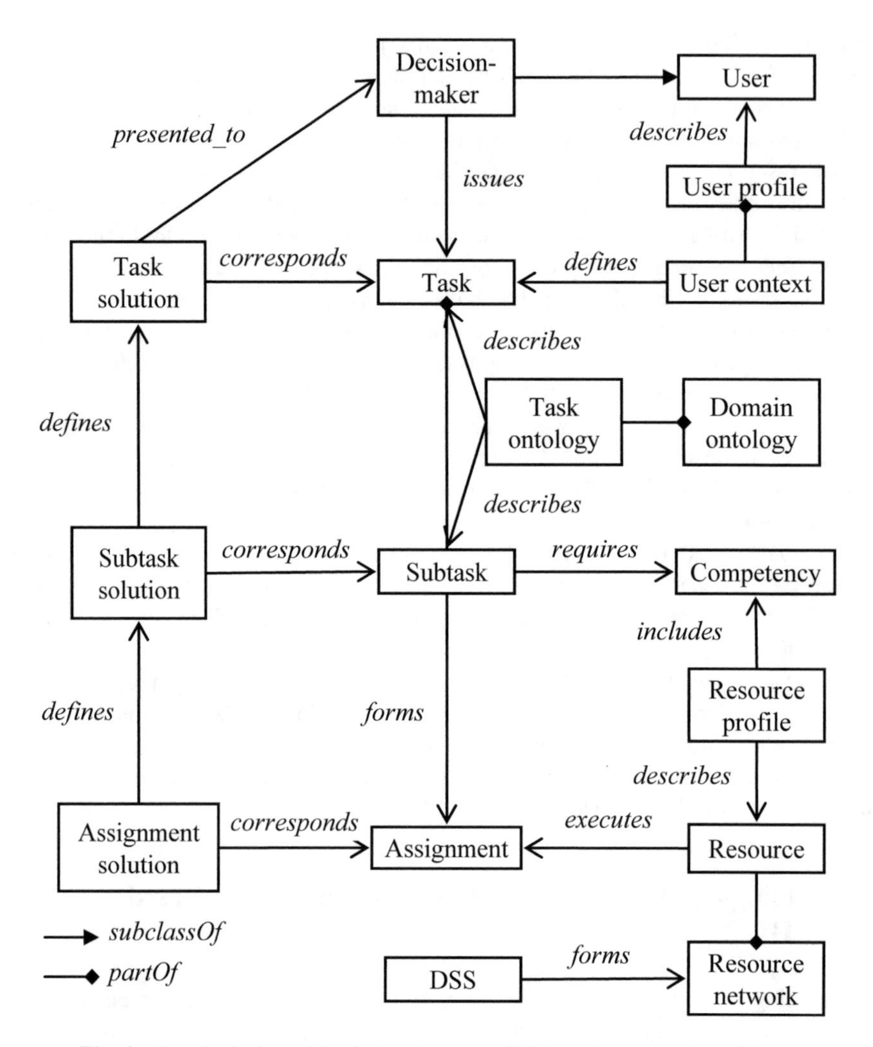

Fig. 2. Ontological model of DSS based on HCC environment (simplified)

4 Dynamic Resource Network Configuration

The DSS based on the HCC operates in a dynamic environment, as it uses independent resources, whose load can change (for example, a participant in the environment may at some point interrupt his work and refuse for some time from performing a task). Thus, the formation of a network of resources for executing a task must be carried out dynamically.

The proposed model of dynamic configuration of the resource network is based on the usage of task ontology, allowing to reuse the knowledge about solving different problem domain tasks by encoding them using a standard dictionary [16].

For the purposes of this project, a structure of the task ontology is proposed. The ontology consists of a set of tasks and subtasks, sets of input and output parameters for tasks, sets of admissible parameter values, as well as a set of constraints connecting subtasks to a task in the hierarchy of task decomposition, parameters to tasks and subtasks, and valid values to the parameters:

$$O = (A, IP, OP, D, E),$$

A – a set of tasks (and subtasks), IP – a set of input parameters for tasks, OP – a set of output parameters of tasks, D – a set of allowable parameter values, E – constraints reflecting task-subtask hierarchy, relations between parameters and tasks, and parameters and their domains (sets of allowable values).

Tasks that are not decomposable into subtasks are proposed to be considered subtasks themselves. Such tasks are described by a set of input and output parameters. Also, within the framework of this ontology, it is possible to describe the relationship of tasks with the competences required to solve them.

This ontology of tasks is either created by experts in the problem area, or completely or partially, on the basis of existing task ontologies or their fragments.

4.1 Decision-Making Using Task Ontology

According to the proposed ontological model a task (or subtask) a is described by four sets:

$$a = (A', IP', OP', C), \text{where}$$

$A' \subset A$ – a set of subtasks of a, $IP' \subseteq IP$ – a set of input parameters of a, $IP'' \subseteq IP$ – a set of output parameters of a, C – a set of competencies, required to solve the subtask (defined in terms of application competency ontology).

According to the developed model, the task to be solved is determined by the output parameter (for example, the task of planning a tourist route is determined by the output parameter "tourist route") or by several output parameters. The proposed method of task decomposition consists in determining, on the basis of the task ontology, possible configurations of a network of subtasks that allow one to obtain the values of the output parameter (parameters) by selecting and constructing chains of corresponding subtasks. These chains determine the necessary input parameters for obtaining the solution of the initial task. Possible combinations of input parameters are offered to the user, who can make a choice based on the information available to him/her. If the automatic configuration of the subtasks network for accomplishing the original task is impossible, human resources are involved in the decomposition process. The proposed model allows choosing different ways of executing the task depending on the available initial data.

Since the arrival of new tasks to the system is not predetermined in advance, nor is the time for solving the subtasks, the resource network can change in the process of task execution, i.e. it is dynamic, and the process of its configuration can be repeated until all subtasks are solved. Thus, when the state of the system changes (the appearance of a new task, the change of context, the execution of a task by a resource), a dynamic change in its configuration occurs. Namely, the sequence of the rest subtasks execution is determined and they are distributed among the available resources of the HCC environment.

The algorithm for subtask distribution is based on the multi-agent modeling with the elements of the theory of coalition games (using ideas from earlier work of the authors [12]). Each resource that can accomplish some subtasks is represented by an agent in this algorithm. Initially, each agent creates its own coalition (having some solution structure) assigning itself the role of a leader of that coalition. Then, a leader of each coalition analyses which agent from another coalition can extend the number of required subtasks from the workflow solved by the coalition and/or improve the value of the objective function (reflecting matching of resource competencies and subtask requirements) and sends proposals via shared information space. Each agent choses the coalition having most potential (able to solve more required subtasks or having a better "match" competence-wise). Winning coalition implements the workflow and its resources receive the specified reward.

Such dynamic configuration of the resource network allows not only to efficiently use the resources of the human-computer cloud environment, dynamically evaluating their load, but also to re-use the subtask solutions to solve different tasks, thereby reducing the overall load of resources.

4.2 Decision-Making in Problem Domain "e-Tourism"

This section illustrates the proposed approach applied to building a tourist itinerary. Typical context of this task is that a tourist would like to see the most interesting attractions in the area in some constrained time and in most convenient way. Itinerary planning requires not only information about the popular attractions in the area of interest, but also user preferences as well as transportation options. There are currently various approaches to build itineraries both by software services and by humans. The proposed HCC offers a way to implement and deploy itinerary planning that leverages both software and human (contributor) resources for different subtasks.

Formally, input parameters for the itinerary planning are (simplified): tourist (user) preferences (d_1), location of a tourist itinerary (d_2), time for the itinerary being planned (d_3), context information (d_4), set of attractions (d_5), ranked list of attractions (d_5'), ranked plans of tourist itineraries (d_6). Parameters d_5 (set of attractions) and d_5' (ranked list of attractions) have the same structure, but different semantics.

Subtasks for the itinerary construction, as well as their input and output parameters are shown in Table 1. Note, the task a_5 has the same type of parameter (d_6) both as input and output. It makes sense as the workflow is evaluated not only based on the number of subtasks, but on time, cost and artifacts quality.

According to the model, the task of planning a tourist itinerary is determined by the output parameter "tourist itinerary". In the proposed example, the user can choose, for example, a variant with the input parameter "ranked list of attractions" from the possible subtasks, which will lead to the need to solve only one subtask ($a4$, Create candidate itineraries, respecting local context information), or, a variant in which to solve the task it is necessary to solve all the subtasks defined above (discussed here).

Based on the analysis of input and output parameters, it can be concluded that the subtasks a_1 (Collect context information) and a_2 (Search for attractions based on preferences) are independent and can be solved in parallel. The solution of the subtask a_3 (Improve the list of attractions based on context information) should be preceded by

Table 1. Subtasks for itinerary construction.

Symbol	Subtask description	Input parameters	Output parameters
a_1	Collect context information	d_2, d_3	d_4
a_2	Search for attractions based on preferences	d_1, d_2	d_5'
a_3	Improve the list of attractions based on context information	d_4, d_5'	d_5'
a_4	Create candidate itineraries, respecting local context information	d_5'	d_6
a_5	Refine candidate itineraries	d_6	d_6

the solution of subtasks a_1 and a_2, and the solution of the subtask a_4 (Create candidate itineraries, respecting local context information) is the solution of subtask a_3 (Fig. 3). The subtask a_5 (Refine candidate itineraries) can be performed only the last.

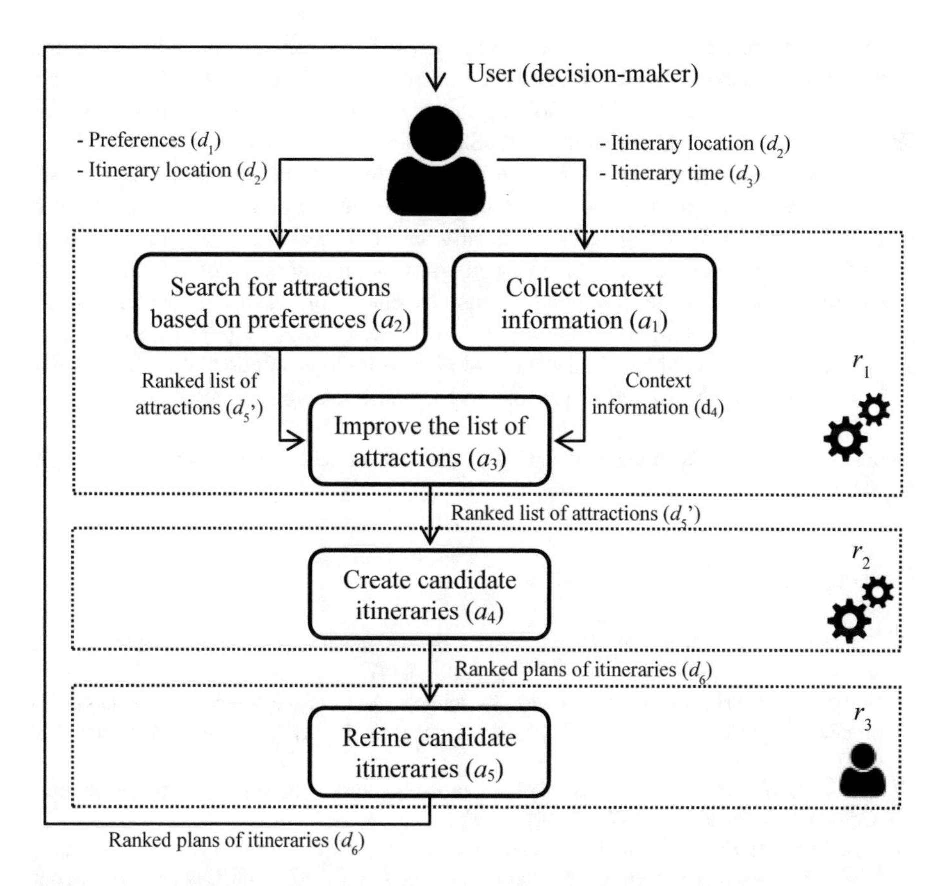

Fig. 3. Model of tourist itinerary creation task and an example resource configuration for it.

After the decomposition of the task, resources (represented by agents – see Sect. 4.1) start to negotiate which resource will accomplish what subtasks. If the process of the negotiations succeeds, each task will be performed by one resource in the resulting schedule. In the example shown in Fig. 3, tasks a_1–a_4 are performed by two software resources r_1 and r_2, while subtask a_5 is performed by human-based resource r_3.

5 Conclusion

The paper describes the set of ontological models powering the proposed human-computer cloud environment as well as principles of dynamic workflow building decision-support systems based on this environment.

The general ontological model of the human-computer cloud environment is based on mOSAIC ontology, that has been extended by (1) classes that allow human contributors to be represented as one of the types of cloud components and (2) classes necessary to configure the resource network, taking into account the competencies of the resources.

An ontological model of the DSS based on the environment of human-machine cloud computing has also been developed. According to the model, decision-making is carried out based on the results of solving the user task by the DSS resource network. The proposed ontological model of the DSS combines the concepts needed to configure the network of resources in accordance with the task of the decision-maker (DM). A formal description of the task is contained in the ontology of the problem domain. A resource network is configured dynamically based on resource competencies. At the core of the proposed model of dynamic configuration of a network of resources, is the task ontology, enabling the knowledge reuse by encoding it using a standard vocabulary. The proposed model allows to choose different ways of performing the task depending on the available initial data, and also to build a solution plan determining which subtasks can be solved in parallel and which ones are sequential.

Acknowledgements. The research is funded by the Russian Science Foundation (project # 16-11-10253).

References

1. Franzoni, C., Sauermann, H.: Crowd science: the organization of scientific research in open collaborative projects. Res. Policy **43**(1), 1–20 (2014)
2. Jollymore, A., Haines, M., Satterfield, T., Johnson, M.: Citizen science for water quality monitoring: data implications of citizen perspectives. J. Environ. Manag. **200**, 456–467 (2017)
3. Faulkner, M.: Community sense and response systems: your phone as quake detector. Commun. ACM **57**(7), 66–75 (2014)
4. Ushahidi. http://www.ushahidi.com/. Accessed 20 Apr 2018
5. Meier, P.: How Crisis Mapping Saved Lives in Haiti. http://voices.nationalgeographic.com/2012/07/02/crisis-mapping-haiti/. Accessed 20 Apr 2018

6. Smirnov, A., Ponomarev, A., Levashova, T., Teslya, N.: Human-computer cloud for decision support in tourism: Approach and architecture. In: Proceedings of the 19th Conference of Open Innovations Association (FRUCT), pp. 226–235 (2016)
7. Smirnov, A., Ponomarev, A., Shilov, N., Kashevnik, A., Teslya, N.: Ontology-based human-computer cloud for decision support: architecture and applications in tourism. Int. J. Embedded Real-Time Commun. Syst. **9**(1), 1–19 (2018)
8. Euzenat, J., Shvaiko, P.: Ontology Matching, 2nd edn. Springer, Heidelberg (2013)
9. Sun, Y.L., Harmer, T., Stewart, A.: Specifying cloud application requirements: an ontological approach. In: Cloudcomp 2012. LNICST, vol. 112, pp. 82–91 (2013)
10. Liu, F., et al.: NIST Cloud Computing Reference Architecture. Recommendations of the National Institute of Standards and Technology [Electronic resource]. National Institute of Standards and Technology Special Publication 500-292 (2011)
11. Bellini, P., Cenni, D., Nesi, P.: Smart cloud engine and solution based on knowledge base. Procedia Comput. Sci. **68**, 3–16 (2015)
12. Smirnov, A., Shilov, N., Kashevnik, A.: BTO supply chain configuration via agent-based negotiation. Int. J. Manuf. Technol. Manag. **17**(1/2), 166–183 (2009). Inderscience
13. Moscato, F., Aversa, R., Martino, B., Rak, M., Venticinque, S., Petcu, D.: An ontology for the cloud in mOSAIC. In: Cloud Computing. Methodology, Systems, and Applications, pp. 467–485. CRC Press, Boca Raton (2011)
14. Moscato, F., Aversa, R., Martino, B.: An analysis of mOSAIC ontology for cloud resource annotation. In: Proceedings of the Federated Conference on Computer Science and Information Systems, pp. 973–980. IEEE (2011)
15. Nyrén, R., Edmonds, A., Papaspyrou, A., Metsch, T., Parák, B.: Open Cloud Computing Interface – Core [Electronic resource]. GFD-R-P.221. Open Grid Forum (2016). https://www.ogf.org/documents/GFD.221.pdf. Accessed 20 Apr 2018
16. Chandrasekaran, B., Josephson, J.R., Benjamins, V.R.: Ontology of tasks and methods. In: Proceedings of the Eleventh Workshop on Knowledge Acquisition, Modeling and Management (KAW 1998). http://ksi.cpsc.ucalgary.ca/KAW/KAW98/chandra/index.html. Accessed 20 Apr 2018

Enterprise Total Agentification as a Way to Industry 4.0: Forming Artificial Societies via Goal-Resource Networks

Valery B. Tarassov(✉)

Bauman Moscow State Technical University, Moscow, Russia
vbulbov@yahoo.com

Abstract. The concept of Industry 4.0 together with its components and tools is discussed. An increasing role of agent-oriented and social technologies for developing smart enterprises, cyberphysical systems, Internet of things, collaborative robots is shown. The arrival of these technologies opens new frontiers in growing and studying artificial societies. A representation of networked enterprise as a mixed society of natural, software and hardware agents is suggested. Some fundamentals of agent theory and multi-agent systems are considered, and GRPA architecture is analyzed. The resource-based approach to enterprise modeling is taken and the principle of dependence of formal apparatus on both agent specification and architecture is formulated to justify the need in new network models extending conventional resource networks. Three basic agent types are introduced and a formalism of goal-resource networks to visualize and simulate communication between agents and formation of both multi-agent systems and artificial societies is presented. Colored goal-resource networks (CGRN) to represent these basic agent types are proposed. Some examples of communication situations and behavior strategies for «robot-robot» and «human-robot» interactions are given.

Keywords: Industry 4.0 · Agent · Industrial agentification · Agent architecture Networked enterprise · Artificial society · Multi-agent system · Weighted graph Resource network · Goal-resource network · Colored goal-resource network

1 Introduction

Recently, the Fourth Industrial Revolution called briefly Industry 4.0 has started to unfold in Germany and other developed countries [12, 28]. The term Industry 4.0 was coined by Walster and first appeared at Hannover Messe in 2011 as an outline of German industry perspectives. Nowadays, the initiative of Industry 4.0 has spread far beyond Germany and is widely used around the world: its counterparts are Industrial Internet in the USA [18], High Value Manufacturing Catapult in the United Kingdom, Usine du Futur in France, Fabbrica del Futuro in Italy, Smart Factory in Netherlands,

This work is supported by Russian Science Foundation project №16-11-00018 and Russian Foundation for Basic Research project №17-07-01374.

A. Abraham et al. (Eds.): IITI 2018, AISC 874, pp. 26–40, 2019.
https://doi.org/10.1007/978-3-030-01818-4_3

Made Different in Belgium, Made in China 2025, National Technology Initiative in Russia, and so on.

In a wide sense, Industry 4.0 encompasses both a new industrial enterprise vision and its keynote mission in the age of digital economy. In a narrow sense, Industry 4.0 is a name for the current trend of information exchange and automation in manufacturing technologies that fuse the physical and digital worlds. It includes Cyberphysical Systems [31], the Internet of Things [20], both cloud [29] and cognitive computing [34], as well as, advanced Artificial Intelligence and Robotic Systems, to enable Smart Factory [26]. While Industry 3.0 faces the problem of automating single machines and technological processes with using computers and electronic devices, Industry 4.0 focuses on the end-to-end digitization of all physical assets and their integration into digital ecosystems with value chain partners. Digitization means the process of converting information in a digital form; the result is called digital representation or, more specifically, a digital image. Here a keyword is «Ubiquitous Digitization», i.e. digitization and integration of both vertical and horizontal value chains, digitization of product and service offerings, digitization of business models and customer access, and so on.

So in 2015, McKinsey [2] defined Industry 4.0 as "the next phase in the digitization of the manufacturing sector, driven by four basic factors: the astonishing rise in data volumes, computational power, and connectivity, especially new low-power wide-area networks; the emergence of analytics and business-intelligence capabilities; new forms of human-machine interaction such as touch interfaces and augmented-reality systems; and improvements in transferring digital instructions to the physical world, such as advanced robotics and 3-D printing."

In our opinion, some fundamental problems closely related to Industry 4.0 technologies still remain neglected. Among them we can mention the synthesis of hybrid and synergetic manufacturing systems, development of pervasive agentification in enterprise, formation of artificial (societies of things) and mixed human-artificial industrial societies. A few ways of their solving will be considered in the paper.

2 Industry 4.0: Smart Enterprises, Cyberphysical Systems, Internet of Things, Collective and Collaborative Robotics – New Challenges for Growing Artificial Societies

The development of future manufacturing enterprises supposes the implementation, refinement and consolidation of different enterprise platforms such as Digital Factories, Smart Factories and Virtual Factories. Here *Digital Factory* represents a network of digital models, methodologies, and applications used to integrate and improve the product lifecycle. It relates the planning and design of manufacturing facilities with the manufacturing process itself. The digital factory concept focuses on an integrated planning and monitoring process that includes product design, process planning, and planning and implementation of the operation, making the manufacturing process more efficient and responsive to change.

In its turn, the concept of *Virtual Factory* makes emphasis on virtual simulation of real enterprise operation that allows us to replicate various manufacturing scenarios and helps in designing and implementing modern networked enterprises. In the latter context, virtual factories encompass the resources of different interacting enterprises in a concrete industrial micro-environment to create added value.

Finally, *Smart Factory* is often viewed as a keynote technological outcome of Industry 4.0 that combines advanced production, information, and communication technologies, with the potential for integration across the entire manufacturing supply chain. Basic components and tools of smart factory are depicted in Fig. 1.

Thus, the main building blocks are smart sensors and instruments, intelligent monitoring and control systems, advanced robotics and friendly human-machine interfaces. In particular, smart sensors and networks enable big data gathering and processing just in the course of manufacturing. Here such tools as cognitive sensors and sensor networks [33] capable both to perform measurements and «understand» obtained results by providing appropriate visualizations, as well as reason by taking into considerations contextual information in the form of evaluations and norms are of special concern.

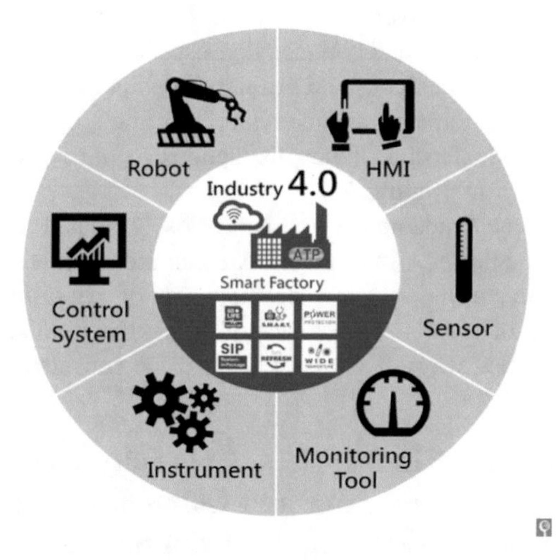

Fig. 1. Basic tools for building smart factory in the context of Industry 4.0

An important component of Industry 4.0 is the fusion of the physical and the virtual worlds [12] provided by *cyberphysical systems*. The emergence of cyberphysical systems (CPhS) means the inclusion of computational resources into physical-technical processes. Embedded computers and networks monitor and control the physical-technical processes with feedback loops where physical processes affect computations and vice versa. A virtual copy of physical world is created and well-timed decentralized decisions are made.

Let us note that CPhS are descended from mechatronic systems enhanced by advanced tools of data/knowledge acquisition, communication and control. Their components continuously interact, providing CPhS self-adjustment and adaptation to changes. It is obvious that CPhS are crucial for production digitization and agentification. Here work pieces, devices, equipment, production plant and logistics components with embedded software are all talking to each other. Smart products know how they are made and what they will be used for. Thus, both production machines and equipment and products become cognitive agents involved into manufacturing and logistics processes.

Internet of Things (IoT) plays the role of nervous system for Industry 4.0. It can be viewed as the ubiquitous and global network that provides the functionality of integrating the physical world. This is done through the collection, processing and analysis of data generated by IoT sensors, which will be present in all things and will be integrated through the public communication network. In industrial branches, it will allow people and things (enterprise employees and equipment) to be connected anytime, anyplace, with anyone and anything, ideally using corporate network.

So this rapidly growing paradigm means establishing connectivity of smart devices by which objects (things) can sense one another and communicate. According to International Data Corporation (IDC), IoT is a network of networks of uniquely identifiable objects that communicate without human interaction using IP. Nowadays, such communication and network technologies as IPv6, web-services, Radio Frequency IDentification (RFID) and high speed mobile 4G Internet networks are employed.

It is worth stressing that recently social dimension of CPhS and IoT has appeared crucial: now both *Cyberphysical Social Systems* [4, 30, 35] and *Socio-Cyberphysical systems* [10] are of special concern. Besides, the concept of the *Social Internet of Things*, the meeting point of IoT with social networks [1], is worth mentioning.

The most revealing examples of using social technologies for Industry 4.0 are related to Collective Robotics and Collaborative Robotics. The state-of-the-art in the field of *Collective Robotics* was presented in [14]; some basic problems of Collective Robotics and *Synergetic Artificial Intelligence* were considered in our plenary talk at IITI'17 (see [13]). Below we shall pay a special attention to Collaborative Robotics (briefly, Cobotics).

Cobotics is a neologism formed by crossing the words «collaborative» and «robotics» [23]. Collaborative robot is intended to physically interact with humans in a shared workspace. This is the difference with respect to conventional industrial robots, that operate autonomously or with a limited guidance.

A cobotic system includes a human and a robot collaborating in synergy to perform a required task. In order to do such a joint work hand in hand with human beings, any cobot needs to be equipped with powerful onboard computer and complex sensor system, including an advanced computer vision and learning facilities. It allows prevent the collisions of robot with human partners and obstacles, as well as operate in case of software crash.

To differ from classical master-slave relations, human-robot partnership in cobotic systems is based on collaboration via interactive information management, where the robot partner can initiate the dialogue with human partner to precise the task, request additional data or obtain his evaluation of learning results. Here, some new opportunities for cobotic applications in Industry 4.0 are open by a strategy of direct teaching «do as I do» by showing the necessary motions to the robot.

Therefore, the main requirements for cobots are focused on safety, light weight, flexibility, versatility, and collaborative capacity.

Without the need of robot's isolation, its integration into human workspace makes the cobotic system more economical and productive, and opens up many new opportunities to compare with classical industrial robots. On the one hand, cobots increase information transparency via their ability to collect data and pass it on to other systems for analysis, modeling and so on. On the other hand, they provide technical assistance, in the sense that they "physically support humans by conducting a range of tasks that are unpleasant, too exhausting, or unsafe for their human co-workers" [8, 27].

According to ISO 10218, the following classes of industrial cobots can be viewed: (a) robots-manipulators sharing a workspace with humans (for instance, on the assembly line) to facilitate their workload (as a first interactive industrial robot Baxter); (b) mobile transportation robots, as well as mobile robots working in production rooms together with people; (c) industrial multi-robot systems. All these robots have the status of artificial cognitive agents.

Furthermore, it is worth employing in Russian counterpart of Industry 4.0 earlier national theoretical investigations related to Technetics and Technocenosis theory [15, 25]. The term «Technetics» stands for the theory of technosphere evolution. Here a holistic approach to techniques, technologies, materials, products, waste is taken. An important part of technetics is technogenetics that encompasses the problems of creation and transfer of hereditary information by design and technological documentation and other means.

The term «cenosis» usually means a community of organisms. In case of product agentification in technetics four basic levels of cenological approach are: (1) product; (2) population of products; (3) technocenosis; (4) technosphere. Population of products – a group of artificial objects of the same kind –is viewed as an elementary unit of technoevolution. By a technocenosis we mean various communicating populations of artificial objects (things). Generally such artificial populations are sets of weakly connected products limited in time and space. Like biological communities, technocenoses force «artificial organisms» (agents) entering into them, to live according to the laws dictated by the whole community.

The analysis carried out in Sect. 2 shows that we need considering both agentification and socialization issues for Industry 4.0 components and tools. By industrial agentification in a weak sense we mean a process of integration of software and hardware equipment already existing in the enterprise into a multi-agent system. Moreover, industrial agentification in a strong sense supposes the formation of networked enterprises and growing of mixed societies of both natural and artificial agents. A representation of networked enterprise as a society of natural, software and hardware agents is given in Fig. 2.

In Industry 4.0 driven by IoT the simulation of Artificial Societies (see [7]) becomes a top priority. It allows understanding how does the heterogeneous behavior of artificial agents generate the global macroscopic regularities of enterprise social phenomena, studying group formation, transformation and activities for agentified things, as well as dependence of social structures on the agent type.

In the next sections we will consider the above mentioned components of Industry 4.0 on the basis of agent-oriented approach, multi-agent systems and resource exchanges in networks. Primarily, we give the necessary information about networks, and show the difference between classical flow networks and resource networks. Secondarily, we present some details from the theory of agents to justify the need in modifying and extending the resource networks formalisms in order to understand both agent's behavior and communication between agents. We will use *the principle of dependence* of formal apparatus on both agent specification and architecture.

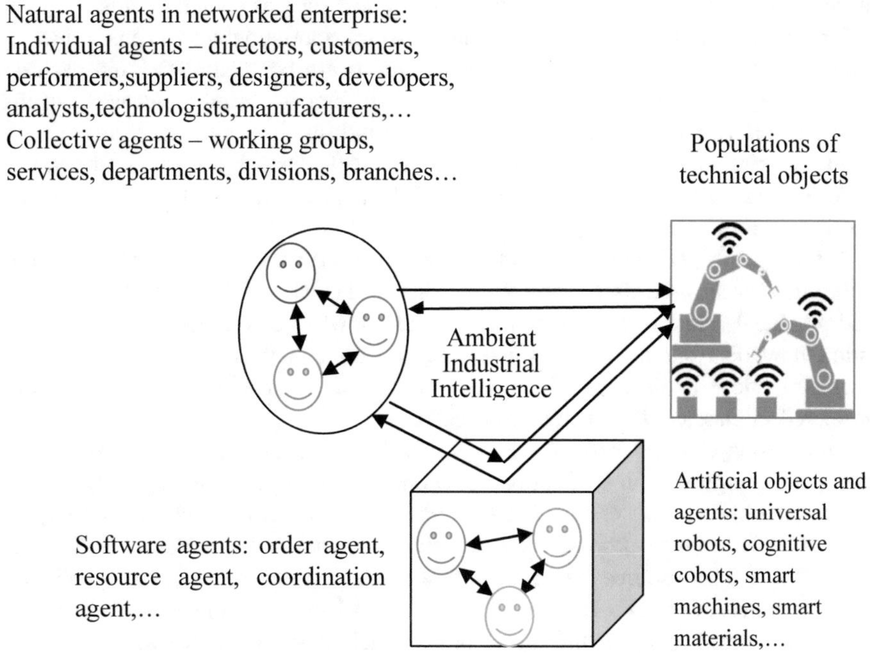

Natural agents in networked enterprise:
Individual agents – directors, customers, performers,suppliers, designers, developers, analysts,technologists,manufacturers,...
Collective agents – working groups, services, departments, divisions, branches...

Populations of technical objects

Ambient Industrial Intelligence

Software agents: order agent, resource agent, coordination agent,...

Artificial objects and agents: universal robots, cognitive cobots, smart machines, smart materials,...

Fig. 2. Networked enterprise as a society of natural and artificial agents

3 From Flow Networks and Flow Graphs to Resource Networks

Graphs and networks are the best tools to visualize and understand various relationships between agents and develop multi-agent systems and artificial societies. Here graph labeling refers to the assignment of labels to the arcs and/or vertices of a graph, by using certain rules depending on the situation. A *weighted* graph associates a special label called *weight* with every arc in the graph. In practice the interpretation of weights is of primary concern (for instance, strength of relation, cost or arc conductivity).

Weights are usually real numbers; in particular, they may be restricted to rational numbers or integers. Some algorithms require further restrictions on weights; for instance, Dijkstra's algorithm operates properly only for positive weights.

In graph theory, any network *N* is seen as a directed graph equipped with a function that makes in correspondence a non-negative real number to each arc [3, 11]. Formally it is given by a triple

$$N = \langle X, R, W \rangle, \tag{1}$$

where *X* is a set of nodes, *R* is a set of arcs, and $W\colon R \rightarrow \mathbf{R}^{+}$, $\mathbf{R}^{+}$ is a set of non-negative real numbers. A network may contain special nodes, such as source or sink.

For example, a flow network (also known as transportation network) is a directed graph without loops, where each arc has a capacity (or conductivity) *w*and each arc receives a flow. Here the amount of flow on an arc cannot exceed the capacity of this arc. A network can be used to model road traffic, fluids in pipes, currents in an electrical circuit, or anything similar in which something travels through a network of nodes.

A classical network method is the Ford-Fulkerson algorithm which computes the maximum flow in a network [9].

A flow graph [22] and an extended flow graph [19] can be employed for representing, analyzing and discovering knowledge in databases. They may be seen as a good graphical framework for data mining and knowledge discovery based on information flow distribution. Flow graphs are used as a mathematical tool to analyze information flow in decision algorithms, in contrast to material flow optimization considered in classical flow network analysis.

Among non-classical networks Pospelov's deeds (action frames) are worth mentioning in the context of linking actions and communications between agents (see [24]). In more detail graph theoretic methods for multi-agent systems are considered in [21].

In 2009 Kuznetsov proposed a new network model called a Resource Network (RN) [16, 17]. The resource network is a flow model represented by oriented weighted graph, in which every two vertices are either not adjacent or connected by a pair of oppositely directed arcs. To differ from conventional transportation networks and Ford-Fulkerson model, where the resources flowing from source to sink nodes are located in the arcs, in RN all resources are assigned to the nodes, and the weights of arcs indicate their capacities. The nodes exchange resources, following the definite rules. The basic problem is to analyze the exchange processes and their stabilization.

Definition 1 [16]. A Resource Network is a directed graph whose vertices v_i are assigned nonnegative numbers $q_i(t)$ called *resources* varying in discrete time *t* and whose arcs (v_i, v_j) are assigned time invariant positive numbers r_{ij} called capacities.

A pair of arcs (v_i, v_j) and (v_j, v_i) is called bidirectional. Any network whose vertices are connected only by bidirectional pairs is called a bidirectional network.

A bidirectional network is said to be complete if any two vertices are connected by a bidirectional pair and is said to be symmetric if the capacities in each bidirectional pair are identical.

Let *n* be the number of vertices. The state of the network at a given time *t* is defined by the vector $Q(t) = (q_1(t), q_2(t), \ldots, q_n(t))$. At each time, resources are transferred from the vertices along the outgoing arcs with the resource amounts depending on the arc capacities. The rules for resource transfer satisfy the following conditions: (a) the network is closed; i.e., no resources are supplied from the outside; (b) a resource amount sent out of a vertex is subtracted from its total resource, while a resource amount arriving at a vertex is added to its total resource, i.e. the total resource W is conserved. The capacity matrix of a network is defined as: $R = \|r_{ij}\|_{n \times n}$

A state $Q(t)$ is called stable if $Q(t) = Q(t + 1) = Q(t + 2) = \ldots$

A state $Q^* = (q_1^*, q_2^*, \ldots, q_n^*)$ is said to be asymptotically reachable from the state $Q(0)$ if for any $\varepsilon > 0$ there exists t_ε, such that $t > t_\varepsilon |q_i^* - q_i(t)| < \varepsilon$, $i = 1,2, \ldots. n$ for all $t > t_\varepsilon$. A network state is called a limit state if it is either stable or asymptotically reachable.

Now let us adapt the RN-model above for agent-oriented modeling (see also [5]).

Definition 1*. A Resource Network for MAS is a bilateral (multi-lateral) weighted directed graph

$$RN = \langle A, R, RES, W \rangle, \tag{2}$$

where the set of nodes is associated with the agent set A, the set of arcs is seen as the set of relations R between agents from A, RES stands for resource set (each agent $a_i \in A$ has a resource amount $res(a_i) \in RES$) and W is the conductivity set. Each arc $r_{ij} \in R$ is equipped witha non-negative number called capacity (or conductivity) from a_i to a_j. In other words, conductivity in MAS specifies a maximum amount of resources that an agent can give to other agent in a time t.

The resource network is homogeneous if all its conductivities are equal. Generally, when MAS includes agents of different types, resource networks are not homogeneous, because both resource amount and conductivity values depend on agent type. Moreover, the agent typology restricts the possibility of MAS formation.

4 Communication, Goals, Resources of Agents in MAS

Establishing communication between agents is the first prerequisite for building MAS and artificial societies. Such a communication means deploying interactions aimed at forming bilateral and multilateral dynamic relations. According to synergetic methodology [32], interactions are primary and the structures they generate are secondary, i.e. communications are the sources of organizations. Multi-agent systems are polystructural by their nature and ensure the fusion of extensive and intensive structures. From the viewpoint of graph theory, the amplification of extensive structures supposes adding new nodes to the initial graph, whereas the deployment of intensive structures is provided with adding new arcs. The necessary conditions to start interactions between agents are: (1) *goal* sharing or imposition; (2) agents relations to *resources*, an amount of available resources and a need in additional resources; (3) *commitments* and agreements between agents.

The reason for any agent's activity is the need seen as a distance between desired and current agent's state. The interpretation of this need is referred as motivation; the role of motivation in activity theory is the same as the role of force in mechanics. The development of motive leads to generating goal as a model of agent's wanted future. Goal-driven activity and autonomy are basic agent's features. Agent's autonomy is enabled by available resources, as well as by their periodic acquisition from the environment (including other agents).

Often the pair «autonomy – intelligence» is viewed as agent's main attributes. Here *intelligence* is mainly related to the analysis of sensor data, knowledge elicitation and reasoning, whereas *autonomy* is associated with decision-making, planning and acting, agent's resource allocation and re-allocation.

Various resources are required to achieve a goal. The term «resources» is viewed here in a wide sense as any means useful for achieving goals of agent or multi-agent system. For instance, the resources of manufacturing systems are workers, machines, raw materials, robots, etc. We specify two main resource attributes: (1) resource amount that is its spatial measure, and (2) resource action that is its temporal measure. It is evident that the same amount of resources can bring about different resource actions.

Agent's communication is often associated with resource allocation, exchange and shared use. Generally resources have two basic parameters: (a) location; (b) accessibility. The idea of resource agentification [6] consists in providing each resource with the knowledge about its own structure, location and state to facilitate resource employment.

A general agent architecture GRPA (Goal – Resources – Perception – Action) was proposed in [5] (Fig. 3). In this architecture goals and resources are seen as key attributes of agent's internal world, whereas perception and action are basic processes of interacting with external environment.

Let us introduce basic agent's types by taking two criteria: (a) a capacity to generate individual goals and allocate the appropriate resources; (b) a capacity to form shared goals and organize the exchange of resources to achieve such collective goals. Then three basic agent types are straightforwardly specified in Table 1: (1) benevolent agent a_b; (2) self-interested or egoist agent a_e; (3) altruistic agent a_a.

So the total agent set is partitioned into the subsets of benevolent, egoist and altruistic agents: $A = A_b \cup A_e \cup A_a$, where $A_b \cap A_e = \varnothing$, $A_b \cap A_a = \varnothing$,$A_e \cup A_a = \varnothing$.

Let us give a short goal-resource description of each agent type. The benevolent agent a_b has both individual goal and the capacity to form/take collective goal. He takes an active part in exchanging resources and saves some resources for proper needs.

The egoist agent a_e strives to achieve only his goals and ignores the goals of other agents. He imposes his goal and initiates resource sharing only for his own benefit; in fact the resource exchange is substituted here by the flow of resources to the a_e. At last, the altruistic agent a_a takes anyone else's goal as his own goal and is ready to the resource exchange even if he loses from this.

The own agent's goal is a decisive factor for his self-preservation by resources, whereas the goal sharing is a necessary condition to obtain some coalition.

Such a specification of agent types opens the opportunity to build various multi-agent organizations or exclude some of them. For instance, multi-agent systems that include only egoist agents, as well as MAS of altruistic agents cannot exist. In the first case we observe the inability of A_e to goal sharing and mutually beneficial exchange of

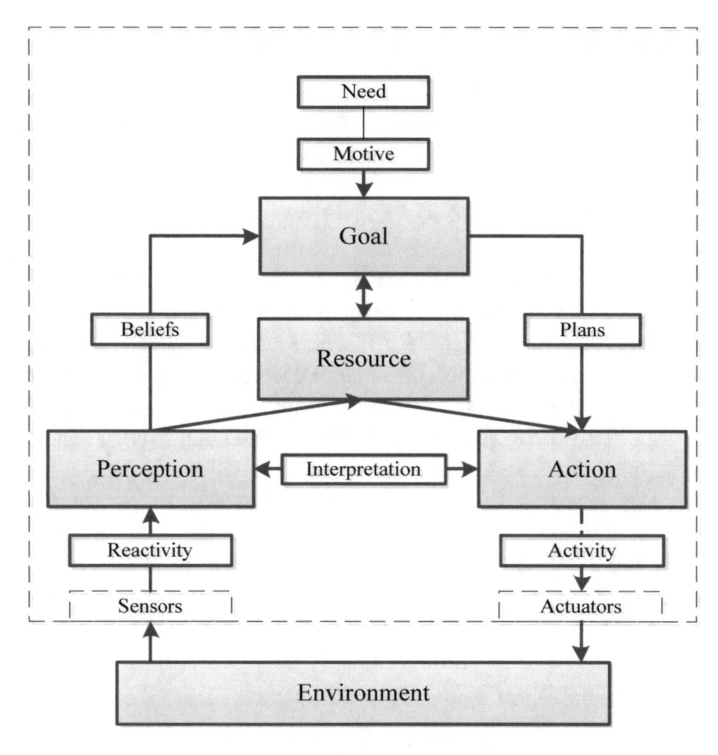

Fig. 3. An illustration of GRPA model

Table 1. Definition of agent types

Agent type	Individual goal	Collective goal	Colored agent denotation
Benevolent agent a_b	+	+	Green
Egoist agent a_e	+	–	Red
Altruistic agent a_a	–	+	Deep Blue

resources. In the second case only altruistic agents cannot form individual goals (and spent the appropriate resources) and hence generate collective goals.

The availability of resources can influence the agent type. For example, a significant inflow of resources can turn the benevolent agent a_b into the egoist agent a_e. Inversely, the lack of resources can transform the benevolent agent a_b into the altruistic agent a_a. The necessary conditions for MAS formation can be formulated a follows: (1) $res(a_a(t)) \geq res_{min}$ (a living minimum condition for the altruistic agent); (2) $res(a_e(t)) < res_{max}$ (a social ability preservation by the egoist agent).

The above mentioned considerations show the need in introducing both goal-driven and resource-shared structures into classical Resource Networks and taking into account the agent type, for instance, by using nodes of different colors.

5 From Goal-Resource Networks to Colored GRN

Definition 2. *A Goal-Resource Network* (*GRN*) for MAS is a *bilateral weighted* (twice) directed graph

$$GRN = \langle A, R, K, RES, W, t \rangle, \tag{3}$$

where a collection of nodes A is viewed as a set of agents, and a set of arcs R is partitioned into two non-overlapping subsets: a subset of goal arcs R_G and a subset of resource arcs R_{RES}: $R = R_G \cup R_{RES}$, $R_G \cap R_{RES} = \varnothing$. Let us denote resource arcs by solid lines and goal arcs – by dotted lines. Here each node $a \in A$ is characterized by a type $k \in K$, K = {benevolent, self-interested, altruistic},$|K| = 3$ and a non-negative value of resource $res \in RES$. Each arc $r \in R$ has some conductivity value $w \in W$ and W is also partitioned into a subset of resource conductivities W_{RES} and a subset of goal conductivities W_G. Finally a discrete time set is given, $t = 0, 1, 2, \ldots, n$. Generally, the values of input and output conductivities can be different.

Let n be the number of agents in A. In any discrete time t the state of MAS is given by the state vector $RES_A(t) = (res(a_1), \ldots, res(a_n))$, where n is the number of agents in GRN. A state of MAS is called stable, if $RES_A(t)$ = const. A state of MAS is called asymptotically reachable for an infinitesimal ξ, $\xi > 0$, if $\forall i = 1, \ldots, n$ the following inequality takes place $|res(a_i(t + 1)) - res(a_i(\mathrm{t}))| < \xi$.

Definition 3. A GRN is called colored, if is includes colored nodes corresponding to various types of agents. Below we will denote benevolent agents by green color, self-interested (egoist) agents – by red color, and altruistic agents –by deep blue color.

Let us consider main communication situations and behavior strategies for «robot – robot» and «human-robot» interactions. We begin with pairwise communication.

A. Communication of two benevolent agents a_b. Let us denote two benevolent agents by a_{bi} and a_{bj} (Fig. 4). Here an equitable information exchange leads to a formation of shared goal and resource exchange.

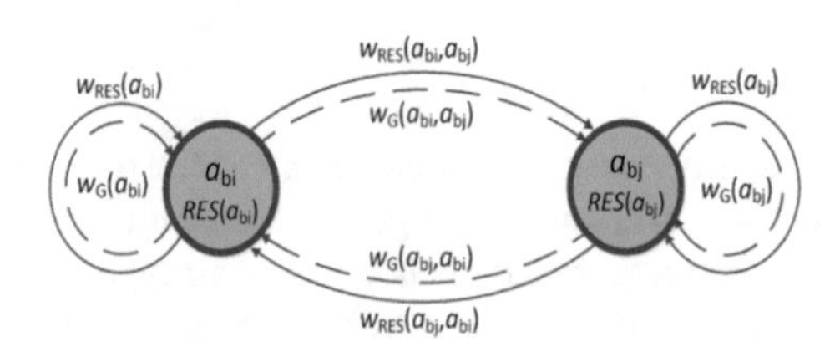

Fig. 4. A goal-resource network for the communication of two benevolent agents

A multi-agent system that includes n benevolent agents seems to be the most effective architecture to implement the strategies of Decentralized AI when a complete graph structure appears (see Fig. 5). Such a communication structure contributes to growing «democratic» artificial society.

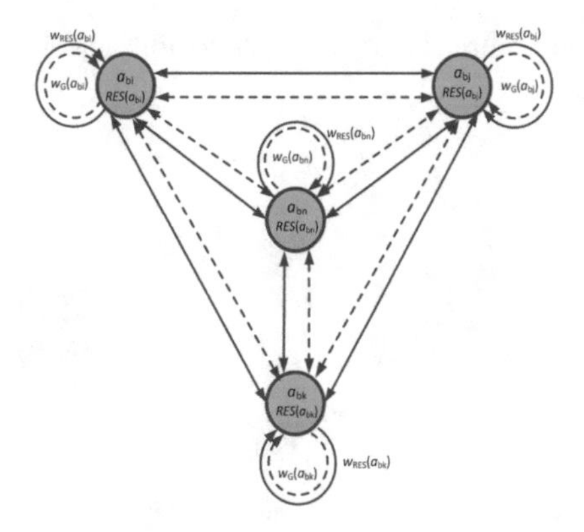

Fig. 5. A goal-resource network with a coordinator agent

B. Communication of egoist agent a_e and altruistic agent a_a. Here an egoist agent a_e imposes his goal to an altruistic agent a_a and employs the a_a resources to achieve it (see Fig. 6). In fact, it is a transfer of resources from a_a to a_e that can lead to the death of the altruistic agent, if $res(a_a(t)) < res_{min}$. A basic star structure (1 egoist and n altruistic agents) degenerates with time and becomes an isolated vertex – the resource monopoly.

C. Communication of egoist agent a_e and benevolent agent a_b. Such a communication can only happen when the agent a_b accepts the goal of a_e (Fig. 7). In fact, an illusion of the resource exchange appears. The egoist agent a_e needs some resources for achieving his goal, but, in response, he tries not to give anything away. Usually, the benevolent agent a_b avoids such a communication and participates in it only in case of emergency. Here any communication between a_b and a_e stops, if $res(a_b(t))$ is close to res_{min} (the instinct of self-preservation in a_b).

The communication schemes *B* and *C* lead to the formation of «autocratic» artificial societies.

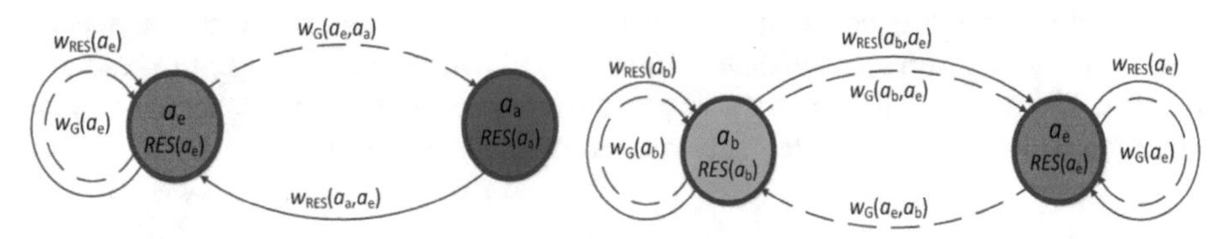

Fig. 6. A goal-resource network for the communication between egoist agent and altruistic agent

Fig. 7. A goal-resource network for the communication between egoist agent and benevolent agent

D. Communication of benevolent agent a_e and altruistic agent a_a. Here effective resource exchanges take place, because a_a shares the goal of a_b (see Fig. 6) and the latter does not allow the situation of resource depletion in the former. In case of one benevolent agent a_b and n altruistic agents a_a a natural wheel or star structure of GRN is formed (see Fig. 8).

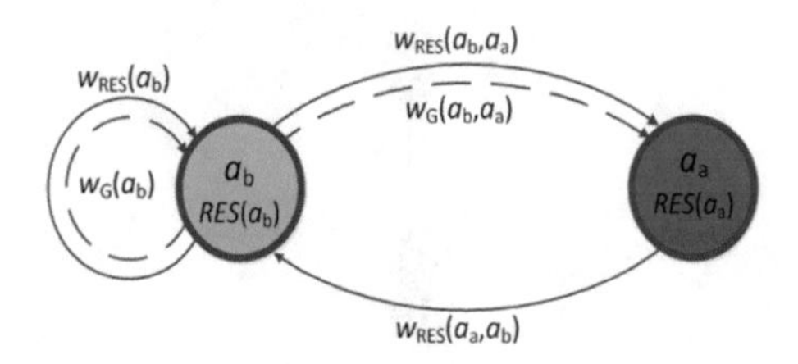

Fig. 8. A goal-resource network for the communication between benevolent agent and altruistic agent

In all the considered examples of GRN, some constant types of agents are analyzed that do not change in the process of communication. Besides, interesting situations of modifying agent type by changing the amount of available resources or goal settings are of special concern, for instance, a transmutation of benevolent agent a_b to egoist agent a_e or an acquisition of purposefulness by an altruistic agent a_a.

Generally the agent's type is not crisp, and we have to give a triple of membership functions μ_{ab}, μ_{ae}, μ_{aa} on the set $K = \{a_b, a_e, a_a\}$ (the degrees of benevolence, egoism and altruism respectively).

It is well-known from graph theory that the degree of the vertex is determined by the number of arcs incident to this vertex. For goal-resource networks we should take into consideration both the number of outgoing arcs and their total conductivity. Thus, the potential influence of agent in a multi-agent system is given by a triple

$$I(a_i(t)) = \langle res(a_i(t)), w_S(a_i(t)), m(a_i(t)), (t)\rangle,$$

where $res(a_i(t)$ stands for the amount of resources in the agent a_i at the time t, $w_S(a_i(t))$ is the sum number of conductivities related to a_i at the time t and $m(t)$ is the number of agents interacting with a_i at the time t. The main influence criterion is agent-to-resource relation. The more are the agent resources, the more he can influence other agents and the more is the amount of potentially reachable resources for him.

6 Conclusion

An idea of total agentification for networked enterprise has been put forward that means its consideration as a family of various heterogeneous MAS forming artificial societies. The generation and simulation of AS on the basis of goal-resources networks seems to be a useful approach to investigating social dimension in Industry 4.0 and its components and tools. The concept of adequacy between agent architecture and required formal representations has been discussed. An original agent's architecture GRPA (Goal – Resources – Perception – Action) has been proposed. The CGRN (Colored Goal-Resource Networks) have been introduced in the paper and the appropriate archetypes of communications between agents have been constructed, visualized and investigated.

In our further work, we hope to simulate and compare various types of AS in the context of Industry 4.0 and develop an ontological system for mutual understanding and joint work of artificial agents in such societies.

References

1. Atzori, L., Iera, A., Morabito, G., Nitti, M.: The social internet of things (SIoT) – when social networks meet the internet of things: concept, architecture and network characterization. Comput. Netw. **56**(16), 3594–3608 (2012)
2. Baur, C., Wee, D.: Manufacturing's Next Act. McKinsey & Company Operations, June 2015. http://www.mckinsey.com/businessfunctions/operations/our-insights/manufacturings-next-act
3. Diestel, R.: Graph Theory. Springer, Heidelberg (2005)
4. Dressler, F.: Cyber physical social systems: towards deeply integrated hybridized systems. In: 2018 International Conference on Computing, Networking and Communications (ICNC), pp. 420–424 (2018)
5. Dyundyukov, V.S., Tarassov, V.B.: Goal-resource networks and their application to agents communication and co-ordination in virtual enterprise. In: Proceedings of the 7th IFAC Conference on Manufacturing Modelling, Management and Control, MIM 2013, 19–21 June 2013, St. Petersburg, Russia, IFAC Proceedings, vol. 46, no. 9, pp. 347–352 (2013)
6. Emelyanov, V.V.: A multi-agent model of decentralized resource management. In: Proceedings of the International Conference on Intelligent Control – New Intelligent Technologies in Control Problems, ICIT 1999, 6–9 December 1999, Pereslavl-Zalesski, Yaroslavl Region, Russia. PhysMathLit, Moscow, pp. 121–126 (1999). (in Russian)
7. Epstein, J., Axtell, R.L.: Growing Artificial Societies: Social Science From the Bottom Up. MIT Press, Cambridge (1996)
8. Faneuff, J.: Designing for Collaborative Robotics. O'Reilly, Sebastopol (2016)
9. Ford Jr., L.R., Fulkerson, D.R.: Flows in Networks. Princeton University Press, Princeton (2010)
10. Frazzon, E.M., Hartmann, J., Makuschewitz, Th., Scholtz, B.: towards socio-cyber-physical systems in production networks. Procedia CIRP **7**, 49–54 (2013)
11. Harary, F.: Graph Theory. Addison–Wesley, Reading (1994)
12. Kagermann, H., Helbig, J., Hellinger, A., Wahlster W.: Recommendations for Implementing the Strategic Initiative Industrie 4.0: Securing the Future of German Manufacturing Industry. Final Report of the Industrie 4.0 Working Group (2013)
13. Karpov, V.E., Tarassov, V.B.: Synergetic artificial intelligence and social robotics. In: Abraham, A., Kovalev, S., Tarassov, V., et al. (eds.) Proceedings of the 2nd International Scientific Conference «Intelligent Information Technologies for Industry», IITI 17. Advances in Intelligent Systems and Computing, 14–16 September 2017, Varna, vol. 679, pp. 3–15. Springer, Cham (2018)
14. Kernbach, S. (ed.): Handbook of Collective Robotics: Fundamentals and Challenges. Pan Stanford Publishing Pte. Ltd., Singapore (2013)
15. Kudrin, B.I.: Introduction to Technetics. TGU Press, Tomsk (1993). (in Russian)
16. Kuznetsov, O.P.: Homogeneous resource networks. 1. complete graphs. Autom. Remote Control **11**, 136–147 (2009). (in Russian)

17. Kuznetsov, O.P., Zhilyakova, L.Yu.: Flows and limit states in bidirectional resource networks. In: Proceedings of the 18th IFAC World Congress, 28 August - 2 September 2011, Milano, Italy, pp. 14031–14035 (2011)
18. Li, J.Q., Yu, F.R., Deng, J., et al.: Industrial internet: a survey of the enabling technologies, applications and challenges. IEEE Commun. Surv. Tutorials **19**(3), 1504–1526 (2017)
19. Liu, H., Sun, J., Zhang, H., Liu, L.: Extended Pawlak's flow graphs and information theory. In: Transactions on Computational Science V. LNCS, vol. 5540, pp. 220–236. Springer, Heidelberg (2009)
20. McEven, A., Cassimally, H.: Designing the Internet of Things. Wiley, Chichester (2014)
21. Mesbahi, M., Egerstedt, M.: Graph Theoretic Methods in Multiagent Networks. Princeton University Press, Princeton (2010)
22. Pawlak, Z.: Flow graphs and data mining. In: Transactions on Rough Sets III, pp. 1–36. Springer, Berlin (2005)
23. Peshkin, M., Colgate, J.: Cobots. Indus. Robots **26**(5), 335–341 (1999)
24. Pospelov, D.A.: Modeling of deeds in artificial intelligence systems. Appl. Artif. Intell. **7**(1), 15–27 (1993)
25. Pospelov, D.A., Varshavsky, V.I.: The Orchestra Plays without a Conductor. Reflections on Evolutionand Control of Some Technical Systems. Librokom Publishing House, Moscow (2009). (in Russian)
26. Radziwon, A., Biberg, A., Bogers, M., Madsen, E.S.: The smart factory: exploring adaptive and flexible manufacturing solutions. Procedia Eng. **69**, 1184–1190 (2014)
27. Ronzhin, A., Rigoll, G., Meshcheryakov, R. (eds.): Interactive collaborative robotics. In: Proceedings of the First International Conference (ICR 2016). LNAI, vol. 9812. Springer, Berlin (2016)
28. Schwab, K.: The Fourth Industrial Revolution. World Economic Forum, Geneva (2016)
29. Sehgal, N.K., Bhatt, P.Ch.: Cloud Computing. Springer, Heidelberg (2018)
30. Smirnov, A.V., Levashova, T.V., Kashevnik, A.M.: Ontology-based cooperation in cyber-physical social systems. In: Industrial Applications of Holonic and Multi-Agent Systems. LNCS, vol. 10444, pp. 66–79 (2017)
31. Suh, S.C., Tanik, U.J., Carbone, J.N., Eroglu, A.E. (eds.): Applied Cyber-Physical Systems. Springer, Heidelberg (2014)
32. Tarassov, V.B.: From Multi-Agent Systems to Intelligent Organization. Editorial URSS, Moscow (2002). (in Russian)
33. Tarassov, V.B.: Cognitive sensors: interpreting measurement results. Soft Meas. Comput. **1**, 133–140 (2017)
34. Wang, Y.: On cognitive computing. Int. J. Softw. Sci. Comput. Intell. **1**(3), 1–15 (2009)
35. Zeng, J., Yang, L.T., Lin, M., Ning, H., Ma, J.: A survey: cyber-physical-social systems and their system-level design methodology. Future Gen. Comput. Syst. **56**, 504–522 (2016)

Data Mining and Knowledge Discovery in Intelligent Information and Control Systems

Context-Dependent Guided Tours: Approach and Technological Framework

Alexander Smirnov[1], Nikolai Shilov[1](✉), and Oleg Gusikhin[2]

[1] SPIIRAS, 39, 14th line, St. Petersburg 199178, Russian Federation
{smir,nick}@iias.spb.su

[2] Ford Motor Company, 2101, Village Rd., Dearborn, MI, USA
ogusikhi@ford.com

Abstract. Development of information and communication technologies causes appearance of new various concepts (Internet of Things, e-business, e-government, etc.). The paper proposes an on-going research aimed at integration of the concepts of connected car as a product-service system (system that integrates physical components and electronic services) with e-tourism. The concept and of a system aimed at context-dependent planning and dynamic adaptation of guided tourist rides in a car are described. The system is based on the usage of car connectivity technologies and cloud-based services and is built around the previously developed by the authors tourist support system TAIS.

Keywords: Infomobility · Smart city · Connected car · E-tourism
Ad-hoc tour planning

1 Introduction

Modern technologies substantially change our lives and drive appearance of new various concepts (Internet of Things, e-business, e-government, etc.). "Infomobility" infrastructure is one of the key elements of the smart city. It stands for operation and service provision schemes whereby the use and distribution of dynamic and selected multi-modal information to the users, both pre-trip and, more importantly, on-trip, play a fundamental role in attaining higher traffic and transport efficiency as well as higher quality levels in travel experience by the users [1]. Infomobility is a new way of service organization appeared together with the development of personal mobile and wearable devices capable to present user multimodal information at any time. Infomobility plays an important role in the development of efficient transportation systems, as well as in the improvement of the user support quality.

In-vehicle electronic systems are developing fast and accumulating new and new features related to connectivity, information support and entertainment. Such systems have transformed from simple audio players to complex solutions, referred to as "infotainment systems". They are not only capable to connect to smartphones but also can share information from different vehicle sensors and present various information through in-vehicle screen (visual information) or stereo system (audio information) [2]. Such systems make it possible to treat a car not as a single product but as a

A. Abraham et al. (Eds.): IITI 2018, AISC 874, pp. 43–50, 2019.
https://doi.org/10.1007/978-3-030-01818-4_4

product-service system [3, 4], i.e. system that tightly integrates physical products and software services.

Usage of infotainment systems does not only improve the driver's experience but also opens a wide range of possibilities through the "connected car" technologies (e.g., [5]). "Connected car" or "connected vehicle" is a relatively new term originating from the Internet-of-Things vision. It stands for the vehicle's connectivity with its surroundings on a real time basis for providing the safety and expedience to the driver [6].

The paper extends the previously presented work [7, 8] aimed at introduction of application of the concept of infomobile systems to support customized on-demand tours based on the "Connected car" technology. The paper is structured as follows. Section 2 introduces e-tourism, main directions of its development and its connection to transportation services. The concept of the developed system is presented in Sect. 3. It is followed by the description of the approach to context-driven tour generation and its evaluation. Main results are summarized in the conclusion.

2 E-Tourism

Due to availability of information through Internet, tourists are becoming more active and wanting to explore new sights on their own. The sights, in turn, e.g. museums, palaces, parks, etc. have been upgrading their offerings with interactive displays, audio tours, and other [e.g., 9, 10]. It had become more interesting to spend time at the sights than just getting a quick view when passing by. The experiments show that usage of mobile applications allows tourists to discover more sights and stay at them longer than usage of conventional means (paper maps and books) [11, 12]. The internet allows tourists to do their own research and planning before arriving, allowing them to use public transportation or other means to see the main sights.

Transport service providers have been reacting to this shift introducing various services for individual tourists. For example, local commuter networks had introduced 2- and 3-day travel passes, directly targeting the tourists' transportation needs. Hop-on-hop-off bases are available in many tourist-oriented cities. They provide for multi-day ticket that allow customers unlimited use of the sightseeing buses, entering and exiting at convenient stops (sights, hotels and train stations), at their own schedule. This makes it possible to increase the utilization by each customer, spread the use over the whole day, and allow the operator to get more utilization out of each bus, instead of optimizing for peak periods.

Uber has launched UberTour in some cities that integrates the concept of Uber (ad-hoc car reservation using a mobile app) and hop-on-hop-off buses that drive the tourist through a number of pre-defined attractions.

Analysis of existing at the moment apps [1, 13] in the market shows that there is a trend towards providing proactive tourist support based on his/her location, preferences, and current situation in the area (weather, traffic jams, and etc.) [14]. Development of such systems is still an actual task that attracts researchers from all over the world (e.g., [15–18]). Such systems are aimed to solve the following tasks:

- generate recommended attractions and their visiting schedule based on the tourist and region contexts and attraction estimations of other tourists. The tourist context characterizes the situation of the tourist, it includes his/her location, co-travelers, and preferences; the region context characterizes the current situation of tourist location area, it includes such information as weather, traffic jams, closed attraction, etc.
- collect information about attractions from different sources and recommend the tourist the best for him/her attraction images and descriptions;
- propose different transportation means for reaching the attraction;
- update the attraction visiting schedule based on the development of the current situation.

3 The Concept

The system being described assumes ad-hoc generation of a tour route and support of the tour (guiding, narration, presentation of additional information) based on the personal schedule analysis, current situation monitoring and prediction, and integration of the on-board infotainment system with personal smartphone/tablet and external services via Ford's SYNC Applink API (Fig. 1). It is based on the earlier developed by the authors personalized tourist assistance service (TAIS) [19].

The tour is generated based on the preferences stored in the tourist's profile, situation in the area and its possible development (e.g., regular traffic jams during rush hours can be easily predicted), and available information about attractions. It is proposed to the tourist, who has a possibility to modify the tour.

Based on the number of tourists a car of required type (standard, minivan, van, etc.) is selected automatically. The driver is notified about the tour order and, if he/she accepts the tour, picks up the tourist(s) at the time and location indicated in the on-board navigation system. Then, the driver just follows the tour route loaded into the car's navigation system.

Tourist(s) can use their smartphones/tablets during the tour for narration, imagery and video synchronized with the vehicle's location, speed and orientation. The guiding can be done through an on-board infotainment system (this is especially convenient when a minivan is used for a group of tourists). Guiding information is extracted from accessible in the smart city services and predefined libraries.

The tourist is also able to communicate in some extent through the SYNC system with the driver if he/she doesn't speak the person' language (e.g., ask for a stop near an attraction) and control some car elements as opening/closing windows or adjusting climate control.

The system takes into account the person's preferences described in his/her profile (preset and revealed) and the current situation at the location (season, weather, traffic jams, etc.) to anticipate what the passenger would want and need.

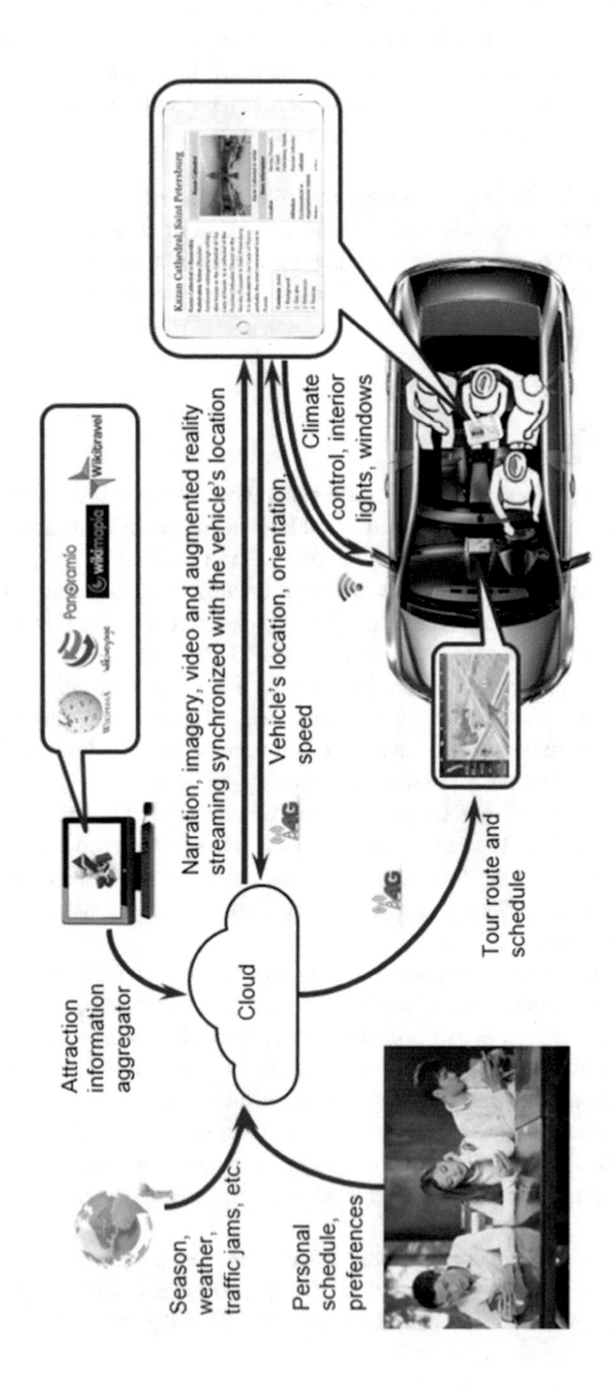

Fig. 1. The overall scheme of the on-board dynamic tour support system

4 Adaptive Guidance

For adaptive guidance an approach to creating a context-aware personalized narrative from fragments similar to an assembly line is proposed (Fig. 2). It assumes that there is a library of (relatively) short fragments (called "clips") each described with subject, running time, interests (keywords), and additional contextual parameters (appropriate weather, location, direction). Additionally, constraints are introduced that define alternative (mutually excluding) clips, clip restrictions (one clip cannot be played without another one), and clip sequences (one clip cannot be played after another one). The carried out analysis has shown that text to speech generation is not an appropriate solution for this task since long listening of monotonous narration is not comfortable. As a result, it was decided that clips had to be pre-recorder by human narrators. Besides, the cars tested are equipped with enough storage space to store audio clips for a region (e.g., St. Petersburg, Russia).

The system analyses the current context and builds a sequence of clips of the given duration that is the most appropriate to the current situation and tourist's preferences. It can be seen in the figure that the narration for a given duration can be assembled from available fragments in a number of ways. As a result, there is a task of selecting those, which do not only fit the given duration but also match the context (both the tourist context and the region context) in the best way.

This task can be formalized as follows:

$o_i = (t_i, R_i, C_i, W_i)$, where

o_i – is a narrative fragment (clip) i, $i = 1 \ldots n$,

t_i – duration of clip i,

R_i – is a set of constraints related to clip i,

C_i – is a set of contextual parameters of clip i,

W_i – is a set of associated weighted keywords.

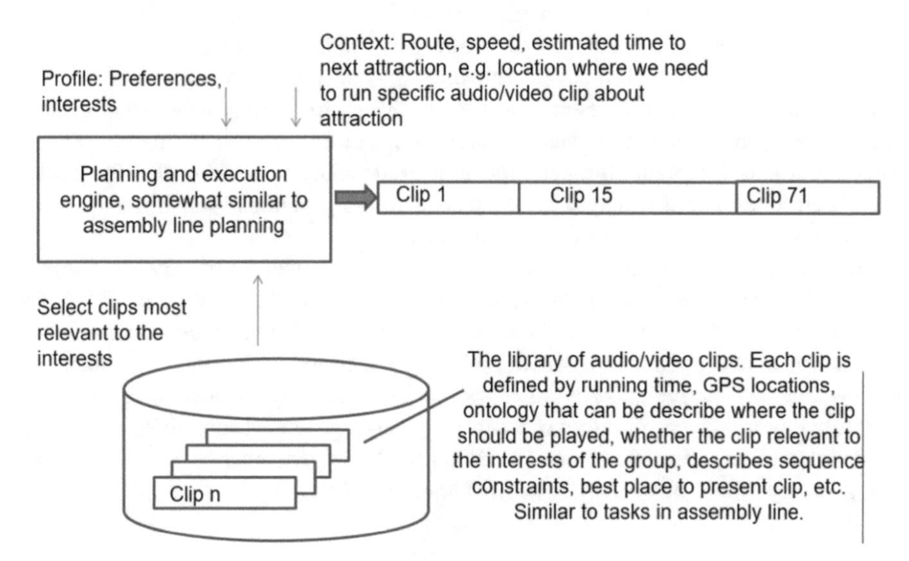

Fig. 2. Adaptive audio guidance

The vector $X = (x_1, x_2, \ldots, x_n)$ is a solution of the task, where x_i is a Boolean variable denoting that clip i is used in the solution ($x_i = 1$) or not ($x_i = 0$).

The duration constraint is formalized as $|\Delta t| \leq t_t k_t$, where k_t is an acceptable deviation from the target duration (e.g., 0.2), and t_t is the target duration.

$\Delta t = t_t - t_{sum}$, where $t_{sum} = \sum_{i=1}^{n} x_i t_i$ is the actual duration of the built narrative.

The usefulness of the clip for the tourist is based on the matching of the weighted keywords associated with the clip (W_i) and weighted keywords associated with the tourist (W_j): $u_i = f(W_i, W_j)$. This can be done, for example, based on the cosine function.

The contextual parameters are processed individually, and then they are aggregated into a weighted sum (c_i). For example, the value of the clip can be decreased if it is aimed at watching at something remoted and the weather is rainy or foggy.

The overall goal function is $u_{sum} = \sum_{i=1}^{n} u_i t_i c_i \rightarrow max$.

5 Evaluation

With the purpose of proving the concept an experiment has been conducted aimed to estimate (1) the acceptance of the generated narrations by people as well as (2) the capability of the system implementing the approach to generate narrations similar to those prepared by people.

The Evangelical Lutheran Church of St. Michael located in St. Petersburg, Russia[1], was chosen as an attraction since (i) it is not very well known and the information about it would be new for the participants, and (ii) it is located near the place where the experiment was conducted, so the participants would be interested to get to know something new about it. The narration fragments were created based on the Wikipedia description as well as other descriptions found in the Internet, with total amount of fragments being equal eight. Each of fragments was annotated with keywords, importance and length. This fragments were read by a human narrator and recorded.

The experiment was conducted in two stages. First, 13 participants selected their interests among the proposed seven keywords, and were given a task to create a tour out of available fragments (presented as text) of the approximately given length. At the second stage, the system generated an audio tour out of recorded fragments with the same input parameters and the participants were listening to it. After that the participants were interviewed if they were satisfied with the generated tour, if the generated tour was better than one created by them and why.

It was observed that in most cases the system generated tours were the same as those created by human participants (precision 73%, recall 79%, F-Measure 76%). The interviews showed that deviations were mainly due to different understanding of the importance of the fragments by them and by the team preparing the experiment. As well as different interpretation of the keywords. Besides, people tended to keep the tour length below the given time frame, when the system generated tours both longer and shorter. However, all participants agreed that the tours generated by the system were fine and they would be satisfied when listening for such tours.

[1] https://www.google.com/maps/@59.9444108,30.2832951,20z.

6 Conclusion

The paper describes the concept and approach to creating a system aimed at context-dependent planning and dynamic adaptation of guided tourist rides in a car based on the usage of car connectivity technologies and cloud-based services. The system is based on the integration of the previously developed by the authors tourist support system TAIS with Ford SYNC Applink. The paper concentrates on the approach to automatic generation of tours based on the context and tourist's preferences. The approach is based on the "assembly line" principle, when the tour is generated based on available fragments. The carried out experimentation confirmed that the clips generated are not significantly different from those created by people and are well accepted by the experiment participants.

Acknowledgements. The paper is partially due to the project sponsored by the Ford University Research Program, State Research # 0073-2018-0002, and projects funded by grants ## 18-07-01201, 18-07-01272, 17-29-03284 of the Russian Foundation for Basic Research.

References

1. Ambrosino, G., Nelson, J.D., Bastogi, B., Viti, A., Romazzotti, D., Ercoli, E.: The role and perspectives of the large-scale flexible transport agency in the management of public transport in urban areas. In: Ambrosino, G., Boero, M., Nelson, J.D., Romanazzo, M. (eds.) Infomobility Systems and Sustainable Transport Services, pp. 156–165. ENEA, Rome (2010)
2. Taramov, A., Shilov, N.: A systematic review of proactive driver support systems and underlying technologies. In: Balandin, S., Levina, A., Tyutina, T. (eds.) Proceedings of the 20th Conference of Open Innovations Association FRUCT, St. Petersburg, Russia, 3–7 April 2017, pp. 448–459. ITMO University, St. Petersburg (2017)
3. Vezzoli, C., Kohtala, C., Srinivasan, A., Xin, L., Fusakul, M., Sateesh, D., Diehl, J.C.: Product-Service System Design for Sustainability. Routledge, Abingdon (2017)
4. Tukker, A., Tischner, U. (eds.): New Business for Old Europe: Product-Service Development, Competitiveness and Sustainability. Routledge, Abingdon (2017)
5. Cherubini, S., Iasevoli, G., Michelini, L.: Product-service systems in the electric car industry: critical success factors in marketing. J. Clean. Prod. **97**, 40–49 (2015)
6. Hashimoto, N., Kato, S., Minobe, N., Tsugawa, S.: Automated vehicle mobile guidance system for parking assistance. In: IEEE Intelligent Vehicles Symposium, pp. 630–635 (2007)
7. Shim, H.B.: The technology of connected car. J. Korea Inst. Inf. Commun. Eng. **20**(3), 590–598 (2016)
8. Smirnov, A., Shilov, N., Gusikhin, O.: Connected car-based customised on-demand tours: the concept and underlying technologies. In: Galinina, O., Balandin, S., Koucheryavy, Y. (eds.) Internet of Things, Smart Spaces, and Next Generation Networks and Systems, 16th International Conference NEW2AN 2016 and 9th Conference ruSMART 2016. Lecture Notes in Computer Science, St. Petersburg, Russia, 26–28 September 2016, vol. 9870, pp. 131–140. Springer, Heidelberg (2016)

9. Smirnov, A., Shilov, N., Gusikhin, O.: Cyber-physical-human system for connected car-based e-tourism. In: Proceedings of 2017 IEEE Conference on Cognitive and Computational Aspects of Situation Management, Savannah, GA, USA, 27–31 March 2017. IEEE Communications Society (2017)
10. Giuseppe, M., Michela, B., Matteo, F., Chiara, F., Alessandro, M., Stefano, M., Fabiana, R.: QRCODE and RFID integrated technologies for the enhancement of museum collections. In: Euro-Mediterranean Conference, pp. 759–766. Springer, Cham, November 2014
11. Li, R.Y.C., Liew, A.W.C.: An interactive user interface prototype design for enhancing on-site museum and art gallery experience through digital technology. Mus. Manag. Curatorship **30**(3), 208–229 (2015)
12. Modsching, M., Kramer, R., Ten Hagen, K., Gretzel, U.: Effectiveness of mobile recommender systems for tourist destinations: A user evaluation. In: 4th IEEE Workshop on Intelligent Data Acquisition and Advanced Computing Systems: Technology and Applications, IDAACS 2007, pp. 663–668. IEEE, September 2007
13. Price, E.: Travel apps that can replace your tour guide (2015). http://www.cntraveler.com/stories/2015-02-24/travel-apps-that-can-replace-your-tour-guide
14. Cowen, B.: A personal tour guide – almost everywhere – for $9.99 or less! (2015). http://www.johnnyjet.com/2015/01/a-personal-tour-guide-almost-everywhere-for-9-99-or-less/
15. Smirnov, A., Kashevnik, A., Ponomarev, A., Shilov, N., Teslya, N.: Proactive recommendation system for m-tourism application. In: Johansson, B., Andersson, B., Holmberg, N. (eds.) Perspectives in Business Informatics Research, Proceedings of the 13th International Conference on Business Informatics Research, BIR 2014. LNBIP, vol. 194, pp. 113–127. Springer, Heidelberg (2014)
16. Gerhardt, T.: 3 Ways Multi-Modal Travel is Tricky for App Developers (2015). http://mobilitylab.org/2015/03/10/3-ways-multi-modal-travel-is-tricky-for-app-developers/
17. Staab, S., Werthner, H., Ricci, F., Zipf, A., Gretzel, U., Fesenmaier, D.R., et al.: Intelligent systems for tourism. IEEE Intell. Syst. **17**(6), 53–64 (2002)
18. Hasuike, T., Katagiri, H., Tsubaki, H., Tsuda, H.: Interactive approaches for sightseeing route planning under uncertain traffic and ambiguous tourist's satisfaction. In: Eto, H. (ed.) New Business Opportunities in the Growing E-Tourism Industry, pp. 75–96. Business Science Reference, Hershey (2015)
19. Smirnov, A.V., Kashevnik, A.M., Ponomarev, A.: Context-based infomobility system for cultural heritage recommendation: Tourist Assistant—TAIS. Pers. Ubiquit. Comput. **21**(2), 297–311 (2017)

Retention to Describe Knowledge of Complex Character and Its Formalization in Category Theory

A. V. Zhozhikashvily[1] and V. L. Stefanuk[1,2(✉)]

[1] Institute for Information Transmission Problems,
Bolshoy Karetny per. 19, 127051 Moscow, Russia
stefanuk@iitp.ru
[2] Peoples' Friendship University of Russia, Miklucho-Maklaya str. 6,
117198 Moscow, Russia

Abstract. Knowledge based computer systems may be designed for many environments, demonstrating different patterns of behavior. Though the inferences obtained may be similar, the use of the inferences may require some supplementary information directly related to the subject domain properties. For such complex cases a concept of retention is proposed in this paper intended for applications to a wide variety of situations, where an intelligent system may depend upon complex external conditions. Examples of systems designed by the authors are provided that support this new concept. In conclusion, an attempt to formalize this concept in the language of Category Theory is provided.

Keywords: Retention · Environment · Dynamic systems
Dynamic and Static Knowledge · Autonomic Systems · Lattice
Monoid · Defeasible reasoning

1 Introduction

Rule-based computer systems are designed for many environments, demonstrating different patterns of behavior. Though the inferences obtained may be similar in some systems, the use of the inferences may require some supplementary information directly related to the subject domain properties. For such complex cases a concept of retention is proposed in this paper intended for applications to a wide variety of situations, where an intelligent system may depend upon various external conditions.

The behavior of an intelligent system operating in dynamic environments may serve as a bright example of such an altered behavior. These systems were first introduced in publications [1, 2]. In the proposed dynamic intellectual system, the data that supporting the inference may change in time, became outdated and lead to outdating the inferences that were already made. The present authors worked on the dynamic intellectual systems for many years and they worked out the following approach.

With any facts, both original one and the one obtained in the process of logical inference, a certain supplementary parameter was connected. This parameter describes

A. Abraham et al. (Eds.): IITI 2018, AISC 874, pp. 51–58, 2019.
https://doi.org/10.1007/978-3-030-01818-4_5

some additional properties of the result of inference. Say this parameter may show if the truth-value of the obtained fact is valid until a certain moment in time, and it should re-checked afterwards [3]. As we found several types of such parameters, we decided to give it a special name - retention[1]. Some examples will be demonstrated below showing that the retention is not necessary numerical value.

Several references might be found that treat subjects that are rather close to retention concept. Thus, in the serious of important publications by Pollock [4] it was noted that there are decisions of two types – defeasible and non-defesible ones, where defeasible reasoning turns to be close to non-monotonic reasoning, and non-defesible to be close to "strict logic".

This line of thinking continues in [5, 6] where defeasible reasoning is used to resolve logical contradictions.

Actually all above and our retention concept follow some previous studies of the Frame Problem, where the question of reconsideration of inferences was first formulated [7]. Actually in our practical mobile intelligent system [1, 2 and 3] this problem found a practical solution in Seismology.

Nevertheless we decided to introduce a concept of retention to be able to formalize this notion to be able to build a strong theoretical base for it with the further goal to use it within the category language [8] that we introduced before to make intelligent rule-based systems.

Before going further let us note that the properties described via concept of retention are especially significant for autonomic intelligent systems. The role of retention may be the most important for such systems, as there is no an "external owner", i.e. the author or the designer, who is able to correct the performance of intellectual system if the environment is changing considerably.

The paper is organized in the following way.

In the part 2 "Retention" the concept of retention is discussed from the point of view of ordering on the set of retention.

In the part 3 "Operations with retentions" the operations over the set of retention are described that allows calculating the retentions inference from the retentions of facts used to infer it.

In the part 4 "Retention Monoid" the concept of monoid of retention is introduced. In the frames of this algebraic structure the operation mentioned above are implemented.

In the part 5 "Retentions in Theory of Category Language" a multiple category is defined, which is being built using a Category and the Monoid.

2 Retention

The simplest kind of retention in dynamic systems would be the information if the established truth of a certain fact might change in time, or the fact is always true and never requires verification. The facts of the first type have been called as static facts, others were referred to as dynamic ones.

---

[1] Retention means storing, keeping, preserving and etc. (from latin *Retentio*).

In more complex cases the retention pointed not only to the information that the truth-value of the fact might change, but also informs on the conditions showing when the truth-vale should be verified. For example, with each fact there may be related it's inherent "life time". The retention in this case might report until what time the truth-value of the fact is preserved. After this moment the truth of the fact must be verified.

The retention may describe *certain conditions* under which the truth of the fact is remained. Or even some Boolean expression may be used that should be true for the preservation of the fact. The last two cases demonstrate examples when the retention is not a numerical value.

The set of retentions may be naturally ordered, considering that retentions p and q are connected with inequality $p \leq q$, if in the situation where the truth-value of the fact with retention p is preserved, the truth-value of fact with the retention q is also preserved. In general such an ordering is not necessary linear one. In the examples listed above the ordering is defined in the following way.

If facts may be considered as static or dynamic and static factors are denoted by the number 1, while dynamic ones are denoted with 0, then the usual ordering on the set $\{0, 1\}$ is considered. If retention is a moment in time, and until this moment there is no need in reconsidering of the fact, then the order is usual ordering on the numerical scale of the moments in time.

If the retention is the set of some conditions, then the inequality $p \leq q$ means an inclusion $q \subset p$. At last, if retention is a Boolean, then inequality $p \leq q$ means the logical implication $p \Rightarrow q$.

3 Operations with Retentions

In order to be able to describe the retentions for inferred facts we need a collection of operations with retentions. Of the first importance, however, are operations AND-compositions and OR-compositions.

In such systems if the fact f_n was inferred with the help of some rule from facts $f_1, \cdots, f_{n-1}$, then the retention of the fact f_n is defined as AND-composition of retentions of facts $f_1, \cdots, f_{n-1}$. It means that the truth-value of the fact f_n is preserved when the truth-values of all facts $f_1, \cdots, f_{n-1}$ are preserved.

OR-compositions is used in the case when a certain fact may be obtained from various rules and its retention is obtained as an OR-compositions of retentions that are obtained from retentions obtained from each of the rules. It means that the truth-value of the fact is preserved if the activity of at least one of the rules is preserved.

In our researches we assumed that the partially ordered set of retentions presents a lattice [9]. This is quite reasonable convention. It fulfilled in the above examples. In the case of lattice it is natural to consider the AND-composition and OR-composition to be $\wedge$ and $\vee$ i.e. the operations of computation of the least upper bound and the greatest lower bound.

The next example of the informational environment for the systems designed by present authors was a fuzzy environment. We applied rather popular in the Expert System Theory method. Namely with each of the facts it was associated a certainty factor with respect to the truth-value of the fact. Such an approach was proposed by Shortliffe for the system MYCIN [10], and it was widely used in Expert Systems area.

Later, especially with penetration of some ideas from Category Theory (that will be discussed in the current paper) it became clear that the certainty factors also may be considered as retentions. And despite of the fact that the certainty factors say nothing about the time interval when truth-value is preserved, we decided to apply the term "retention" also to them.

Like the previous retentions certainty factors in dynamic systems are ordered, and for them such operations as AND-composition and OR-composition may be defined as well.

The other environments different from fuzzy and dynamic ones, where the technique of retentions may be successfully applied may be also considered.

Let's take the system where Knowledge Base is created by the experts from different scientific areas who represent different scientific schools and approaches to science. In result some of its inferences may be recognized by all the experts, yet some other inferences may be rejected by a group of experts.

Let S be a set of all possible scientific schools. As the set of retentions P we will take the set of all subsets of the set S. As the ordering over P we will take the set-theory relation *inclusion*. And as the operations *AND*-composition and *OR*-composition we will take set-theoretical *union* and *intersection*.

By this we will achieve that the truth-value of fact f_3 under the rule $f_1, f_2 \rightarrow f_3$ will be admitted by those experts that agreed both with f_1 and f_2.

Similar model with somewhat different goals have been used to study joint behavior of a collective if intellectual systems. The designed system, using in its activity some definite facts, was able to take into account whether some other members of the collective able to influence on the truth-values of the original facts. When it is the case the facts under considerations and the inferences based on these facts must be reconsidered each time, when it became known that some important influencing system has made some changes to the Knowledge Base. (Let S be the set of all the systems included into the collective. Let P will be the set of all the subsets of the set S as it was in previous example.)

In many of the cases listed above the set of retentions is an algebraic lattice, and lattice operation may be used as the operations over retentions. However there are some exceptions, which, by the way, are very significant. In the publication [11] by using an axiomatic approach there were obtained formulas for combining of certainty factors. The built operations do not have idempotent property, and as a result they may not be represented as operations in some lattice. This fact forced us to refrain from the use of a lattice and look for more general algebraic object than algebraic lattice.

To build a multiple category that will be discussed below it would be enough if the set of retentions represent a mooned [12]. However, the monoid is too very general a concept, that does not include some important for the theory of retentions elements such as ordering.

4 The Retention Monoid

To include all such necessary elements let us introduce the concept of *Monoid of Retentions*. By the monoid of retentions we understand a commutative ordered monoid with involution that satisfies to an additional condition for its elements $p \leq q$ if and only if $p = q \wedge r$ for some element r of the monoid.

Let us provide some more detailed definition.

The Retention Monoid is a partially ordered set P with a binary operation $\wedge$ and with a unitary operation $\sim$ satisfying the next conditions:

(1) $(P, \wedge)$ is commutative monoid;
(2) $p \leq 1$ for every $p \in P$;
(3) If $p \leq q$, then $p \wedge r \leq q \wedge r$ for any $p, q, r \in P$;
(4) If $p \leq q$, then $p = q \wedge r$ for some $r \in P$;
(5) if $p \leq q$, then $\sim q \leq \sim p$;

From (2) and (3) it follows that relation $p \wedge q \leq q$ is valid for any $p, q \in P$.

And also in reverse direction: starting with the assertion $p \wedge q \leq q$, it is possible to obtain the assertion (2) assuming $q = 1$. It means that the conditions (2) and (4) in above definition may be replaced with the following one: $p \leq q$ if and only if the relation $p = q \wedge r$ is valid for some $r \in P$. The last step leads to the brief formulation of the definition that we started with.

In our retention monoid the operation $\vee$ may defined with the expression $x \vee y = \sim(\sim x \wedge \sim y)$. In this case $(P, \vee)$ will be also commutative monoid, where the role of unit would be played with $0 = \sim 1$.

Let us give examples for retention monoids corresponding to various environments.

1. P is a distributive complemented lattice, $\leq$ и $\wedge$ are the common ordering and lower – common ordering and greatest lower bound of the lattice., $\sim x$ – дополнение к x.

In this case the operation $\vee$ will be least upper bound of the lattice. Such lattices were considered by present authors previously in studies of their dynamic intellectual systems and have been referred to as context lattices. The distributivity is required in order to provide the uniqueness of complements.

2. $P = [0, 1]$, common ordering, $x \wedge y = \min(x, y)$, $\sim x = 1 - x$.

In this case one has $x \vee y = \max(x, y)$. With such definition of operations $\wedge$ and $\vee$, it represents the popular choice of operations AND and OR as the compositions of certainty factors. Note that current example is not a special case of example 1 as thought the line [0,1] with operations min and max is a lattice, yet the operation $\sim x = 1 - x$ is not a complementation.

3. $P = [0, 1]$, usual ordering, $x \wedge y = xy$, $\sim x = 1 - x$.

In this case $x \vee y = x + y - xy$. Thus the operations $\wedge$ and $\vee$ correspond to proposed in [11] formulas for *AND*-composition and *OR*-composition of certainty factors.

The above examples demonstrate that the retention monoid may be used both in fuzzy and dynamic situations. By the way, in both cases with monoid instruments it is possible to describe various ways for arranging AND and OR compositions.

Retention monoid is a very unusual algebraic structure. It has a lot in common with the lattice: a pair of binary operations related to the ordering in a similar manner. However in the common case it is not a lattice. The main difference consists in the obstacle that the operations are not idempotent. In any lattice there are valid the distribution inequality:

$$p \wedge (q \vee r) \geq (p \wedge q) \vee (p \wedge r),$$

$$p \vee (q \wedge r) \leq (p \vee q) \wedge (p \vee r).$$

In the retention monoid, where operations are defined with the expressions suggested in [11], $p \wedge q = pq$, $p \vee q = p + q - pq$, one has

$$p \wedge (q \vee r) = pq + pr - pqr,$$

$$(p \wedge q) \vee (p \wedge r) = pq + pr - p^2qr.$$

or

$$p \wedge (q \vee r) \leq (p \wedge q) \vee (p \wedge r),$$

the first inequality differs from that shown above as it has changed its sign to the opposite one.

5 Retentions in Theory of Category Language

In what follows it will be shown how to include the concept of retention in the previously built our theory of category model of production AI system [8]. The basic element of this category model was the notion of a pattern and associated with it some matching and concretization operations. In the developed theory of category language the patterns correspond to some morphisms and the mentioned operations consist in search of morphisms to make certain diagrams commutative [8].

The required separation of the subject domain knowledge from the environmental knowledge is achieved here with the use of the following two categories applied simultaneously. One category describes subject domain. The other category describes the character of information environment. Used for construction of knowledge base resulting category is a product of these two categories. This arrangement allows transposing the resulting category system to various other environments by replacing the second category.

The product of two categories in CT is built in the following way.

Let **C** and **D** be two categories. A pair (C, D) of two objects belonging to categories **C** and **D** becomes an object in the product category. If C_1, C_2 are the objects of the category **C**, and D_1, D_2 are the objects of category **D**, then the morphism (C_1, D_1)

(C_2, D_2) is a pair of morphisms in the corresponding categories (f, g), where f: $C_1 \to C_2$, g: $D_1 \to D_2$ are morphisms in categories **C** and **D**. Hence if f_i: $C_i \to C_{i+1}$ and g_i: $D_i \to D_{i+1}$, $i = 1, 2$ are morphisms in the corresponding categories, then composition of the morphisms (f_1, g_1):$(C_1, D_1) \to (C_2, D_2)$ and (f_2, g_2):$(C_2, D_2) \to (C_3, D_3)$ is defined with an expression (f_1, g_1) $(f_2, g_2) = (f_1f_2, g_1g_2)$. Such an operation on the two categories is easy for implementation in a computer program.

In this paper we will consider a special case of proposed construction where the second category is a single object category. The definition of single object category is in fact the definition of the monoid of endomorphism of this object. This monoid in our theory is the example of retention monoid.

A category that presents a product of category **C** and a single object category that was built using our retention monoid will be referred to as *P*-multiple category. The objects of *P*-multiple category that was built using category **C** and monoid *P* are the same as the objects in **C.** If C_1 и C_2 are some objects of the category **C** and consequently the objects of the multiple category, then the morphism $C_1 \to C_2$ of multiple category in a the pair (f, p), where $f : C_1 \to C_2$ is a morphism of category **C**, and $p \in P$. The composition of morphisms of multiple category is defined with expression $(f_1, p_1)(f_2, p_2) = (f_1f_2, p_1p_2)$. Thus, the restrictions on the use of composition of morphisms are naturally related to restriction on the use of the constituents: the use of the morphism f_1f_2 is admissible, if admissible the use of both f_1 and f_2.

6 Conclusion

So, the theory of category language opens a possibility to study the behavior of intelligent systems in various environments. It is valuable observation is useful for applications as it shows the possibility to transpose the system to various environments by changing only the programming implementation of the retention monoid.

The proposed in this paper theory of retentions has let us give a uniform description of different systems such as fuzzy, dynamic, and social systems. Embedding the retention theory in our theory of category model and built previously technology of knowledge based systems opens a possibility to adjust a system to work with knowledge containing various kinds of retentions. For this it would be enough to change the implementation of the retention monoid.

Altogether the proposed retention theory will permit to consider some new environments with *a priori* unknown retentions.

Acknowledgements. We wish to express many thanks to anonymous reviewers for their valuable remarks that let us to improve the final paper.

The research was partially supported with Russian Fond for Basic Research, grants N 18-07-00736 and N 17-29-07053.

References

1. Stefanuk, V.L.: The behavior of quasi-static expert systems in changing fuzzy environment. In: Proceeding of 4th National Conference with International Participation Artificial Intelligence 1994, Rybynsk, vol. 1, pp. 199–203 (1994). (in Russian)
2. Stefanuk, V.L.: Dynamic expert systems. Kybernetes. Int. J. Syst. Cybern. **29**(5), 702—709 (2000). MCB University Press
3. Zhozhikashvily, F.V., Stefanuk, V.L.: Multiple categories for dynamic production system description. Theory Syst. Control **5**, 71–76 (2008). (in Russian)
4. Pollock, J.L.: Defeasible reasoning. In: Adler, J.E., Rips, L.J. (eds.) Reasoning: Studies of Human Inference and its Foundations, p. 31. Cambridge University Press, Cambridge (2006)
5. Vagin, V.N., Morosin, O.L.: Argumentation in intelligent decision support systems. J. Inf.-Measur. Control Syst. **11**(6), 29–36 (2013)
6. Vagin, V.N., et al.: Exact and Plausible Reasoning in Intelligent Systems, p. 704. Fizmatlit, Moscow (2004). (in Russian). Edited by Vagin, V.N., Pospelov, D.A
7. Patrick, H.: The frame problem and related problems in artificial intelligence. University of Edinburgh (1971)
8. Stefanuk, V.L., Zhozhikashvily, F.V.: Collaborating Computer: Problems, Theory, Applications. Nauka, Moscow (2007). (In Russian)
9. Birkhoff, G.: Lattice Theory, 3rd edn. American Mathematical Society, Col Pub, Providence (1967)
10. Rule-Based Expert Systems: The MYCIN Experiments of the Stanford Heuristic Programming Project. Addison-Wesley, Boston (1984). Edited by Buchanan, B.G., Shortliffe, E.H
11. Stefanuk, V.L.: Some aspects of expert system theory. Sov. J. Comp. Sci. (Tech. Kibernet.) **2**, 85–91 (1987). Moscow
12. Jacobson, N.: Basic Algebra 1, 2nd ed., Dover (2009). ISBN 978-0-486-47189-1

Proximity of Multi-attribute Objects in Multiset Metric Spaces

Alexey B. Petrovsky[1,2,3,4(✉)]

[1] Federal Research Center "Informatics and Control", Russian Academy of Sciences, Prospect 60 Letiya Octyabrya, 9, Moscow 117312, Russia
pab@isa.ru
[2] Belgorod State National Research University, Belgorod, Russia
[3] V.G. Shukhov Belgorod State Technological University, Belgorod, Russia
[4] Volgograd State Technical University, Volgograd, Russia

Abstract. The paper considers new classes of metric spaces of finite, bounded, measurable multisets. We discuss the possibilities to use new types of metrics, pseudometrics, quasimetrics, symmetrics for estimating proximity of objects with many numerical and/or verbal attributes that are represented as multisets. New indexes of similarity and dissimilarity of multi-attribute objects are used in new methods of group multiple criteria decision making.

Keywords: Multiset · Space of multisets · Metric · Pseudometric
Quasimetric · Symmetric · Multi-attribute objects
Similarity and dissimilarity of objects · Group multiple criteria decision making
Group verbal decision analysis

1 Introduction

In various theoretical and practical problems of decision making, artificial intelligence, pattern recognition, data processing, and other subject areas, it is often necessary to assess a similarity or dissimilarity of objects (alternatives, options) by their properties. There are known a lot of attribute spaces with various metrics for characterizing proximity of objects. We mention the spaces of vectors, sequences, functions, sets, fuzzy sets, and so on [3, 5, 6]. Metrics in spaces of multisets or sets with repeating elements were introduced firstly by the author [7, 9–12] and studied in a small number of other works [4, 13].

This paper describes new classes of multiset spaces. We discuss how to use new types of metrics, pseudometrics, quasimetrics, and symmetrics for estimating the similarity and dissimilarity of objects with many numerical and/or verbal attributes that exist in several different exemplars. We demonstrate that the proposed concepts allow representing and comparing multi-attribute objects. The new indexes of similarity and dissimilarity of objects, which generalize the known indexes for multisets, are used for solving problems of group multiple criteria choice.

A. Abraham et al. (Eds.): IITI 2018, AISC 874, pp. 59–69, 2019.
https://doi.org/10.1007/978-3-030-01818-4_6

2 Notions of Multiset Theory

Consider briefly the basic notions of multiset theory [2, 8, 12, 14]. *A multiset* $\boldsymbol{A}$ drawn from an ordinary (crisp) set $X = \{x_1, x_2, \ldots\}$ with different elements, which is called *a domain*, is a collection of groups of identical elements

$$\boldsymbol{A} = \{k_{\boldsymbol{A}}(x) \circ x \,|\, x \in X, k_{\boldsymbol{A}}(x) \in Z_+\}. \quad (1)$$

Here $k_A : X \to Z_+ = \{0, 1, 2, 3, \ldots\}$ is *a multiplicity function* that marks the number of occurrence of an element $x \in X$ in the multiset $\boldsymbol{A}$. It is denoted by the symbol $\circ$. A multiset becomes a set when $k_{\boldsymbol{A}}(x) = \chi_A(x)$, where $\chi_A(x) = 1$ if $x \in A$, $\chi_A(x) = 0$ if $x \notin A$.

The set Supp$\boldsymbol{A} = \{x \in \boldsymbol{A} | \chi_{\text{Supp}\boldsymbol{A}}(x) = \min[k_{\boldsymbol{A}}(x), 1]\}$ is named *a support set* or *carrier* of a multiset $\boldsymbol{A}$. The maximum value $\max_{x \in X} k_{\boldsymbol{A}}(x)$ of the multiplicity function is called *a height* alt$\boldsymbol{A}$ of the multiset $\boldsymbol{A}$. *The cardinality* of a multiset $\boldsymbol{A}$ is defined as a total number of all multiset elements: card$\boldsymbol{A} = |\boldsymbol{A}| = \sum_x k_{\boldsymbol{A}}(x)$, and *the dimensionality* of a multiset $\boldsymbol{A}$ is defined as a total number of all distinct elements: dim$\boldsymbol{A} = /\boldsymbol{A}/ = \sum_x \chi_{\boldsymbol{A}}(x)$,. For instance, the cardinality of a n-dimensional multiset $\boldsymbol{A} = \{k_{\boldsymbol{A}}(x_1) \circ x_1, \ldots, k_{\boldsymbol{A}}(x_n) \circ x_n\}$ over a n-element set $X = \{x_1, \ldots, x_n\}$ is equal to $|\boldsymbol{A}| = k_{\boldsymbol{A}}(x_1) + \ldots + k_{\boldsymbol{A}}(x_n) = m$, and the dimensionality of a multiset $\boldsymbol{A}$ is equal to $/\boldsymbol{A}/ = \chi_{\boldsymbol{A}}(x_1) + \ldots + \chi_{\boldsymbol{A}}(x_n) = n = |\text{Supp}\boldsymbol{A}| = |X|$.

Multisets $\boldsymbol{A}$ and $\boldsymbol{B}$ are said to be *equal* $(\boldsymbol{A} = \boldsymbol{B})$ when $k_{\boldsymbol{A}}(x) = k_{\boldsymbol{B}}(x)$. A multiset $\boldsymbol{B}$ is said to be *included* in a multiset $\boldsymbol{A}(\boldsymbol{B} \subseteq \boldsymbol{A})$ when $k_{\boldsymbol{B}}(x) \le k_{\boldsymbol{A}}(x), \forall x \in X$. Then the multiset $\boldsymbol{B}$ is named *a submultiset* or *multisubset* of the multiset $\boldsymbol{A}$. As for sets, the fulfillment of the inclusions $\boldsymbol{A} \subseteq \boldsymbol{B}$ and $\boldsymbol{B} \subseteq \boldsymbol{A}$ implies the equality of the multisets $\boldsymbol{A} = \boldsymbol{B}$.

A multiset is called *the constant multiset* $N_{[h]}$if $k_{\boldsymbol{N}[h]}(x) = h = \text{const}, h \in Z_+$, *the empty multiset* $\varnothing$ if $k_{\varnothing}(x) = 0$; *the maximal multiset* $\boldsymbol{Z}$ if $k_{\boldsymbol{Z}}(x) = \max_{\boldsymbol{A} \in \mathsf{A}} k_{\boldsymbol{A}}(x), \forall x \in X$, that is all multisets $\boldsymbol{A}$ of the multiset family A are submultisets of the multiset $\boldsymbol{Z}$.

Define the following operations on multisets:

union $\boldsymbol{A}_1 \cup \ldots \cup \boldsymbol{A}_n = \cup_{i \in I} \boldsymbol{A}_i = \{k_{\bigcup i\boldsymbol{A}i}(x) \circ x | k_{\bigcup i\boldsymbol{A}i}(x) = \max_{i \in I} k_{\boldsymbol{A}i}(x), x \in X\}$;

intersection $\boldsymbol{A}_1 \cap \ldots \cap \boldsymbol{A}_n = \cap_{i \in I} \boldsymbol{A}_i = \{k_{\bigcap i\boldsymbol{A}i}(x) \circ x | k_{\bigcap i\boldsymbol{A}i}(x) = \min_{i \in I} k_{\boldsymbol{A}i}(x), x \in X\}$;

arithmetic addition $\boldsymbol{A}_1 + \ldots + \boldsymbol{A}_n = \sum_{i \in I} \boldsymbol{A}_i = \{k_{\sum i\boldsymbol{A}i}(x) \circ x | k_{\sum i\boldsymbol{A}i}(x) = \min [\sum_{i \in I} k_{\boldsymbol{A}i}(x), k_{\boldsymbol{Z}}(x)], x \in X\}$;

arithmetic subtraction $\boldsymbol{A} - \boldsymbol{B} = \{k_{\boldsymbol{A}-\boldsymbol{B}}(x) \circ x | k_{\boldsymbol{A}-\boldsymbol{B}}(x) = \max[k_{\boldsymbol{A}}(x) - k_{\boldsymbol{B}}(x), k_{\varnothing}(x)], x \in X\}$;

symmetric subtraction $\boldsymbol{A}\Delta\boldsymbol{B} = \{k_{\boldsymbol{A}\Delta\boldsymbol{B}}(x) \circ x | k_{\boldsymbol{A}\Delta\boldsymbol{B}}(x) = |k_{\boldsymbol{A}}(x) - k_{\boldsymbol{B}}(x)|, x \in X\}$;

complement $\overline{\boldsymbol{A}} = Z - \boldsymbol{A} = \{k_{\overline{A}}(x) \circ x | k_{\overline{A}}(x) = k_{\boldsymbol{Z}}(x) - k_{\boldsymbol{A}}(x), x \in X\}$;

multiplication by a scalar (reproduction) $b \bullet \boldsymbol{A} = \{k_{b \bullet \boldsymbol{A}}(x) \circ x | k_{b \cdot \boldsymbol{A}}(x) = \min [b \times k_{\boldsymbol{A}}(x), k_{\boldsymbol{Z}}(x)], b \in Z_+, x \in X\}$;

arithmetic multiplication $\boldsymbol{A}_1 \bullet \ldots \bullet \boldsymbol{A}_n = \prod_{i \in I} \boldsymbol{A}_i = \{k_{\prod_i \boldsymbol{A}_i}(x) \circ x | k_{\prod_i \boldsymbol{A}_i}(x) = \min \left[\prod_{i \in I} k_{\boldsymbol{A}_i}(x), k_{\boldsymbol{Z}}(x)\right], x \in X\}$;

raising to the arithmetic power $A^n = \{k_A n(x) \circ x | k_A n(x) = \min[(k_A(x))^n, k_Z(x)], x \in X.\}$

direct multiplication $A_1 \times \ldots \times A_n = \{k_{A_1 \times \ldots \times A_n}(x_{g_1}, \ldots, x_{gn}) \circ \langle x_{g_1}, \ldots, x_{g_n} \rangle | k_{A_1 \times \ldots \times A_n}(x_{g_1}, \ldots, x_{gn}) = \min[\prod_i k_{A_i}(x_{g_i}), \prod_i k_Z(x_{g_i}),] x_{g_i} \in A_i, g_i \in G_i, i = 1, \ldots, n;$

raising to the direct power $(\times A)^n = \{k_{(\times A)} n(x) \circ x | k_{(\times A)} n(x) = \min[\prod_i k_A(x_i), \prod_i k_Z(x_i)], x_i \in A, i = 1, \ldots, n\}$

Determine some special types of families of set and multisets.

A family of all subsets of a set A that includes the set A and the empty set $\emptyset$ is called the *power set* or *Boolean* [6] of the set A and denoted by $\mathsf{P}(A)$.

A family of all different submultisets of a multiset A over a domain X is called *the macroset* or *Boolean* [12] of the multiset A and denoted by $\mathsf{P}(A)$. In the family $\mathsf{P}(A)$, different submultisets are always present in single exemplars, including the multiset A, which is the maximal multiset for the family $\mathsf{P}(A)$, the set Supp$A$, and the empty multiset $\emptyset$. The Boolean $\mathsf{P}(A)$ is a set of submultisets of a multiset A.

A family of all possible submultisets of a multiset A over a domain X is called *the power multiset* [2, 14] or *multiboolean* [12] of the multiset A and denoted by $\mathsf{Q}(A)$. The multiboolean $\mathsf{Q}(A)$, in contrast to the Boolean $\mathsf{P}(A)$, is a multiset where several identical exemplars of a multiset A are allowed. The Boolean $\mathsf{P}(A)$ of a multiset A is the support Supp $\mathsf{Q}(A)$ of the multibulean $\mathsf{Q}(A)$. If $A \subseteq A$, then $P(A) \subseteq \mathsf{P}(A) \subseteq \mathsf{Q}(A)$. Some of elements of the Booleans $\mathsf{P}(A), \mathsf{P}(A)$ and the multiboolean $\mathsf{Q}(A)$ are themselves sub(multi)sets of other (multi)sets of these families.

A family of multisets over the set $X = \{x_1, x_2, \ldots\}$ is called *a semiring* E of multisets if it contains the empty multiset $\emptyset$, intersection $A_i \cap A_j$ of the multisets $A_i, A_j \in \mathsf{E}$, and any multiset $A \in \mathsf{E}$ is representable as a finite union $\bigcup_{s=1}^{r} A_s = A$ of disjoint submultisets $A_s \subseteq A$; *a ring* K of multisets if it contains the empty multiset $\emptyset$, union $A_i \cup A_j$, sum $A_i + A_j$ and difference $A_i - A_j$ of the multisets $A_i, A_j \in K$; *an algebra* S of multisets if it is a ring that includes the maximal set, named a unit of the algebra, and is closed under finite unions, additions and complements of multisets; *σ-ring* K_σ of multisets if it is a ring and contains a countable union $\bigcup_{s=1}^{\infty} A_s$ and sum $\sum_{s=1}^{\infty} A_s$ of the multisets $A_s \in \mathsf{K}_\sigma$; *δ-ring* K_σ of multisets if it is a ring and contains a countable intersection $\bigcap_{s=1}^{\infty} A_s$ of the multisets $A_s \in \mathsf{K}_\sigma$; *σ-algebra* S_σ of multisets if it is a σ-ring, contains a unit of algebra, and is closed under countable unions, additions and complements of multisets. In particular, every ring of multisets is a semiring of multisets, every σ-algebra of multisets is a δ-algebra of multisets, and vice versa.

Introduce the notion of a measure of a multiset as follows [7, 12]. A non-negative real-valued function $m : A \to \mathrm{R}$, defined on a family $\mathsf{A} = \{A_i\}_{i \in I}$ of multisets, is called *a strongly additive* or *strongly finitely additive measure*, and *a strongly σ-additive* or *strongly countably additive measure* of a multiset if the equalities

$$m(\sum_{i=1}^{n} A_i) = \sum_{i=1}^{n} m(A_i), m(\sum_{i=1}^{\infty} A_i) = \sum_{i=1}^{\infty} m(A_i) \tag{2}$$

hold correspondingly for any finite and any countable collection A of multisets. A family A is a semiring E, a ring K, σ-ring K_σ, or σ-algebra S_σ of submultisets A_i of multiset $A = \{k_A(x_1) \circ x_1, k_A(x_2) \circ x_2, \ldots\}$ over a set $X = \{x_1, x_2, \ldots\}$. The condition

$m(\varnothing) = 0$ follows from the equalities (2) and an identity of the empty multiset $\varnothing + \varnothing = \varnothing$.

For a family A of disjoint multisets ($\boldsymbol{A}_i \bigcap \boldsymbol{A}_j = \varnothing, i \neq j, \boldsymbol{A}_i, \boldsymbol{A}_j \in$ A), a sum of multisets is equal to union: $\sum_{i=1}^{n} \boldsymbol{A}_i = \bigcup_{i=1}^{n} \boldsymbol{A}_i$. Then the expressions (2) can be written as

$$m(\bigcup_{i=1}^{n} \boldsymbol{A}_i) = \sum_{i=1}^{n} m(\boldsymbol{A}_i), \quad m(\bigcup_{i=1}^{\infty} \boldsymbol{A}_i) = \sum_{i=1}^{\infty} m(\boldsymbol{A}_i). \tag{3}$$

A measure m of a multiset satisfying the above conditions (3) is called correspondingly *weakly additive* or *weakly finitely additive*, and *weakly σ-additive* or *weakly countably additive*. The strongly additive measure of a multiset is also weakly additive. The converse is not true. The weak additivity (3) of a multiset measure coincides with the finite and countable additivity of a set measure [6]. For sets, the operation of addition is not feasible, and the strong additivity of a measure is absent.

Let the measure of a singleton $\{x_i\}$ be equal to $m(\{x_i\}) = w_i$, $0 \leq w_i < \infty$. Then, according to (2), the function $m(\boldsymbol{A}) = m(\sum_i \boldsymbol{A}_i) = \sum_i m(\boldsymbol{A}_i) = \sum_i [k_{\boldsymbol{A}}(x_i) m(\{x_i\})] = \sum_i w_i k_{\boldsymbol{A}}(x_i)$, defined on the family A of multisets, is a strongly additive or σ-additive measure of a multiset. If we put all the numbers $w_i = 1$, then we obtain: $m(\boldsymbol{A}) = \sum_i k_{\boldsymbol{A}}(x_i) = |\boldsymbol{A}|$. Thus, the multiset cardinality is also a multiset measure.

3 Multiset Metric Spaces

Recall some notions of metric spaces [3, 5, 7, 12].

A real-valued function d_X, defined on the direct product $X \times X$ of a set X, is called *a metric* if it satisfies: (1^0) the axiom of symmetry $d_X(x,y) = d_X(y,x)$; (2^0) the axiom of identity $d_X(x,y) = 0 \Leftrightarrow x = y$; ($3^0$) the axiom of a triangle $d_X(x,y) \leq d_X(x,z) + d_X(z,y)$ for all $x,y \in X$. The non-negativity $d_X(x,y) \geq 0$ follows from the conditions (10)–(3^0). A set X with a metric d_X given on a set X is called *a metric space* and denoted by (X,d_X).

A real-valued function d_X, defined on the direct product $X \times X$ of a set X, is called *a quasimetric* if it satisfies the axiom (1^0) of symmetry and the coincidence condition (4^0) $d_X(x,x) = 0$ for all $x,y \in X$. A set X with a quasimetric d_X is called *a quasimetric space*. A quasimetric d_X satisfying the axiom (3^0) of a triangle is called *a pseudometric* on a set X, and a set X with a pseudometric d_X is called *a pseudometric space*. In the quasi- and pseudometric spaces, the axiom (2^0) of identity for d_X is not satisfied, in general, that is the condition $d_X(x,y) = 0$ does not imply the equality of the elements x and y. Note that any metric is also a quasimetric and a pseudometric. The converse is not generally true. A real-valued function d_X defined on the product $X \times X$ of a set X, satisfying the axioms (1^0) of symmetry and (2^0) of identity, is called *a symmetric* on a set X, and a set X with a symmetric d_X is called *a proximity space*. A symmetric d_X, satisfying the inequality (5^0) $d_X(x,y) \leq \max_{z \in X}[d_X(x,z), d_X(z,y)]$ for any elements of a set X, is called *an ultrametric* on a set X. Note that any ultrametric is also a metric. The converse is not generally true. The numerical value of any function $d_X(x,y)$ is called *a distance* between the elements x and y of a set X.

Consider now possible approaches to form spaces on families of multisets.

The Boolean $\mathsf{P}(\boldsymbol{X})$ of different submultisets $\boldsymbol{A}_s \subseteq \boldsymbol{X}$ of an n-dimensional multiset $X = \{k_X(x_1) \circ x_1, \ldots, k_X(x_n) \circ x_n\}$ over a finite set $X = \{x_1,\ldots,x_n\}$ is equivalent to the set $\mathbb{R}^n$ of n-dimensional vectors $\mathbf{y}_s = (y_{s_1}, \ldots, y_{s_n})$ with real components $y_{si} = k_{A_s}(x_i) \in Z_+, i = 1, \ldots, n$, and forms *the metric space of finite multisets* [4, 11, 12]:

$\boldsymbol{P}_1 = (\mathbf{P}(\boldsymbol{X}), d_{P1})$ with the Hamming-type metric

$$d_{P1}(\boldsymbol{A}, \boldsymbol{B}) = \sum\nolimits_{i=1}^{n} |k_A(x_i) - k_B(x_i)|; \tag{4}$$

$\boldsymbol{P}_2 = (\mathbf{P}(\boldsymbol{X}), d_{P2})$ with the Euclid-type metric

$$d_{P2}(\boldsymbol{A}, \boldsymbol{B}) = \left[\sum\nolimits_{i=1}^{n} |k_A(x_i) - k_B(x_i)|^2\right]^{1/2}; \tag{5}$$

$\boldsymbol{P}_p = (\mathbf{P}(\boldsymbol{X}), d_{Pp})$ with the Minkowski-type metric

$$d_{Pp}(\boldsymbol{A}, \boldsymbol{B}) = \left[\sum\nolimits_{i=1}^{n} |k_A(x_i) - k_B(x_i)|^p\right]^{1/p}, p \geq 1 \text{ is an integer}; \tag{6}$$

$\boldsymbol{P}_\infty = (\mathbf{P}(\boldsymbol{X}), d_{P\infty})$ with the Chebyshev-type metric

$$d_{P\infty}(\boldsymbol{A}, \boldsymbol{B}) = \max_i |k_A(x_i) - k_B(x_i)|. \tag{7}$$

The σ-algebra $\mathbf{S}_\sigma(\boldsymbol{X})$ of different submultisets $\boldsymbol{A}_s \subseteq \boldsymbol{X}$ of a multiset $\boldsymbol{X} = \{k_X(x_1) \circ x_1, k_X(x_1) \circ x_2, \ldots\}$ over an arbitrary set $X = \{x_1, x_2, \ldots\}$, where any submultiset $\boldsymbol{A}_s$ satisfies the restriction $k_{A_s}(x_i) \leq k_s < \infty$ or $\sum_{i=1}^{\infty} (k_{A_s}(x_i))^p \leq k^p < \infty$ for all integers $p \geq 1$, is equivalent to the set R^N of bounded numerical sequences $Y_s = \{y_{si}\} = \{y_{s1}, y_{s2}, \ldots\}$ with real terms $y_{si} = k_{A_s}(x_i) \in \mathbb{Z}_+$ satisfying the condition $|y_{si}| \leq k_s < \infty$ или $\sum_{i=1}^{\infty} |y_{si}|^p \leq k^p < \infty$ for integers $p \geq 1$, and forms *the metric space of bounded multisets* [11, 12]:

$\boldsymbol{S}_1 = (\mathbf{S}_\sigma(\boldsymbol{X}), d_{S1})$ with the Hamming-type metric

$$d_{S1}(\boldsymbol{A}, \boldsymbol{B}) = \sum\nolimits_{i=1}^{\infty} |k_A(x_i) - k_B(x_i)|; \tag{8}$$

$\boldsymbol{S}_2 = (\mathbf{S}_\sigma(\boldsymbol{X}), d_{S2})$ with the Euclid-type metric

$$d_{S2}(\boldsymbol{A}, \boldsymbol{B}) = \left[\sum\nolimits_{i=1}^{\infty} |k_A(x_i) - k_B(x_i)|^2\right]^{1/2}; \tag{9}$$

$\boldsymbol{S}_p = (\mathbf{S}_\sigma(\boldsymbol{X}), d_{Sp})$ with the Minkowski-type metric

$$d_{Sp}(\boldsymbol{A}, \boldsymbol{B}) = \left[\sum\nolimits_{i=1}^{\infty} |k_A(x_i) - k_B(x_i)|^p\right]^{1/p}; \tag{10}$$

$\mathbf{S}_\infty = (\mathbf{S}_\sigma(\boldsymbol{X}), d_{S\infty})$ with the Chebyshev-type metric

$$d_{S\infty}(\boldsymbol{A},\boldsymbol{B}) = \sup_i |k_{\boldsymbol{A}}(x_i) - k_{\boldsymbol{B}}(x_i)|. \tag{11}$$

The space $(\mathbf{S}_\sigma(\mathbf{Z}), m)$ of multisets with a measure, where $\mathbf{S}_\sigma(\mathbf{Z})$ is σ-algebra of the maximal multiset $\mathbf{Z} = \{k_{\mathbf{Z}}(x_1) \circ x_1, k_{\mathbf{Z}}(x_2) \circ x_2, \ldots\}$ over a set $X = \{x_1, x_2, \ldots\}$, multisets $\boldsymbol{A}_s \subseteq \mathbf{Z}$ are measurable, a multiset measure m is strongly σ-additive and completely σ-finite, forms *the metric spaces of measurable multisets* $\boldsymbol{Z}_{qp} = (\mathsf{S}_\sigma(\mathbf{Z}), d_{Z_{qp}})$, q = 1, 2, 3, 4, $p \geq 1$ is an integer, with the Petrovsky-type metrics [7, 11, 12]:

$$d_{Z1_p}(\boldsymbol{A},\boldsymbol{B}) = [m(\boldsymbol{A}\Delta\boldsymbol{B})]^{1/p}, \tag{12}$$

$$d_{Z2_p}(\boldsymbol{A},\boldsymbol{B}) = [m(\boldsymbol{A}\Delta\boldsymbol{B})/m(\mathrm{Z})]^{1/p}, \tag{13}$$

$$d_{Z3_p}(\boldsymbol{A},\boldsymbol{B}) = [m(\boldsymbol{A}\Delta\boldsymbol{B})/m(\boldsymbol{A} \cup \boldsymbol{B})/]^{1/p}, \tag{14}$$

$$d_{Z4_p}(\boldsymbol{A},\boldsymbol{B}) = [m(\boldsymbol{A}\Delta\boldsymbol{B})/m(\boldsymbol{A} + \boldsymbol{B})]^{1/p}. \tag{15}$$

The functions d_{Z3_p} and d_{Z4p} are not defined for $\boldsymbol{A} = \boldsymbol{B} = \emptyset$, therefore it is assumed by definition that $d_{Z3_p}(\emptyset, \emptyset) = d_{Z4_p}(\emptyset, \emptyset) = 0$. The function d_{Z1p} gives the mapping $d_{Z1_p} : \mathbf{S}_\sigma \times \mathbf{S}_\sigma \to \mathrm{R}_+$, and the functions d_{Z2p}, d_{Z3p}, d_{Z4p} give the mapping $d_{Z_{qp}} : \mathsf{S}_\sigma \times \mathsf{S}_\sigma \to \mathrm{R}_{01} = [0, 1]$.

The functions $d_{Zqp}(\boldsymbol{A},\boldsymbol{B})$, q = 1, 2, 3 are pseudometrics, and the function $d_{Z4p}(\boldsymbol{A}, \boldsymbol{B})$ is a quasimetric. In general, the equality $\boldsymbol{A} = \boldsymbol{B}$ does not follow from the condition $m(\boldsymbol{A}\Delta\boldsymbol{B}) = 0$. For multisets that differ by a multiset of measure zero, the condition $m(\boldsymbol{A}\Delta\boldsymbol{B}) = 0$ implies the so-called m-equalities of multisets $\boldsymbol{A} = {}_m\boldsymbol{B}, \boldsymbol{A}_1 \cup \boldsymbol{A}_2 = {}_m\boldsymbol{B}_1 \cup \boldsymbol{B}_2, \boldsymbol{A}_1 \cap \boldsymbol{A}_2 =_m \boldsymbol{B}_1 \cap \boldsymbol{B}_2$, which are valid almost everywhere. Then the axiom (2^0) of identity holds for the functions $d_{Zqp}(\boldsymbol{A},\boldsymbol{B})$, q = 1–4. So, the functions d_{Z1p}, d_{Z2p}, d_{Z3p} become metrics, and the function d_{Z4p} becomes a symmetric almost everywhere on the corresponding space of measurable multisets [14, 15].

We shall call the pseudometric $d_{Z1p}(\boldsymbol{A},\boldsymbol{B})$ on the space $(\boldsymbol{S}_\sigma(\boldsymbol{Z}),m)$ of measurable multisets *general* or *basic*, the pseudometric $d_{Z2p}(\boldsymbol{A},\boldsymbol{B})$ *completely averaged*, the pseudometric $d_{Z3p}(\boldsymbol{A},\boldsymbol{B})$ *locally averaged*, the quasimetric $d_{Z4p}(\boldsymbol{A},\boldsymbol{B})$ *averaged*. The general pseudometric $d_{Z1p}(\boldsymbol{A},\boldsymbol{B})$ marks the proximity of two multisets $\boldsymbol{A}$ and $\boldsymbol{B}$ in the original space. The completely averaged pseudometric $d_{Z2p}(\boldsymbol{A},\boldsymbol{B})$ marks the proximity of two multisets $\boldsymbol{A}$ and $\boldsymbol{B}$ reduced to the maximal possible distance in the original space. The locally averaged pseudometric $d_{Z3p}(\boldsymbol{A},\boldsymbol{B})$ marks the proximity of two multisets $\boldsymbol{A}$ and $\boldsymbol{B}$ reduced to the joint "common part" of these two multisets in the original space. The averaged quasimetric $d_{Z4p}(\boldsymbol{A},\boldsymbol{B})$ marks the proximity of two multisets $\boldsymbol{A}$ and $\boldsymbol{B}$ reduced to the maximal possible "common part" of these two multisets in the original space.

When we replace the multiplicity function $k_{\boldsymbol{A}}(x)$ of a multiset by the characteristic function $\chi_A(x)$ of a set in (4)–(11), we obtain the metric spaces of the finite and bounded sets. Replacing the multiplicity function $k_{\boldsymbol{A}}(x)$ of a multiset by the characteristic function $\chi_A(x)$ of a set in (12)–(15), we define the metric spaces of the measurable sets. The general metrics $d_{X11}(A,B) = m(A\Delta B)$ and $d_{X11}(A,B) = m|A\Delta B|$ are called the Fréchet distance or a measure metric. The locally averaged metric $d_{X31}(A,B) = m(A\Delta B)/m(A\cup B)$ is called the Steinhaus distance, and the metric $d_{X31}(A,B) = m|A\Delta B|/m|A\cup B|$ is called the biotopic distance [3]. The spaces of measurable sets and measurable multisets with metrics (12)–(15) were introduced firstly by the author.

4 Group Multiple Criteria Decision Making

The problem of multiple criteria decision making is formulated generally as follows. There is given a collection of objects (alternatives, options) $O_1,\ldots,O_q$, which are described by many attributes $K_1,\ldots,K_n$ and can exist also in several exemplars (versions, copies) with different values of attributes. Each attribute K_l has a discrete numerical and/or verbal scale $X_l = \{x_l^1,\ldots,x_l^{h_l}\}, l = 1,\ldots,n$, grades of which are ordered or not. Various exemplars of the object arise, for instance, in cases, when an object is evaluated by several experts on many criteria, or the characteristics of object are calculated several times by different methods, or measured several times by different tools. Based on the knowledge of experts and/or preferences of decision makers (DM), it is required: to select one or several best objects; to order all objects; and to assign all objects into several classes (categories). Examples of such tasks are group ordering and classification of multi-attribute objects, recognition of graphic symbols, processing text documents.

Discuss possible ways of representing and comparing multi-attribute objects [9–11]. Introduce a combined scale (hyperscale) of attributes – the set $X = X_1\cup\ldots\cup X_n = \{x_1^1,\ldots,x_1^{h_1};\ldots;x_n^1,\ldots,x_n^{h_n}\}$, which consists of n groups of object characteristics and combines all gradations of estimates on all rating scales. Correspond a multiset

$$\boldsymbol{A}_i = \{k_{\boldsymbol{A}_i}(x_1^1)\circ x_1^1,\ldots,k_{\boldsymbol{A}_i}(x_1^{h_1})\circ x_1^{h_1};\ldots;k_{\boldsymbol{A}_i}(x_n^1)\circ x_n^1,\ldots,k_{\boldsymbol{A}_i}(x_n^{h_n})\circ x_n^{h_n}\} \tag{16}$$

to an object O_i. Here, the value of the multiplicity function $k_{\boldsymbol{A}_i}\left(x_l^{el}\right)$ shows how many times the estimate $x_l^{e_l}\in X_l, e_l = 1,\ldots,h_l$ of the attribute K_l is present in the description of an object O_i. Note that a multiset (16) can be easily written in the "usual" form (1) as $\boldsymbol{A}_i = \{k_{\boldsymbol{A}_i}(x^1)\circ x^1,\ldots,k_{\boldsymbol{A}_i}(x^h)\circ x^h\}$ if make a change: $x_1^1 = x^1,\ldots,x_1^{h1} = x^{h1}$, $x_2^1 = x^{h1+1},\ldots,x_2^{h2} = x^{h1+h2},\ldots,x_n^{hn} = x^h, h = h_1+\ldots+h_n$.

The variety of operations over multisets provides the ability to aggregate multi-attribute objects in different ways. It is common to combine objects into groups (classes) using the operation of addition of vectors or union of sets. The group D_f of objects described by multisets can be formed as the sum $\boldsymbol{C}_f = \sum_i \boldsymbol{A}_i, k_{Cf}(x^j) = \sum_i k_{\boldsymbol{A}_i}(x^j)$, union $\boldsymbol{C}_f = \bigcup_i \boldsymbol{A}_i, k_{Cf}(x^j) = \max_i k_{\boldsymbol{A}_i}(x^j)$, intersection $\boldsymbol{C}_f = \bigcap_i \boldsymbol{A}_i, k_{Cf}(x^j) = \min_i k_{\boldsymbol{A}_i}(x^j)$, or as a linear combination $\boldsymbol{C}_f = \sum_i b_i\bullet\boldsymbol{A}_i, \boldsymbol{C}_f = \bigcup_i b_i\bullet\boldsymbol{A}_i, \boldsymbol{C}_f = \bigcap_i b_i\bullet\boldsymbol{A}_i$ of multisets corresponding to objects. While adding multisets, all properties (all values of all attributes) of all members of the group are aggregated. While uniting or intersecting multisets, the best properties (the maximum values of all attributes) or the worst properties (the minimum values of all attributes) of individual members of the group are increased.

Present now a s-th version $O_i^{<s>}, i = 1,\ldots,q$, $s = 1,\ldots,t$ of multi-attribute object O_i, as a multiset $\boldsymbol{A}_i^{<s>} = \{k_{\boldsymbol{A}_i}^{<s>}(x_1^1) \circ x_1^1, \ldots, k_{\boldsymbol{A}_i}^{<s>}(x_1^{h1}) \circ x_1^{h1}; \ldots; k_{\boldsymbol{A}_i}^{<s>}(x_n^1) \circ x_n^1, \ldots, k_{\boldsymbol{A}_i}^{<s>}(x_n^{hn}) \circ x_n^{hn}\}$ over the $X = X_1 \cup \ldots \cup X_n = \{x_1^1, \ldots, x_1^{h1}; \ldots; x_n^1, \ldots, x_n^{hn}\}$ of estimates. Let form the multiset $\boldsymbol{A}_i$ representing the object O_i by the weighted addition of multisets describing the versions of this object: $\boldsymbol{A}_i = c^{<1>}\boldsymbol{A}_i^{<1>} + \ldots + c^{<t>}\boldsymbol{A}_i^{<t>}$. The multiplicity function is equal to $k_{\boldsymbol{A}_i}(x^j) = \sum_s c^{<s>} k_{\boldsymbol{A}_i}^{<s>}(x^j)$, $j = 1,\ldots,h$. The integer coefficient $c^{<s>}$ marks a significance of the s-th version of the object (expert competence, measurement accuracy).

Consider a multi-attribute object O_i as a point of metric space (A, d) of multisets, $\mathsf{A} = \{\boldsymbol{A}_1,\ldots,\boldsymbol{A}_q\}$, d is a metric of type (4)–(15). In practical problems, a multiset measure can be given by a linear combination: $m(\boldsymbol{A}) = \sum_{l=1}^n w_l k_{\boldsymbol{A}}(x^l) = \sum_{l=1}^n w_l \sum_{el=1}^{h_l} k_{\boldsymbol{A}}(x_l^{e_l})$, where $w_l > 0$ is a significance (weight) of the element x_l or the attribute K_l, $l = 1,\ldots,n$. A proximity of the analyzed objects in the attribute space is characterized either by dissimilarity or similarity of object properties. So, a dissimilarity between the objects O_i and O_j is marked by one of the metrics (12)–(15), which we rewrite as follows:

$$d_{Z1p}(O_i, O_j) = [H_{ij}]^{1/p}; \; d_{Z2p}(O_i, O_j) = [H_{ij}/W]^{1/p}; \tag{17}$$

$$d_{Z3p}(O_i, O_j) = [H_{ij}/M_{ij}]^{1/p}; d_{Z4p}(O_i, O_j) = [H_{ij}/N_{ij}]^{1/p}. \tag{18}$$

A similarity of the objects O_i and O_j can be estimated by the following indicators:

$$s_{0p}(O_i, O_j) = [L_{ij}]^{1/p}; s_{1p}(O_i, O_j) = [L_{ij}/W]^{1/p}; \tag{19}$$

$$s_{2p}(O_i, O_j) = [1 - d_{21}(O_i, O_j)]^{1/p} = [1 - (H_{ij}/W)]^{1/p}; \tag{20}$$

$$s_{3p}(O_i, O_j) = [1 - d_{23}(O_i, O_j)]^{1/p} = [L_{ij}/M_{ij}]^{1/p}; \tag{21}$$

$$s_{4p}(O_i, O_j) = [1 - d_{24}(O_i, O_j)]^{1/p} = [2L_{ij}/N_{ij}]^{1/p}. \tag{22}$$

Here the factors

$$H_{ij} = m(\boldsymbol{A}_i \Delta \boldsymbol{A}_j) = \sum_{l=1}^n w_l \left|k_{\boldsymbol{A}_i}(x^l) - k_{\boldsymbol{A}_j}(x^l)\right| = \sum_{l=1}^n w_l \sum_{el=1}^{hl} \left|k_{\boldsymbol{A}_i}(x_l^{el}) - k_{\boldsymbol{A}_j}(x_l^{el})\right|;$$

$$L_{ij} = m(\boldsymbol{A}_i \cap \boldsymbol{A}_j) = \sum_{l=1}^n w_l \min\left[k_{\boldsymbol{A}_i}(x^l), k_{\boldsymbol{A}_j}(x^l)\right] = \sum_{l=1}^n w_l \sum_{el=1}^{hl} \left[\min k_{\boldsymbol{A}_i}(x_l^{el}), k_{\boldsymbol{A}_j}(x_l^{el})\right];$$

$$M_{ij} = m(\boldsymbol{A}_i \cup \boldsymbol{A}_j) = \sum_{l=1}^n w_l \max\left[k_{\boldsymbol{A}_i}(x^l), k_{\boldsymbol{A}_j}(x^l)\right] = \sum_{l=1}^n w_l \sum_{el=1}^{hl} \max\left[k_{\boldsymbol{A}_i}(x_l^{el}), k_{\boldsymbol{A}_j}(x_l^{el})\right];$$

$$N_{ij} = m(\boldsymbol{A}_i + \boldsymbol{A}_j) = \sum\nolimits_{l=1}^{n} w_l[k_{\boldsymbol{A}_i}(x^l) + k_{\boldsymbol{A}_j}(x^l) = \sum\nolimits_{l=1}^{n} w_l \sum\nolimits_{el=1}^{hl} [k_{\boldsymbol{A}_i}(x_l^{el}) + k_{\boldsymbol{A}_i}(x_l^{el})];$$

$$W = m(Z) = \sum\nolimits_{l=1}^{n} w_l \max_{\boldsymbol{A}\in\mathbf{A}} k_{\boldsymbol{A}}(x^l) = \sum\nolimits_{l=1}^{n} w_l \sum\nolimits_{el=1}^{hl} \max_{\boldsymbol{A}\in\mathbf{A}} k_{\boldsymbol{A}}(x_l^{el}).$$

The relations $N_{ij}= M_{ij}+ L_{ij}$, $H_{ij}= M_{ij}\text{-}L_{ij}$ hold.

The expressions s_{01}, s_{11}, s_{21}, s_{31}, s_{41} (19)–(22) for p = 1 generalize to the multisets the known non-metric indexes of objects' similarity, such as, respectively, the measure of absolute similarity, the Russell-Rao similarity measure, the simple matching coefficient, the Jacquard coefficient or the Rogers-Tanimoto measure, the Sorensen index [1, 11, 15].

We used the proposed indexes of dissimilarity (17), (18) and similarity (19)–(22) between objects, which exist in several copies with many different, in particular, contradictory values of quantitative and/or qualitative attributes, in the new methods for group verbal decision analysis, based on the multiset space theory [9, 10].

The method ARAMIS (Aggregation and Ranking Alternatives nearby the Multi-attribute Ideal Situations) focuses on group ordering objects with many verbal attributes. This method estimates a proximity of objects to any "ideal" object in the multiset metric space and allows us to rank multi-attribute objects without pre-constructing individual rankings.

The methods CLAVA-HI and CLAVA-NI (CLuster Analysis of Verbal Alternatives – HIerarchical and NonhIerarchical) allow to generate classes (clusters) of several copies of objects in the verbal attribute space, when a number of classes is fixed or not fixed in advance. In hierarchical clustering, one combines step by step the nearest pairs of objects and/or clusters such that $d(\boldsymbol{C}_u,\boldsymbol{C}_v) = \min_{p,q} d(\boldsymbol{C}_p,\boldsymbol{C}_q)$, and forms new clusters as a sum, union or intersection of multisets describing objects/clusters. Finally, one builds two or more groups of objects.

The method MASKA (abbreviation of the Russian words Multi-Attribute Consistent Classification of Alternatives) focuses on a group distribution of objects with many verbal attributes between several classes. This method allows us to find collective rules for sorting multi-attribute objects that aggregate a lot of the manifold (and may be inconsistent) individual rules.

The developed methods were applied for solving practical problems of multiple criteria choice. Among them there are the multi-indicator evaluation of company activity, competitive selection of research projects, multi-attribute classification of credit cardholders, multi-aspect evaluation of research efficiency, multiple criteria selection of the advanced computing complex, and so on.

Rating of Russian companies on information and telecommunication technologies was constructed with the ARAMIS method. 50 experts estimated activity of companies by dozen criteria with numerical or verbal scales as follows: K_1. Total profit; K_2. Gross sales; K_3. Net sales; K_4. Number of projects; K_5. Projects' importance; K_6. Number of employees; K_7. Professional skills of staff, and others. For instance, the criterion K_7. 'Professional skills of staff' had the following verbal scale:

x_7^1 – a very high level of staff qualification and experience;
x_7^2 – a high level of staff qualification and experience;
x_7^3 – a medium level of staff qualification and experience;
x_7^4 – a low level of staff qualification and experience

Quantitative continuous scales had been transformed into qualitative discrete scales with a small number of grades as follows: x_s^1 – very high (large); x_s^2 – high (large); x_s^3 – medium; x_s^4 – low (small). Each verbal grade of a scale corresponds to any determined interval of numerical values of a criterion. 30 companies were selected as the leading high-tech companies, 10 as the mostly dynamic developing companies, and 10 companies as the leading developers of software.

A competitive selection of research projects for the State Research Program on High-Temperature Superconductivity was produced with the MASKA method. Three experts estimated every application by 6 criteria with verbal scales as follows: K_1. Project contribution to the program goals; K_2. Long-range value of the project; K_3. Novelty of approach to the tasks' solution; K_4. Qualification of team; K_5. Resources available for the project realization; K_6. Profiles of the project results. Each criterion has a nominative or ordered scale of verbal estimates. For instance, the scale of criterion K_3.' Novelty of approach to the tasks' solution' was as follows:

x_3^1 – a new original approach;
x_3^2 – a modernized approach;
x_3^3 – a known traditional approach

Experts also made one of the following conclusions: r_1 – to approve the project; r_2 – to reject the project; r_3 – to consider the project later after improving. The individual expert recommendations for a project selection may coincide or not as well.

For selecting the approved projects the following collective decision rules were built. The first rule: "The project is to be included in the Program if a qualification and experience of team are the best or sufficient for the project realization". (Criterion estimates x_4^1 or x_4^2). The second rule: "The project is to be included in the Program if a qualification and experience of team are the best or sufficient for the project realization, the project is very important or important for an achievement of the major program goals, and available resources are fully sufficient or sufficient for the project realization". (Criteria estimates x_4^1 or x_4^2, and x_1^1 or x_1^2, and x_5^1 or x_5^2). The last decision rule coincided with the empirical sorting rule in the real-life situation when more than 250 applications were considered and about 170 projects were approved.

5 Conclusions

The paper describes new indicators of the objects' proximity that can be used in various methods of group classification, sorting, ranking multi-attribute objects. The representation of such objects as points of multiset metric space allows us to solve new problems never being sold earlier and known traditional problems in more simple and effective ways.

The new methods of group verbal decision analysis expand the range of considered problems and have no analogues. These methods operate with objects existed in several copies and described by many numerical, symbolic and/or verbal attributes. In these methods, verbal attributes are not transformed in or replaced by any numerical ones as, for instance, in MAUT, TOPSIS and fuzzy methods. These methods take into account diversity and contradictions of preferences of many decision makers and/or knowledge of experts without searching for a consistency of actors' judgments.

Acknowledgments. This work was supported by the Russian Foundation for Basic Research (projects 16-29-12864, 17-07-00512, 17-29-07021, 18-07-00132, 18-07-00280).

References

1. Anderberg, M.R.: Cluster Analysis for Applications. Academic Press, New York (1973)
2. Blizard, W.D.: Multiset theory. Notre Dame J. Formal Logic **30**, 36–65 (1989)
3. Deza, M.M., Deza, E.: Encyclopedia of Distances. Springer, Berlin (2016)
4. Kosters, W.A., Laros, J.F.J.: Metrics for multisets. In: Bramer, M., Coenen, F., Petridis, M. (eds.) Research and Development in Intelligent Systems XXIV. Proceedings of AI-2007, pp. 294–303. Springer, London (2008)
5. O'Searcoid, M.: Metric Spaces. Springer, London (2007)
6. Oxtoby, J.: Measure and Category. Springer, New York (1971)
7. Petrovsky, A.B.: Metric spaces of multisets. Doklady Acad. Sci. (Doklady Akademii Nauk). **344**(2), 175–177 (1995). (in Russian)
8. Petrovsky, A.B.: Operations with multisets. Doklady Math. **67**(2), 296–299 (2003)
9. Petrovsky, A.: Group verbal decision analysis. In: Adam, F., Humphreys, P. (eds.) Encyclopedia of Decision Making and Decision Support Technologies, vol. 1, pp. 418–425. IGI Global, Hershey (2008)
10. Petrovsky, A.B.: Methods for the group classification of multi-attribute objects. Sci. Tech. Inf. Process. **37**(5), 346–356 (2010). 357–368
11. Petrovsky, A.B.: Indicators of similarity and dissimilarity of multi-attribute objects in metric spaces of sets and multisets. Artificial Intell. Decis. Making (Iskusstvennyiy intellekt i prinyatiye resheniy) **4**, 78–94 (2017). in Russian
12. Petrovsky, A.B.: Theory of Measurable Sets and Multisets. Nauka, Moscow (2018). in Russian
13. Rebai, A.: Canonical fuzzy bags and bag fuzzy measure as a basis for MADM with mixed non cardinal data. Eur. J. Oper. Res. **78**, 34–48 (1994)
14. Singh, D., Ibrahim, A.M., Yohana, T., Singh, J.N.: An overview of the applications of multisets. Novi Sad J. Math. **37**(2), 73–92 (2007)
15. Sneath, P.H.A., Sokal, R.R.: Numerical Taxonomy. The Principles and Practice of Numerical Classification. Freeman, San Francisco (1973)

Evidence Theory for Complex Engineering System Analyses

Boris Palyukh[1], Vladimir Ivanov[1(✉)], and Alexander Sotnikov[2]

[1] Tver State Technical University, Tver, Russia
{pboris,mtivk}@tstu.tver.ru
[2] Joint Supercomputer Centre of the RAS, Moscow, Russia
asotnikov@jscc.ru

Abstract. The evolution of manufacturing systems with internal multi-stage processes takes place in a fuzzy dynamic environment. One of the instruments of system component state and property modelling is a mathematical theory of evidence. The article examines some examples of the evidence theory applications which deal with the two different stages of the system development. In particular, the problem of simulation, diagnostics, and assessment of complex engineering system states is considered. Necessity and sufficiency conditions of critical component states are shown to be relaxed. The other example of the evidence theory application illustrates the quantitative assessment approach to technical system component innovation. The coprocessing of primary innovation data is taken up, with the data being retrieved from various sources with different measurement and expert appraisal completeness and reliability.

Keywords: Theory of evidence · Multi-stage process
Dynamic environment · Diagnostics · Innovation · Assessment

1 Introduction

Decision-making problems can, in some cases, be characterized by uncertainty, i.e. practically null information on object states. On the other hand, the exact knowledge of object state probabilities can be available which allows decision making with only risk limitations considered. There is also some intermediate situation where the information about object or its component states is available but inaccurate and/or incomplete. Such problems are solved with a theory of evidence also known as Dempster-Shafer theory.

The article outlines the evidence theory application domains for simulating, diagnosing and assessing complex engineering system states. The authors provide the two examples of applying the theory. The first example examines the problem of complex engineering system elements durability. The diagnostics model of the object is described as a multiagent system whose agents (system elements) cooperate with each other and the control area. The evidence theory is shown to enable the accurate recognition of agents critical situations. The other example

A. Abraham et al. (Eds.): IITI 2018, AISC 874, pp. 70–79, 2019.
https://doi.org/10.1007/978-3-030-01818-4_7

considers the quantitative assessment approach to the innovation of different objects. The assessment model based on the object data retrieval from various databases is proposed. The specific application of the evidence theory including the combination of different indicator measurements is introduced.

2 Fundamentals of Evidence Theory

So, the evidence theory (see [1] and [2]) provides a mathematical basis for simulating and computing the event probability after single parts of the event information retrieved from different sources are combined. The source information represents, as a rule, imperfect (interval) values of expert appraisals, measurements or observations. The significant theory concepts are considered to be functions of belief and plausibility.

A belief function shows the degree of confidence in that fact that objects have the given property:

$$Bel(A_i) = \sum_{A_j \subseteq A_i} m(A_i) \tag{1}$$

where A_i is an event proving an object (object sets) has the given property; $A_i \subseteq C$; C is an exhaustive event set (a problem solution result); $m(A_j)$ is a basic distribution of event probabilities A_j; $m(A_j) \in [0,1]$. In addition, $Bel(\emptyset) = 0$, $Bel(A_i) \in [0,1]$ and $Bel(C) = 1$.

A plausibility function shows the degree of plausibility in that fact that objects have the given property:

$$Pl(A_i) = 1 - Bel(\overline{A}_i) = 1 - \sum_{A_j \cap A_i = \emptyset} m(A_i) \tag{2}$$

On the whole, supposing the existence of $p(A_i)$ to be true probability of the given property of the objects from $A_i \subseteq C$ we have:

$$Bel(A_i) \leq p(A_i) \leq Pl(A_i) \tag{3}$$

where $Bel(A_i)$ and $Pl(A_i)$ are lower and upper limits of $p(A_i)$.

3 Related Works

Since the appearance of the first papers (see [1] and [2]), thousands of works on the evidence theory have been published in the world. Classical works which constitute exhaustive fundamental basis of the evidence theory are collected in [3]. Also, we would like to mention the BELIEF conference organized by Belief Functions and Applications Society (BFAS) (http://www.bfasociety.org). The articles in [4] describe the latest information about theoretical issues and applications in various aspects of the evidence theory.

Research and developments with the use of the evidence theory is rapidly evolving around the world. The analysis of publication activity over the previous 25 years shows the evident nonlinear growth of publications dedicated to the evidence theory tendencies and registered in one of the worlds largest databases of scientific and technical papers and articles IEEE Xplore® Digital Library (http://ieeexplore.ieee.org), see Fig. 1.

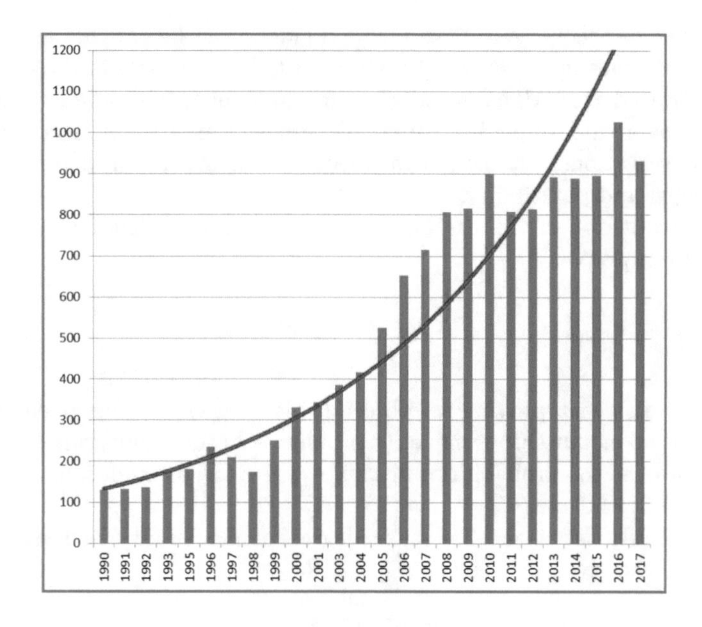

Fig. 1. Number of publications related to Dempster-Shafer theory

4 Evidence Theory Application Domains

Here is the list of the evidence theory applications for solving decision-making problems with imperfect (interval) values of expert appraisals, measurements, and observations: diagnostic systems, manufacturing systems, expert appraisal analysis, remote sensing technology, structural state analysis and evaluation.

Diagnostic system research and development focus on the following main areas: diagnostics of chemical-technological processes, weld defect analysis, ultrasonic and eddy current product testing, quality control of water distribution systems, flutter probability and risk estimate, engine diagnostics, operating unit fault diagnostics, diagnostic systems of multistage decision making, wireless sensor networks, diagnostics of high-voltage circuit breakers.

In the area of manufacturing systems the following trends are in the focus: the choice of product data management system for a machine-building enterprise, CAD/CAE solution of problems, robotics (a stand-alone robots decision-making with probability models), support of production system reliability.

The expert appraisal analysis deals with: conflict analysis of group expert appraisals, research analysis of expert appraisals, expert appraisal analysis in managing high-tech projects (e.g. in forming work performer teams, clustering and ranging of group expert appraisals.

The relevant technologies of remote sensing are the following: object and object group classification in multi- and hyperspectral aerospace images, remote sensing image enhancement by detecting changes in initially defective images, geolocation systems (wireless sensor-driven navigation systems, incident prevention).

Research and development in the area of structural state analysis and evaluation deals with the following: structural reliability analysis, pipeline, joint and manhole state evaluation and defect analysis.

5 Evidence Theory for Diagnosing Complex Engineering System States

5.1 Basic Concepts

There is a problem of analyzing complex engineering system states in the diagnostic systems [5]. Solutions of this or other similar problems are provided in the intelligent optimal control system for multi-stage process evolutions in a fuzzy dynamic environment, the system being under development (see [6] and [7]).

Normal operation of a product means the absence of crises and performance degradations X_k of its components (agents) $A_i \in A_k$. The indicator of crisis or performance degradation is the fact that $X_k = [(\underline{x_k}), (\overline{x_k})]$ exceeds allowed normative values $D_k = [(\underline{d_k}), (\overline{d_k})]$. Diagnostic hypotheses are checked and give the following error types: (a) type I errors or "false alarm" and (b) type II errors or "alarm absence in performance degradations".

Let A^* be a set of agents being in crisis or having performance degradations. Then a diagnostic procedure means the search for solutions of the following Boolean equation:

$$F : P^* \to A^* \tag{4}$$

where F is a predicate function of $F = \vee(\wedge A_i)$, $P_k = 1$ and $P^* = \{P_k \mid P_k = 1\}$ are sets of registered performance degradations, P_k is an indicator function.

Indicator function P_k transforms from Boolean function as a performance degradation indicator into a continuous function defined at $[0, 1]$ and showing a degree of performance degradation.

Provided $X_k > \overline{d_k}$, function P_k is given by:

$$P_k = \begin{cases} 0, & \text{if } \overline{x_k} \leq \overline{d_k} \\ 1, & \text{if } \underline{x_k} \geq \overline{d_k} \\ \frac{\overline{x_k} - \overline{d_k}}{\overline{x_k} - \underline{x_k}}, & \text{if } \underline{x_k} < \overline{d_k} < \overline{x_k} \end{cases} \tag{5}$$

Provided $X_k > \underline{d_k}$, function P_k is given by:

$$P_k = \begin{cases} 0, & \text{if } if\ \underline{x_k} \geq \underline{d_k} \\ 1, & \text{if } if\ \overline{x_k} \leq \underline{d_k} \\ \frac{\underline{d_k} - \underline{x_k}}{\overline{x_k} - \underline{x_k}}, & \text{if } \underline{x_k} < \underline{d_k} < \overline{x_k} \end{cases} \tag{6}$$

If $P_k \neq 0$, the fact is interpreted as the suspicion of performance degradation by k-th indicator. At that, a fuzzy set of probably critical agents A_k is activated. The diagnostic critical conditions $s \in S$ of agent A_i are as follows:
$(\forall s \in S)(P_k = 0) \leftrightarrow (A_k \cap A^* = \emptyset)$ – a necessary condition,
$(\forall s \in S)(P_k = 1) \leftrightarrow (A_k \cap A^* = A^*)$ – a sufficient condition.

5.2 Methodology and Example

Imagine that the degree of X_k influence on a critical state of agent A_i is assessed by experts with a number in [0; 100]. Table 1 presents the expert appraisal results.

Table 1. Expert appraisal results

	A_1	A_2	A_3	A_4	A_5
X_1	20	40			60
X_2	15		40		60
X_3	90	50	30	30	

Assume $X_1 = [42,8\%; 52,4\%]$ and $X_3 = [53,0\%; 58,0\%]$ with normative interval S=[50%; 55%]. Indicator degradation probabilities X_1 and X_3 are:
$P_{x1} = (50.0 - 42.8)/(52.4 - 42.8) = 0.75$ and $P_{x3} = (58.0 - 55.0)/(58.0 - 53.0) = 0.6$.
The table gives the following activated sets:
$A_{x1} = \{(a_1; 0.2), (a_2; 0.4), (a_5; 0.6)\}$ and
$A_{x3} = \{(a_1; 0.9), (a_2; 0.5), (a_3; 0.3), (a_4; 0.3)\}$.
The normative distributions of critical state probabilities are as follows:
$m_x = \langle a_1; a_2; a_5 \rangle = \langle 0.17; 0.33; 0.50 \rangle$ and
$m_x = \langle a_1; a_2; (a_3 \vee a_4) \rangle = \langle 0.53; 0.30; 0.15 \rangle$.
The redistributed distributions of the probabilities are computed in the following way:

$$m_x^{'} = m_x * P_k, \ \ m_x^{'}(A) = 1 - P_x. \tag{7}$$

Thus, $m_{x1}^{'} = \langle a_1; a_2; a_5; A \rangle = \langle 0.13; 0.25; 0.48; 0.25 \rangle$ and
$m_{x3}^{'} = \langle a_1; a_2; (a_3 \vee a_4); A \rangle = \langle 0.32; 0.18; 0.10; 0.40 \rangle$.
In accordance with Dempster's rule different evidences (events) with probability distributions m_1 and m_2 are combined as follows:

$$m(C_N) = \frac{1}{(1-K)} \sum_{A_i \cap A_j \neq \emptyset} m_1(A_i) m_2(A_j) \tag{8}$$

Table 2. The combinations of different m_{xi} in terms of Dempster's rule

		m'_{x1}			
		a_1	a_2	a_5	A
m'_{x3}	a_1	a_1	$\emptyset$	$\emptyset$	a_1
	a_2	$\emptyset$	a_2	$\emptyset$	a_2
	$a_3 \vee a_4$	$\emptyset$	$\emptyset$	$\emptyset$	$a_3 \vee a_4$
	A	a_1	a_2	a_5	A

where $m(C_N)$ is a function of combining events A_i and A_j corresponding to object properties C_N; $K = \sum_{A_i \cap A_j = \emptyset} m_1(A_i)m_2(A_j)$.

Based on the data of the example under consideration we have: $K = 0.08 + 0.12 + 0.02 + 0.07 + 0.01 + 0.02 + 0.04 = 0.35$. The combinations of different m_{xi} in terms of Dempster's rule are presented in Table 2.

For the critical agent hypotheses: $m(a_1) = (0.04 + 0.08 + 0.05)/0.65 = 0.26$; $m(a_2) = (0.04 + 0.04 + 0.10)/0.65 = 0.28$; $m(a_3 \vee a_4) = 0.02/0.65 = 0.03$; $m(a_5) = 0.19/0.65 = 0.29$; $m(A) = 0.10/0.65 = 0.15$.

The final probability distribution is:

$m(\emptyset) = \langle a_1; a_2; (a_3, a_4); a_5; A\rangle = \langle 0.26; 0.28; 0.03; 0.29; 0.15\rangle$.

5.3 Conclusions

The conclusions of the given example results are presented in Table 3.

Table 3. Example solution results

Conventional methods (with fuzzy sets)	Evidence theory of agent crisis recognition
The failure of the necessity condition results in type II errors, the failure of sufficiency condition results in type I errors	The relaxation of necessity and sufficiency conditions to $(\forall s \in S)(P_k = 0) \leftrightarrow P^* \neq \emptyset$
The solution result: $A^* = (a_1; 0.2), (a_2; 0.4)$	The solution result: $a_1[0.26; 0.41], a_2[0.28; 0.43]$, $a_3[0.03; 0.18], a_4[0.03; 0.18]$, $a_5[0.29; 0.44]$, $A[1.0; 1.0]$
Rejecting the hypothesis of agent A_5 crises gives type II error	Accepting the hypothesis of agent A_5 crises, determining the crisis cause

6 Evidence Theory for Assessing Object Innovation

6.1 Basic Concepts

The analysis of the literature dedicated to different innovation issues (see, e.g., [8] and [9]) shows that the notion innovation always includes such connotations as new, efficiency-increasing, profit-making. The object innovation is considered within its system relations, a degree of the object influence on a subject and external objects. Consider the following key terms:

- Innovation is an object (such as an invention, an engineering system component, a technology, a method, etc.) having some properties determining its technological novelty, relevance, and implementability.
- Technological novelty means significant improvements, a new way of using or granting an object or a technology. Novelty subjects are potential users or producers themselves.
- Relevance is a potential producers awareness of the object necessity as a demand.
- Search pattern is a target object description (a linguistic model) consisting of a marker and key terms. A search pattern is used to generate queries.
- Marker is a key term determining an object application (action) field.
- Query is a set of key terms and a marker used by a search engine to search object information in a databases.

6.2 Methodology and Example

Suppose that an objects linguistic model consists of N key terms. Suppose there is a universal set of term numbers $\Omega = |K|$ and the marker has number $0 \notin \Omega$. Queries $A_{qk} \subseteq \Omega$, $k = [1, N]$ are constructed with the terms and marker is added to each query. For example, $A_{q1} = \{1\}$, $A_{q2} = \{1, 2\}$, etc. Queries A_{qk} serve as observable subsets Ω. They are executed in different search engine, retrieve data from various databases including the Internet. The number of new object relevant results are supposed to be less than that of old and well-known ones. We assume that object novelty Nv is determined as follows:

$$Nv = 1 - \frac{1}{S} \sum_{k=1}^{S} R_k^{01} \tag{9}$$

where $R_k^{01} = 1 - \exp\left(1 - \frac{R_k}{R_0^{01}}\right)$ is the value of R_k standardized for $[0; 1]$; R_k is a number of documents resulted from the execution of k-th query in a database containing the information about the data domain; R_0^{01} is the number of documents standardized for $[0; 1]$ and resulted from the execution of a query containing one term-marker; S is a query number. Object relevance is estimated as follows:

$$Rv = \frac{1}{S} \sum_{k=1}^{S} F_k^{01} \tag{10}$$

where $F_k^{01} = 1 - \exp(1 - \frac{F_k}{F_0^{01}})$ is the value of F_k standardized for $[0; 1]$; F_k is a frequency of execution by users queries similar to k-th query; F_0^{01} is the value of F_k^{01} for the query containing one term-marker.

The number of relevant object search results being applicable to a search pattern are supposed to be larger than that of the objects with demand decline. So, estimated indicators of query A_{qk} executions are $p(Nv)$ – the probability of an object having technological novelty and $p(Rv)$ – the probability of an object having relevance (in demand). And measured indicators of query A_{qk} executions are R – the number of documents retrieved with the search engine and F – the frequency of query executions by search engine users.

Let us consider the order of computation for innovation indicator $p(Nv)$ as an example. The number of the retrieved documents R_k is computed for each query and specified in accordance with (9). The number of group intervals is determined as $I = \sqrt{S}$. When $N = 15$, $I = 3.873 \approx 4$. In case of the equal intervals $A_1 = [0.0000; 0.2500]$, $A_2 = [0.2501; 0.5000]$, $A_3 = [0,5001; 0,7500]$, $A_4 = [0.7501; 1.0000]$. In terms of measuring the mentioned intervals correspond to the nominal scale *It is novel* (IN), *It is evidently novel* (IEN), *It is evidently not novel* (IENN), *It is not novel* (INN).

Suppose that query executions A_{q1}–A_{q15} in Search Engine 1 (e.g. Yandex) give the following results: four queries have the results from the interval A_1 i e. ($q_1 = 4$), $q_2 = q_3 = 5$, and $q_4 = 1$. According to [2] base probability $m(A_{ki}) = q_k/S$ where q_k is a number of observed subsets. For our example we have $m(A_1) = q_1/S = 4/15 = 0.27$, $m(A_2) = 0.33$, $m(A_3) = 0.33$, $m(A_4) = 0.07$. The estimation results for $Bel(A)$ and $Pl(A)$ for the interval $A = [0.0; 0.5]$ (IN or IEN) are the following (see formula (1) and (2): $Bel(A) = m(A_1) + m(A_2) = 0.60$, $Pl(A) = 0$. Thus, the observed object is new with probability $p(Nv) = 0.60$.

Suppose that Search Engine 2 (e.g., Google) gives the measurement results for indicator R_k while executing queries A_{qk}. In the end, $q_1 = q_2 = 2$, $q_3 = 6$, $q_4 = 5$. Thus, $m(A_1) = 0.13$, $m(A_2) = 0.13$, $m(A_3) = 0.40$, $m(A_4) = 0.34$. The interval $A = [0,0; 0,5]$ (IN or IEN) gives the following: $Bel(A) = m(A_1) + m(A_2) = 0.26$, $Pl(A) = 0$. That is, $p(Nv) = 0.26$ which differs from the results retrieved with Search Engine 1.

Table 4 presents data intersections of the examples for Search Engine 1 and 2 used for calculating a combined base probability.

Table 4. Data intersections Search Engine 1 and 2

		A_j^1 (Google)			
		{1,4}	{3,7}	{2, 5, 6, 9, 10, 13}	{8, 11, 12, 14, 15}
A_i^1 (Yandex)	{1, 2, 3, 4}	{1,4}	{3}	{2}	∅
	{5, 7, 8, 9, 10}	∅	{7}	{5, 9, 10}	{8}
	{6, 11, 12, 13, 15}	∅	∅	{13}	{11, 12, 15}
	{15}	∅	∅	∅	{15}

In accordance with formula (8) $K = 0.267$ and $m_{12}(A_1) = 0.243$, $m_{12}(A_2) = 0.392$, $m_{12}(A_3) = 0.333$, $m_{12}(A_4) = 0.032$. For the interval $A = [0,0;0.5]$ (IN or IEN): $Bel_{12}(A) = m_{12}(A_1) + m_{12}(A_2) = 0.243 + 0.392 = 0.635$, $Pl_{12}(A) = 0$. Thus, $p(Nv) = 0.635$.

The source credibility can be considered with the introduction of discount factor α for base probability $m(A)$. Let the expert appraisal are $\alpha_1 = 0.1$, and $\alpha_2 = 0.2$. Hence, the results of the Search Engine 1 are more credible than those of the Search Engine 2. Discounted base probabilities are estimated as $m^{\alpha}(A) = (1 - \alpha)m(A)$. With Dempster's rule and expression (8) we have $K^{\alpha} = 0.193$, $m^{\alpha}_{12}(A_1) = 0.219$, $m^{\alpha}_{12}(A_2) = 0.329$, $m^{\alpha}_{12}(A_3) = 0.291$, $m^{\alpha}_{12}(A_4) = 0.037$. For the interval A = $[0,0;0,5]$ (IN or IEN): $Bel^{\alpha}_{12}(A) = m^{\alpha}_{12}(A_1) + m^{\alpha}_{12}(A_2) = 0.219 + 0.329 = 0.548$; $Pl^{\alpha}_{12}(A) = 0$. Thus, $p(Nv) = 0.548$.

6.3 Experiments and Discussion

Figure 2 illustrates the results of the experiments conducted to determine novelty of an engineering solution from the patent data base http://www1.fips.ru. The solution novelty was determined individually with the database subsets corresponding to patent registration years (a ten-year period was used). The adequate trend lines verify the hypothesis of the object novelty reduction in time. The polynomial approximation (a trend wavy line) shows the novelty recurrence of the analyzed object that requires the check of the hypothesis of innovation cycles in this application area. It should be noted that the concepts describing the society development in general and economy in particular as a consequence of recurring cycles have long been recognized [10]. Our research only deals with the hypothesis generation which needs to be followed up.

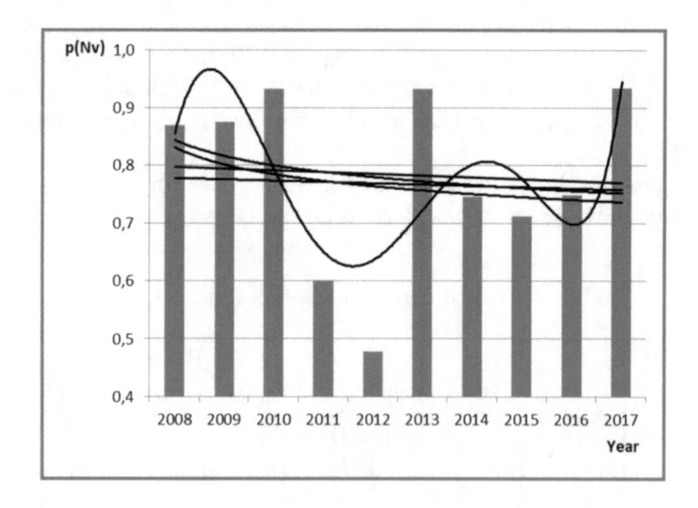

Fig. 2. Recurrence and general reduction of an object novelty in time

7 Conclusions

The use of the evidence theory for considering some uncertainty in diagnosing and analyzing complex engineering system states gives a number of advantages. The measurements are combined to transform the results of many assessments in order to make them consistent and to combine them with mathematical modeling results. So, necessity and sufficiency conditions of critical states are relaxed.

The approach to determining object innovation indicators is one of the outcomes of the research Data Warehousing Based on Search Agent Intellectualization and Evolutionary Model of Target Information Selection. Carrying on this research we are planning to substantiate and formalize the generation procedure of an objects linguistic model and include algorithms based on the examined techniques in a search agent composition.

Acknowledgment. The work was supported by RFBR (Projects No. 18-07-00358 and No. 17-07-01339).

References

1. Dempster, A.: Upper and lower probabilities induced by a multivalued mapping. Ann. Math. Stat. **38**(2), 325339 (1967). https://doi.org/10.1214/aoms/1177698950
2. Shafer, G.: A Mathematical Theory of Evidence. Princeton University Press, Princeton (1976)
3. Yager, R., Liu, L.: Classic Works of the Dempster-Shafer Theory of Belief Functions. Springer, London (2010)
4. Vejnarov, J., Kratochvl, V.: Belief Functions: Theory and Applications. Springer International Publishing, Prague (2016). 4th International Conference (Czech Republic, Prague, 21–23 September 2016)
5. Horst, C. (ed.): Handbook of Technical Diagnostics. Springer, Heidelberg (2013)
6. Palyukh, B.V., Vetrov, A.N., Yegereva, I.A.: Architecture of intelligent optimal control system for multi-stage processes evolution in fuzzy dynamic environment. Softw. Prod. Syst. **30**(4), 619624 (2017)
7. Palyukh, B.V., Ivanov, V.K., Yegereva, I.A.: Innovation intelligent search and production system evolution control. In: 8th International Research and Practice Conference of AI Integrated Models and Soft Computing, Kolomna, 18–20 May 2015, vol. 1, p. 418427 (2015)
8. Oslo Manual: Guidelines for Collecting and Interpreting Innovation Data. The Measurement of Scientific and Technological Activities, 3rd Edition (2005). https://www.oecd-ilibrary.org/science-and-technology/oslo-manual_9789264013100-en
9. Tucker, R.B.: Driving Growth Through Innovation: How Leading Firms are Transforming Their Futures, 2nd edn. Berrett-Koehler Publishers, San Francisco (2008)
10. Schumpeter, J.A.: Business Cycles: A Theoretical, Historical, and Statistical Analysis of the Capitalist Process, New York(1939)

Analysis of Software Development Process in Respect to Anomaly Detection

Denis Zavarzin(✉) and Tatyana Afanaseva

Ulyanovsk State Technical University, Ulyanovsk, Russia
dzavarzin91@gmail.com, tv.afanasjeva@gmail.com

Abstract. The problem of detecting anomalous states or trends of changing the key design metrics is an actual problem in large software companies, which use software project management systems. The relevance of this task is conditioned by such factors: application of Agile software development methods; the amount of the project code; duration of the project (long-term projects); continuous improvement of the project functionality; stabilization of the system; improving the quality of the development process and the quality of the end product throughout all its life cycle. In this paper, a new method was proposed for searching for anomalous TS values through k-means clustering, using primary preprocessing - fuzzy transform (F-transform), which allows detecting the outliers not only in stationary TS but also in nonstationary TS extracted from software project management systems. This method is able to identify anomalies in the TS that characterized by strong oscillatory changes in the trend behavior or identify single atypical TS values. This method can be used for quickly localizing sections of TS trends atypical behavior for excluding such values in further analysis. In addition, this method can be applied iteratively, until the complete exclusion of values that clearly do not correspond to the behavior of TS tendencies (elementary, local, general).

Keywords: Anomaly detection · Time series · Granular representation
Fuzzy models · Design metrics

1 Introduction

Many large software companies currently use software project management systems. The problem of detecting anomalous states or trends of changing in the key design metrics is an actual problem in such systems.

The relevance of this task is conditioned by such factors: application of Agile software development methods; the amount of the project code; duration of the project (long-term projects); continuous improvement of the project functionality; stabilization of the system; improving the quality of the development process and the quality of the end product throughout all its life cycle.

In view of these factors, the project management takes place in the partial uncertainty, even if the development process was planned.

The relevance of this task is confirmed by the Russian State standard GOST R ISO/IEC 25010-2015, which emphasizes the dependence of the final software

A. Abraham et al. (Eds.): IITI 2018, AISC 874, pp. 80–88, 2019.
https://doi.org/10.1007/978-3-030-01818-4_8

product's quality on the quality of the design process, which, in turn, depends on the key design metrics [1]. We can predict, identify and track unusual situations dependent on the time events of the design process in order to increase the success of the project. We can also identify the types of key project metrics that can be easily recovered, represented as a TS and can be comprehensively analyzed.

We will extract metrics from the existing software systems for project management: there are project management software (including software repositories, issue tracking tools, etc.). These systems have the basic analytic tools: burnout diagrams, comparisons with planned values of design metrics, reports in tabular form, etc. Despite the multitude of analytic tools, they are sufficiently limited in their capabilities for complex analysis and taking stock of the result of analysis; also, this is a proprietary software. In addition, the design process can be built uniquely for each project. We based on the Atlassian Jira as one of the most popular, functional and understandable issue-tracking product.

We create special risk triggers of the project's failure for tracking the unusual situation on the project. Such triggers linked to the metrics of project software repositories: variability of requirements, software errors, total development time, human re-sources and their quality. Time series, as a mathematical interpretation of the process of changing these metrics can be analyzed for the presence of anomalous behavior patterns. There are examples of the successful application of the design metrics data analysis to solve the detecting anomalous events problem in the large project software repositories (for example, IDE NetBeans) [2].

The tasks of developing and designing software based on Agile will be clarified in the second chapter; will be given practical examples of using analytic tools from the Atlassian Jira. The third chapter will describe the FCL algorithm implementation and consider advantages and disadvantages of the algorithm for finding anomalies. The fourth chapter presents an experiment on project data.

2 Related Works

There are several approaches to the finding and detection of the anomalies in the TS of the process with the normal/anomalous behavior occurring under unknown models.

Statistical Approach. In this approach, hypotheses about the similarity of statistical attributes in consistent fragments of stationary TS are checking, and if they differ significantly, it is concluded that anomalies are possible in the corresponding fragments [3]. However, in this approach there are a number of problems, for example: which TS fragment is taken as an anomalous fragment; how to detect a significant difference in the observed TS values. For example, if a single atypical change has occurred, it is difficult to determine the corresponding time point exactly.

Machine Learning. Inter alia, there is a solution for the problem of classifying a set of numerical TS containing previously known types of anomalies. This is considered in a number of papers, for example, in [4, 5]. This approach has the following lacks: in case of the absence of experts, it is difficult to form a training set (the sets of TS with assigned classes of anomalies and attributions); the inaccuracies and duration of retraining classification methods, that are trained on known types of normal/anomalous TS. The algorithm training must be repeated whenever a new class of anomaly arises; the classification methods are not applicable under conditions where it is required to identify specific anomalies for one small TS in conditions where information about possible anomalous/normal TS fragments classes is unknown.

Clustering Methods. The cluster number can be used as a label of the TS class in the variety of these methods. In this case, TS with significant outliers can form a separate cluster with a small number of elements. However, this approach contains the following lacks: it is necessary to set the specific parameters of the clustering method (the length of the TS segment, the number of clusters, the initial coordinates of the cluster centers, etc.), which is impossible without the involvement of an expert; we need to solve the clustering problem several times: for finding outliers in values and outliers in the first differences; we need the additional analysis of TS, which form small clusters - this leads to significant time costs and problems with diagnosing the type and the time of the anomalous event.

Modern techniques for the detecting anomalies are insufficient automated to identify anomalous patterns under conditions of partial uncertainty. They mainly detect one type of anomalies (outlier type) and require training set /model tuning for each unique case; or based on visual analysis.

Turning to the software project management systems, in particular Atlassian Jira as one of the most popular system, we can divide the following popular visual analytic tools:

1. Burn down charts.
2. Gantt charts.
3. Charts of created/completed tasks, etc.

Russian and foreign scientists, such as Yarushkina, Afanasyeva, Vagin, Kovalev, Antipov, Fomina, Sukhanov, Cheboli, Wei, Song, Goldberg, Chandola, Ghahramani, Cunningham, Khan, Ma, Protopapas, Chan, Keogh, Lin and others, made a significant contribution to the development into area of the anomalies detecting in the processes presented in the form of TS.

The classification of the anomalous types of presentation granules of TS information was given based on the granular representation of TS definition [6]. In this case, the problem of finding anomalies consists in the dividing the components of the TS analysis objects into valid and invalid (anomalous) without information about this at each level of granularity. The classes of anomalies can be divided into a class of numerical values anomalies, a class of fuzzy labels anomalies, a class of fuzzy trends and its components anomalies [7].

In this paper, we present a new method for finding anomalies that allows finding anomalies for stationary and nonstationary TSs without using additional TS. The TS decomposition into the trend item and residue item is used to solve this problem. TS decomposition used the F-transform of TS because of its excellent filtering properties, its quick and easy calculation [8].

3 Mathematical Description of TS Anomaly Detection Technic

3.1 Fuzzy Smoothing of Time Series

Fuzzy smoothing of TS based on fuzzy transformation (F-transform) - a technique developed by Irina Perfileva, which could be classified as method of fuzzy approximation. The F-transform technic is presented for continuous functions and functions on a limited set of points. It implies specifying a fuzzy partition of the universal set [9].

Suppose that the value of TS is defined as a function $X(t)$, and its values known in points $p_1, \ldots, p_n \in w$. We divide the interval w to N_k equidistant nodes of fixed length $x_k = v_L + h(k-1)$, $k = 1, \ldots, N$ where $h = \frac{v_R - v_L}{N-1}$, $v_L = t_1, v_R = t_N$.

$$A_1(t) = \begin{cases} 1 - \frac{(t-t_1)}{h_1}, t \in [t_1, t_2] \\ 0 \end{cases} \tag{1}$$

$$A_k(t) = \begin{cases} 1 - \frac{(t_k - t)}{h_{k-1}}, t \in [t_{k-1}, t_k] \\ 1 - \frac{(t-t_k)}{h_k}, t \in [t_k, t_{k+1}] \\ 0 \end{cases} \tag{2}$$

$$A_k(t) = \begin{cases} 1 - \frac{(t-t_{k-1})}{h_{k-1}}, t \in [t_{k-1}, t_k] \\ 0 \end{cases} \tag{3}$$

Using basic functions, we transform $X(t)$ function into a tuple of K real numbers $[F_1, \ldots, F_k]$ which defined by a weighted average formula:

$$F_k = \frac{\sum_{t=1}^{N} X(t) A_k(t)}{\sum_{t=1}^{N} A_k(t)}, k = 1, \ldots, N \tag{4}$$

The resulting set of numbers is a direct F-transform of TS values. The inverse F-transform allows obtaining values of the time series based on the direct transformation, using the formula:

$$f(t) = \sum_{k=1}^{N} F_k A_k(t) \tag{5}$$

To evaluate the accuracy of the inverse transformation, the MAPE and SMAPE can be calculated.

3.2 Decomposition of TS

The paper [10] describes the decomposition of TS components into the trend item and the residual item is represented by the formula below:

$$y_t = f(t) + x(t) \tag{6}$$

Here $f(t)$ – the trend item of the source TS; $x(t)$ – the residual vector of TS which was obtained as the result of the F-transformation.

3.3 TS Clustering

In this paper, we will use the k-means clustering method to determine the anomalous values in the residue vector. The k-means method is a cluster analysis method, the purpose of which is division of m-observations (from space R^n) to k clusters where each observation referring to that cluster which the center (centroid) is closest.

As the object of clustering, we select the vector of the residual values $x(t)$, which was obtained by decomposition of source TS. It takes time to read off the TS of the residue vector into groups so that in each cluster there will be similar values of the TS of the residues.

The Euclidean distance is used as a metric:

$$\rho(x, y) = \|x - y\| = \sqrt{\sum_{p=1}^{n} (x_p - y_p)^2}, где\, x, y \in R^n \tag{7}$$

When the proximity measure for the centroid was determined, then the partitioning of objects into clusters is reduced to determining the centroids of these clusters. The researcher in advance (in our case we choose two clusters, potentially anomalous and normal values for residual values TS) sets the number of clusters k.

The k-means method is to recalculate at each step the centroid for each cluster obtained in the previous step and divides the observations into k groups (or clusters) $(k \leq m)$ $S_i = \{S_1, S_2, \ldots, S_k\}$ to minimize the total quadratic deviation of cluster points from the centroids of these clusters:

$$\min\left[\sum_{i=1}^{k} \sum_{x^{(j)} \in S_i} \left\|x^{(j)} - \mu_i\right\|^2\right], где\, x^{(j)} \in R^n, \mu_i \in R^n \tag{8}$$

After determining the clusters, we classify the membership of this cluster to anomaly set by introducing a border of the size of the anomaly p cluster.

$$\begin{cases} 1, k \leq p \cdot n \\ 0, k > p \cdot n \end{cases}, \tag{9}$$

where p – where is the cluster border size anomaly threshold; k – the number of elements found in the cluster S_i; n – number of elements in the residual vector of the source TS.

4 Algorithmic Description of FCL Anomaly Detection Technic

In this paper, we propose a new TS anomaly detection technic – FCL method (see Fig. 1). The initial data for this algorithm are as follows: the number of basic functions = 3; number of clusters = 2; the cluster size anomaly border is 0.1. The algorithm of the method is describing in the steps below:

Step 1. Construction of the set of basic functions according to formulas (1), (2), (3).
Step 2. Calculation of fuzzy direct TS F-transform, according to formula (4) and obtaining of residual TS vector, according to formula (6).
Step 3. Normalization of the residue vector by taking modulo values of the residual vector of the original TS.
Step 4. Clustering of the obtained normalized residual vector by the k-means method [8], as described in Sect. 3.3.
Step 5. Separation of the obtained values into clusters according to the anomaly border value of the ratio of clusters of anomalous and normal values, according to formula (9). Formation for each value of TS mark 0, if there is no anomaly, and 1 if there are anomalies.

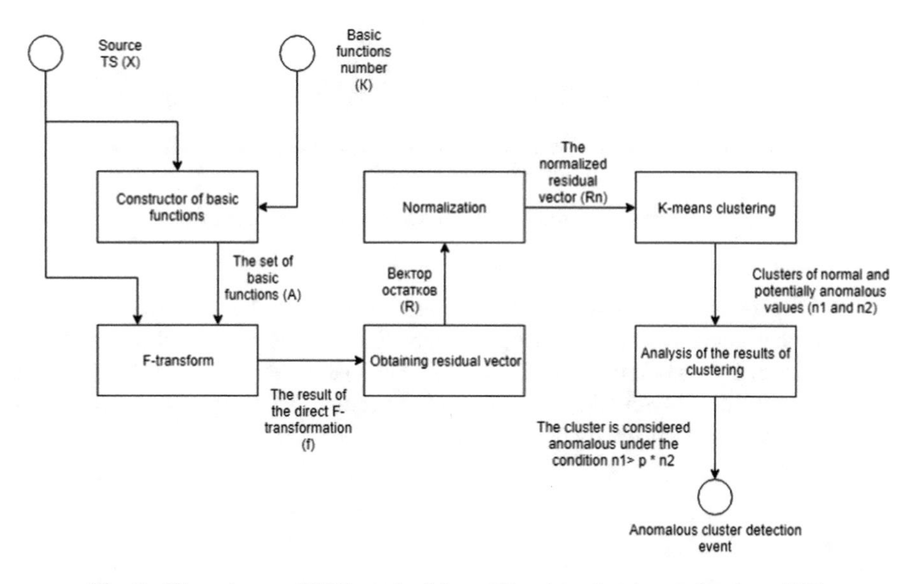

Fig. 1. The schema of FCL method for outlier detection in number-based TS.

5 Experiments

Experiments were conducted on TS from the open data bank of the CIF-2015 TS forecasting competition. On Figs. 2, 3, 4 and 5 show the application of the FCL method to search for anomalies in the numerical values of TS CIF-2015 (ts3, ts19, ts29, ts40). The description for the experimental TS is given in Table 1.

Table 1. Experiment results table

№ TS	Total values number	TS behavior type	Total number of anomalies	Number of anomalies found by FCL method
ts3	518	Stationary	–	18
ts19	30	Non-stationary	5	4
ts29	26	Non-stationary	3	1
ts40	51	Non-stationary	5	2

The TS ts3 is stationary and characterized by huge increases in the trend. Visually, it is not possible to determine the number of anomalies in such TS type. The FCL–method identified 18 anomalies that characterize exceptionally strong growth and fall between pairs of TS values.

The time series ts19 is characterized by rare oscillatory trend changes that are not constant with respect to the root-mean-square deviation of the TS values. The FCL-method detected 4 out of 5 anomalies detected by visual analysis.

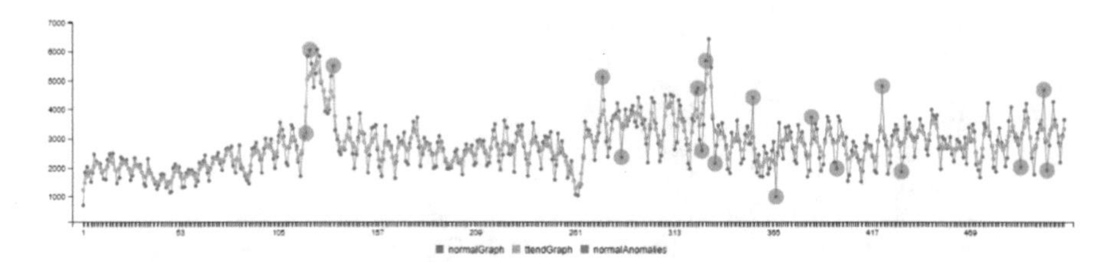

Fig. 2. Experiment for TS ts3, which contains many anomalies.

The TS ts29 contains more complex anomaly types. It is characterized by pairs of points of local trends with minor changes in direction and intensity. In this experiment the FCL algorithm worked worse (1 out of 3 visually determined anomalies was found), because it does not take into account the peculiarities of changes in the trends of local subsequences in the dynamics of TS behavior.

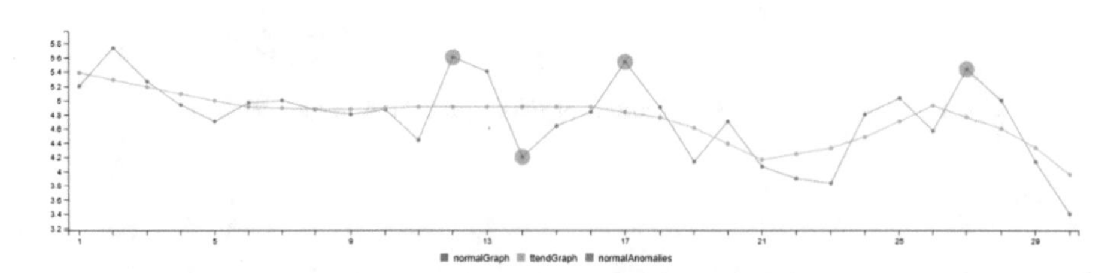

Fig. 3. Experiment for TS ts19, which contains many anomalies.

The TS ts40 is represented by the vibrational component. At this TS, the FCL-method showed the worst result (2 out of 5 hard-to-detect visual anomalies). This is explained by the fact that it is difficult to distinguish the ejection from the vibrational TS, which also contains the trend component (growth direction).

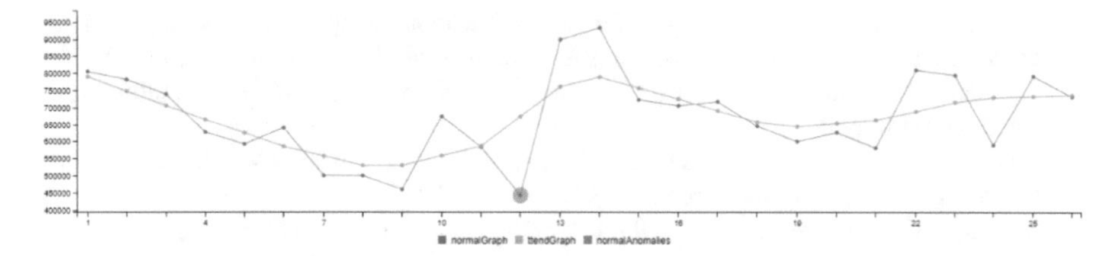

Fig. 4. Experiment for TS ts29, which contains single anomaly.

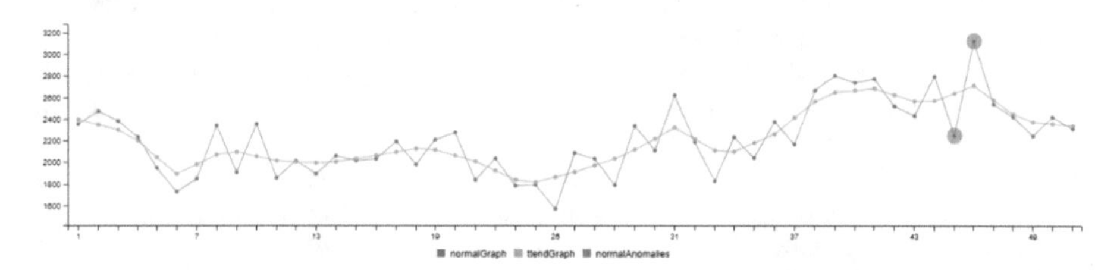

Fig. 5. Experiment for vibrational TS ts40, which contains two anomalies.

6 Conclusions

A new method was proposed for searching for anomalous TS values through k-means clustering, using primary preprocessing - fuzzy transform (F-transform), which allows to detect the outliers not only in stationary TS, but also in nonstationary TS.

The advantages of this method:

1. The identification of the anomalies in the TS that characterized by strong oscillatory changes in the trend behavior.
2. The identification of the single atypical TS values that differ from nearby trend values in the dynamics of its change.

The disadvantages of this method:

1. The multiple anomalies in the chaotic TS trends behavior are found poorly.
2. The anomalies, characterized by a gradual change in TS trends and their properties (type and intensity), are found poorly.

Thus, this method can be used for quickly localizing sections of TS trends atypical behavior for excluding such values in further analysis. In addition, this method can be applied iteratively, until the complete exclusion of values that clearly do not correspond to the behavior of TS tendencies (elementary, local, general).

Acknowledgements. The authors acknowledge that this paper was supported by the project no. 16-07-00535 and by the project no. 16-47-730715 of the Russian Foundation of Basic Research.

References

1. GOST R ISO/IEC 25010-2015 Information Technology (IT): System and software engineering. Requirements and assessment of the quality of systems and software. Models of quality systems and software products. http://docs.cntd.ru/document/1200121069. Accessed 06 May 2018
2. Tunnell, J.W.: Using time series models for defect prediction in software release planning . Electronic Thesis Depository, p. 90 (2015)
3. Marchuk, V.I., Tokareva, S.V.: Methods for detecting anomalous values in the analysis of nonstationary random processes. In: Marchuk, S.V. (ed.) Tokarev. - Mines: State Educational Establishment of Higher Professional Education of the Southern Eurasian State University (2009). 209 c
4. Fomina, M., Antipov, S., Vagin, V.: Methods and algorithms of anomaly searching in collections of TimeSeries. In: Proceedings of IITI 2016, pp. 63–74 (2016)
5. Cózar, J., Puerta, J.M., Gámez, J.A.: An application of dynamic bayesian networks to condition monitoring and fault prediction in a sensored system: a case study. Int. J. Comput. Intell. Syst. **10**, 176–195 (2017)
6. Zavarzin, D.V., Afanasyeva, T.V.: Detection of anomalous states in the behavior of processes in organizational and technical systems. In: Proceedings of the Conference "Informatics. Modeling. Automation of design (IMAP-2016)", pp. 69–75. UlSTU, Ulyanovsk (2016)
7. Zavarzin, D.V., Afanasyeva, T.V.: Multi-model description of anomalous patterns in the problem of finding anomalies of complex systems. Fuzzy systems, soft calculations and intelligent technologies (NSMVIT-2017) Proceedings of the VII All-Russian Scientific and Practical Conference, St. Petersburg, 3–7 July 2017, 220 p. in 2 tons. T.1. -Pb. Politechnica-service (2017)
8. Yarushkina, N.G. The integral method of fuzzy modeling and analysis of fuzzy trends. In: Yarushkina, N.G., Afanasyeva, T.V., Perfilieva, I.G. (eds.) Automation of Control Processes, vol 2, no. 20, pp. 59–64. UlSTU, Ulyanovsk (2010)
9. Romanov, A.A.: Application of the F-transformation method for the forecast of the trend and the numerical representation of the time series, 7 p. UlSTU, Ulyanovsk (2011)
10. Afanasyeva, T.V.: Modeling of fuzzy tendencies of time series, 215 p. UlSTU, Ulyanovsk (2013). TV Afanasyeva

A Model of Multiagent Information and Control System Distributed Data Storage

Eduard Melnik[1] and Anna Klimenko[2(✉)]

[1] Southern Scientific Center of the Russian Academy of Science, 41 Chekhov St., Rostov-on-Don 344006, Russia
[2] Scientific Research Institute of Multiprocessor Computer Systems of Southern Federal University, 2 Chekhov St., GSP-284, Taganrog 347928, Russia
Anna_klimenko@mail.ru

Abstract. This paper deals with the information and control system dependability issue. Firstly, the "criterion delegating" approach is presented briefly. This approach, developed earlier, allows to improve the system elements reliability by the criteria number reducing within the configuration forming problem. As such approach needs a data storage to distribute up-to-date monitoring and control tasks context data through the system, the one's models are developed and presented. We develop two model types, centralized (based on Viewstamped Replication protocol) and fully decentralized. The models are considered and discussed in terms of communication environment workload on the operation and reconfiguration stages of the system.

Keywords: Information and control system · Dependability · Reliability Distributed data storage · Data replication · Multiagent control system

1 Introduction

Information and control systems (ICS) are likely to be the key component of the automated and intelligent mechatronic complexes, e.g., gas and oil production systems, aircraft, transport and many others, including such important concept as the Industrial Internet of Things.

As the failure of such systems can cause disasters, casualties and economical losses, the dependability issue becomes quite topical. It must be mentioned, that the dependability [1, 2] of a complex object depends on all integral subsystems. Yet, as was said in [3], the mechatronic complex mission success is provided by the ICS up to 50%. In other words, the ICS failure is critical for the object controlled. So, ICS must be as dependable as possible.

The current paper deals with the particular class of ICSs, described in [4–6]. The peculiarity of the ICS considered is the performance redundancy as a way to obtain fault-tolerance and the decentralized monitoring and control based on the principles of multiagent systems. The performance redundancy usage affects the system recovery procedure in the following way: instead of switching the spare unit on, in case of element failure only monitoring and control tasks are relocated to the operable units. This makes the term "configuration" to be rather important. In the current context

A. Abraham et al. (Eds.): IITI 2018, AISC 874, pp. 89–98, 2019.
https://doi.org/10.1007/978-3-030-01818-4_9

"configuration" means the binding of the monitoring and control tasks to the particular computational units. In addition to this, the fully decentralized ICS monitoring and control implemented with intelligent agents allows to avert the situation, when the main control unit fails and the system is ruined.

As was mentioned in our previous work, which contains the description of the Reconfigurable Distributed Information and Control System Multiagent Management Approach, we affect the ICS dependability by fault-tolerance attribute and reliability function [7].

The configuration forming problem is one of the most important here because of its impact on the ICS units reliability functions. Our previous work presented a kind of draft of "Criterion delegating approach" (CDA) [7], which allowed to improve the configuration quality and so the reliability function. Yet the base of the CDA is the distributed data storage, and it's model and architecture was not developed and observed properly.

The current paper presents a new model of multiagent ICS decentralized data storage, whose usage is the essential for the CDA. We'll present the previous model of the data storage, the new model implemented in terms of viewstamped replication protocol and the fully decentralized one. All models will be compared in terms of network workload, stability and the time needed for the general functions delivery.

The following sections of the paper contain:

- a brief description of the ICS considered architecture;
- a brief review of the CDA;
- models of the distributed data storage;
- discussion and conclusion.

2 Information and Control Systems with Performance Redundancy and Decentralized Monitoring and Control

The key idea of the performance redundancy is presented in details in [5].

We can avert the structural redundancy and its disadvantages (cost, weight, etc.) by using of computational units with performance redundancy. As was mentioned in [5] such reservation method gives the reliability function enhancement in particular configurations.

In Fig. 1 the scheme of the ICS with the performance redundancy and decentralized monitoring and control is presented.

It is seen that every computational unit (CU) is controlled by its own software agent. Agents are equal and have equal knowledge about the system. The agent's tasks are: to send and receive heartbeats, maintaining their knowledge about the system integrity, and to perform the reconfiguration procedure when it is needed. Besides, every agent has the list of configurations which is equal to other agents' lists. Reconfiguration is carried out according to the configurations in configuration lists, or, if it were considered, through the on-the-fly configuration forming.

In case of CU failure the reconfiguration takes place, and after recovery the system will be as is shown in Fig. 1.

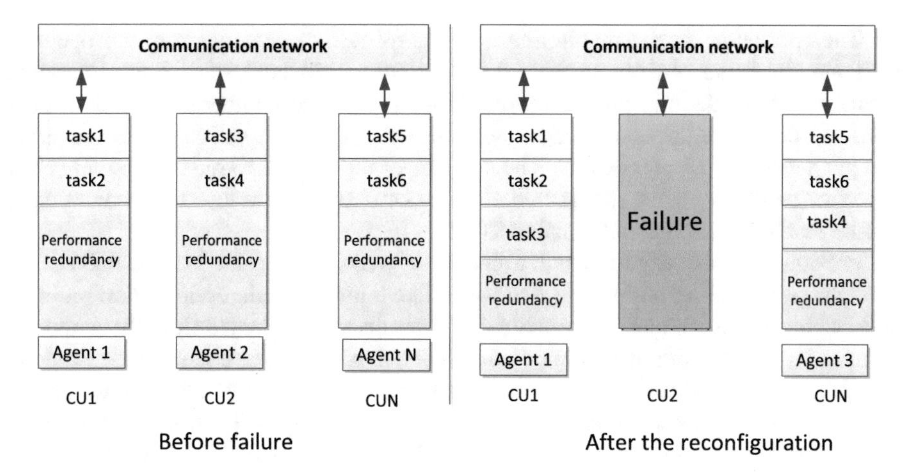

Fig. 1. The general scheme of the system with performance redundancy and decentralized monitoring and control

So, the monitoring and control tasks from the inoperable CU are distributed among the operable ones. The drawback of such reservation method is the CU reliability function degradation in cases of CU overloading [8]. So, the configurations must be formed with mandatory objective to make the CU workloads as equal as possible.

3 Criterion Delegating Approach

The model of configuration forming problem was presented and described in details in [7]. To recall the problem in general, the objective functions of configuration forming problem are presented below:

$$F_1 = \sum_{i=1}^{N} (a_i - a_i^{'}) \rightarrow MIN; \tag{1}$$

$$F_2 = |A| - |A'| \rightarrow MIN; \tag{2}$$

$$F_3 = \sum_{k=1}^{K} u_{kj}^{'} - \sum_{l=1}^{L} u_{lq}^{'} \rightarrow MIN,\ \forall j, q; \tag{3}$$

where F_1 is the number of monitoring and control tasks, which are relocated from the operable CUs, F_2 – the number of non-critical monitoring and control tasks to be removed from the system, F_3 is the workload dispersion on the CUs after the reconfiguration procedure, a_i is the tuple $< j,\ u_{ij},\ t_i >$, which describes the link between task$_i$ and CU j. Here u_{ij} is the CU resources utilization percentage. A determines the configuration before the recovery, and A' – the configuration after the recovery, assuming $A = \{a_i\}$.

The restriction on the monitoring and control tasks removing from the operable CUs has the quite obvious cause: all context data will be lost. Yet the important assumption was made in previous work: if we reduce the number of objective functions, we can solve the optimization problem with reduced objective function number and get a result of a better quality. The key idea of the CDA is to reduce the number of objective functions, while the desired aim (in our case – not to lose any context data) can be reached by some additional software.

Previously some experimental estimations were made (Fig. 2). We solved the configuration forming problem in two ways: like a three criteria optimization problem and like a two criteria one. The quality of the results was evaluated from the workload point of view. The optimization problems were solved with the "quenching" technique of simulated annealing [9]. For this simulation a random set of 25 monitoring and control tasks with computational complexity of 10–40 conventional units was generated. Monitoring and control tasks were assigned to the 10 CUs with equal performance.

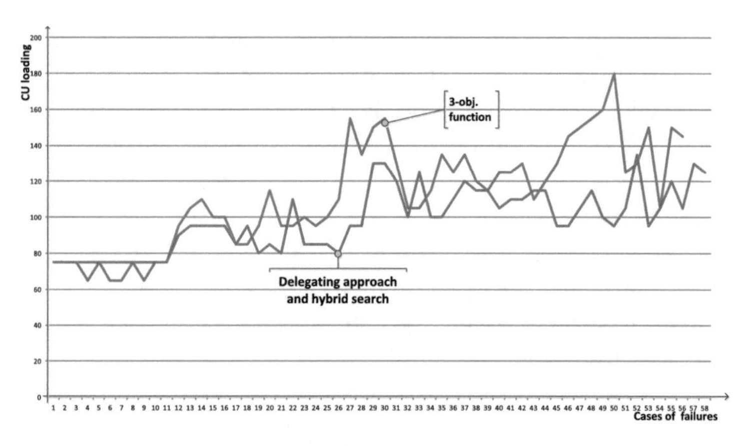

Fig. 2. The CU maximum loadings comparing the 3-criteria objective function and 2-criteria one

The CU workloads are shown in Fig. 2 when configurations were made with three-criteria objective function and two-criteria one for one (cases 1–10) and two (11–58) arbitrary failures. Combining search via simulated annealing with different criteria number, we get the CU workloads as is shown in Fig. 2.

So, the elimination of the tasks relocation criterion allowed to obtain the configurations of a better quality in terms of CU workload.

Yet the model of software and the distributed data storage were presented and described in general, so, the work on it was continued, and in the next section the results will be presented.

4 Models of Multiagent Information and Control System Decentralized Data Storage

To obtain the possibility to relocate monitoring and control tasks from the operable CU and not to lose task context data, the distributed and decentralized data storage must be implemented. Before the presentation of the data storage models, some important assumptions about the monitoring and control tasks are to be made.

Let's assume the monitoring and control task as a determined state machine, so as each state can be described by the values of input data and the ones of internal variables. The simplified scheme of such presentation is shown in Fig. 3.

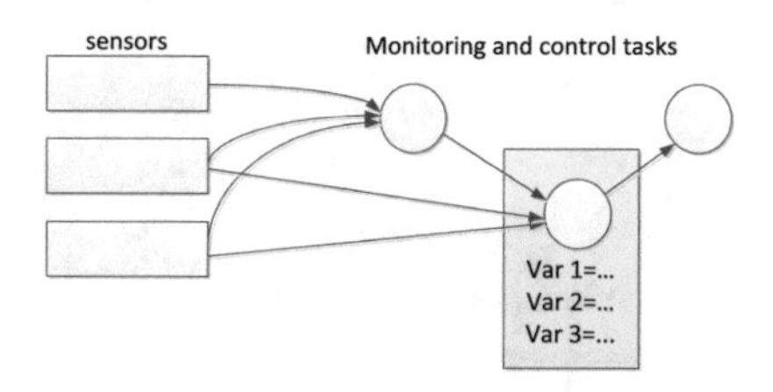

Fig. 3. The scheme of the monitoring and control task presentation

Assuming this, we can determine the portions of input data and internal variables as a kind of transactional data, which is generated periodically by portions. Each portion is a result of the following:

- input data received from sensors and/or another monitoring and control tasks;
- internal variable values.

Also we assume that transitions between states are instantenious, and each state generates a data portion for replication and further storage.

Previously [7] we presupposed the model of the distributed data storage as is presented in Fig. 4.

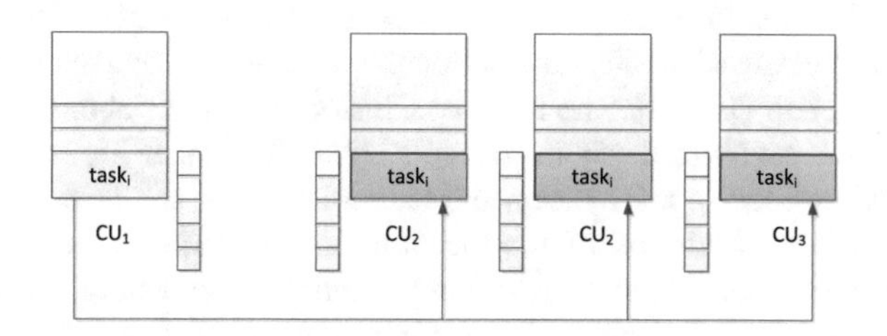

Fig. 4. The distributed data storage model with one active task

In such model of a data storage there is an active monitoring and control task, which replicates context data to CUs where this task can be relocated according to the configuration lists. In case of CU failure – and according to the general system dependability providing mechanism [5–7] - the agents initialize the reconfiguration, and the copy of $task_i$ is launched with appropriate context data.

The model described above allows to eliminate the objective of the active task relocation number minimization, yet it does not handle the case of $task_i$ failure. As is seen, if the active task fails (yet the CU does not), the context data distribution is stopped, and if there is a CU failure after that, the system will not recover properly.

So, the decision was made to improve the data storage model by means of redundancy involving (Fig. 5).

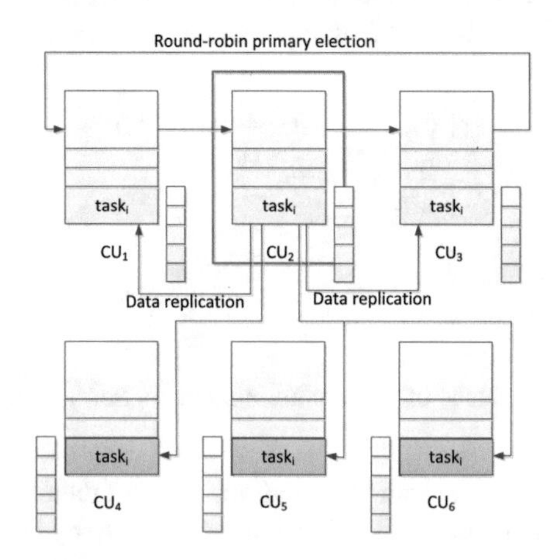

Fig. 5. The distributed data storage model with replicated active tasks and the leader elected

Here we presuppose that $task_i$ is launched on the 2f + 1 CUs, where f – is the number of failures to be handled [10]. Besides, we presuppose the fair behavior of the tasks without the byzantine one. The system is asynchronous, and every replica of the $task_i$ receives input data from sensors in the same order but, may be, not in the same time moments. Also we assume that the running task can not consume the new context data.

The general aim is to reach the data consensus in case of reconfiguration.

Firstly, the most used replication protocols were analyzed: Paxos, Viewstamped replication, Raft, Zab [11–14]. The main conclusion was made about those protocols: all of them use the centralized scheme of replication. The leader (or primary) is elected every ballot (view/epoch), and, except the particular cases, the leader establishes the up-to-date data. The discipline of the election can be various, for example, Viewstamped replication uses simple round-robin discipline dependent on the view number.

In our particular case the copies of the monitoring and control task have to elect the leader, which would replicate its context data to those CUs where this task can be launched after the reconfiguration. It must be mentioned, that the relations between active tasks and their replicas can be determined as a passive replication [14]. Context data is presupposed to be stored in chain-like structure of the fixed size with FIFO discipline. So, the simplified model of data storage elements interaction was developed on bases of the Viewstamped replication principles.

Within this interactional model, every data portion is "stamped" with <view_number, package_number>, where view_number is the number of view (epoch) and initially is 0, and the package_number is the number of context data portion generated by active task. The data chains are supposed to be identical if <view_number, package_number> of the first chain are equal to the <view_number, package_number> of the second chain.

Leader Election

1. The leader is elected by round-robin discipline.
2. Once the leader has been elected, the replicas exchange their context data, sending it to leader.
3. The leader chooses the chain with a MAX(view_number, package_number) as up-to-date and sends this chain to other active task replicas. The choice of MAX (view_number, package_number) is expedient due to possible failures of active task replicas as is used in ViewStamped Replication.

Operational State

1. Only leader sends its context data to passive replicas
2. Followers just collect their own context data, marking it with <view_number, package_number>.

View Change

1. view_number++;
2. Leader election.

Task Failure and Recovery

1. If the leader is inoperable, the view change is initialized;
2. If the follower is inoperable, it can be restore by the operation system software. After recovery a follower can request the context data from leader to update its data storage, or it can be done on the View_change stage.

Reconfiguration

1. The passive task replica is active now and consumes all context data from its store.
2. According to the configuration list the communication between the replicated tasks is established.
3. View change.
4. Go to the Operational state.

The model based on the Viewstamped replication protocol has the strong advantage, reducing the communication environment workload, as well as the space needed for data storage. Yet if the primary fails, the change view procedure must be initialized. Besides, change view procedure seems to be slow due to the context data exchange and distribution. So, we decided to develop the fully decentralized model of a context data storage (Fig. 6).

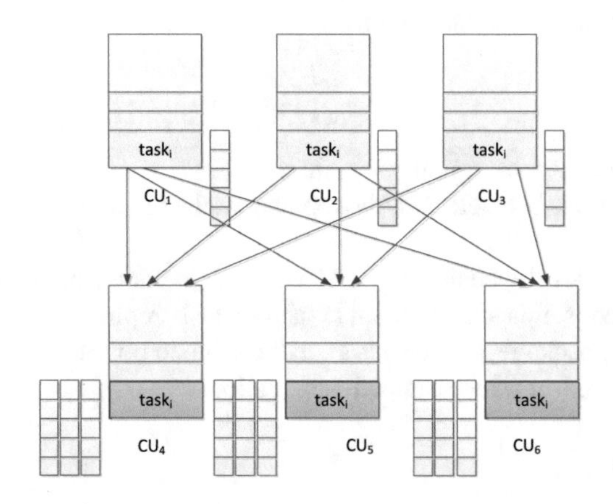

Fig. 6. The decentralized data storage model

In the model proposed there is no primary at all, all active tasks are identical. The "stamp" is changed: instead of "view_number" (there are no views), the "replica_id" is used. So, each data portion is marked with <replica_id, package_number>.

The operational stage begins from the reconfiguration: each task sends its context data to each inactive replica. Then, if any replica fails, nothing happens with data storage of inactive tasks. So, there will be just two stages: operational stage and recovery stage after the reconfiguration.

Operational State

1. Active tasks send their context data, marked with <replica_id, package_number>, to the inactive task replicas.
2. Inactive task replicas collect context data (it can be done also by the agent of the CU).

Recovery

1. New active task replica takes the context data from the storage with max (package_number). There can be several data portions with the same package_number, but identified by the <replica_id, package_number>.
2. Active task replica selects the context data by random selection from the set of data portions.
3. Active task replica begins to distribute its own context data to the passive replicas.

So, it is seen that the fully distributed model is much more simple, yet generates more communication environment workload. In the following section we will discuss centralized and decentralized model in terms of possible workloads and time consumption.

5 Discussion and Conclusion

In previous sections of this paper three models of a multiagent information and control system decentralized data storage were presented and considered

The first model, "non-replicated" is quite simple but unstable in cases of active task failure. The following models were developed with usage of Viewstamped Replication protocol principles and in a fully decentralized manner. These models have pros and cons, so it is expedient to consider them. Onward, the centralized and decentralized models will be discussed in terms of time and resources consumption.

It must be mentioned, that ICS are the real-time systems, and often the maximum time of task performing is estimated.

Centralized model is much less resource consumptive then the decentralized one. In centralized model only the primary distributes the data portions though the network, and the inactive tasks collect only data portions from the primary. It can be quite useful when the communication environment is of a low bandwidth and the time of the operations is measured by minutes, hours, days. In case of primary failure, view change and – what is notably important - reconfiguration procedure – consumes weighty time and resources due to the consensus finding and data exchange.

Decentralized model is greedy for the communicational environment and memory space due to the distributing data redundancy and the storing of all context data from all active replicas. Yet every active task replica can fail arbitrary, and there is no need to do anything at all except the task recovery, what is done by the operation system of the CU. Besides this, in cases of reconfiguration the recovery procedure of the decentralized model is rather fast in comparison with the centralized one because of a small number of operations.

So, the following conclusion can be made: the models presented implement the "criterion delegating approach" data storage and so the configurations of a higher quality can be obtained. It is extremely important for the reliability enhancement of ICS. Yet, centralized and decentralized models should be used in different operational conditions. The centralized model offloads the network and the memory space, yet the decentralized one allows to reduce recovery and reconfiguration time significantly.

The future work presupposes the formal modelling and precise estimation of the centralized and decentralized models drawbacks, as well as the reliability increase estimation in case of decentralized model implementation.

Acknowledgement. The paper has been prepared within the GZ SSC RAS N GR project 01201354238 and RFBR projects 17-08-01605 A, 18-29-03229 mk, 18-05-80092 and SO SSC RAS N GR project 01201354238, the program of RAS presidium fundamental research I.30 "Theory and technologies of multilevel decentralized group management in conflict and cooperation conditions" (project № AAAA-A18-118011290099-9).

References

1. Crestani, D., Godary-Dejean, K.: Fault tolerance in control architectures for mobile robots: fantasy or reality? In: 7th National Conference on Control Architectures of Robots (2012). http://hal.archives-ouvertes.fr/docs/00/80/43/70/PDF/2012_CAR_FTRobotic-FantasyOrReality.pdf
2. Avizienis, A, Laprie, J.C., Randell, B.: Fundamental concepts of dependability. Technical report, Seriesuniversity of Newcastle Upon Tyne, Computing Science, vol. 1145, pp. 7–12 (2001). https://doi.org/10.1.1.24.6074
3. Carlson, J., Murphy, R.R.: How UGVs physically fail in the field. IEEE Trans. Robot. **21**(3), 423–437 (2005). https://doi.org/10.1109/TRO.2004.838027
4. Melnik, E., Korovin, I., Klimenko, A.: Improving dependability of reconfigurable robotic control system. Lecture Notes in Computer Science (including subseries Lecture Notes in Artificial Intelligence (LNAI) and Lecture Notes in Bioinformatics), vol. 10459 (2017). https://doi.org/10.1007/978-3-319-66471-2_16
5. Melnik, E.V., Klimenko, A.B., Schaefer, G., Korovin, I.S.: A novel approach to fault tolerant information and control system design. In: 2016 5th International Conference on Informatics, Electronics and Vision, ICIEV 2016 (2016). https://doi.org/10.1109/ICIEV.2016.7760182
6. Korovin, I., Melnik, E., Klimenko, A.: A recovery method for the robotic decentralized control system with performance redundancy. Lecture Notes in Computer Science (LNCS) (including subseries Lecture Notes in Artificial Intelligence and Lecture Notes in Bioinformatics), vol. 9812 (2016). https://doi.org/10.1007/978-3-319-43955-6_2
7. Melnik, E., Klimenko, A., Korobkin, V.: Reconfigurable distributed information and control system multiagent management approach. In: Advances in Intelligent Systems and Computing, vol. 680 (2018). https://doi.org/10.1007/978-3-319-68324-9_10
8. Strogonov, A.: Dolgovechnost Integralnih schem I proizvodstvenniye metody ee prognozirovaniya. ChipNews, № 6, pp. 44–49 (2002)
9. Ingber, L.: Simulated annealing: practice versus theory. Math. Comput. Model. **18**(11), 29–57 (1993). https://doi.org/10.1016/0895-7177(93)90204-C
10. Liskov, B.: From viewstamped replication to Byzantine fault tolerance. Lecture Notes in Computer Science (LNCS) (including subseries Lecture Notes in Artificial Intelligence and Lecture Notes in Bioinformatics), vol. 5959, pp. 121–149 (2010). https://doi.org/10.1007/978-3-642-11294-2_7
11. Kirsch, J., Amir, Y.: Paxos for system builders. In: Proceedings of the 2nd Workshop on Large-Scale Distributed Systems and Middleware - LADIS 2008, p. 1 (2008). https://doi.org/10.1145/1529974.1529979
12. Liskov, B., Cowling, J.: Viewstamped replication revisited. In: IEICE Transactions on Information and Systems, (MIT-CSAIL-TR-2012-021), pp. 1–14 (2012). http://18.7.29.232/handle/1721.1/71763
13. Van Renesse, R., Schiper, N., Schneider, F.B.: Vive La Différence: Paxos vs. Viewstamped Replication vs. Zab. IEEE Trans. Depend. Secure Comput. **12**(4), 472–484 (2015). https://doi.org/10.1109/TDSC.2014.2355848
14. Junqueira, F.P., Reed, B.C., Serafini, M.: Zab: high-performance broadcast for primary-backup systems. In: Proceedings of the International Conference on Dependable Systems and Networks (pp. 245–256). https://doi.org/10.1109/DSN.2011.5958223

Representation and Use of Knowledge for the Reconfiguration of the Mechanical Transport System

Stanislav Belyakov[1(✉)], Marina Savelyeva[1], and Igor Rozenberg[2]

[1] Southern Federal University, Taganrog, Russia
beliacov@yandex.ru, marina.n.savelyeva@gmail.com
[2] Research and Development Institute of Railway Engineers, Moscow, Russia
I.kudreyko@gismps.ru

Abstract. The problem of controlling the reconfiguration of the mechanical transport system is considered with the aim of minimizing the total transportation costs. The features of decision making about the choice of the reconfiguration option in the conditions of the lack of information on the dynamics of the state of the external environment are analyzed. The approach to decision-making based on the use of experience is described. The distinctive feature of the approach is the presentation of knowledge about the precedents of reconfiguration in the form of images. The model of the image is given. The principle of obtaining logical conclusions on the basis of image analysis is described. The operations of comparing and transforming images are discussed. The features of the implementation of the transformation operator in the mechanical transport system are considered.

Keywords: Mechanical transport system · Reconfiguration · Intelligent system Image analysis of precedents

1 Introduction

Mechanical transport systems (MTS) are a kind of transport systems related to the Internet of things (IoT). In accordance with known technologies (for example, RFID) labels are placed on the transported objects, allowing to track the position and dynamics of the object's movement. The means of transportation is a system of conveyors and switches for the direction of flow of transported objects. Each object moves from the departure point to the delivery point, passing through a sequence of conveyors and switches. Examples of MTS are baggage handling systems at airports, technological systems at industrial enterprises, and service systems at logistics centers.

The quality of transportation of MTS is determined by the parameters of the delivery time of the cargo, the cost of transportation by the network, the risk of damage to the transported object. External environment has significant effect on parameter quality values of transportation in MTS. This influence is difficult to predict. Instability of cargo flow, unacceptable variation of its geometric measurements, weight and strength of package can lead not only to emergency suspension, but also to equipment

A. Abraham et al. (Eds.): IITI 2018, AISC 874, pp. 99–106, 2019.
https://doi.org/10.1007/978-3-030-01818-4_10

failure of separate conveyors. In this situation, it is necessary to make quick decisions on reconfiguring the MTS, which involves both replacing damaged equipment and adding new sections of the network with conveyors and switches. Reconfigured MTS is a new version of the system, the work of which is poorly researched and generates unaccounted risks of failures of the elements. In conditions of a large number of variants of MTS reconfiguration, choosing the best of them, becomes a difficult problem.

Assessing the current state of the MTS, the state of the cargo flow and predicting the behavior of MTS after reconfiguration is a multifactorial task, to obtain an analytical solution of which is rather problematic. The possible way to build solutions can be the use of the reconfiguration experience of MTS, which experts have. Any decision taken is based on deep knowledge of the mutual influence of external factors, the response to their impact, the significance of individual operations of the reconfiguration process. Any made decision is based on deep knowledge of the mutual influence of external factors, the response to their impact, the significance of separate operations of the reconfiguration process. The reuse of experience in the problem situation is attractive that, it provides a higher reliability of the made decision. To implement this approach, tasks of a different kind should be solved: how is the experience of MTS reconfiguration and how to apply it in the changed operating conditions.

In this paper, the method is analyzed that allows to accumulate and transform experience into given conditions for the emergence of problem situations.

2 Review of Reconfiguration Control Methods

The cost of delivering cargo in MTS is an integral measure of the quality of transportation. Any unforeseen situation leading to the suspension of the work of MTS can be estimated through cost increase in value. Thus, in general view, the task of minimizing the transportation cost of MTS, which is solved when operating the MTS, can be formulated as follows:

$$\begin{cases} E_T(t) + E_R(f) \rightarrow min \\ t < t^* \\ f \subseteq F \end{cases} \quad (1)$$

where E_T is the average cost of delivery of the consignment, determined by the amount of time t the cargo is in the MTS; t^* is restriction on the time of transportation. The function $E_R(f)$ expresses the cost of losses from emergency situations in which reconfiguration was applied. F denotes the set of known defects that contribute to the emergence of emergency situations, and through f denotes a subset of actual defects during the transportation of the consignment. Statement of the problem (1) reflects a number of features of transportation by means MTS. First, the reconfiguration procedure takes a considerable amount of time, commensurate with the time of finding the consignment in the network. This should be noted when, as a possible reconfiguration method a parametric setting of routing in the network is considered. Changing the routing tables in the direction switches, no doubt, affects the status of the cargo flow in the network and should be used to prevent accidents. But as the way of reconfiguring, setting up the routing algorithm has limited capabilities and takes a little time. Therefore, we will further assume that the cost of this type of reconfiguration is included in the value of $E_T(t)$. Secondly, the reconfiguration of MTS is implemented within the set of defects, studied in essence and included by MTS developers in the operational documentation. Thus, the variety of defects is narrowed by the operating procedure, and the choice of the reconfiguration option is the ability of technical staff to adequately assess the problem situation. Third, predicting the quality of transportation after reconfiguration considers the actual set of functioning defects f.

Let us consider the methods for managing the reconfiguration of networks according to problem formulation (1).

Traditionally used analytical tool for the research of the behavior of networks is the analysis of their capacity [1]. The emergency is associated with the achievement by capacity of some limit value. The difficulty of practical application of this method consists in the dynamics of the values of the network capacity and the intensities of the input flows. The reason for changing the capacity of the conveyor can be power decrease in the supply voltage network, the development of the working resource by the actuator, the change in the geometry of the network. The intensity of the input cargo flow, in turn, varies due to changes in the mass-dimensional characteristics of the objects, the way of loading the consignment in MTS. For these reasons, the estimate of the capacity balance is necessary, but not sufficient for the rational choice of the reconfiguration method.

Researches of the most general principles and methods of reconfiguring of geographically distributed networks are of undoubted interest [2]. Analyzing the reconfiguration mechanisms and decision-making strategies for choosing the best system configuration, the authors note the importance of applying high-level conceptual models of knowledge representation about reconfiguration [3]. Analysis of the work of this direction suggests the need for a narrow specialization of conceptual models [4]. The universal models are not very effective [5].

A special role in the solution of the problem of reconfiguration management is played by the use of intelligent methods for detecting pre-emergency situations [6]. Identification of the pre-emergency situation is possible with a high degree of reliability only on the critical values of the parameters that were observed during operation [7]. Since this problem can be solved by a classification method, neural networks are used to find solutions [8]. The disadvantage of this approach in this case should be considered a significant role of the logical conclusions of the expert: slightly different data samplings generate significantly different estimates depending on the logic of reasoning.

Complex relationships of data and the logic of their evaluation lead to the idea of modeling the expert's imaginative thinking [9]. To the problem (1), methods of image description and comparison of situations can be applied [10–12]. Such opportunity requires an analysis of the structure of knowledge about the reconfiguration of MTS.

3 Features of Control of Reconfiguration Using Knowledge

The task of controlling reconfiguration in the event of an emergency situation is to select from among many known reconfiguration options R such r that minimizes the $E_R(f)$ component in the task (1). Given the considerable dimensionality of MTS state space and the large number of possible scenarios after the MTS structure is changed, it is advisable to use the knowledge of experts to choose the reconfiguration option. This knowledge arises as a result of the experts' analysis of the consequences of the real precedents of reconfiguration. It can be argued that knowledge is formed as a mental image in the mind of the expert [13]. The image integrally reflects both the concrete situation, and a lot of situations that are close to it in meaning. The Intelligent Reconfiguration System (IRS) should use the appropriate image description formalism.

Modeling of image thinking is proposed to be built on the operations of comparison and display. Comparison of images consists in comparison of their semantic proximity [14]. For this, the metric of semantic proximity must be introduced, reflecting the emerging situations of reconfiguration in a special space. Note that the use of proximity metrics of situations in MTS state space is unsatisfactory, it involves comparing unique situations. The real world is dynamic and a deep understanding of the behavior of the observable system is difficult to achieve through comparison of single events. It is reasonable to compare images with knowledge of possible transformations that saves the meaning of the observed situations.

The displaying operation is designed to transfer the known image to a given real-world area. The need to use this operation is determined by the desire to increase the reliability of the comparison of images. The formal comparison of the proximity of precedents does not assume the assessment of whether the previously known precedent can arise in the considered area of space. Nevertheless, the topology of the real world can completely exclude this phenomenon, which will lead to unreliable conclusions. Thus, the use of these operations increases the reliability of the conclusions about the semantic equivalence of images. It should be noted that the IRS should have the knowledge to transform the observed critical parameters of MTS state into the image of the situation. The actual parameters values should be provided with the description of their possible changes. This operation reminds fuzzification in fuzzy systems, but in conceptual relation it is characterized by the presence of a space-time reference and the need to observe topological restrictions.

4 Transformation of Images

Transformation of images is the basis of the displaying operation necessary to transfer the reconfiguration experience. The essence of the transformation is the reproduction of the known precedent in given conditions. Most often conditions determine the space-time area and the values of the MTS state parameters. We can assume that the MTS state space and the real-world space form the space of images.

The result of the transformation is determined by the topology of the image space, i.e. relations and rules for preserving the continuity of displays in the space of images.

Rules and relations define the transformation invariant. Formally, the topology of the image space (A) is represented as

$$A = \bigcup_i A_i, A_i \cap A_j \neq \emptyset,$$

$$\forall c \in A : H(c) \subseteq A_i.$$

From the informative point of view, a new (transformed) image J_{TR} must save an intuitively understood meaning by the expert as transformations corresponding to the topology of space. The center of the transformed image $\bar{c}$ displays the actual situation, the decision of which can be re-used. Formally, the transformation is described as

$$< \bar{c}, H(\bar{c}) > \; = F_{TR}(< c, H(c) >),$$

where F_{TR} is transformation operator, $\bar{c}$ is center of the transformed image, $H(\bar{c})$ is admissible transformations of the center of the transformed image.

The transformed image exists under the following conditions:

$$\bar{c} \neq \emptyset \tag{2}$$

$$h_i(\bar{c}) \neq \emptyset, \; i = \overline{1, |H(\bar{c})|} \tag{3}$$

Condition (2) guarantees the reliability of the transfer of experience, which is confirmed by the reproduction of the known precedent in a given area of image space. The known situation can be reproduced, if the topological limitations of space are not broken. For example, if the space of images is flat, then the accident on a linear part of the conveyor cannot be reproduced on a spiral.

Condition (3) ensures meaningful transfer of experience. In view of the fact that the essence of the observed precedent is concluded in its admissible transformations, the impossibility to reproduce at least one transformation indicates a loss of meaning. The total number of transformations is arbitrary. It can be argued that an increase in this number indicates a greater study of the precedent. The more admissible transformations are known, the less probability it is to make a mistake in assessing the meaning of the situation.

Let us consider an example of the transformation of experience. Let there be an emergency situation in some subnetwork of MTS, the reason for which was the high intensity of the input flow due to the tight schedule of the arrival of external freight transport. There was an overload of one of the conveyors, which led to a partial failure of the equipment. The speed of transportation on this conveyor sharply decreased. In accordance with the recommendations for operation, it was required to stop the segment of the MTS, which includes a defective section, and to make repairs. In the situation that arose, this could lead to significant losses. To reduce losses, one of the admissible solutions was accepted: a mobile conveyor was connected and the cargo flow was redirected to neighboring sections of the conveyors. This allowed repairs to be performed without shutting down the defective conveyor. Analyzing the results of the reconfiguration operation, the expert concluded that it is not always possible to apply the decision once again, since it makes sense in conditions of intensive cargo traffic, the cost of potential damage for which is high, and the compulsory availability of a driveway for the mobile conveyor. The attempt to transfer experience to another emergency situation must take into account the specified conditions. The transformation operation gives the chance to estimate the feasibility of such conditions.

5 About the Implementation of the Transforming Operator for MTS

The transformation operator F_{TR} constructs the objects of admissible transformations, controlling compliance with topological restrictions of the state space. If $c = <n, m, v_{in}, v_{out}, V, r, g>$ is the center of the situation images, the operator performs the "carry" procedures using the specified parameters. As already noted, the number of admissible transformations does not coincide with the number of parameters and must be more than 1.

Let us consider an example of topological restrictions of MTS. They are formulated, based on the operating regulations of the MTS, its territorial location, temporary modes of arrival of cargo. For example,

- subnetwork n is the set of no more than 15 conveyors connected by direction switches;
- the number of cargo units m on subnet n should not exceed 120 units;
- the intensity of the incoming and outgoing flow (v_{in}, v_{out}) should not exceed 0.25 s^{-1};
- linguistic estimates of the parameters V, r, g have a subset of bounding values from the set of possible ones. For example, for the parameter g of possible values {excellent, good, satisfactory, bad}, the bounding subset can be {excellent, good}.

If at least one restriction is not executed, reproduce the well-known precedent in the given area, i.e. for the given set of parameters $n, m, v_{in}, v_{out}, V, r, g$ is impossible, i.e. condition (2) is not satisfied.

Consider an example of the admissible transformation that uses the parameters n and m for some reconfiguration precedent. Suppose that the analysis showed the importance of the territorial position of the reconfigured subnet. In particular, the network should not be near sources of energy supply and stands of baggage reception, but at the same time adjacent to technical drives in locations. Proceeding from this, on the geographic map or the MTS scheme, the expert draws the boundaries of the possible zones of the network elements' location relative to the subnet and indicates the spatial relations that must be observed between the boundary of the zone and the objects on the map. Any changes in the network within the specified boundaries of zones and relations are admissible and do not affect the essence of the made decision. Thus, the transformation operator must build the area around the subnet in the given area of the map and monitor the proximity of the placement of the above objects. The algorithm for constructing the boundary of the zone is deterministic. For example, a buffer construction algorithm used by geoinformation systems can be used.

The admissible conversion for the value m can be specified, for example, as follows: the number of cargo units can be more on 3–5 units, provided that the subnet is located near the unloading terminal and transportation happens in the morning. In case of transformation in this example, the admissible deviation on 3–5 units should be accompanied by observance of the relation "is near" and "in the morning".

6 Discussion of Results

Increasing the reliability of decisions about reconfiguring MTS is the main advantage of the proposed method of using knowledge. The increase in reliability is achieved by using the reconfiguration precedents, which are the necessary element of the image. The above operations on the use case images do not use any complicated logic, relying on the knowledge developed by the expert as a result of a posteriori case-based reasoning. Undoubtedly, such approach means an extensional accumulation of knowledge that requires significant amounts of resources. Nevertheless, this is the payment for the quality of the work of the IRS.

Note also that the basis of machine learning IRS in this problem are not samples of data, but instances of complex structured images. Like ontologies, they describe the concepts of knowledge area. This approach to machine learning complicates the procedures for preparing data for the computer system, but it guarantees a more reliable and stable result by describing no single events and situations, but areas and sets. In this case, "insignificant" differences in the data in the sample cannot lead to significant semantic errors. As for the complexity of the description of knowledge, then the information system with an advanced visualization system can be an instrument for overcoming it. An example can be geoinformation systems. The ability to visualize transformations and relationships on maps, plans and schemes greatly facilitates the description of knowledge and their verification.

7 Conclusion

Reconfiguration of MTS with the use of knowledge gives the more effect, the more the number of elements includes the MTS and the more complex the relationship between the reconfiguration process and the external environment. The variety of situations and conditions of their occurrence increases. In such cases, it is difficult for applications engineers of MTS to confine themselves to operational documentation. Knowledge is needed to make decisions based on practical experience. The method of representation and use of kind of knowledge described in this paper is a promising way to solve the problem.

Further research, from the point of view of the authors, should be directed to the research of the mechanism of experience transformation in MTS, differing in scale, coverage of the territory, topology of the space of images.

Acknowledgment. This work has been supported by the Russian Foundation for Basic Research, projects № 17-01-00119.

References

1. Cormen, T.H., Leiserson, C.E., Rivest, R.L., Stein, C.: Introduction to Algorithms, 3rd edn. MIT Press, Cambridge (2009)
2. Youjie, M., Feng, L., Xuesong Z., Zhiqiang G.: Overview on algorithms of distribution network reconfiguration. In: 2017 36th Chinese Control Conference (CCC), pp. 10657–10661. IEEE Press, New York (2017)
3. Shariatzadeh, F., Kumar, N., Srivastava, A.K.: optimal control algorithms for reconfiguration of shipboard microgrid distribution system using intelligent techniques. IEEE Trans. Ind. Appl. **53**(1), 474–482 (2017)
4. Srivastava, I., Bhat, S.S.: Soft computing techniques applied to distribution network reconfiguration: a survey of the state-of-the-art. In: 2016 8th International Conference on Computational Intelligence and Communication Networks (CICN), pp. 702–707. IEEE Press, New York (2016)
5. Brennan, R.W., Vrba, P., Tichy, P., Zoit, A., Sunder, C., Strasser, T., Marik, V.: Developments in dynamic and intelligent reconfiguration of industrial automation. Compt. Ind. **59**(6), 533–547 (2008)
6. Makarova, I., Pashkevich, A., Mukhametdinov, E., Mavrin, V.: Application of the situational management methods to ensure safety in intelligent transport systems. In: Proceedings of the 3rd International Conference on Vehicle Technology and Intelligent Transport System, pp. 339–345. (2017)
7. Makarova, I., Khabibulli, R., Mukhametdinov, E., Pashkevich, A., Shubenkova, K.: Efficiency management of robotic production processes at automotive industry. In: Proceedings of the 2016 17th International Conference on Mechatronics Mechatronika, pp. 1–8. (2016)
8. Azab, A., ElMaraghy, H., Nyhuis, P., Pachow-Frauenhofer, J., Schmidt, M.: Mechanics of change: a framework to reconfigure manufacturing systems. CIRP J. Manuf. Sci. Technol. **6**, 110–119 (2013)
9. Kuznetsov, O.P.: Kognitivnaya semantika i iskusstvennyy intellekt. Iskusstvennyy intellekt i prinyatie resheniy **4**, 32–42 (2012)
10. Belyakov, S., Bozhenyuk, A., Rozenberg I.: The intuitive cartographic representation in decision-making. In: World Scientific Proceeding Series on Computer Engineering and Information Science, vol. 10, pp. 13–18 (2016)
11. Belyakov, S., Belyakova, M., Savelyeva, M., Rozenberg, I.: The synthesis of reliable solutions of the logistics problems using geographic information systems. In: 10th International Conference on Application of Information and Communication Technologies (AICT), pp. 371–375. IEEE Press, New York (2016)
12. Belyakov, S., Savelyeva, M.: Protective correction of the flow in mechanical transport system. In: Proceedings of the 6th Computer Science On-line Conference 2017 (CSOC 2017), pp. 180–185. (2017)
13. Kaplan, R., Schuck, N.W., Doeller, C.F.: The role of mental maps in decision-making trends in neuroscience. Trend Neurosci. **40**(5), 256–259 (2017)
14. Lenz, M., Bartsch-Spörl, B., Burkhard, H.D.: Case-Based Reasoning Technology: From Foundations to Applications. Springer, Heidelberg (2003). https://doi.org/10.1007/3-540-69351-3

Synthesis of Adaptive Algorithms for Estimating the Parameters of Angular Position Based on the Combined Maximum Principle

Andrey Kostoglotov(✉), Sergey Lazarenko, Anton Penkov, Igor Kirillov, and Olga Manaenkova

Rostov State Transport University, Rostov-on-Don, Russian Federation
{kostoglotov, lazarenkosv}@icloud.com,
penchal285@yandex.ru,
{kirillov8084, manaenkova_o}@mail.ru

Abstract. The problem of synthesis of adaptive dynamic filter is presented in the form of problem of quasi-optimal control. The solution is obtained based on the theorem of the maximum of the function of the generalized power and analysis of the Lagrangian along characteristic trajectories in phase space. This allows to construct a model of controlled motion that can be represented in a quasilinear form. The obtained equation of the adaptive filter of the dynamic estimation of the motion parameters differs from the known equations by its feedback structure. On the basis of mathematical modeling it is shown that estimations of the proposed filter provide an increase in accuracy with less computational costs.

Keywords: Adaptation · The combined maximum principle · Estimation
The Hamilton-Ostrogradsky principle

1 Introduction

The process to increase the efficiency of information control systems is directly related to the level of intellectualization of information processing algorithms. The algorithms become "intelligent" when the developer provides for the possibility for the standard solutions to adapt via parameters or even by structure. The paper presents a new approach to the use of the possibility of adaptation to the model of the process under study as the basis for constructing the production rules for intelligent information processing systems.

The basis of algorithms for estimating the state of controlled dynamical systems are their mathematical models, which follows from the laws of motion, represented in the form of differential equations or variational principles. The law of motion establishes the dependence of the object state on the control action. In case of random impact the methods of statistical synthesis are traditionally used to solve the problem of estimating the trajectory parameters [1]. Among the solutions obtained on this basis the most common algorithms are of the Kalman structure [1, 2]. Typical disadvantages are

A. Abraham et al. (Eds.): IITI 2018, AISC 874, pp. 107–115, 2019.
https://doi.org/10.1007/978-3-030-01818-4_11

relatively high computational complexity and weak dependence of the feedback coefficients on observations in a steady state [3–5]. As a result, quasi-optimal estimation algorithms have become more widely used [1, 4]. In this case, the error in selecting the model class of the object dynamics can lead to unacceptably high estimation errors. This is one of the reasons for the development of adaptive dynamic filtering methods [5, 6].

In the case when the state vector is assumed to be quasi-deterministic, the estimation problem is posed as the problem of synthesis of optimal control. One of the effective approaches to solving such an extremal problem is based on the use of the theorem of the maximum of the function of the generalized power [7–15]. It allows to obtain the general structure of the mathematical model for controlled dynamic system with adaptation to the observed dynamics [9, 11, 12, 15]. The determination of the parameters of the obtained model requires to investigate the behavior of the kinetic potential on the characteristics in the phase space.

The purpose of the work is the synthesis of the adaptive filter for dynamic estimation of the angular position parameters based on the adaptive model with correction based on observations.

2 Formulation of the Problem

In the observation space we define the target functional [9, 11, 12, 15]:

$$\begin{aligned} J &= \frac{1}{2}\int_0^{t_1} [y(t) - H(q,t)]^T N^{-1} [y(t) - H(q,t)] dt \\ &= \frac{1}{2}\int_0^{t_1} \Delta^T N^{-1} \Delta dt = \int_0^{t_1} F(y,q) dt, \end{aligned} \quad (1)$$

where N^{-1} is the weight matrix characterizing the intensity of interference in the observation channel

$$y(t) = H(q,t) + \zeta(t);$$

Here $H \in R^n$ is the matrix of the projection of the state space onto the space of observations, $q \in R^n$ is the vector of generalized coordinates, $\xi(t) \in R^n$ is the vector of random effects on the observation channel of known intensity, $t \in [0, t_1] \subset R$, n is the number of degrees of freedom of the dynamical system.

The motion of the object satisfies the Hamilton-Ostrogradskii principle [16], and the motion quality is estimated using the extended action functional [7–15]

$$S = \int_0^{t_1} [\lambda(T + A) + F] dt, \quad (2)$$

where $T = T(q, \dot{q}, t) = \frac{1}{2}\dot{q}^T p$ is the kinetic energy, p is the generalized momenta vector, $A = \int_{q(0)}^{q(t_1)} \sum_{s=1}^{n} Q_s dq_s$ is the work of generalized forces $Q_s = Q_s(q, \dot{q}, t)$ at the true trajectory, n is the number of degrees of freedom. Generalized forces can depend on control additively $Q_s = Q_s^A + U_s$ or multiplicatively (parametrically) $Q_s = Q_s^M(q, \dot{q}, t, U)$. The control vector is chosen from a certain valid range

$$U \in \overline{G}_U.$$

The value $\lambda = \lambda(\xi, q, \dot{q})$ is the Lagrange multiplier depending on the vector of random effects $\xi(t)$ and the trajectory parameters $(q, \dot{q}) \in R^{2n}$.

It is required to find the control vector $U(q, \dot{q}, \xi, \lambda)$ as a function of the trajectory parameters $(q, \dot{q}) \in R^{2n}$ and random effects $\xi \in R^n$.

The solution to this problem is based on the methodology of the combined maximum principle [7–15]. To refine the structure of the solution obtained, it is necessary to analyze the behavior of the invariants of motion on the characteristic trajectories in the phase space.

3 Model of the Object Motion with Correction According to Current Observations

To solve the problem, we use the combined maximum principle, where the optimality condition corresponds to the maximum of the generalized power [7–15]

$$\Phi(q, \dot{q}, Q(U), \lambda, \xi) = \max \sum_{s=1}^{n} [\lambda Q_s + V_s]\dot{q}_s,$$

where $V = \operatorname{grad} F$, and we satisfy the transversality conditions for the Hamiltonian H

$$H|_0^{t_1} = \lambda(A + T) + F|_0^{t_1} = 0.$$

The expression for the generalized force Q_s in terms of the minimum of criterion (1) has the form

$$Q_s = \lambda^{-1}[\mu_s p_s - V_s], s = \overline{1, n},$$

where μ_s is synthesizing function [7–15]. Depending on the parameters of the trajectory its structure is determined by the following relation [14]

$$[H_0, Q_0] = \frac{\partial H_0}{\partial p_s}\frac{\partial Q_0}{\partial q_s} - \frac{\partial H_0}{\partial q_s}\frac{\partial Q_0}{\partial p_s} = 0, \tag{3}$$

where q_s, p_s are the Hamilton variables, $H_0 = \lambda T + F$, $Q_0 = \mu_s p_s - V_s = 0$, Eq. (3) is extended expression for the operation of the Poisson bracket.

For this problem in this formulation [7–15] we have the solution of the inverse problem in the form

$$Q_s = \lambda^{-1}\left[-\frac{|\dot{q}_s|p_s}{\lambda^{-1}|V_s|} - V_s\right], s = \overline{1,n}, \tag{4}$$

from which expression the model of motion with correction for observations takes the form

$$\frac{d}{dt}\frac{\partial T}{\partial \dot{q}_s} - \frac{\partial T}{\partial q_s} = Q_s(q, p, \xi, \lambda). \tag{5}$$

4 Adaptation of the Motion Model Based on the Lagrangian Invariance Principle

The Hamiltonian in the action integral (2) can be represented in the form with the explicitly separated Lagrangian $L = \lambda T - F$

$$\begin{aligned} H &= \lambda(T + A) + F = 2F + \lambda A + (\lambda T - F) \\ &= 2\lambda T + \lambda A + (-\lambda T + F). \end{aligned}$$

From the analysis of the structure of the phase space at the junction points of the hyperbolic paraboloid and ellipsoid [14] it follows that we satisfy the condition for Lagrangian

$$L_0 = \lambda T - F = 0.$$

Let us consider the state of the system in the second (fourth) quadrant of the phase plane.

Along the optimal trajectory the Hamilton action has a stationary value, hence with $Q = Q_0 = 0$

$$p_s\dot{q}_s - \lambda^{-1}V_s^2 = 0.$$

Consequently, there exists a kinematic constraint

$$\dot{q}_s = \pm\sqrt{\eta_s^{-1}}V_s,$$

Where the value η_s is found in accordance with equation

$$\sum_{s=1}^{n} a_{sk}(q)\dot{q}_s\dot{q}_k - \frac{\lambda^{-1}}{\eta_s^{-1}}q_s^2 = 0, s = \overline{1,n},$$

where $a_{sk}(q)$ is the inertia coefficient.

The characteristic determinant

$$\left|a_{sk} - \delta_s^k \Lambda_s\right|_{k=1}^{n} = 0, s = \overline{1,n},$$

where δ_s^k is the Kronecker symbol.

The relation between the factors Λ_s, λ and the roots of the characteristic determinant is as follows:

$$\Lambda_s = \frac{\lambda^{-1}}{\eta_s^{-1}}, s = \overline{1,n}.$$

Thus, the feedback law (4) is reduced to the form

$$Q_s = \Lambda_s\left(-\sqrt{\eta_s^{-1}}\dot{q}_s - \eta_s^{-1}V_s\right),$$

and we can write the equation of motion in the Lagrangian variables as

$$\frac{d}{dt}\frac{\partial T}{\partial \dot{q}_s} - \frac{\partial T}{\partial q_s} = Q_s,$$

$$\Lambda_s\ddot{q}_s + \sum_{s,k=1}^{n}\left[\frac{\partial \Lambda_s}{\partial q_k}\dot{q}_k\dot{q}_s - \frac{1}{2}\frac{\partial \Lambda_k}{\partial q_s}\dot{q}_s\dot{q}_k\right] = \Lambda_s\left(-\sqrt{\eta_s^{-1}}\dot{q}_s - \eta_s^{-1}V_s\right),$$

and in the Hamiltonian variables as

$$\frac{dq_s}{dt} = \frac{p_s}{\Lambda_s},$$

$$\frac{dq_s}{dt} = \frac{\partial E}{\partial q_s} + Q_s;$$

$$\frac{dp_s}{dt} = \frac{1}{2}\sum \frac{\partial \Lambda_k}{\partial q_s}\frac{p_k}{\Lambda_k^2} + \left(-\sqrt{\eta_s^{-1}}p_s - \eta_s^{-1}p_s - \eta_s^{-1}\Lambda_s V_s\right),$$

where $T = \frac{1}{2}\sum_{k=1}^{n}\Lambda_k\dot{q}_k^2 = \frac{1}{2}\sum_{k=1}^{n}\frac{p_k^2}{\Lambda_k}$ is the kinetic energy, $p_s = \frac{\partial T}{\partial \dot{q}_s} = \Lambda_s\dot{q}_s$ is the generalized momentum, $\dot{q}_s = \frac{p_s}{\Lambda_s}$ is the generalized velocity.

5 Example of Constructing an Adaptive Model of Motion

Let the kinetic energy of an object with two degrees of freedom is as follows:

$$T = \frac{1}{2}\sum_{s,k=1}^{2} p_s\dot{q}_s = \frac{1}{2}\sum_{s,k=1}^{2} a_{sk}\dot{a}_s\dot{a}_k = \frac{1}{2}\frac{p_k^2}{\Lambda_k}.$$

The characteristic determinant

$$\begin{vmatrix} a_{11}-\Lambda & a_{12} \\ a_{21} & a_{22}-\Lambda \end{vmatrix} = 0. \tag{6}$$

The roots of the characteristic equation are:

$$\Lambda_{1,2} = 0.5\left(S_p \pm \sqrt{S_p^2 - 4D}\right),$$
$$S_p = a_{11} + a_{22};\ D = a_{11}a_{22} - a_{12}a_{21}.$$

The Hamiltonian takes the form

$$H = \lambda\left[T + \sum_{s=1}^{2}\int_{q_0}^{q} Q_s dq_s\right] + F$$

The equation of motion in the Hamiltonian variables are:

$$H = \lambda\left[T + \sum_{s=1}^{2}\int_{q_0}^{q} Q_s dq_s\right] + F\frac{dq_s}{dt} = \frac{\partial H}{\partial p_s} = \dot{q}_s = \frac{p_s}{\Lambda_s},$$
$$\frac{dq_s}{dt} = \frac{\partial T}{\partial q_s} + Q_s = -\frac{1}{2}\sum_{k=1}^{2}\frac{\partial \Lambda_k}{\partial q_s}\frac{p_k}{\Lambda_k^2} + \left(-\sqrt{\eta_s^{-1}}p_s - \eta_s^{-1}\Lambda_s V_s\right).$$

For an object with one degree of freedom

$$T = \frac{1}{2}m\dot{q}^2 = \frac{1}{2}\frac{p_2}{m},$$

we have

$$\ddot{q} = -\sqrt{\eta^{-1}}\dot{q} - \eta^{-1}V$$

in the Lagrangian form, and

$$\dot{q} = \frac{p}{m},$$

$$\frac{dp}{dt} = -\sqrt{\eta^{-1}}p - \eta^{-1}mV$$

in the Hamiltonian form.

6 Synthesis of the Estimation Equations

Now we consider a variant of constructing an adaptive filter of dynamic estimation, when the observation is defined as follows:

$$y = q + \zeta,$$

where ζ is the observation noise, the generalized coordinate.

The motion Eq. (5) with $n = 1$, $T = \frac{1}{2}\dot{q}^2$ has the form

$$\ddot{q} = U.$$

The quality of the estimation is determined by the functional

$$J = \frac{1}{2}N^{-1}\int\limits_0^{t_1} [y - \hat{q}]^2 dt,$$

where the sign $\wedge$ denotes an estimate, and $N-$ is the spectral density of the noise.

It follows from (6) that the equation for the optimal filter has the form

$$\ddot{\hat{q}} = -\sqrt{\lambda^{-1}}\dot{\hat{q}} - \lambda^{-1}N^{-1}(y - \hat{q}).$$

The paper presents the results of mathematical modeling of the process of estimating the readings of the orientation sensors (accelerometer and gyroscope) based on the new algorithm with adaptation via the parameter λ using the kinematic constraints for the angular motion in one variant and the simple integration of measurements in the other variant.

The mathematical modeling results are shown in Figs. 1, 2 and 3.

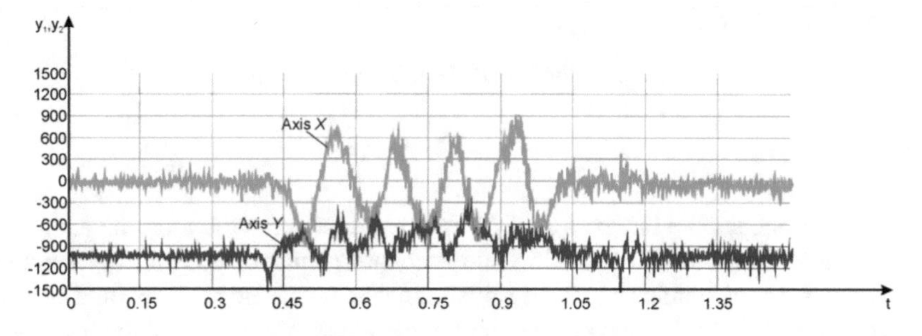

Fig. 1. The initial accelerometer data.

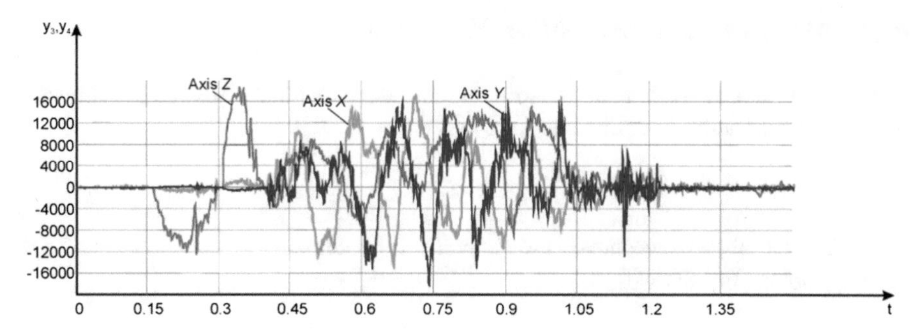

Fig. 2. The initial gyroscope data.

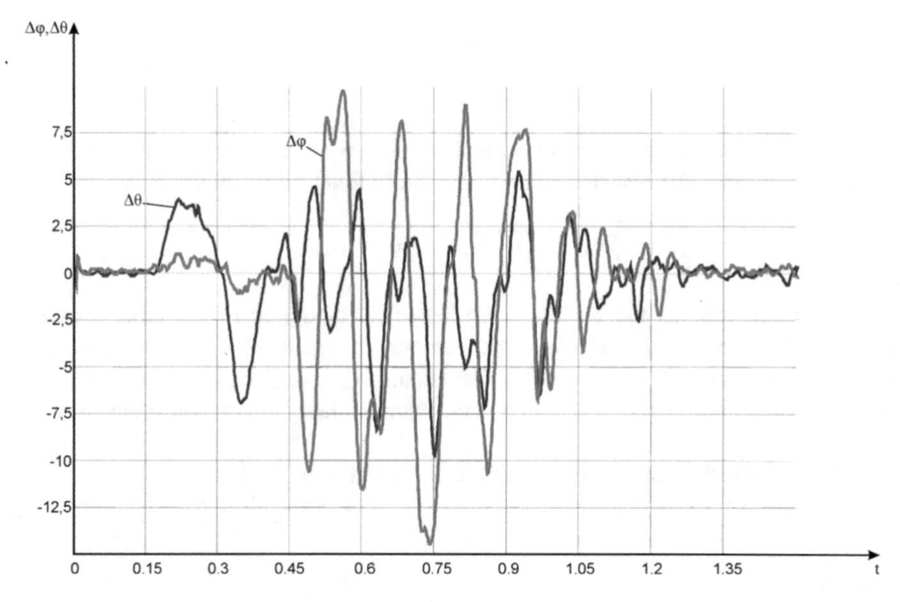

Fig. 3. Estimates of absolute angular errors

7 Conclusion

The results of the simple integration of measurement data in estimating the parameters of the angular position without taking into account kinematic constraints and without adaptation of the dynamics model show significant (on average 5°–6° and at some moments up to 10°–12°) errors at the output of the filters in the mode of the fast change of the angular position.

Adaptation of dynamic filters based on motion models obtained using the combined maximum principle allows to increase the accuracy of the estimation of the angular position parameters.

Acknowledgements. The paper has been accomplished with the support of Russian Federal Property Fund grants No. 18-01-00385 A, № 18-08-01494 A and grant from the Rostov State Transport University.

References

1. Bar-Shalom, Y.: Estimation with Applications to Tracking and Navigation, New York (2001)
2. Jin-long, Y.: A novel robust two-stage extended Kalman filter for bearings-only maneuvering target tracking. Int. J. Phys. Sci. **3**, 987–991 (2011)
3. Rudenko, E.A.: The optimal structure of discrete nonlinear filters of small order. Autom. Remote Control **9**, 58–71 (1999)
4. Schooler, C.C.: Optimal a-b filters for systems with modeling inaccuracies. IEEE Trans. Aerosp. Electron. Syst. **6**, 1300–1306 (1975)
5. Rudenko, E.A.: Analytical-numerical approximations of the optimal recurrent logical-dynamical low order filter-predictor. J. Comput. Syst. Sci. Int. **5**, 691–714 (2015)
6. Kaufman, P.J.: Smarter Trading: Improving Performance in Changing Markets, New York (1995)
7. Derabkin, I.V., Kostoglotov, A.A., Lazarenko, S.V., Lyashchenko, Z.V.: Intellectualization of industrial systems based on the synthesis of a robotic manipulator control using a combined-maximum principle method. Adv. Intell. Syst. Comput. **451**, 375–384 (2016)
8. Kostoglotov, A.A., Deryabkin, I.V., Andrashitov, D.S., Lazarenko, S.V., Pugachev, I.V.: Synthesis of algorithms for estimation of parameters and state of dynamic systems using additional invariants. In: Proceedings of 2016 IEEE East-West Design and Test Symposium, vol. 1, pp. 1–5 (2016)
9. Kostoglotov, A.A., Kuznetcov, A.A., Lazarenko, S.V., Deryabkin, I.V.: The method of structural adaptation of discrete algorithms for the combined maximum principle in problems of estimation of motion parameters. Inf.-Control Syst. **85**, 10–15 (2016)
10. Kostoglotov, A.A., Lazarenko, S.V., Derabkin, I.V., Kuznetcova, O.N., Yachmenov, A.A.: Combined maximum principle as the basis of intellectualization of control systems for a suspension of vehicles. Adv. Intell. Syst. Comput. **679**, 384–393 (2017)
11. Kostoglotov, A.A., Lazarenko, S.V., Deryabkin, I.V., Kuzin, A.P., Pugachev, I.V., Manaenkova, O.N.: Fuzzy control laws in the basis of solutions of synthesis problems of the combined maximum principle. Adv. Intell. Syst. Comput. **679**, 375–383 (2017)
12. Kostoglotov, A.A., Kuzin, A.P., Lazarenko, S.V., Pugachev, I.V.: The combined maximum principle in the problem of synthesis of an adaptive dynamic filter under conditions of disturbances in the measurement process. In: MATEC Web of Conference, vol. 132, pp. 1–5 (2017)
13. Kostoglotov, A.A., Deryabkin, I.V., Andrashitov, D.S., Lazarenko, S.V., Kuznetcov, A.A.: Method of estimation algorithms synthesis of dynamic processes with construction of the reference trajectory in transients disturbances. In: International Conference on Mechanical, System and Control Engineering, vol. 1, pp. 367–371 (2017)
14. Kostoglotov, A.A., Lazarenko, S.V.: Synthesis of adaptive tracking systems based on the hypothesis of stationarity of the hamiltonian on the switching hypersurface. J. Commun. Technol. Electron. **62**, 123–127 (2017)
15. Kostoglotov, A.A., Lazarenko, S.V., Lyaschenko, Z.V.: Intellectualization of measuring systems based on the method of structural adaptation in the construction of tracking filter. In: Proceedings of 2017 20th IEEE International Conference on Soft Computing and Measurements, Saint Petersburg (2017)
16. Lur'e, A.I., Analiticheskaya Mekhanika (Analytical mechanics). Gos. Izd. Fiz.-Matem. Liter. Moscow (1961)

Synthesis of Intelligent Discrete Algorithms for Estimation with Model Adaptation Based on the Combined Maximum Principle

Andrey Kostoglotov[1,2(✉)], Sergey Lazarenko[1,2(✉)], Igor Pugachev[2(✉)], and Alexey Yachmenov[1]

[1] Rostov State Transport University, Rostov-on-Don, Russian Federation
{kostoglotov, lazarenkosv}@icloud.com, yachmenov-aa@mail.ru
[2] Don State Technical University, Rostov-on-Don, Russian Federation
bakut_8536@mail.ru

Abstract. The paper considers the problem to estimate the parameters of the motion of maneuvering targets. The structure of the filter for estimating the parameters of motion is determined by the mathematical model of the motion. At present time the kinematic models are widely used, but they do not fully correspond to the observed dynamics. This may lead to divergence of the estimation process and failure of the computational procedure. New dynamic filters of the combined maximum principle with the dynamic model of motion possess higher accuracy and stability and smaller amount of computational costs in comparison with common filters. The parametric adaptation of the procedure is carried out using fuzzy logic.

Keywords: Adaptation · Combined maximum principle · Estimation
Mathematical model · Lagrange equation of the second kind · Fuzzy logic

1 Introduction

One of the requirements for modern technical systems is to provide a wide range of working conditions that lead to different, often contradictory, modes of operation. They are characterized by a combination of quality indicators, constraints on the dynamics of systems and control actions, the nature and intensity of disturbances, etc. As applied to radar stations, this is the modes of viewing and tracking a target. The dynamics of the observed object at different parts of its trajectory can vary significantly, which lead to a great variety of mathematical models of motion. Among the models, the kinematic motion models described by a polynomial are most widely used. The polynomial coefficients are the parameters of the target trajectory. It should be noted that when observing maneuvering targets, the kinematic model which is the basis for the synthesis procedure, as a rule does not fully correspond to the observed dynamics, which may lead to the disruption of tracking [1, 2].

The well-known results of studies [3–7] allow to state that the use of the combined maximum principle ensures the construction of dynamic mathematical models of

A. Abraham et al. (Eds.): IITI 2018, AISC 874, pp. 116–124, 2019.
https://doi.org/10.1007/978-3-030-01818-4_12

motion. Their application to the problems of estimating the parameters of motion involve the effect of structural adaptation, which, in comparison with traditional filters, increases the accuracy of tracking the maneuvering target with less computational costs. At the same time the problems to adapt the constructed dynamic models to the observed motion via the adjustment of the parameters were not investigated. To address this, under the conditions of insufficient a priori information about the probabilistic properties of the occurring processes, it is reasonable to use methods of intellectualization [8].

The aim of the research is to improve the efficiency of solving the problems of estimating the parameters of the maneuvering targets movement by constructing a dynamic mathematical model of the motion using the combined maximum principle and then by parametric adaptation of the model using fuzzy logic methods [8–10].

2 Formulation of the Problem

In the observation space we define the target functional of the error [1, 3–7, 10]:

$$J = \frac{1}{2}\int_{0}^{t_1} [\mathbf{y}(t) - \hat{\mathbf{q}}(t)]^T \mathbf{N}^{-1} [\mathbf{y}(t) - \hat{\mathbf{q}}(t)]dt = \int_{0}^{t_1} F(\mathbf{y}, \hat{\mathbf{q}}, t)dt, \tag{1}$$

where here $\wedge$ is a sign that denotes an estimate, and $\mathbf{N}^{-1}$ is the weight matrix characterizing the intensity of interference in the observation channel

$$\mathbf{y}(t) = \mathbf{q}(t) + \xi(t), \tag{2}$$

here $\mathbf{q} \in R^n$ - is the vector of generalized coordinates, $\xi(t) \in R^n$ is the vector of random effects on the observation channel of known intensity, $t \in [0, t_1] \subset R$, n is the number of degrees of freedom of the dynamical system.

The observed dynamics satisfies the Hamilton-Ostrogradskii principle [11]. Let us write the extended action functional in the form [3–7, 10]

$$S = \int_{0}^{t_1} [\lambda(T + A) + F]dt, \tag{3}$$

where $T = T(\mathbf{q}, \dot{\mathbf{q}}, t) = \frac{1}{2}\dot{\mathbf{q}}^T\dot{\mathbf{q}}$ is the kinetic energy, $A = \int_{q(0)}^{q(t_1)} \sum_{s=1}^{n} Q_s dq_s$ is the work of generalized forces $Q_s = Q_s(\mathbf{q}, \dot{\mathbf{q}}, t)$ at the true trajectory,

$$\mathbf{Q} \in \overline{G}, \tag{4}$$

here $\overline{G}$ is the set of generalized forces admitting the observed motion, $\lambda = \lambda(\xi, \mathbf{q}, \dot{\mathbf{q}})$ is the Lagrange multiplier depending on the vector of random effects $\xi(t)$ and on the trajectory $(\mathbf{q}, \dot{\mathbf{q}}) \in R^{2n}$.

It is required to find the vector of generalized forces $\mathbf{Q}(\mathbf{q}, \dot{\mathbf{q}}, \boldsymbol{\xi}, \lambda)$ as a function of the trajectory $(\mathbf{q}, \dot{\mathbf{q}}) \in R^{2n}$. and random effects $\boldsymbol{\xi} \in R^n$.

The solution to this problem is based on the methodology of the combined maximum principle [3–7, 10]. To refine the structure of the solution obtained, it is necessary to analyze the behavior of the invariants of motion on the characteristic trajectories in the phase space.

3 The Synthesis of Motion Models Using the Combined Maximum Principle

The condition of the minimum of the target functional can be represented in the form of the condition of the maximum of the generalized power [3–7, 10]

$$\Phi(\mathbf{q}, \dot{\mathbf{q}}, \mathbf{Q}, \lambda, \boldsymbol{\xi}) = \max \sum_{s=1}^{n} [\lambda Q_s + V_s]\dot{q}_s, \tag{5}$$

where $\mathbf{V} = gradF$; under the transversality condition

$$H|_0^{t_1} = \lambda(A + T) + F|_0^{t_1}, \tag{6}$$

The use of Eq. (5) allows to establish a feedback structure up to a synthesis function μ_s

$$Q_s = \lambda^{-1}[\mu_s \dot{q}_s - V_s], s = \overline{1, n}. \tag{7}$$

Taking into account the expression for the kinetic energy [12] we get

$$\mu_s = -\lambda \left| \frac{d\dot{q}_s}{dq_s} \right|, \; s = \overline{1, n}. \tag{8}$$

For the target of unit mass the Hamilton-Ostrogradskii principle yields the Lagrange equations of the second kind [11]

$$\ddot{q}_m = \sum_{k=1}^{3} g^{mk} Q'_r - \sum_{s,k=1}^{3} \Gamma^m_{sk} \dot{q}_s \dot{q}_r = \lambda^{-1} \left[-\lambda \left| \frac{\dot{q}_m}{L q_m} \right| \dot{q}_m - V_m \right], \; m = \overline{1, 3}, \tag{9}$$

where g^{mr} are the contravariant components of the metric tensor $\hat{g}$,

$$\sum_{m=1}^{3} \left[g_{mr} \left(\ddot{q}_m + \sum_{s,k=1}^{3} \Gamma^m_{sr} \dot{q}_s \dot{q}_r \right) \right] = Q'_r, \; \text{m} = \overline{1, 3}, \tag{10}$$

Γ^m_{sr} are the Christoffel symbols of the second kind, g_{sr} are the covariant components of the metric tensor $\hat{g}$. Then in accordance with (8) we get:

$$\ddot{q}_s = -\left|\frac{d\dot{q}_s}{dq_s}\right|\dot{q}_s + \lambda^{-1}N_{ss}[y_s - q_s]. \tag{11}$$

Let the equations of lines passing through the terminal point of the phase space have the form

$$q_s^a - k\dot{q}_s^b = 0 \tag{12}$$

where a_s, b_s, k – *const*. Since

$$\frac{d\left(q_s^a - k\dot{q}_s^b\right)}{d\dot{q}_s} = aq_s^{a-1} - kb\dot{q}_s^{b-1}\frac{d\dot{q}_s}{dq_s} = 0, \tag{13}$$

then

$$\frac{d\dot{q}_s}{dq_s} = \frac{a}{b}\frac{q_s^{a-1}}{k\dot{q}_s^{b-1}} = \frac{a}{b}\frac{q_s^{a-1}\dot{q}_s}{k\dot{q}_s^b} = \frac{a}{b}\frac{\dot{q}_s}{q_s}. \tag{14}$$

In accordance with the chosen form of the kinetic energy

$$\mu_s = -\tilde{\lambda}\left|\frac{\dot{q}_s}{q_s}\right|,\ \tilde{\lambda} = \lambda\frac{b}{a} = \lambda L \geq 0,\ s = \overline{1,n}. \tag{15}$$

Approximations of the synthesis function lead to the Lagrange equation of the second kind

$$\ddot{q}_s = \tilde{\lambda}\left[-\lambda^{-1}\frac{\dot{q}_s^2}{|q_s|}sign(\dot{q}_s) + L^{-1}N_{ss}[y_s - q_s]\right]. \tag{16}$$

It follows from (11) that after the expansion of the state space [4, 5] we get

$$\dot{\mathbf{x}} = \mathbf{A}\mathbf{x} - \mathbf{\Delta}, \tag{17}$$

where

$$\mathbf{A} = \begin{bmatrix} 0 & 1 & 0 & 0 & \cdots & 0 & 0 \\ 0 & -\frac{\tilde{\lambda}}{\lambda}\frac{|x_2|}{|x_1|} & 0 & 0 & \cdots & 0 & 0 \\ 0 & 0 & 0 & 1 & \cdots & 0 & 0 \\ 0 & 0 & 0 & -\frac{\tilde{\lambda}}{\lambda}\frac{|x_4|}{|x_3|} & \cdots & 0 & 0 \\ \vdots & \vdots & \vdots & \vdots & \ddots & \vdots & \vdots \\ 0 & 0 & 0 & 0 & \cdots & 0 & 1 \\ 0 & 0 & 0 & 0 & \cdots & 0 & -\frac{\tilde{\lambda}}{\lambda}\frac{|x_{2n}|}{|x_{2n-1}|} \end{bmatrix}, \tag{18}$$

$$\mathbf{x} = \begin{bmatrix} x_1 \\ x_2 \\ \vdots \\ x_n \\ x_{2n} \end{bmatrix}, \ \boldsymbol{\Delta} = \begin{bmatrix} 0 \\ \lambda^{-1} N_{11} [y_1 - x_1] \\ \vdots \\ 0 \\ \lambda^{-1} N_{nn} [y_n - x_{2n-1}] \end{bmatrix}. \tag{19}$$

Passing to the discrete time we obtain from (17):

$$\mathbf{x}(k+1) = \boldsymbol{\Phi}\left(\Delta T, \lambda, \tilde{\lambda}, k\right)\mathbf{x}(k) + \tilde{\Delta}, \tag{20}$$

where ΔT is the data sampling period, k is the current time,

$$\tilde{\Delta} = \begin{bmatrix} 0 \\ \lambda^{-1} N_{11} [y_1(k) - x_1(k)] \\ \vdots \\ 0 \\ \lambda^{-1} N_{nn} [y_n(k) - x_{2n-1}(k)] \end{bmatrix}, \tag{21}$$

and the transition matrix

$$\boldsymbol{\Phi}(\Delta T, M, k) = \begin{bmatrix} 1 & \Delta T & 0 & 0 & \cdots & 0 & 0 \\ 0 & -\left[\frac{\tilde{\lambda}\Delta T}{\lambda}\frac{|x_2(k)|}{|x_1(k)|} - 1\right] & 0 & 0 & \cdots & 0 & 0 \\ 0 & 0 & 1 & \Delta T & \cdots & 0 & 0 \\ 0 & 0 & 0 & -\left[\frac{\tilde{\lambda}\Delta T}{\lambda}\frac{|x_4(k)|}{|x_3(k)|} - 1\right] & \cdots & 0 & 0 \\ \vdots & \vdots & \vdots & \vdots & \ddots & \vdots & \vdots \\ 0 & 0 & 0 & 0 & \cdots & 1 & \Delta T \\ 0 & 0 & 0 & 0 & \cdots & 0 & -\left[\frac{\tilde{\lambda}\Delta T}{\lambda}\frac{|x_{2n}(k)|}{|x_{2n-1}(k)|} - 1\right] \end{bmatrix}. \tag{22}$$

4 Discrete Dynamic Filter of the Combined Maximum Principle

In accordance with the synthesis procedure [7], Eq. (20) yields the following equations of the discrete dynamic filter:

$$\begin{aligned} &\hat{\mathbf{x}}(k+1) = \boldsymbol{\Phi}\left(\Delta T, \lambda, \tilde{\lambda}, k\right)\hat{\mathbf{x}}(k) + \mathbf{H}_1\left(\Delta T, \lambda^{-1}, \mathbf{N}\right)[\tilde{\mathbf{y}}(k) - \mathbf{H}_2\hat{\mathbf{x}}(k)], \\ &\hat{\mathbf{x}}(0) = \hat{\mathbf{x}}_0, \end{aligned} \tag{23}$$

where $\hat{\mathbf{x}}_0$ is the initial conditions vector,

$$
\mathbf{H}_1\left(\Delta T,\lambda^{-1},N_{ss}\right)=\begin{bmatrix}
0 & 0 & 0 & 0 & \cdots & 0 & 0\\
0 & \Delta T\lambda^{-1}N_{11}^{-1} & 0 & 0 & \cdots & 0 & 0\\
0 & 0 & 0 & 0 & \cdots & 0 & 0\\
0 & 0 & 0 & \Delta T\lambda^{-1}N_{22}^{-1} & \cdots & 0 & 0\\
\vdots & \vdots & \vdots & \vdots & \ddots & \vdots & \vdots\\
0 & 0 & 0 & 0 & \cdots & 0 & 0\\
0 & 0 & 0 & 0 & \cdots & 0 & \Delta T\lambda^{-1}N_{nn}^{-1}
\end{bmatrix}, \tag{24}
$$

$$
\tilde{\mathbf{y}}=\mathbf{H}_3\mathbf{y}=\begin{bmatrix}
0 & 0 & \cdots & 0\\
1 & 0 & \cdots & 0\\
0 & 0 & \cdots & 0\\
0 & 1 & \cdots & 0\\
\vdots & \vdots & \ddots & \vdots\\
0 & 0 & \cdots & 0\\
0 & 0 & \cdots & 1
\end{bmatrix}\begin{bmatrix} y_1\\ y_2\\ \vdots\\ y_n\end{bmatrix}, \tag{25}
$$

$$
\mathbf{H}_2=\begin{bmatrix}
0 & 0 & 0 & 0 & \cdots & 0 & 0\\
1 & 0 & 0 & 0 & \cdots & 0 & 0\\
0 & 0 & 0 & 0 & \cdots & 0 & 0\\
0 & 0 & 1 & 0 & \cdots & 0 & 0\\
\vdots & \vdots & \vdots & \vdots & \ddots & \vdots & \vdots\\
0 & 0 & 0 & 0 & \cdots & 0 & 0\\
0 & 0 & 0 & 0 & \cdots & 1 & 0
\end{bmatrix}. \tag{26}
$$

5 Synthesis of an Intelligent Discrete Algorithm for Estimating the Movement Parameters of Maneuvering Target

The measurement process is defined by equation

$$
y(k)=\mathbf{H}\mathbf{x}(k)+\xi(k), \tag{27}
$$

where $\mathbf{x}=\begin{bmatrix} x_1\\ x_2\end{bmatrix}$, $x_1(k)=q(k\Delta T), x_2(k)=\dot{x}_1(k)$ are the generalized coordinate being measured and velocity of its change, respectively, $\mathbf{H}=[1\quad 0]$ is the matrix of the projection of the state space onto the space of observations.

The intellectualization of a discrete dynamic filter of the combined maximum principle is provided by the adaptation of the parameter $\tilde{\lambda}$ characterizing the maneuver intensity to the observed motion.

The initial conditions of the synthesis are as follows: the non-maneuvering target moves at a constant speed; an increase in the norm of the discrepancy vector of observations is due to a change in the character of the motion, caused by the inadequacy of the mathematical model to the observed motion.

We consider the linguistic variables "the modulus of the current value of the observation discrepancy $\tilde{\Delta}$" and the "controlled parameter $\tilde{\lambda}$"; as well the terms "small" and "large". The membership functions $\tilde{\mu}_1$ and $\tilde{\mu}_2$ are shown in Fig. 1.

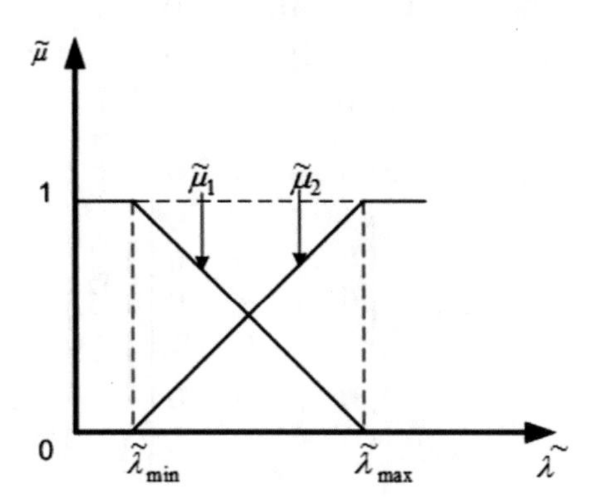

Fig. 1. The membership functions

Then the fuzzy instruction to adapt the mathematical model of the combined maximum principle is reduced to a simple production rule: for a "small" observations discrepancy $\tilde{\Delta}$ the target does not maneuver and the controlled parameter $\tilde{\lambda}$ is "small", and for a "large" observations discrepancy $\tilde{\Delta}$ there is a change in the nature of the target motion under the influence of some force and the controlled parameter $\tilde{\lambda}$ becames "big".

Performing dephasing by analogy with [9] we obtain

$$\tilde{\lambda}(k) = \frac{\tilde{\lambda}_{\min}\tilde{\mu}_1\left(\|\tilde{\Delta}\|\right) + \tilde{\lambda}_{\max}\tilde{\mu}_2\left(\|\tilde{\Delta}\|\right)}{\tilde{\mu}_1\left(\|\tilde{\Delta}\|\right) + \tilde{\mu}_2\left(\|\tilde{\Delta}\|\right)}, \tag{29}$$

where $\tilde{\lambda}_{\min}, \tilde{\lambda}_{\max}$ are the minimum and maximum values of the controlled parameter.

For this case the equations of an intelligent discrete dynamic filter can be written as follows:

$$\begin{gathered}
\hat{\mathbf{x}}(k+1) = \mathbf{\Phi}(\Delta T, k)\hat{\mathbf{x}}(k) + \mathbf{H}_1[\mathbf{H}_3 y(k) - \mathbf{H}_2\hat{\mathbf{x}}(k)], \\
\mathbf{\Phi}\left(\Delta T, \lambda, \tilde{\lambda}, k\right) = \begin{bmatrix} 1 & \Delta T \\ 0 & -\left[\frac{\tilde{\lambda}(k)\Delta T}{\lambda}\frac{|\hat{x}_2(k)|}{|\hat{x}_1(k)|} - 1\right] \end{bmatrix}, \\
\mathbf{H}_1 = \begin{bmatrix} 0 & 0 \\ 0 & \Delta T\tilde{\lambda}(k)L^{-1}N^{-1} \end{bmatrix}, \mathbf{H}_3 = \begin{bmatrix} 0 \\ 1 \end{bmatrix}, \mathbf{H}_2 = \begin{bmatrix} 0 & 0 \\ 1 & 0 \end{bmatrix}.
\end{gathered} \tag{30}$$

6 Conclusion

The use of the maximum condition for the function of generalized power provides a set of dynamic motion models. Their structure is determined by the method of approximation of the synthesizing function. This allows to obtain relatively simple equations for estimating the parameters of motion of controlled objects. Adaptation of the obtained solution to the observed dynamics is carried out using fuzzy logic. As the exact input data for the intelligent algorithm we choose the residual observations module, and the parameter of the synthesized dynamic mathematical model characterizing the intensity of the maneuver is used as the output exact value.

The paper has been accomplished with the support of Russian Federal Property Fund grants No. 18-01-00385 A, No 18-08-01494 A, No 18-38-00937 mol_a, Umnik No 0033574.

References

1. Bar-Shalom, Y., Li, X.R., Kirubarajan, T.: Estimation with Applications to Tracking and Navigation. Wiley, New York (2001)
2. Li, X.R., Jilkov, V.P.: Survey of maneuvering target tracking. Part I: dynamic models. IEEE Trans. Aerosp. Electron. Syst. **39**, 1333–1364 (2003)
3. Kostoglotov, A.A., Kostoglotov, A.I., Lazarenko, S.V.: Joint maximum principle in the problem of synthesizing an optimal control of nonlinear systems. Autom. Control. Comput. Sci. **41**, 274–281 (2007)
4. Derabkin, I.V., Kostoglotov, A.A., Kuznetcov, A.A., Lazarenko, S.V., Losev, V.A.: The stochastic synthesis of the adaptive filter for estimating the controlled systems state based on the condition of maximum of the generalized power function. In: MATEC Web of Conference, vol. 77, pp. 1–4 (2016)
5. Kostoglotov, A.A., Kuznetcov, A.A., Lazarenko, S.V., Deryabkin, I.V.: The method of structural adaptation of discrete algorithms for the combined maximum principle in problems of estimation of motion parameters. Inf.-Control. Syst. **85**, 10–15 (2016)
6. Kostoglotov, A.A., Kuzin, A.P., Lazarenko, S.V., Pugachev, I.V.: The combined maximum principle in the problem of synthesis of an adaptive dynamic filter under conditions of disturbances in the measurement process. In: MATEC Web of Conference, vol. 132, pp. 1–5 (2017)
7. Kostoglotov, A.A., Lazarenko, S.V.: Synthesis of adaptive tracking systems based on the hypothesis of stationarity of the hamiltonian on the switching hypersurface. J. Commun. Technol. Electron. **62**, 123–127 (2017)
8. Eliseev, A.V., Muzychenko, N.Yu.: Method of adaptive Kalman filter settings in the task of tracking for a dynamic object with unknown acceleration. J. Radio Eng. **8**, 39–44 (2014)
9. Muzychenko, N.Yu.: Synthesis of an optimum linear meter for observations in the presence of correlated interferences on the basis of fuzzy-logic algorithms. J. Commun. Technol. Electron. **55**, 755–758 (2014)
10. Kostoglotov, A.A., Lazarenko, S.V., Lyaschenko, Z.V.: Intellectualization of measuring systems based on the method of structural adaptation in the construction of tracking filter. In: Proceedings of 2017 20th IEEE International Conference on Soft Computing and Measurements, Saint Petersburg (2017)

11. Lur'e, A.I., Analiticheskaya Mekhanika (Analytical mechanics). Gos. Izd. Fiz.-Matem. Liter., Moscow (1961)
12. Derabkin, I.V., Kostoglotov, A.A., Kuzin, A.P., Lazarenko, S.V., Manaenkova, O.N., Pugachev, I.V.: Fuzzy control laws in the basis of solutions of synthesis problems of the combined maximum principle. In: Advances in Intelligent Systems and Computing, vol. 679, pp. 322–329 (2018)

Ontological Modeling, Semantic Technologies and Knowledge Engineering

The Approach to Extracting Semantic Trees from Texts to Build an Ontology from Wiki-Resources

Nadezhda Yarushkina, Aleksey Filippov, Vadim Moshkin(✉), and Ivan Dyakov

Ulyanovsk State Technical University, Ulyanovsk, Russia
{jng, al.filippov, v.moshkin, i.dyakov}@ulstu.ru

Abstract. The article describes the developed method of extracting semantic trees from text resources. This method is based on the use of a sequence of linguistic algorithms in constructing a syntactic sentence tree. The basis of the developed method is the algorithm for translating syntax trees of text fragments into the structures of semantic trees using a set of rules. A formal model of the rules is presented. The resulting semantic trees can be combined into a domain ontology taking into account the built-in relations between objects in the wiki resource. An example of our approach is also presented.

Keywords: Domain ontology · Semantic analysis · Linguistics Text resources

1 Introduction

Currently, methods of artificial intelligence are used to solve various problems in the field of business process automation. The use of methods of artificial intelligence allows intelligent systems to solve intellectual tasks at a level close to a human. Intelligent systems must have knowledge about the PrA to successfully solve the intellectual tasks. The methods of knowledge engineering allow to describe the features of the PrA in the form of a domain ontology [1–6].

At present experts in the problem area (PrA) form ontologies. The expert must have skills in the field of ontology engineering and have a good understanding of the specifics of a particular PrA. Building an ontology is a long and complex process.

The main lack of domain ontologies is the need for their development and updating due to PrA change. Knowledge extraction is carried out to extend the ontology. Knowledge extraction is carried out using semi-automatic methods for transforming unstructured, semi-structured and structured data into conceptual structures.

Now there are several directions for building the ontology:

1. extraction of knowledge from Internet resources (in particular, wiki-resources);
2. analysis of dictionaries and thesaurus;
3. merging of different ontological structures;

A. Abraham et al. (Eds.): IITI 2018, AISC 874, pp. 127–137, 2019.
https://doi.org/10.1007/978-3-030-01818-4_13

4. extraction of terminology in the process of text processing using statistical and linguistic methods.

Thus, the task of automatically building ontologies based on the analysis of the contents of text resources is currently relevant.

A large number of researches are devoted to the automatic building of the domain ontology on the basis of the analysis of the content of wiki-resources. Wiki-resource – a website whose structure and content can be modified by using a special markup language. User do not need additional tools and IT skills to work with wiki-resources. So different wiki-resources may be used as data sources for the building of ontologies as they contain knowledge of various PrAs and freely available for use.

There are various approaches to the automatic generation of ontologies based on the analysis of the contents of wiki-resources:

1. Formation of classes and relations of ontology on the basis of analysis of the structure of wiki-resources [7–11].
2. Formation of objects and relations of ontology on the basis of analysis of the structure of wiki-resources [7, 12–15].
3. Formation of an ontology in the process of combining several ontologies [16–20].

For example, in the YAGO project for automatic building of the domain ontology, data from Wikipedia and data from the semantic WordNet network were used. The ontology was built on the basis of a hierarchy of Wikipedia pages and information from info-boxes, and then expanded based on WordNet data. As you can see, the contents of the pages of wiki-resources are almost not taken into account, instead, various widely available thesauri are used.

We believe that the analysis of the content of the wiki-resources will increase the completeness of the description of the PrA in the form of a domain ontology. Also an ontology can be built on the basis of an analysis of the contents of a set of text documents. The idea of our approach is to use the existing methods of linguistic analysis to construct a syntactic tree of sentence. Further, using a set of rules, you can translate a syntax tree into a semantic tree. Semantic representation of the text on native language (NL) is the most complete of those that can be achieved only by linguistic methods. The domain ontology can be build from the semantic trees extracted from content of text resources and then merge the semantic trees with the built-in relations between the objects in the wiki-resources.

2 A Method of Translating a Syntactic Tree into a Semantic Tree

It is necessary to determine the syntactic structure of the sentence on NL for constructing the semantic tree. There are several parsing tools of texts in Russian, for example [21–24]:

1. Lingo-Master;
2. Treeton;
3. DictaScopeSyntax;

4. ETAP-3;
5. ABBYY Compreno;
6. Tomita-parser;
7. AOT etc.

In our work, for constructing a syntactic tree the results of the AOT project were used. Consider the application of the algorithm of translating a syntactic tree into a semantic tree using the example of test sentence in Russian: "*Онтология в информатике - это попытка всеобъемлющей и подробной формализации некоторой области знаний с помощью концептуальной схемы*" (in English: "*Ontology in informatics is an attempt at comprehensive and detailed formalization of a certain field of knowledge with the help of a conceptual scheme*").

Formally the function of translating a syntactic tree into a semantic tree

$$F^{Sem}:\left\{N_{li}^{Synt},P_j\right\}\rightarrow\left\{N^{Sem},R^{Sem}\right\},$$

where N_{li}^{Synt} – i-th node of l-th level of the syntactic tree. For example, the first node of the first level is the node "ontology", the second - "pg", the third - "is", etc. for the parse syntactic in Fig. 1. The node of the syntactic tree can be a member of the sentence, for example, the node "ontology", or also can be a syntactic label that defines the constituent members of the sentence, for example, "pg" (the prepositional group); P_j – j-th rule for translating the nodes of the syntactic tree. The nodes of the syntactic tree will be translated into nodes and relations of the semantic tree. The resulting syntactic tree of test sentence is shown in the Fig. 1.

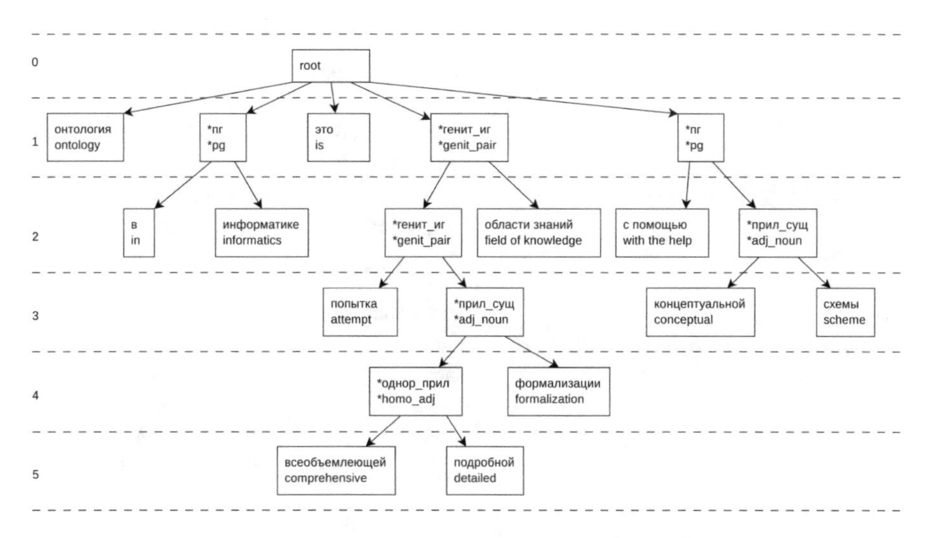

Fig. 1. Example of a syntactic tree of test sentence.

The rule is a collection of several words (units) united according to the principle of semantic-grammatical-phonetic compatibility. Formally rule:

$$\{N_1^{Synt}, N_2^{Synt}, \ldots, N_k^{Synt}\} \rightarrow \{N^{Sem}, R^{Sem}\},\ k = \overline{1, K},$$

where $N_1^{Synt}, N_2^{Synt}, \ldots, N_k^{Synt}$ is the set of units of the rule corresponding to the set of nodes of the syntactic tree. The rule only works if all the units match. Examples of rules and the results of their use are presented in Table 1. K is number of units in the rule; $\{N^{Sem}, R^{Sem}\}$ is set of nodes N^{Sem} and relations R^{Sem} of the semantic tree, obtained as a result of translation of the syntactic tree into a semantic tree.

The expert should check the consistency of the rules for obtaining the correct semantic tree. Otherwise, the relationship between the nodes of the tree will have an incorrect meaning. In this case, the resulting ontology will not reflect the semantics of the analyzed wiki-resource.

Table 1. Examples of rules for translating nodes of syntactic tree into nodes of a semantic tree and the results of their application.

Initial data	Rule	Result
attempt-***genit_pair**-formalization	node1-***genit_pair** -node2 → node1-associateWith-node2	*attempt*-associateWith-*formalization*
in-***pg**-*informatics*	node1-***pg** -node2 → prevNode-***dependsOn***(node)-node2	lastNode-***dependsOn***-*informatics*
is	**is** → prevNode-nextNode	lastNode-isA-nextNode
conceptual-***adj_noun**-*scheme*	node1-***adj_noun**-node2 → node2-hasAttribute-node1	*scheme*-hasAttribute-*conceptual*
comprehensive **-*homo_adj**-*formalization* *detailed*-***homo_adj**-*formalization*	node1-***homo_adj–**node2 → node2-hasAttribute-node1	*formalization* -hasAttribute-*comprehensive* *formalization*-hasAttribute- *detailed*

$$R^{Sem} = \left\{ R^{Sem}_{isA}, R^{Sem}_{partOf}, R^{Sem}_{associateWith}, R^{Sem}_{dependsOn}, R^{Sem}_{hasAttribute} \right\},$$

where R^{Sem}_{isA} – set of transitive relations of hyponymy;
R^{Sem}_{partOf} – set of transitive relations «part/whole» ;
$R^{Sem}_{associateWith}$ – set of symmetrical relations of association
$R^{Sem}_{dependsOn}$ – set of asymmetric relations of associative dependence;
$R^{Sem}_{hasAttribute}$ – set of asymmetric relations describing the attributes of nodes.

3 The Algorithm of Translating a Syntactic Tree into a Semantic Tree

The algorithm of translating a syntactic tree into a semantic tree consists of the following steps:

1. Go to the first level of the syntactic tree.
2. Select the next node of the current tree level. If there are no unprocessed nodes, go to step 12.
3. If the node is marked as processed, go to step 2.
4. If the node is not a syntax label (not starts with "*"), go to step 10.
5. If the node is a syntax label (starts with "*") and does not have child elements, go to step 10.
6. If the node is a syntax label (starts with "*") and all its child nodes are not syntax labels, go to step 10.
7. If there is a temporary parent node, then replace it, otherwise create a temporary node.
8. If there is no connection between the nodes, create a temporary relationship between them and go to step 2.
9. If both nodes are not temporary and there is no connection between them, create an "associateWith" relationship between them and go to step 2.
10. Apply the rule for translation.
11. Mark the nodes as processed and go to step 2.
12. Go to the next level of the syntactic tree, and then go to step 2.

4 Example of the Algorithm of Translating a Syntactic Tree into a Semantic Tree

Let's consider an example of translating the syntactic tree of test sentence presented above into a semantic tree. The following nodes of syntactic tree (syntactic units) were identified in the first level of the syntactic tree of the test sentence (see Fig. 1):

- *ontology*;

- ***pg** (*informatics*);
- *is*;
- ***genit_pair**(***genit_pair**(*attempt*, ***adj_noun**(***homo_adj**(*comprehensive*, *detailed*), *formalization*)), *field of knowledge*);
- ***pg**(*with the help*, ***adj_noun**(*conceptual*, *scheme*)).

Figure 2 shows the semantic tree of test sentence at the beginning of the algorithm.

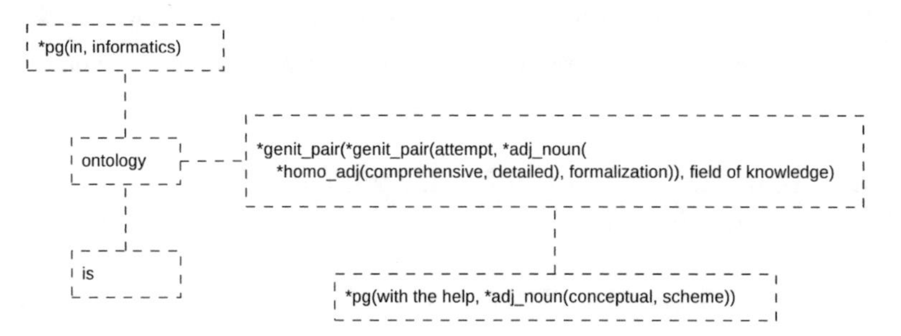

Fig. 2. Example of a semantic tree of test sentence at the beginning of the algorithm.

Figure 3 shows the semantic tree of test sentence at the first iteration of the algorithm.

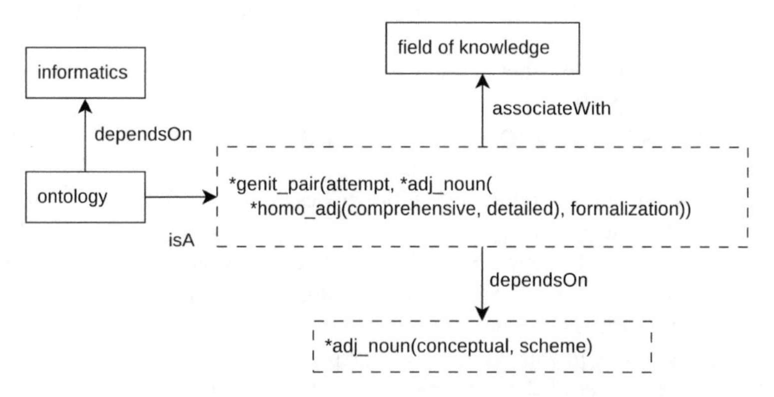

Fig. 3. Example of a semantic tree of test sentence at the first iteration of the algorithm.

As you can see from Fig. 3, all syntactic units of the first level of the syntactic tree of the test sentence were processed. After applying the translation rules:

- the syntactic unit "ontology" was included in semantic tree;
- from the syntactic unit "\is" the relation "isA" was formed between the node "ontology" and the temporary node "***genit_pair**(…)";
- from the syntactic unit "***pg**(*in, informatics*)" the node "informatics" and relation "dependsOn" between the nodes "informatics" and "ontology" were formed;

- from the syntactic unit "***genit_pair**(***genit_pair**(…)), *field of knowledge*)" the temporary node "***genit_pair**(…))" and the node "field of knowledge" were formed that are connected by the relation "associateWith";
- from the syntactic unit "***pg**(*with the help*, …)" the temporary node "***adj_noun** (*conceptual*, *scheme*)" and relation "dependsOn" between that node and the temporary node "***genit_pair**(…))" were formed.

All syntactic units of the first level and all syntactic units of the second level that are related to the syntactic units of the first level were marked as processed in the syntactic tree of test sentence.

Figure 4 shows the semantic tree of test sentence at the second iteration of the algorithm.

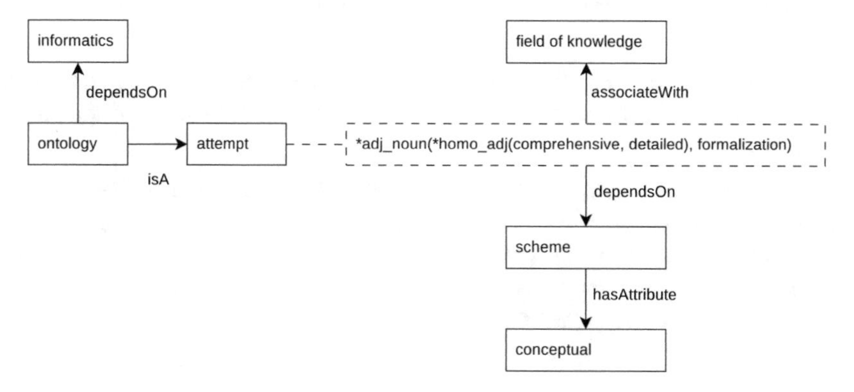

Fig. 4. Example of a semantic tree of test sentence at the second iteration of the algorithm.

As you can see from Fig. 4, all syntactic units of the second level of the syntactic tree of the test sentence that not marked as processed were processed. After applying the translation rules:

- from the syntactic unit "***genit_pair** (*attempt*, ***adj_noun**(***homo_adj** (*comprehensive*, *detailed*), *formalization*))" the node "attempt" and temporary node "***adj_noun**(…)" were formed that are connected by relation "associateWith". In the genitive pair, the second node is the main node, so the existing relationships refers to the second node;
- from the syntactic unit "***adj_noun**(*conceptual*, *scheme*)" nodes "conceptual" and "scheme" and relation "hasAttribute" between them were formed.

All syntactic units of the second level and all syntactic units of the third level that are related to the syntactic units of the second level were marked as processed in syntactic tree of test sentence.

Figure 5 shows the semantic tree of test sentence at the third iteration of the algorithm.

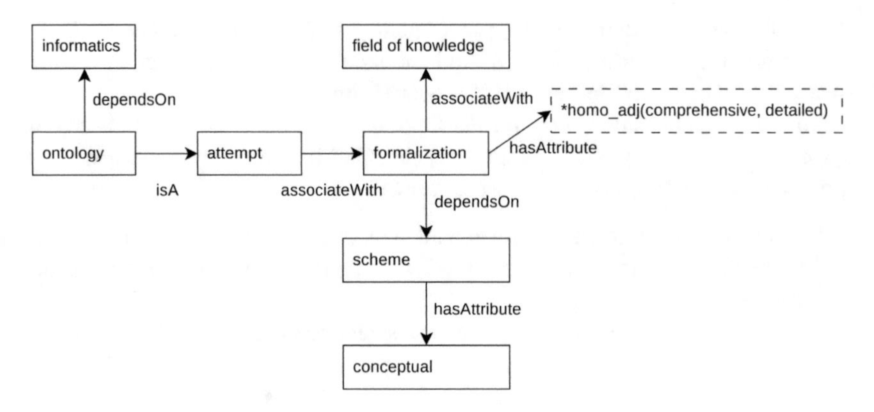

Fig. 5. Example of a semantic tree of test sentence at the third iteration of the algorithm.

As you can see from Fig. 5, all syntactic units of the third level of the syntactic tree of the test sentence that not marked as processed were processed. After applying the translation rules:

- form the syntactic unit "***adj_noun**(***homo_adj**(*comprehensive*, *detailed*), formalization)" the node "formalization" and the temporary node "***homo_adj**(…)" were formed that are connected by the relation "hasAttribute". In a pair adjective-noun a noun is the main node, so the existing relationships refers to a noun;
- also between the nodes "attempt" and "formalization" a relation "associateWith" was created.

All syntactic units of the third level and all syntactic units of the fourth level that are related to the syntactic units of the third level were marked as processed in syntactic tree of test sentence.

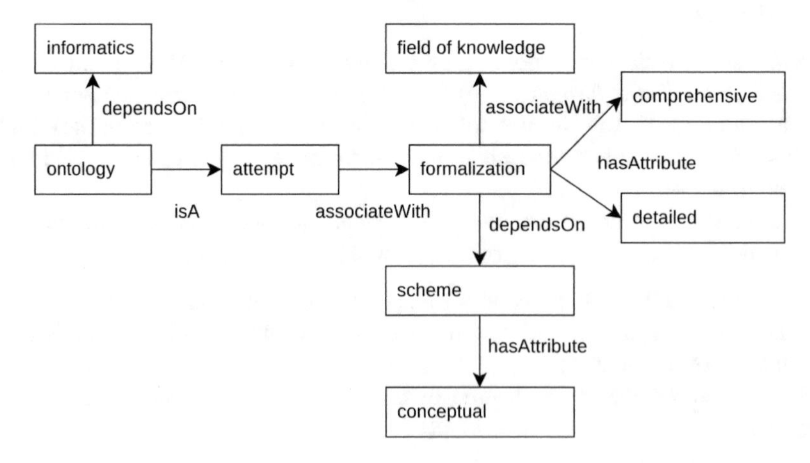

Fig. 6. Example of a semantic tree of test sentence at the fourth iteration of the algorithm.

Figure 6 shows the semantic tree of test sentence at the fourth iteration of the algorithm.

As you can see from Fig. 6, all syntactic units of the fourth level of the syntactic tree of the test sentence that not marked as processed were processed. After applying the translation rules form the syntactic unit "***homo_adj** (*comprehensive*, *detailed*)" the nodes "comprehensive" and "comprehensive" of semantic tree were formed that are connected by relation "hasAttribute" with node "formalization".

All syntactic units of the fourth level and all syntactic units of the fifth level that are related to the syntactic units of the fourth level were marked as processed in syntactic tree of test sentence.

At the fifth iteration of the algorithm, the process of building the semantic tree of the test sentence is complete. The resulting semantic tree for the test fragment is shown in Fig. 6. The resulting semantic tree can be merged with other semantic trees in a text resource. In addition, this semantic tree can be merged with the domain ontology created by the expert.

5 Conclusions and Future Work

We have described a modular pipeline that can be used for translating a syntactic tree of sentence into a semantic tree. This approach can be used to automatically build a domain ontology from wiki-resources. Manually building an ontology is a long and complex process. The idea of our approach is to use the existing methods of linguistic analysis to construct a syntactic tree of sentence. Further, using a set of rules, you can translate a syntax tree into a semantic tree. The domain ontology can be build from the semantic trees extracted from text content of wiki-resource taking into account the built-in relations between wiki objects.

Also we have described the algorithm of translating a syntactic tree into a semantic tree. An example of the proposed approach of translating the syntactic tree of test sentence into a semantic tree is considered in detail.

In the future work we plan to use methods of deep learning to translating the syntactic tree of sentence into a semantic tree. Comparison of the two approaches to solving problem of automatically build a domain ontology will allow us to understand when you need to use the semantic approach and when you need to use the methods of deep learning.

Also we plan to extend the set of rules for translating the syntactic tree into a semantic tree to cover a greater number of types of semantic relationships between objects of PrA. We also plan to develop an algorithm for automatically checking the consistency of these rules. In addition, we plan to develop an algorithm for evaluating the quality of the resulting ontology.

Acknowledgments. This work was financially supported by the Russian Foundation for Basic Research (Grants No. 16-47-732054 and 18-37-00450) and Ministry of Education and Science of Russia in framework of project № 2.4760.2017/8.9 and Russian Foundation of base Research in framework of project № 17-07-00973 A.

References

1. Konstantinova, N.S., Mitrofanova, O.A.: Ontology as a knowledge storage system. Portal "Information and Communication Technologies in Education". http://www.ict.edu.ru/ft/005706/68352e2-st08.pdf. Accessed 21 Mar 2018
2. Hepp, M.: Products and services ontologies: a methodology for deriving OWL ontologies from industrial categorization standards. Int. J. Semant. Web Inf. Syst. (IJSWIS) **2**(1), 72–99 (2006)
3. Damljanovic, D., Agatonovic, M., Cunningham, H.: Natural language interfaces to ontologies: combining syntactic analysis and ontology-based lookup through the user interaction. Lecture Notes in Computer Science, vol. 6088, pp. 106–120 (2010)
4. Pazienza, M., Pennacchiotti, M., Zanzotto, F.: Terminology extraction an analysis of linguistic and statistical approaches. Studies in Fuzziness and Soft Computing, vol. 185, pp. 255–279 (2004)
5. Kovacevic, A., Konjovic, Z., Milosavljevic, B., Nenadic, G.: Mining methodologies from NLP publications: a case study in automatic terminology recognition. Comput. Speech Lang. **26**, 105–126 (2012)
6. Zarubin, A., Koval, A., Filippov, A., Moshkin, V.: Application of syntagmatic patterns to evaluate answers to open-ended questions. In: CITDS-2017. Communications in Computer and Information Science, pp. 150–162. Springer (2017)
7. Zarubin, A.A., Koval, A.R., Moshkin, V.S., Filippov, A.A.: Construction of the problem area ontology based on the syntagmatic analysis of external wiki-resources. http://ceur-ws.org/Vol-1903/paper26.pdf. Accessed 03 Mar 2018
8. Shestakov, V.K.: Development and maintenance of information systems based on ontology and Wiki-technology. In: 13-and All-Russian Scientific Conference, RCDL 2011, Voronezh, pp. 299–306 (2011)
9. Bizer, C., Lehmann, J., Kobilarov, G., Auer, S., Becker, C., Cyganiak, R., Hellmann, S.: DBpedia – a crystallization point for the web of data. J. Web Semant. Sci. Serv. Agents World Wide Web **7**, 154–165 (2009)
10. Suchanek, F.M., Kasneci, G., Weikum, G.: YAGO: a core of semantic knowledge unifying WordNet and Wikipedia. In: Proceedings of the 16th International Conference on World Wide Web, Banff, Alberta, Canada, 8–12 May 2007, pp. 697–706. ACM Press, New York (2007)
11. Suchanek, F.M., Kasneci, G., Weikum, G.: YAGO: a large ontology from Wikipedia and WordNet. Web Semant. Sci. Serv. Agents World Wide Web **6**(3), 203–217 (2008)
12. Subkhangulov, R.A.: Ontological search for technical documents based on the intelligent agent model. Autom. Manag. Process. **4**(38), 85–91 (2014)
13. Astrakhantsev, N.A., Fedorenko, D.G., Turdakov, D.Y.: Automatic enrichment of informal ontology by analyzing a domain–specific text collection. In: Materials of International Conference "Dialog", vol. 13, no. 20, pp. 29–42 (2014)
14. Cui, G.Y., Lu, Q., Li, W.J., Chen, Y.R.: Corpus exploitation from Wikipedia for ontology construction. In: Proceedings of the Sixth International Language Resources and Evaluation (LREC 2008), Marrakech, pp. 2125–2132 (2008)
15. Hepp, M., Bachlechner, D., Siorpaes, K.: Harvesting wiki consensus – using Wikipedia entries as ontology elements. In: Proceedings of the First Workshop on Semantic Wikis – From Wiki to Semantics, Annual European Semantic Web Conference (ESWC 2006), pp. 124–138 (2006)
16. McGuinness, D.L., Fikes, R., Rice, J., Wilder, S.: An environment for merging and testing large ontologies. In: KR 2000, pp. 483–493 (2000)

17. Noy, N.F., Musen, M.A.: The PROMPT suite: interactive tools for ontology merging and mapping. Int. J. Hum.-Comput. Stud. **59**(6), 983–1024 (2003)
18. Pottinger, R.: Mapping-based merging of schemas. In: Schema Matching and Mapping, pp. 223–249. Springer, Berlin (2011)
19. Raunich, S., Rahm, E.: Automatic target-driven ontology merging with ATOM. In: Information Systems, Pergamon, vol. 42, pp. 1–14
20. Raunich, S., Rahm, E.: Target-driven merging of taxonomies. Cornell University Library (2010). https://arxiv.org/ftp/arxiv/papers/1012/1012.4855.pdf. Accessed 18 Feb 2018
21. Sokirko, A.V.: Semantic words in automatic processing. Dissertation, Ph.D. (05.13.17); State Committee of the Russian Federation for Higher Education Russian State University for the Humanities, p. 120 (2001)
22. Boyarskiy, K.K., Kanevskiy, Ye.A.: Semantico-syntactic parser semsin, scientific and technical herald of information technologies. Mech. Opt. **5**, 869–876 (2015)
23. Artemov, M.A., Vladimirov, A.N., Seleznev, K.Ye.: Survey of natural text analysis systems in Russian. Sci. J. Bull. Voronezh State Univ. http://www.vestnik.vsu.ru/pdf/analiz/2013/02/2013-02-31.pdf. Accessed 22 Mar 2018
24. Automatic text processing. http://aot.ru. Accessed 22 Mar 2018

Ontology-Based Semantic Models for Industrial IoT Components Representation

Nikolay Teslya(✉) and Igor Ryabchikov

SPIIRAS, 14th line 39, 199178 St. Petersburg, Russia
teslya@iias.spb.su, i.a.ryabchikov@gmail.com

Abstract. The concept of Industry 4.0 is related to the transition of the industry to new ways of production organizing. New technologies provide basis for creating socio-cyberphysical systems for production also known as smart factories. The main components of smart factories are smart objects, including machines (production robots), people, software services, processed materials and manufactured products. Their interaction requires a model of each component also known as digital twin that reflects the properties of real object in virtual world and could be processed by other components. The paper proposes to use ontologies to develop the model for each type of objects. Five main ontologies for components and two additional is developed to describe production process. In addition, the interrelations between the developed ontologies is described.

Keywords: Socio-cyberphysical system · Industry 4.0 · Industrial IoT Digital twin · Ontology model

1 Introduction

The concept of Industry 4.0 (also known as Industrial Internet of Things) is related to the transition of the industry to new ways of production organizing. It is associated with the development of robotics, the concept of cyberphysical systems and, in particular, the Internet of Things, the development of big data processing technologies, the automation of production and the use of distributed digital ledgers [1, 2]. These technologies provide not only production automation, but also a basis for the creation of so-called "smart factories", which are regarded as cyberphysical systems (or socio-cyberphysical systems considering the staff), possessing complete autonomy and knowledge of the production process. Such systems are able to interact with each other in the physical and virtual world. This approach allows to reconfigure production to the needs of each user, which in turn allows the production of highly customized products that meet the needs of an individual user, without the need for a profound reconfiguration of the production base.

The main components of the industrial socio-cyberphysical systems are smart objects, including machines (production robots), people, software services, processed materials and manufactured products. The term "smart" in this case means that in the production process, the machines, software services and people are able to analyze the current situation and make decisions based on available information about the state of

A. Abraham et al. (Eds.): IITI 2018, AISC 874, pp. 138–147, 2019.
https://doi.org/10.1007/978-3-030-01818-4_14

production and production facilities [3–6], and the processed materials as well as manufactured products are capable to accumulate the history of interaction and to analyze their own state either independently or using external services. The interaction of individual objects in socio-cyberphysical systems is due to the development of software services (agents), which are digital twins of real objects that dynamically display their properties in a virtual (cyber) space and provide information exchange between them through a common information space, forming smart factory. Unlike an automated factory, a smart factory is characterized by a complete interaction of all elements of production, which makes it possible to create a flexible system that can independently optimize productivity, adapt to new conditions and be trained in real time, and also autonomously execute the production process [7]. In this process, one to several smart factories can be involved. In the latter case, a network of smart factories is viewed that can adapt to complex production conditions, including the production of individual parts of the product, logistics and assembly in a specialized department or factory [7].

Within the scope of a smart factory, as well as in case of their cooperation, a large number of heterogeneous components (machines, people, software services) are interact with each other, which requires interoperability and trust between them. Considering that each component of a smart factory from an individual machine to the whole factory can be represented by its own software agent, their integration can be viewed as a multi-agent system in which each agent has certain characteristics, competencies and requirements [4, 8]. In multi-agent industrial systems, interoperability is most often provided by using ontologies [9, 10].

The purpose of this paper is to identify the main components of the socio-cyberphysical system and, the industrial Internet of Things in particular, as well as to develop of their ontologies, on the basis of which digital twins can be created. The description of the digital twin using ontology will allow creating a dynamically updated semantic model, available for processing both machine and human components of the socio- cyberphysical system. An ontology of the interrelationships between the components of socio- cyberphysical components will also be presented, and their links to existing ontologies.

2 Related Work

One of the main tasks within the Industry 4.0 context is to organize the interaction and collaboration of various smart factories, as well as individual components within the factories. This problem is often solved using the concept of the Internet of Things, which allows to unite a lot of heterogeneous components in the common information space, providing facilities for information exchange between them [5, 11]. With reference to industry, the concept of the Internet of Things is concretized by considering the components and the nature of their interrelations from the point of view of the production process, switching to the Industrial Internet of Things (IIoT) and providing interaction of physical, virtual and social components of production in the common information space.

To provide interoperability between components connected to the common information space, ontologies and ontology mapping mechanisms are used [12–14]. These mechanisms are developed and successfully used in a large number of projects related to socio-cyberphysical systems [15, 16], the Internet of Things and the industrial Internet of Things, with the goal of creating a semantic description of information accessible to all components of socio-cyberphysical systems [17–19].

The analysis of publications and projects devoted to the fourth industrial revolution, production automation and application of modern technologies in production, allowed to outline the following main approaches and scenarios where ontologies provide significant impact.

2.1 Multi-agent Systems Concept Application in Smart Factories

The concept of multi-agent system includes the creation of software agents for production components (production machines, schedulers, process analyzers) interacting with each other in a common environment for coordinating actions and exchanging information [20, 21]. To interact with agents, a common language or way of representing knowledge is needed. The ontology is one of the possible ways for representing such an information. The interaction environment can be a common information space in which agents can publish their knowledge in the form of ontology and ask for knowledge of others. The application of this concept allows to provide modularity and simplify the implementation of changes in production processes, thus providing the necessary flexibility. Examples of application of the concept of a multi-agent system in production are presented in [22–24].

2.2 Collection and Analysis of Information About Products Usage

One aspect of the Industry 4.0 concept is the closer integration of consumers with manufacturers and feedback on the use and current condition of the products. By analyzing this information, manufacturers can improve the quality of their future products, as well as support users in the event of improperly functioning products.

The paper [25] presents a scenario for collecting information on the use of production equipment by its supplier to predict possible malfunctions and to seek services for its repair or replacement. This approach can reduce or completely prevent possible downtime. Repair service can be provided by the organization's own forces or by third-party organizations. A solution for industrial IoT that allows for the universal monitoring of industrial machines and products to provide predictive maintenance is presented in [23].

Thus, the problem arises of storing information about the state and functioning of products. Product description can be divided into classes that have common and individual indicators. In addition, products may consist of parts supplied by various organizations (for assembling the final product or replacing part of the product during repair), and indicators of interest can be considered for both the individual component and the entire product based on all information. The use of ontology in this scenario provides a unified structure for describing the characteristics of production equipment, as well as manufactured products, containing, in addition to the characteristics themselves, their semantic relationships. Representation of characteristics in the form of ontology allows to carry out not only automated, but also manual processing, and also to give access to them to any software services capable of processing semantic structures of knowledge.

2.3 Collecting and Providing Complete Information on Products Along the Supply Chain

The availability of complete information on the supply chain of products (on materials used in the production process, its components, production methods, storage and transportation methods) can increase the confidence of supply chain participants in the current state and location of their assets (for example, in the case of long-distance deliveries), allowing to identify errors in transportation or production as well as fraud. For consumers, it is possible to guarantee the quality of the delivered products, for example, to confirm the license (so that the product is not falsified, stolen and delivered illegally) or, in the case of poor materials or defects in components, as well as malfunctions of production machines, to track which products have been affected, and where their current position is.

This approach has become especially relevant with the development of blockchain technology [26], which allows the creation of a decentralized distributed information space for independent heterogeneous components, in which the trust between all components is achieved without the need for a certification center, and such space is resistant to the failure or malfunctioning of a part of the components. Ontology in this case is used to represent product characteristics, providing interoperability between heterogeneous components of the information space when tracking product progress along the supply chain.

The benefits of increasing transparency and awareness in supply chains are described in many works, including [27–29]. In [27], an overview is given of projects successfully using blockchain technology to support supply chains, including tracing fish in Indonesia to counteract illegal activities; tracking the supply of products to counter delivery of substandard products; tracking supplies Walmart, etc.

2.4 Interaction Automation of Enterprises to Participate in Common Manufacturing Processes and the Dynamic Supply Chain

Services provided by enterprises (supply of materials, production, transportation of goods, etc.) can be presented in the form of web services with program interfaces for the orders formation. This allows to automate the interaction between enterprises and even provide an automatic search and use of services. This approach is especially relevant in the context of the availability of interchangeable service providers and competition between them. In turn, for service providers, closer cooperation between industries can improve the efficiency of the use of production resources and capacities. Especially if production systems are flexibly adjusted and can be used to produce a large class of products (for example, for additive manufacturing) [30–32].

Description of such services must involve a type of service, conditions provision, access interface (for example, REST (Representational State Transfer) or WSDL (Web

Services Description Language)), as well as support auto search and logical reasoning that requires use of ontologies.

The services themselves can be implemented as REST, WSDL or other types of Web services, but for their automatic search a common registry containing up-to-date information maintained by the providers themselves is required. Such a register can be implemented based on a common decentralized information space, created, for example, on the basis of blockchain technology.

3 Ontologies for Digital Twins of Socio-Cyberphysical System Components

As a result of the analysis of scenarios for the application of industrial socio-cyberphysical systems, ontological models of components were developed that describe their main characteristics for the creation of digital twins. The following are specific ontologies that can be used as a basis for developing digital twins of components.

The ontology needed for scenario 2.1 (application of the concept of multi-agent system in smart factories) should describe raw materials, products and components produced, production machines and their functionality.

The ontology needed for scenario 2.2 (collection and analysis of information on the use of products) should describe products and components, their state and functioning indicators.

The ontology required for scenario 2.3 (gathering and providing complete information about products and the supply chain to stakeholders) should describe products and components, as well as information about operations performed at different stages of the life cycle (for production, transportation, storage and so on.).

The ontology needed for scenario 2.4 (automation of the interaction of enterprises participating in the general production process and the formation of dynamic supply chains) should describe services, in particular services themselves (raw materials supply, food production, storage, transportation), their conditions providing (cost, geographic location) and interfaces for creating automatic queries.

Thus, the developed ontologies of components can be broken down into:

- Ontology of products and materials. Contains an ontological description of the types of products and raw materials. Allows to specify the type of material, its quantitative characteristics (dimensions, weight, quantity), life time, as well as the current position in the supply chain;
- Ontology of indicators of the state and functioning of products. This ontology is designed to track the current state of the product. Includes the time from the first use, repair marks (which component, when it was replaced), the consumables used and their value, the date of the last replenishment of consumables;
- Ontology of production machines and their capabilities. Ontology of machines, parameters of functioning and other elements necessary for the development of multi - agent systems of smart factories;

– Ontology of software services (type of service, operations performed: production of goods, delivery of materials, transportation, storage, time of operation, cost of the operation), with a description of program interfaces for the formation of orders.
– Ontology of human resources. It is used to describe workers profile (name, age, education, social id, etc.) as well as their competences (competence name and description, rating of competence for worker) and preferences (type of work, work duration, salary, etc.).

In addition, auxiliary ontologies were developed that provide a description of the relationships of the components:

– Ontology of structural relations. The ontology of structural relations between products and materials (structural components, materials used) makes it possible to describe from which materials or structural components a particular product was created;
– Ontology of supply chain operations. The ontology of the operations that were performed with the products during the supply chain (from which materials it was produced, how, what machines were used in production, how it was transferred and transported, in what environmental conditions it was stored (temperature, humidity, lighting)).

The ontology of products and materials is central and is included in all other ontologies (Fig. 1). In particular, the ontology of indicators of state and functioning, as well as the ontology of structural relations, uses it as a basic one, since it requires the description of material and structural components. The ontology of software services includes the ontology of products and materials for describing input and output data. The ontology of production machines uses the ontology of products and materials to describe production processes. In the ontology of supply chain operations, it is used to describe the objects of operations. In addition, the ontology of operations in the supply chain uses the ontology of production machines and their capabilities and the ontology of software services to describe objects and specify the ontology of structural relations (it describes the operations by which relationships were formed). When creating ontologies, it is proposed to use existing ontologie to improve compatibility with existing software agents built on their basis, in particular, the following ontologies can be used: GoodRelations [33], OWL-S [34], eCl@ss [35], eClassOWL [36], and the IEC 62264 standard [37].

GoodRelations is an ontology for describing commercial offers, for example, on the sale or lease of assets, or on the provision of services. The ontology allows to describe the information about the supplier, the terms of the offer (cost, quantity, available methods of delivery, payment, etc.), as well as the proposed assets or services. To describe specific assets or services, the ontology is based on third-party standards, for example eCl@ss. However, this ontology is not sufficient for automatic acceptance of sentences, since it does not describe the program interfaces, the expected sequence of calls (protocol), and possible answers.

OWL-S is an ontology for describing services in the semantic web. The purpose of this ontology is to provide the ability to automatically search, call, and composition Web services.

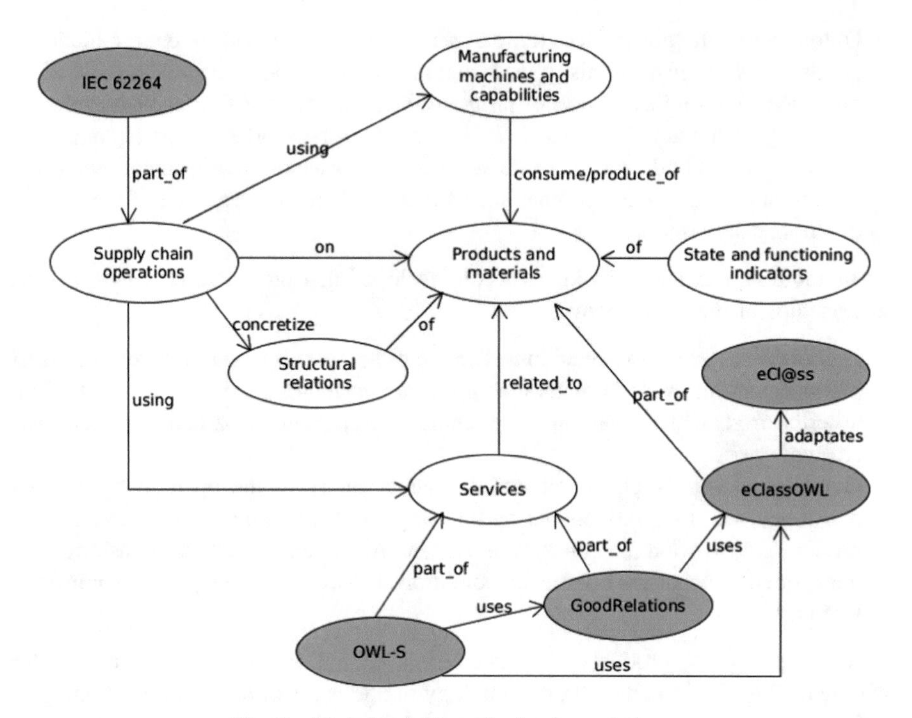

Fig. 1. Digital twins' ontologies interrelations

In addition to the standard ontologies, there are many works on automation of production that describe their own ontologies that can also be used to develop machine ontologies and production capabilities, supply chain operations and ontologies of structural relationships, for example [38–40].

4 Conclusion

The article describes the use of ontologies for creation of digital twins of socio-cyberphysical systems components used in scope of industrial internet of things. Ontologies solve the problems of interoperability between components and digital twins dynamically reflects current state of the components. In addition, digital twins can describe not only separate components but also a whole smart factory in case of joint production by several factories.

In the work, four scenarios of using ontologies in the industrial Internet of things were revealed. All scenarios are related to different aspects of the interaction between production components and the support of final products: the interaction of devices within a single factory; collecting telemetry of products usage to identify possible malfunctions; collection of the complete productions information to support supply chain participants; and automation of production and distribution within several smart factories.

The individual ontologies of components are connected to each other, providing interoperability between socio-cyberphysical systems' components and creating a link between scenarios. Based on the proposed ontologies flexible production can be created, which allows to ensure high customization of the final product.

Acknowledgements. The presented research was partially supported by the projects funded through grants # 16-29-04349, 17-29-07073 and 17-07-00327 of the Russian Foundation for Basic Research.

References

1. Schwab, K.: The fourth industrial revolution. World Economic Forum (2016)
2. Lasi, H., Fettke, P., Kemper, H.G., Feld, T., Hoffmann, M.: Industry 4.0. Bus. Inf. Syst. Eng. **6**, 239–242 (2014)
3. Bedenbender, H., Bentkus, A., Epple, U., Hadlich, T.: Industrie 4.0 plug-and-produce for adaptable factories: example use case definition, models, and implementation, pp. 1–68 (2017)
4. Silva, J.R., Nof, S.Y.: Manufacturing service: from e-work and service-oriented approach towards a product-service architecture. IFAC-PapersOnLine **48**, 1628–1633 (2015)
5. Pisching, M.A., Junqueira, F., Filho, D.J.S., Miyagi, P.E.: Service composition in the cloud-based manufacturing focused on the industry 4.0. In: IFIP Advances in Information and Communication Technology, pp. 65–72 (2015)
6. Zhang, D., Wan, J., Hsu, C.H., Rayes, A.: Industrial technologies and applications for the Internet of Things. Comput. Netw. **101**, 1–4 (2016)
7. Burke, R., Mussomeli, A., Laaper, S., Hartigan, M., Sniderman, B.: The smart factory (2017)
8. Colombo, A.W., Karnouskos, S., Mendes, J.M., Leitão, P.: Industrial agents in the era of service-oriented architectures and cloud-based industrial infrastructures. In: Industrial Agents, pp. 67–87. Elsevier (2015)
9. Borgo, S., Leitão, P.: Foundations for a core ontology of manufacturing. In: Ontologies, pp. 751–775. Springer, Boston (2007)
10. Garetti, M., Fumagalli, L.: P-PSO ontology for manufacturing systems. IFAC Proc. Vol. **45**, 449–456 (2012)
11. Sun, B., Jämsä-Jounela, S.-L., Todorov, Y., Olivier, L.E., Craig, I.K.: Perspective for equipment automation in process industries. IFAC-PapersOnLine **50**, 65–70 (2017)
12. Gierej, S.: The framework of business model in the context of Industrial Internet of Things. Procedia Eng. **182**, 206–212 (2017)
13. Zhu, T., Dhelim, S., Zhou, Z., Yang, S., Ning, H.: An architecture for aggregating information from distributed data nodes for industrial internet of things. Comput. Electr. Eng. **58**, 337–349 (2017)
14. Tang, H., Li, D., Wang, S., Dong, Z.: CASOA: an architecture for agent-based manufacturing system in the context of industry 4.0. IEEE Access **6**, 12746–12754 (2018)
15. Smirnov, A., Levashova, T., Kashevnik, A.: Ontology-based cooperation in cyber-physical social systems (2017)
16. Smirnov, A., Kashevnik, A., Ponomarev, A., Shilov, N.: Context-aware decision support in socio-cyberphysical systems: from smart space-based applications to human-computer cloud services (2017)
17. Euzenat, J., Shvaiko, P.: Ontology Matching. Springer, Heidelberg (2013)

18. Otero-Cerdeira, L., Rodríguez-Martínez, F.J., Gómez-Rodríguez, A.: Ontology matching: a literature review. Expert Syst. Appl. **42**, 949–971 (2015)
19. Teslya, N., Smirnov, A., Levashova, T., Shilov, N.: Ontology for resource self-organisation in cyber-physical-social systems. In: Klinov, P., Mouromtsev, D. (eds.) Knowledge Engineering and the Semantic Web, KESW 2014, pp. 184–195. Springer-Verlag, Berlin (2014)
20. Zou, Y., Finin, T., Ding, L., Chen, H.: TAGA : using semantic web technologies in multi-agent systems. In: Proceedings of the 5th International Conference on Electronic Commerce, pp. 95–101 (2003)
21. Corkill, D.D.: Blackboard and multi-agent systems & the future. In: Proceedings of the International Lisp Conference, vol. 3, pp. 23–118 (2003)
22. Zhong, R.Y., Xu, X., Klotz, E., Newman, S.T.: Intelligent manufacturing in the context of industry 4.0: a review. Engineering **3**, 616–630 (2017)
23. Civerchia, F., Bocchino, S., Salvadori, C., Rossi, E., Maggiani, L., Petracca, M.: Industrial internet of things monitoring solution for advanced predictive maintenance applications. J. Ind. Inf. Integr. **7**, 4–12 (2017)
24. Upasani, K., Bakshi, M., Pandhare, V., Lad, B.K.: Distributed maintenance planning in manufacturing industries. Comput. Ind. Eng. **108**, 1–14 (2017)
25. Braune, A., Diesner, M., Hüttemann, G., Klein, M., Löwen, U., Thron, M.: Exemplification of the industrie 4.0 application scenario value-based service following IIRA structure (2017)
26. Cong, L.W., He, Z.: Blockchain disruption and smart contracts (No. w24399). National Bureau of Economic Research, 52 p. (2018). https://doi.org/10.3386/w24399
27. Kshetri, N.: Blockchain's roles in meeting key supply chain management objectives. Int. J. Inf. Manage. **39**, 80–89 (2018)
28. Bahga, A., Madisetti, V.K.: Blockchain platform for industrial internet of things. J. Softw. Eng. Appl. **09**, 533–546 (2016)
29. Abeyratne, S.A., Monfared, R.P.: Blockchain ready manufacturing supply chain using distributed ledger. Int. J. Res. Eng. Technol. **05**, 1–10 (2016)
30. Sikorski, J.J., Haughton, J., Kraft, M., Street, P., Drive, P.F.: Blockchain technology in the chemical industry: machine-to-machine electricity market. Appl. Energy **195**, 234–246 (2016)
31. Balta, E.C., Jain, K., Lin, Y., Tilbury, D., Barton, K., Mao, Z.M.: Production as a service: a centralized framework for small batch manufacturing. In: 2017 13th IEEE Conference on Automation Science and Engineering, pp. 382–389 (2017)
32. Moghaddam, M., Silva, J.R., Nof, S.Y.: Manufacturing-as-a-service—from e-work and service-oriented architecture to the cloud manufacturing paradigm. IFAC-PapersOnLine **48**, 828–833 (2015)
33. Hepp, M.: GoodRelations: an ontology for describing products and services offers on the web. Lecture Notes in Computer Science (including subseries Lecture Notes in Artificial Intelligence and Lecture Notes in Bioinformatics). LNAI, vol. 5268, pp. 329–346 (2008)
34. Martin, D., Bursten, M., Hobbs, J., Lassila, O., McDermott, D., McIlraith, S., Narayanan, S., Paolucci, M., Parsia, B., Payne, T., Sirin, E., Srinivasan, N., Sycara, K.: OWL-S: semantic markup for web services. https://www.w3.org/Submission/OWL-S/
35. Gräser, O., Hundt, L., John, M., Lobermeier, G., Lüder, A., Mülhens, S., Ondracek, N., Thron, M., Schmelter, J.: White paper AutomationML and eCl@ss integration (2015)
36. Hepp, M., Leenheer, P., Moor, A., Sure, Y. (eds.): Ontology Management. Springer, Boston (2008)
37. IEC: IEC 62264-1 enterprise-control system integration – part 1: models and terminology (2003)

38. Cheng, H., Zeng, P., Xue, L., Shi, Z., Wang, P., Yu, H.: Manufacturing ontology development based on industry 4.0 demonstration production line. In: Proceeding of 2016 3rd International Conference on Trustworthy Systems and their Applications, TSA 2016, pp. 42–47 (2016)
39. Usman, Z., Young, R.I., Case, K., Harding, J.: A manufacturing foundation ontology for product lifecycle interoperability. In: Enterprise Interoperability IV. Making the Internet of the Future for the Future of Enterprise, pp. 147–155 (2010)
40. Martinez Lastra, J.L., Delamer, I.M.: Ontologies for production automation. Lecture Notes in Computer Science (including subseries. Lecture Notes in Artificial Intelligence and Lecture Notes in Bioinformatics). LNCS, vol. 4891, pp. 276–289 (2008)

An Approach to Optimization of Ray-Tracing in Volume Visualization Based on Properties of Volume Elements

Nikolai Vitiska[1], Vladimir Selyankin[2], and Nikita Gulyaev[3](✉)

[1] Scientific Research Institute of Multiprocessor Computer and Control Systems, Co Ltd., Taganrog, Russian Federation
[2] Institute of Computer Technology and Information Security, Engineering and Technological Academy, Southern Federal University, Taganrog, Russian Federation
[3] Rostov State University of Economics, Rostov-on-Don, Russian Federation
m.yo.da@yandex.ru

Abstract. Application of ray-tracing in volume visualization often requires significant optimization, mostly for performance issues. Known approaches can provide good results in average, however, particular cases are often a problem. One of the reasons may be the lack of consideration of properties of data being rendered. In this paper, an approach to optimization of ray tracing based on properties of volume elements is described. Firstly, an approach to ray separation is proposed. The proposed approach is based on that fact, that each position on the ray can be considered as a separate ray, which value may depend on values of previous rays. Taking this into account, the usage of bounding primitives allows to reduce the rendering process to a sequential computation of consecutively arranged rays, where rendering parameters may vary for each individual ray. Secondly, an approach to optimization is proposed. The proposed approach introduces a new strategy for defining individual rendering parameters, which considers properties of volume elements as an influencing factor. However, in many cases it can be complicated to analyze all volume elements, intersected by the ray, so such values are reduced to properties of region of volume elements, which are approximated by an axis-aligned bounding box.

Keywords: Volume visualization · Ray-tracing · Computer graphics

1 Introduction

Volume visualization is a perspective direction of computer graphics, it finds application in many actual tasks, where in many situations it fits better than analogues, for example, surface visualization. Volume visualization as a visualization paradigm provides specialized models, data structures and rendering methods, which are specialized on describing and processing of amorphous, liquid, gaseous phenomena, as well as highly-detailed objects represented by volume elements. Volume visualization in such cases provides result of higher reliability, than surface visualization, however, application of volume visualization often requires a good optimization scheme.

A. Abraham et al. (Eds.): IITI 2018, AISC 874, pp. 148–158, 2019.
https://doi.org/10.1007/978-3-030-01818-4_15

Ray-tracing rendering technique (which is often used in volume visualization) may demand too much computational resources. Also, its results may vary greatly in terms of performance or quality under virtually identical conditions. Optimization of ray-tracing has always been an actual problem, a lot of research has been done on this topic [1–3], however, ray-tracing in volume visualization is still one of the most computationally complex tasks in computer graphics.

However, there is a certain number of subdivision methods, that help to reduce complexity of ray-tracing. The reduction of complexity is achieved by ignoring empty elements (insignificant volume elements that take time to process, but do not contribute to the result). Currently, there are several fundamental methods of subdivision in volume visualization: regular structures (i.e. octree), hierarchical structures (organized primitive hierarchies), and unrelated structures (separate primitives) [4]. Different methods have different advantages and disadvantages, however, a common drawback of all subdivision methods is the impossibility of separation of sibling ray and, sometimes, complex hierarchical structures require excessive computational costs for initialization [5].

There is also a certain number of optimization methods (in any visualization paradigm), most of them boil down to the same idea: there is some value, which describes the «appropriateness» of the result or a degree of «interest» in achieving such result. On the other hand, there is some antagonistic value, which describes the «difficulty» of achieving that result. In common, this scheme can be described as a «cost-benefit» scheme [6], where the value of benefit somehow describes the quality of generated image: by difference of pixel values (mostly used in shading task optimization) or by some other characteristic like number of rendered triangles or culling distance. And the value of cost describes the amount of computing resources, like elapsed time [7]. For example, in surface rendering such optimization schemes are widely used in different optimization approaches [8–10], such are level-of-detail optimization or culling-orientated optimization. Volume visualization has different approaches to optimization proposed [11–15], which concentrate around optimization of algorithmic component. However, this approaches have a common drawback: the process of optimization does not take into account exact properties of data being rendered.

Both subdivision strategy and optimization strategy define a sampling strategy. Among all the variety of such strategies, the most basic approach is continuous regular sampling [16], which suggests only sequential placement of samples with a constant step. Advanced approaches imply the usage of spatial subdivision to reduce the number of samples that miss any significant volume element [17]. Both basic and advanced approaches may have «analytical» mechanism, which operates on a higher level, identifying optimal settings (rendering parameters). Such mechanism is often defined as a solution to some optimization problem, mostly stated in a «cost-benefit» form, which is then used to determine optimal (approximately optimal) values. However, because of the complexity of solving non-linear optimization problems, the «analytical» part is often reduced to a simple linear model, which may not be adequate in particular cases.

In this paper, we propose an approach to optimization of sampling process in ray-tracing in volume visualization based on estimation of characteristics of properties of data being rendered. Our subdivision method is based on a common spatial subdivision structure – axis-aligned bounding box, however, we add a second level of subdivision, which provides subdivision of the ray by «embedded» rays. Our optimization method is based on a «classic» optimization scheme, which defines an optimization problem as a «cost-benefit» decision, which involves modifying values of rendering parameters. But we propose a new strategy for defining individual rendering parameters for separated ray segments or «embedded» rays, which takes into account specific properties of data being rendered.

2 Subdivision Strategy

Let's assume that there is some ray $r = \left(\vec{o}, \vec{d}\right)$, which has its start point in some virtual pixel of the screen plane, let's denote this a point as $\vec{o}$, and its direction, $\vec{d}$, pointing in some direction in scene space. Then this ray defines a certain set of points in space:

$$S = \left\{\vec{s}|\vec{s} = \vec{o} + \vec{d}\lambda, \lambda \in [0, +\infty)\right\}, \tag{1}$$

Let's suppose, that there is some approximation of the scene – all volume elements of volume phenomena are organized by some spatial subdivision structure. Let's assume, that we're using linear, non-hierarchical, axis-aligned bounding boxes, as described in [18]. So, this structure can be defined as a set of primitives $F_1 \ldots F_k$, then for the ray r, traveling through the scene space, there is a certain possibility of intersection with some of this primitives, so a new set, a set of points of intersection P can be defined (Fig. 1):

$$P = \{\vec{p}|\vec{p} \in B_i \wedge \vec{p} \in S\}, \tag{2}$$

where:

B_i – is the set of points, which lie on the border of bounding primitive F_i.

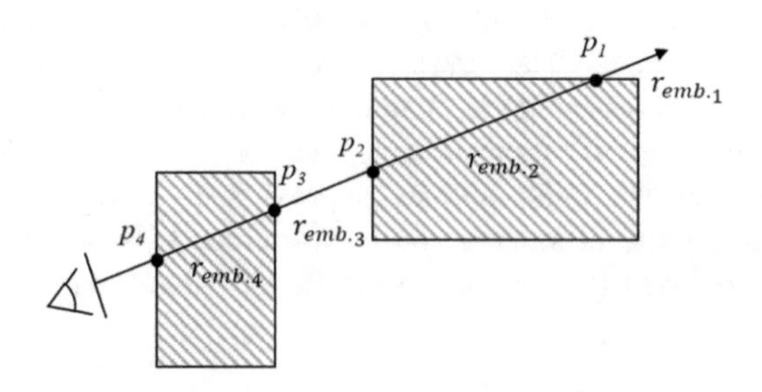

Fig. 1. Separation of the ray into discrete set by bounding primitives

Each point in this set can be considered as a starting point of some new «embedded» ray, $r_{emb.} = \left(\vec{p}, \vec{d}\right)$, which keeps the direction of its «parent» ray. Total amount of embedded rays is equal to the power of the set of points of intersection $|P|$.

Most of rendering techniques in volume visualization consider volume elements as semi-transparent particles or regions filled with such particles. This fact leads to implementation of some opacity accumulation mechanism. For example, in many cases such mechanism can be implemented as a consistent calculation of some value ψ, which can be considered as «accumulated opacity» for current position on some ray:

$$\psi(\lambda) = \int_{\lambda_a}^{\lambda_b} v\left(\vec{o} + \vec{d}\lambda'\right) d\lambda', \tag{3}$$

where:

λ_0 – is the far boundary of ray segment;
λ – is the closest or current boundary of ray segment;
$v(\lambda)$ – is an interpretation function, which returns a certain value of volume element in a given coordinate.

This mechanism is used in many rendering techniques, for example, in volume rendering technique based on absorption model [19], which considers the resulting value as decreased intensity of initial radiation for a certain position on the ray:

$$c = I_0 e^{-\psi(\lambda)}, \tag{4}$$

where:

I_0 – is the initial radiation;
λ – is the position on the ray.

For each ray r, all of its embedded rays $r_{emb.1} \ldots r_{emb.k}$ can be defined by a position on this ray – for each starting point of each embedded ray λ can be found and vice versa. Thus, a certain regularity of corresponding resulting values becomes obvious (Fig. 2). For example, some value ψ_1, which defines total opacity accumulated by $r_{emb.1}$ is equal to zero, because all points on $r_{emb.1}$ lie outside the bounding primitive's inner space, which means that all volume elements intersected by this ray are fully transparent or simply don't exist (out of scene bounds). However, ψ_2, a value of accumulated opacity for the next embedded ray $r_{emb.2}$ is not equal to zero, because some points lie inside the bounding primitive's inner space, where volume elements are not fully transparent.

Next, ψ_3, a value of accumulated opacity for $r_{emb.3}$ is equal to ψ_2, because no new volume elements are intersected by this embedded ray. Finally, ψ_4 can be defined as a sum of accumulated opacity on a segment p_3p_2 and ψ_2. The whole pattern can be described as following: current value of accumulated opacity ψ_i for some embedded ray r_i can be obtained by summation of resulting values of every previous embedded ray and a resulting value for current segment. It can be written as:

$$\psi(\lambda_i) = \int_{\lambda_i}^{\lambda_{i-1}} v\left(\vec{o} + \vec{d}\lambda'\right) d\lambda' + \psi(\lambda_{i-1}). \quad (5)$$

Such interpretation opens a number of opportunities for optimization. As described earlier, it is obvious, that some master ray segments are «empty», like $[\lambda_1, +\infty)$ or $[\lambda_3, \lambda_2]$. It also can be shown by equality of ψ-values of embedded rays $r_{emb.2}$ and $r_{emb.3}$ and «emptiness» of the first ψ-value of $r_{emb.1}$. Thus, an approach to optimization should handle this situation, for example, by lowering rendering parameters for empty segments or even by skipping calculation routine for such segments or rays.

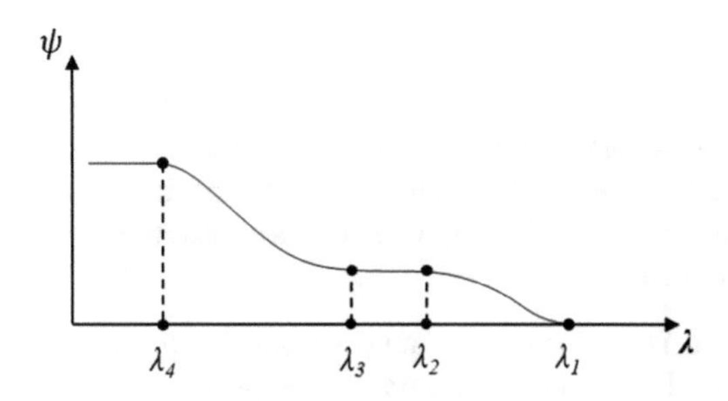

Fig. 2. An example dependence for described case

However, such approach is applicable only in cases, when resulting values are known, which means, that this resulting values should be calculated, though this is not suitable for many situations. Fortunately, this is not the only solution – it is possible to estimate values, or to find some other values, that can provide an informative description of such values. Thus, the proposed approach to optimization is based on described above scheme, but involves the usage of characteristics of volume elements' values.

3 Optimization Strategy

At first, we can define special values for each bounding primitive, that contains some region of volume phenomena, this values can characterize almost any property. In described above case, it is logical to assume, that such values should describe opacity properties. Thus, the first value can be described as a «fullness» coefficient $k_{full.}$ of a certain bounding primitive F_j, which covers this region. This value can be defined based on a binary random variable b, which is 1 when a randomly chosen volume element form inner space of F_i has non-zero opacity and 0 otherwise:

$$k_{full.} = \mathrm{M}[b]. \quad (6)$$

The second value can be described as a «diversity» coefficient $k_{div.}$ and can be defined with the same binary random variable b as following:

$$k_{div.} = \mathrm{D}[b]. \tag{7}$$

Second, we can define a function for individual rendering parameter with $k_{full.}$ and $k_{div.}$. For example, for calculation method (3) we can define:

$$\psi(\lambda) \approx \frac{\lambda_b - \lambda_a}{s} \sum\nolimits_{i=1}^{s} v\left(\vec{o} + \vec{d}\left(\lambda + \frac{\lambda_b - \lambda_a}{2s} i\right)\right), \tag{8}$$

where:

s – is the number of samples along the ray.

Then we can use this scheme for every embedded ray to get sequential calculation back-to-front (5), which makes possible modifications of number of samples s for each embedded ray, leaving each previous embedded ray rendered with the necessary number of samples. Thus, the function for individual (and the only is this case) rendering parameter can be defined as following:

$$s' = s\left(\frac{k_1}{k_{full.}} + k_2 k_{div.}\right). \tag{9}$$

where:

$k_{full.}$, $k_{div.}$ – are «fullness» and «diversity» coefficients for bounding primitive, which is intersected;
k_1, k_2 – are some parameters, that regulate the «significance» of $k_{full.}$ and $k_{div.}$;
s – is the initial number of samples.

After that, the quality function can be defined with s' instead of s. It's obvious, that s' depends on s, but the idea is to transform the original parameter's value, not to substitute it with an independent one. For described above rendering technique, the quality is increased along with the number of samples. Thus, if we have original quality function defined as following:

$$Q(s) = q_0 s + q_1 \tag{10}$$

Then, according to the proposed approach, this function can be redefined trough s' and s for a unit-length segment as following:

$$Q(s) = q_0 s' + q_1 = q_0 s\left(\frac{k_1}{k_{full.}} + k_2 k_{div.}\right) + q_1. \tag{11}$$

In such case, any given value of s gets transformed by both coefficients, which means, that value of Q becomes sensitive to properties that they represent. So, this new Q can be used in different «special» cases, because the estimated quality is lowered by coefficients, that represent real properties of volume elements.

It is obvious, that usage of two properties and only one rendering parameter is not enough for more complex rendering techniques, so this approach can be extended for an arbitrary number of properties $k_{char._1} \ldots .k_{char._t}$, which are actual in terms of a certain task, for an arbitrary rendering parameter, originally defined as x:

$$x' = x(k_1 k_{char._1} + k_2 k_{char._2} + \cdots + k_t k_{char._t}). \quad (12)$$

4 Implementation

Implementation of the proposed approach in volume rendering follows the general idea of optimization in computer graphics – the goal is to achieve maximum quality of the result with minimal computational cost. In volume ray-tracing, there are three basic rendering parameters: number of samples s, number of secondary rays m and the tracing limit l, which contribute to both quality and expenses (however, some rendering techniques may neglect some parameters as described above). In general, each optimization approach should provide a solution, which involves all rendering parameters this technique utilizes. This can be done by defining the quality function $Q(s, m, l)$ and the expense function $P(s, m, l)$ – empirical dependencies or heuristics, which make possible comparison of quantitative values. This functions are used to define the optimization problem: the quality function is used as an objective function, when the expense function is used as a constraint:

$$\begin{matrix} Q(s, m, l) \to max \\ P(s, m, l) \leq \mathcal{P} \end{matrix} \quad (13)$$

The stated optimization problem is being solved on a pre-processing stage during rendering (before processing each new frame). Thus, before we can trace a ray, we have to get optimal values for parameters (s, m and l), which are retrieved by solving the optimization problem. In proposed approach, optimal values of rendering parameters are obtained in a same manner, however the quality function (the objective function) takes a different form: each parameter (s, m and l) is substituted by its «transformed» analogue (s', m' and l'):

$$\begin{matrix} Q(s, m, l) = s(k_1 k_{char._1} + \cdots + k_t k_{char._t}) + m(k_1 k_{char._1} + \cdots + k_t k_{char._t}) \\ + l(k_1 k_{char._1} + \cdots + k_t k_{char._t}) \end{matrix} \quad (14)$$

The casting procedure is divided into two stages: the estimation stage and the sampling stage. During the estimation stage, the ray is cast through the boundary approximation (bounding primitives $F_1 \ldots F_{j.}$) of volume regions, points of intersection and corresponding embedded rays are determined (Fig. 3a). Thus, for every primitive, an embedded ray with a corresponding length is created. Then, the optimization problem (13) for each embedded ray (as well as for whole set of rays) with properties ($k_{char._{1F}} \ldots k_{char._{tF}}$) taken from corresponding primitive is solved. Obtained values then can be normalized in accordance to embedded ray's length as $s \times d$ (because values are initially obtained for a unit-length ray). The process of retrieving volume region's properties is currently trivial, as shown in [20]. The next stage is the sampling stage, during this stage, samples are generated for each embedded ray according to obtained values (Fig. 3b).

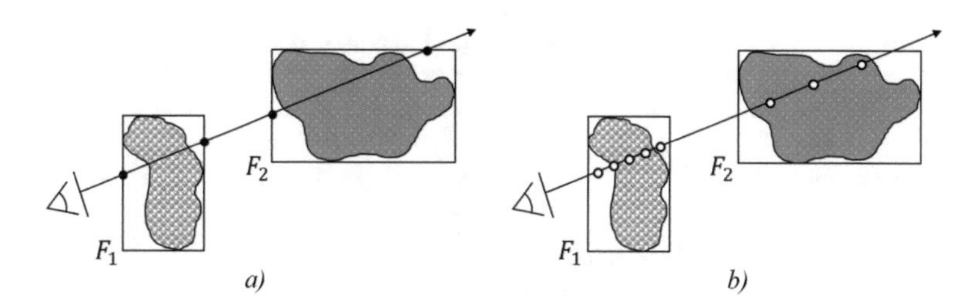

Fig. 3. Stages of the casting procedure

5 Results and Comparison

In order to determine advantages and disadvantages of our approach, we have conducted a number of tests, involving comparison of synthetized images and elapsed rendering time of our approach and most common approaches. The basic approach – continuous regular sampling – is obviously ineffective in terms of performance, however it provides an opportunity to achieve an accurate result (because of over-sampling), which can be used as a reference. Other potentially interesting for comparison approaches are: octree-based approach and bounding box-based approach. To perform a comparison, we firstly synthetize an image with the help of the basic approach. The resulting image is set as a reference, also, rendering parameters (point of view transform) and scene contents are fixed throughout this test. Then, we perform image synthesis with octree-based and bounding box-based approaches. After that, we perform image synthesis with our approach. Finally, we compare images obtained with the help of all approaches (except basic) with the reference image (obtained by basic approach), where the difference between images (plain number of non-matching pixels) is considered as the quality value of approach being tested.

As initial data, we use three voxelized scenes (derived from well-known benchmark 3D models) and three procedurally generated scenes (Table 1). As a visualization model, we use mentioned above absorption model.

Table 1. Initial data used for visualization

Scene	Dimensions (x, y, z)	Source
Stanford Dragon	512 × 2048 × 2048	Voxelized 3D model
Stanford Bunny	2048 × 2048 × 2048	Voxelized 3D model
Happy Buddha	1024 × 1024 × 4096	Voxelized 3D model
Generated scene #1	2048 × 2048 × 1024	Procedurally generated scene
Generated scene #2	1024 × 512 × 4096	Procedurally generated scene
Generated scene #3	2048 × 1024 × 1024	Procedurally generated scene

We use following approach-specific parameters (Table 2), which are fixed throughout all tests. For proposed approach, we use described above function with two characteristics: «fullness» and «diversity», estimated with the following number of samples.

Table 2. Approach-specific parameters

Approach	Parameter	Value
Basic approach	Number of samples per voxel	3
Octree-based approach	Branching threshold (levels)	24
	Bias (opacity)	0.025
Bounding box-based approach	Min. side (voxels)	16
	Min. gap (voxels)	4
	Max. primitive count	256
	Bias (opacity)	0.025
Proposed approach	Min. side (voxels)	64
	Min. gap (voxels)	32
	Max. primitive count	128
	Samples for estimation (% of voxels)	5%
	Bias (opacity)	0.025

All tests were performed through single-threaded CPU rendering on a 2.2 GHz machine with a resolution of 1920 × 1080 with one ray per pixel (no anti-aliasing). The results of our tests are provided in a table below (Table 3), where t denotes elapsed time in milliseconds (t_{init} – subdivision structure initialization time, t_{render} – rendering time) and q denotes quality.

Table 3. Test results

Scene	Ref.	Octree-based			Bounding box-based			Proposed approach		
	t	t_{init}	t_{render}	q	t_{init}	t_{render}	q	t_{init}	t_{render}	q
Stanford Dragon	166	41	30	0.8971	23	29	0.8731	30	14	0.9066
Stanford Bunny	217	54	31	0.9201	27	34	0.8856	31	17	0.9152
Happy Buddha	183	20	23	0.9433	21	20	0.8312	25	12	0.9286
Generated scene #1	197	74	29	0.8625	28	48	0.8551	32	19	0.8979
Generated scene #2	229	82	33	0.8711	24	46	0.8601	28	22	0.9092
Generated scene #3	164	63	30	0.8947	27	44	0.8826	33	18	0.9101

Based on obtained results, we can conclude the following. In comparison with other approaches to visualization on regular data (e.g., a set logically related 3D models), our approach provides a considerable increase in performance – from 10% to 15% compared to octree-based approach and about 10% increase compared to bounding box-based approach. It requires a higher amount of computations on initialization stage (consumed by estimation and optimization procedures) than simple bounding box approach, though. It also provides a slight increase in quality up to 10% compared to bounding box-based approach. However, our approach significantly excels other approaches on irregular and diverse input data (no separate 3D models, variations of color and opacity, what corresponds to demands of most actual tasks in volume rendering), such as procedurally generated scenes, where it avoids typical problems (e.g., bias in «grey» branches), which lead to erroneous sample appearance. It also demands lower primitive parameters (side, gap values), than simple bounding box approach.

6 Conclusions

In many cases, optimal rendering parameters depend on data being rendered – volume elements, or whole regions, which means that in most situations some characteristics of data being rendered should be taken into account. Inclusion of such characteristics can notably improve rendering performance. It may require definition of such characteristics, for every unique task, as well as objective functions and constraints, though. However, such functions are still linear, because characteristics are constant during frame rendering, which allows application of computationally simple linear programming methods.

Acknowledgments. The reported study was funded by RFBR according to the research project № 18-07-00733.

References

1. Wald, I., et al.: State of the art in ray tracing animated scenes. Comput. Graph. Forum **28**(6), 1691–1722 (2009)
2. Blakey, E.: Ray tracing – computing the incomputable? In: Proceedings 8th International Workshop on Developments in Computational Models, Cambridge, UK, pp. 32–40 (2012)
3. Chang, A.: A survey of geometric data structures for ray tracing. Technical report, Polytechnic University, Brooklyn (2001)
4. Reinhard, E., Smits, B., Hansen, C.: Dynamic acceleration structures for interactive ray tracing. In: Rendering Techniques 2000, pp. 299–306. Springer, Vienna (2000)
5. Havran, V., Herzog, R., Seidel, H.P.: On the fast construction of spatial hierarchies for ray tracing. In: Interactive Ray Tracing 2006, pp. 71–80. IEEE (2006)
6. Aliaga, D., Lastra, A.: Automatic image placement to provide a guaranteed frame rate. In: Proceedings of 26th Annual Conference on CG & IT, pp. 307–316 (1999)
7. Funkhouser, T.A., Séquin, C.H.: Adaptive display algorithm for interactive frame rates during visualization of complex virtual environments. In: Proceedings 20th Annual Conference on Computer Graphics and Interactive Techniques, pp. 247–254. ACM (1993)
8. Dong, T., et al.: A time-critical adaptive approach for visualizing natural scenes on different devices. PLoS One **2**(10), e0117586 (2015)
9. Ellul, C., Altenbuchner, J.: Investigating approaches to improving rendering performance of 3D city models on mobile devices. GIS **2**(17), 73–84 (2014)

10. Nijdam, N., et al.: A context-aware adaptive rendering system for user-centric pervasive computing environments. In: 15th IEEE Conference, MELECON 2010, pp. 790–795 (2010)
11. Marmitt, G., Friedrich, H., Slusallek, P.: Interactive volume rendering with ray tracing. In: Eurographics (STARs), pp. 115–136 (2006)
12. Gao, J., et al.: Distributed data management for large volume visualization. In: IEEE Visualization 2005 – VIS 2005, pp. 183–189 (2005)
13. Lee, B., et al.: Fast high-quality volume ray casting with virtual samplings. IEEE Trans. Vis. Comput. Graph. **16**, 1525–1532 (2010)
14. Wang, H., et al.: A parallel preintegration volume rendering algorithm based on adaptive sampling. J. Vis. **19**(3), 437–446 (2016)
15. Wald, I., et al.: Progressive CPU volume rendering with sample accumulation. In: Eurographics Symposium on Parallel Graphics and Visualization, pp. 41–51 (2017)
16. Kaufman, A., Cohen, D., Yagel, R.: Volume graphics. Computer **7**(26), 51–64 (1993)
17. Levoy, M.: Efficient ray tracing of volume data. ACM Trans. Graph. (TOG) **3**(9), 245–261 (1990)
18. Vitiska, N., Gulyaev, N.: An approach to visualization of three-dimensional scenes and objects via voxel graphics for simulation systems. Izvestiya SFedU. Eng. Sci. **4**(165), 77–87 (2015)
19. Max, N.: Optical models for direct volume rendering. IEEE Trans. Vis. Comput. Graph. **1**, 99–108 (1995)
20. Vitiska, N., Gulyaev, N.: A study on modifications of visualization model for volume rendering with ray-tracing. Informatiz. Commun. **3**(8), 30–35 (2016)

Discovering of Part-Whole Relations Used in Architectural Prototyping of Project Tasks

Petr Sosnin(✉) and Anna Kulikova

Ulyanovsk State Technical University, ul. Severny Venets, str. 32, 432027 Ulyanovsk, Russia
sosnin@ulstu.ru, a.push1206@gmail.com

Abstract. The paper presents a way of achieving the necessary architectural understanding in conceptual solving of the project tasks in designing the software intensive systems (SISs). This way should be implemented in the context of automated design thinking, and it is based on discovering of the part-whole relations, their registering in the project ontology and reflecting the initial statement of the perceived task into its graphical architectural description. The suggested way is implemented by designers in the instrumental environment WIQA (Working In Questions and Answers) supporting the conceptual designing of the SISs. Discovering and registering the indicated type of relations are functions that are embedded in the toolkit WIQA, the graphical subsystem of which provides building the architectural prototypes.

Keywords: Architectural prototype · Conceptual designing · Design thinking
Part-whole relation · Project ontology · Question-answering
Software intensive system

1 Introduction

The extremely low degree of success in developing the modern SISs is a reason for the search for innovations in the subject area of software engineering [1]. In any of such searches, innovators should take into account the uniqueness of any SIS-project that is caused by the following its features:

1. The life cycle of the SIS project includes the unpredictable impact on its course of numerous situational factors that it is dangerous not considering at various points in the life cycle (for reference in the useful review [2] about 400 such situational factors are mapped into 48 of their variants distributed across 11 groups).
2. In any of such point, certain designers will react on the base of their unique experience, the use of which will be accompanied by influencing on the reaction both positive and negative manifestations of human factors.
3. Among these points of the life cycle, an essential role will play the new tasks that unpredictably arise in the design process. It is typical that the first of such tasks is a root task of the conceived project.

To deal with any of these situations and react, designers must perceive it, describe appropriate way and understand in a sufficient measure. The main attention of this

A. Abraham et al. (Eds.): IITI 2018, AISC 874, pp. 159–170, 2019.
https://doi.org/10.1007/978-3-030-01818-4_16

paper will focus on the architectural understanding of situations, each of which concerns the appearance of a new project task, beginning with a root task of the corresponding project.

The suggested way to an architectural understanding of the task situations and states of the tasks is oriented on starting the work with any new task with applying the automated design thinking that is very popular in work with innovative projects [3]. The suggested way is embedded in our version of the design thinking approach (DT-approach) that is implemented in a conceptual space formed and studied in the semantic memory of the instrumental environment WIQA [4]. Steps of design thinking include discovering of part-whole relations, combining of which helps to build architectural prototypes that register achieved an understanding of task situations and states of the tasks.

The remainder of the article is structured as follows. Section 2 discloses grounds of our approach to architectural prototyping aimed at controlled understanding in work with project tasks. Necessary prototypes are built with the use of automated design thinking in the frame of the conceptual space. Section 3 points out related works. A way of discovering the part-whole relations is described in Sect. 4. In Sect. 5, we demonstrate the steps of architectural prototyping on a simplified example, and the paper is concluded in Sect. 6.

2 Preliminary Bases

2.1 Phenomenon of Understanding

As told above, interests of the paper are bound with achieving by designers of the necessary architectural understanding in conceptual solving the project tasks in designing the SISs. These interests imply having the constructive specification for the phenomenon of "understanding" and its architectural manifestation.

As an intellectual phenomenon, understanding manifests itself as naturally artificial essence. First of all, a natural side of understanding is responsible for checking the sensible (meaningful) use of natural language, for example, for estimating of the wholeness of describing the perceived situation or state. Such estimating fulfills a managerial role and can be used with different aims, which often requires materializations both as a process of achieving of understanding so its result. Similar additions reflect the artificial side of understanding.

The indicated function of understanding suggests that the process of this phenomenon is initiated in the right hemisphere of the human brain by conceptual processes in the left hemisphere. Both these processes evolve to coordinated outcomes, one of which attempts to confirm the wholeness of another and contrary. The most explicitly, the perceived wholeness of any essence manifests itself through part-whole relations combining its parts, and it corresponds to the system approach to a conceptual presentation of this essence.

The constructive use of the system approach opens the possibility for comparing of the drawn structure of the perceived essence interpreted as a corresponding system with block-and-line scheme extracted from the verbal description of this essence. Having the correspondence between drawn structure (in the context of its environment) and Extracted block-and-line Scheme (ES) can be used as an indicator of understanding. Having the differences is the indicator of misunderstanding or errors in the description or having some details in compared views.

One more version of achieving the necessary degree of understanding is a controlled perception of the ES that is iteratively corrected with the use of interactions of the designer(s) with the accessible experience. Thus, checked artifacts of the ES-type contain indicators of the achieved states of understanding. They express the certain architectural views on the designed system at the language of parts and relations among them. They express architectural understanding that is additional to cause-and-effect understanding.

2.2 Conceptual Space for Designing

In our research and practice of conceptual designing the SYSs, we master the reflection of basic essences of the operational space onto their conceptual representations in forms of nets of question-answer reasoning (QA-nets) implemented and registered in conceptual space CS(t) created in parallel with other actions of designers [4]. Interactions of the designer with nodes of such nets are shown in Fig. 1.

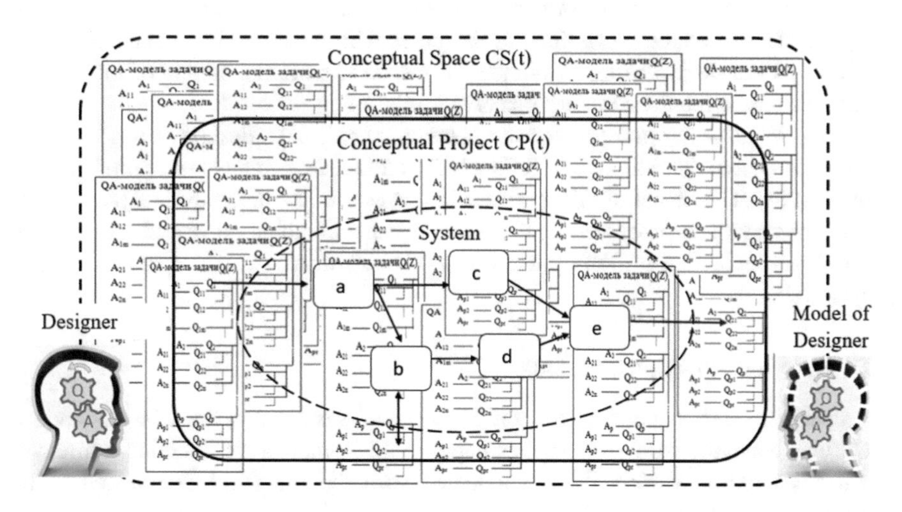

Fig. 1. Interactions of the designer with nodes of QA-nets

The scheme emphasizes that the conceptual project CP(t) has a representation located in CS(t). Moreover, both CP(t) and CS(t) being objectified in the semantic memory in the current state of their QA-nets. By the system approach, CP(t) is a system located in the CS-environment. Also, CP(t) in its current state consists of components (a, b, c, d and e, for example) that were extracted from CS(t) and processed by the necessary way with the use of discovering the part-whole relations. It should be noted; in this process, any component can be interpreted as a system with certain part-whole relations in its structure.

Interacting with objects of the CP(t) or CS(t), the designer interacts with the certain node or nodes of the corresponding QA-nets. Any chosen node has a verbal expression (corresponding basic attribute) that register 'traces' of the designer's access to the natural experience, psychological actions with which are activated in the mental space (MS). During interactions of the designer with the node, we interpret the activated part MS(t) of the mental space MS as an extension of the CS(t) in conditions when such extension is bound with nets by indicated verbal 'traces' of the corresponding node. Similar functions fulfill the images and verbal components of files attached to the node.

Thus, verbal expressions of questions and answers fulfill the roles of intermediaries between CS(t) and MS(t). Signs of questions activate processes in MS(t) among which the central place occupies 'thinking' because "mental space - a space to think" [5]. There are different aims for activating MS(t), for example, to restore the certain state of MS(t) for repeatable activating the corresponding process in MS(t) or to activate the creative potential of the designer for the necessary 'tips' received from the intellect in solving the task.

For our study, MS(t) is a "black box" that manifests itself only through input and output verbal 'traces' objectified in nodes of QA-nets. In the suggested approach, designers use MS(t) as "a suitable space for containing, ruminating and making use of experience" [6]. In the described case, the CS(t) consists of interrelated objects any of which is built with the use of corresponding QA-nets. These objects are formed on the base of facts $\{F_j\}$ about interactions of designers with the accessible experience while solutions of the project tasks. A number of these facts are the source of information about part-whole relations among perceived objects.

In the WIQA-environment, for the creation of QA-nets, designers can fulfill the actions, a system of which we designated as the question-answer approach (QA-approach) [4]. The kernel of this approach is a question-answer analysis (QA-analysis) of the generated and processed textual units that are combine registered facts $\{F_j\}$ in descriptions indicating on the system assemblies of objects in conceptual work with project tasks.

2.3 Automated Design Thinking

The content of this paper focusses on a conceived system that must be designed in conditions when the work of designers begins with an innovative intention. Moreover, initial steps of the work will be conducted with the use the DT-approach in the environment of the toolkit WIQA.

In this case, the first task situation arisen in front of designers is a root task Z*, the statement of which is absent and must be formulated, starting with initial intention as the initial point $P(t_0)$ of the project life cycle. Applying the DT-approach to this task is schematically shown in Fig. 2.

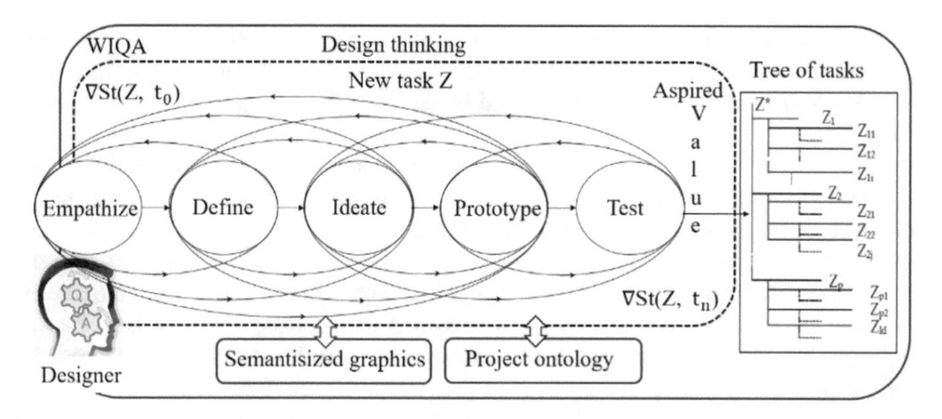

Fig. 2. The iterative process of design thinking

In our version of the DT-approach, the work with the root task Z* and as well with any new project task Z_i begins with a list of keywords registering this task at the level of an initial uncertainty $\nabla U(Z_i, t_0)$. The initial state of the task can be presented with an implication [7]

$$??^{U}U(Z_i) \overset{???^{W}W(Z_i)}{\rightarrow} ?^{V}V(Z_i),$$

where symbols $?^U, ?^V$ and $?^W$ indicates applying of QA- approach [4] to the components of a condition U, value V and potential a solution W of the task Z_i or Z*.

By the QA-approach, questions are objects of interests and applications. In the implication, signs "?", "??" and "???" indicate on uncertainties of a different measure. In this place and below, these signs are used as special indicators required to extract or generate the questions during question-answer analysis of the processed text.

Thus for an innovative project, it is typical to start the life cycle of the project with uncertainty $?^V V(Z^*)$, achieving of which is bound with the project goal. In the described version of work, the list of the certain keywords presents the initial point of the life cycle that corresponds to starting the work with the root project tasks Z* in the frame of design thinking. Below we present elements of such work for the definite project while here we describe the steps of design thinking only schematically. For simplifying the references on specialists considering in thinking, we will count that one designer fulfills all necessary actions.

Looking through keywords, the designer creates some discourses disclosing the essential expectations of the potential users (the Empathize step). These discourses contain information about important requirements, QA-analysis of which helps the designer to formulate the initial statement St(Z*, t2) of the task Z* (The Define step). Analyzing the text St(Z*, t2), the designer tries to invent the idea that can lead to the appropriate solution (the Ideate step), for estimating of which the corresponding prototype should be built (the Prototype step) and tested (the Test step).

In the general case, the described way of working has an iterative character with possible returns to previous steps. Moreover, the first solution is seldom effective, and the search for alternative solutions should be repeated for the future choice of one of them.

The scheme in Fig. 2 also includes some components that uncover conditions of realizing our version of the DT-process. It is implemented in the instrumentally modeling environment WIQA that includes a complex of means for the use of semanticized graphics and project ontology in the analysis and processing of its outcomes [4]. This environment supports agile project management in conditions of multitasking oriented on unpredictable discovering the new task. In the frame of such management, the designer registers the discovered new task with the list of keywords and embeds this task in the tree of the project tasks.

In developing the certain SIS, applications of the described version of the DT-approach are an informational source of registered reasoning of the designer who processes the generated textual units with the following aims that must be achieved in parallel:

1. Creating the structure and content of the space CS(t) for its use as a source of restrictions and requirements for the project to be developed.
2. Creating the project language $L^P(t)$, including the project ontology $O^P(t)$ as a kernel of the language for its controlled use in reasoning and documenting.
3. Creating the project theory $Th^P(t)$ of the substantially evolutionary type for providing the substantiations, prediction, and also modeling and understanding in conceptual designing.
4. Conceptual solving the project tasks, ordering their solutions in the reusable forms and registering them in the current state of the tree of tasks $TT^P(t)$ as the important artifact of the project.

Among these aims, we indicate the creation of the theory $Th^P(t)$ of the substantially evolutionary type that was suggested and investigated by authors of this paper. Such kind of theories is a subclass of Grounded theories [8], the choice of which was caused by the socio-behavioral nature of designing. Constructs and regularities of any applied theory are extracted from the current state of the CS(t) in the real-time work with new tasks any of which is a reason for the next step of the theory evolution. In defining these constructs and regularities, designers apply means of architectural prototyping for experimental confirmation of understanding that can be provided by the theory $Th^P(t)$.

3 Related Works

The first group of related works combines the papers focused on the constructive use of conceptual spaces in the design practice. In this group, we mark the publication [9] defined the ontological viewpoint on the CS, the study [6] suggested and the paper [10] that suggests some templates for documenting the state of the space in the form suitable for decision-making.

The next group of related works concerns the subject area "Design Thinking." The informational source [11] discloses basic features of this activity in developing the SISs. The becoming of the DT-methodology is described in the paper [3]. The interesting analysis of the nature of design thinking is conducted in [7] where K. Dorst connects this nature with two types of abductive reasoning of the designer in interactions with situated problems.

Finally, we disclose some studies that have deals with discovering the part-whole relations in textual units. We have researched the ways to express part-whole relations in both the English and the Russian languages discovered by professional linguists and philologists. The works [12] and [13] contain the most relevant and full information about the ways part-whole relations can be expressed in the Russian language. As far as the English language is concerned, we used the results of the research by Girju, Badulescu and Moldovan [14], who developed a system of patterns and rules to explicate part-whole relations in English texts based on the WordNet thesaurus.

All papers indicated in this section were used as sources of requirements in developing the set of instrumental means provided the work with part-whole relations.

4 Discovering of Part-Whole Relations

In the section, we present automated discovery of part-whole relations (PART (Y, X), where X is a part of Y) from the English texts. We extract special syntactical units – the so-called noun phrases (NP) – from the texts which are linked by part-whole relations detected with the help of various lexical and syntactical patterns [14]:

(1) $\mathbf{NP_X}$ + **P** + $\mathbf{NP_Y}$, where
$\mathbf{NP_X}$ is the noun phrase expressing the part,
$\mathbf{NP_Y}$ is the noun phrase expressing the whole,
P is one of the following prepositions: *of, in, from, throughout, at, on, around, onto, outside.*

(2) $\mathbf{NP_Y}$ + **P** + $\mathbf{NP_X}$, where
P is one of the following prepositions: *with, above.*

(3) $\mathbf{NP_X}$ + *to* + $\mathbf{NP_Y}$.

(4) $\mathbf{NP_X}$ + *all over* + $\mathbf{NP_Y}$.

(5) $\mathbf{NP_X}$ + *and other* + $\mathbf{NP_X}$ + *of* + $\mathbf{NP_Y}$.

(6) $\mathbf{NP_X}$ + *...branch of* + $\mathbf{NP_Y}$.

(7) $\mathbf{NP_Y}$ + *'s/'*(possessive particle) + $\mathbf{NP_X}$.

(8) *in* + $\mathbf{NP_X}$ +, *in* + $\mathbf{NP_Y}$.

(9) $\mathbf{NP_X}$ +, + $\mathbf{NP_Y}$.

(10) $\mathbf{NP_Y}$ + *whose* + $\mathbf{NP_X}$.

(11) $\mathbf{NP_Y}$ + $\mathbf{NP_X}$.

(12) $\mathbf{NP_Y}$ + $\mathbf{NP_X}$ + $\mathbf{NP_X}$.

Apart from that, we consider predicates (verb-connectors) as well as some nouns which mark part-whole relations due to their meaning.

Typical predicates expressing part-whole relations are various verbs and verb-based constructions (collected based on WordNet). They can be divided into two types:

(1) "Whole"-markers: *carry, combine, comprehend, comprise, consist, contain, enclose, feature, have, hold, hold in, house, include, incorporate, inherit, integrate, receive, retain, subsume*;
(2) "Part"-markers: *accommodate, add, admit, affiliate, appertain, be, bear, belong, build in, colligate, compose, compound, confine, constitute, dwell, embrace, encompass, fall in, form, get together, infiltrate, inhere, involve, join, let in, lie, lie in, make, make up, pertain, rejoin, repose, represent, reside, rest, sign up, take.*

Figure 3 shows the overall algorithm of processing any text with the view to discover pairs of syntactical constructions linked by part-whole relations:

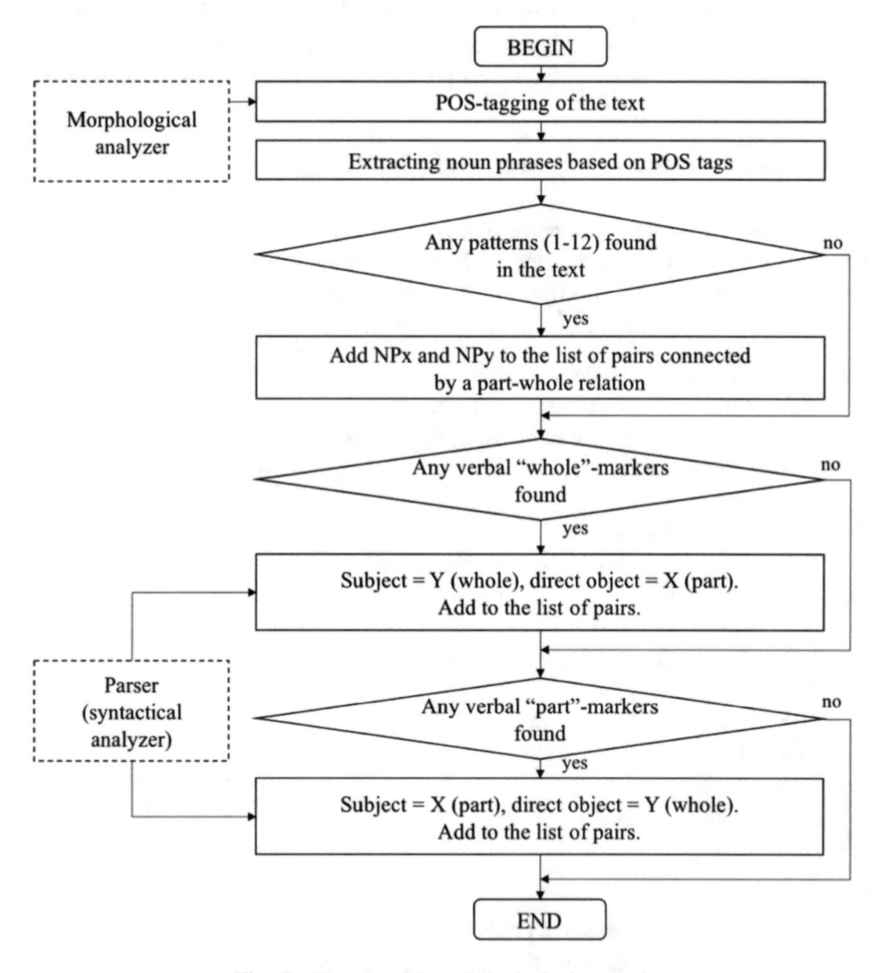

Fig. 3. The algorithm of the lexical control

After discovering any relation PART (Y, X), the designer must include it in the project ontology, where such relations define one of the very important systematization types. Anytime the designer can extract all the registered parts for the chosen whole unit, for example, to extract parts of a system being designed in its current state.

5 Architectural Prototyping

The use of extracted PART (Y, X), we present the example of two architectural prototypes built for the following project task:

Z. Develop a set of tools extracting pairs of lexical items from a text written in the English language linked by a part-whole relation.*

For conceptual solving the task, we applied the DT-approach in the version described above. It needs to note, on the way from discourse D1 to the initial statement of the task Z*, it was generated and analyzed 12 question-answer pairs, discovered 18 semantic errors in their texts, extracted and specified 34 concepts and built seven block-and-line schemes.

The first architectural diagram reflects this task from the viewpoint of the processed data. In automated constructing the prototype presented in Fig. 4, we used concepts chosen from the current state of the project ontology. The two relation types used in this diagram is the part-whole ("consists of") relation and the generic ("is a") relation. The diagram shows that a "text" we analyze consists of "pairs of lexical items linked by a part-whole relation" and "part-whole tags"; it also shows the two components of a "pair of lexical items" and other data types deepening into their relations.

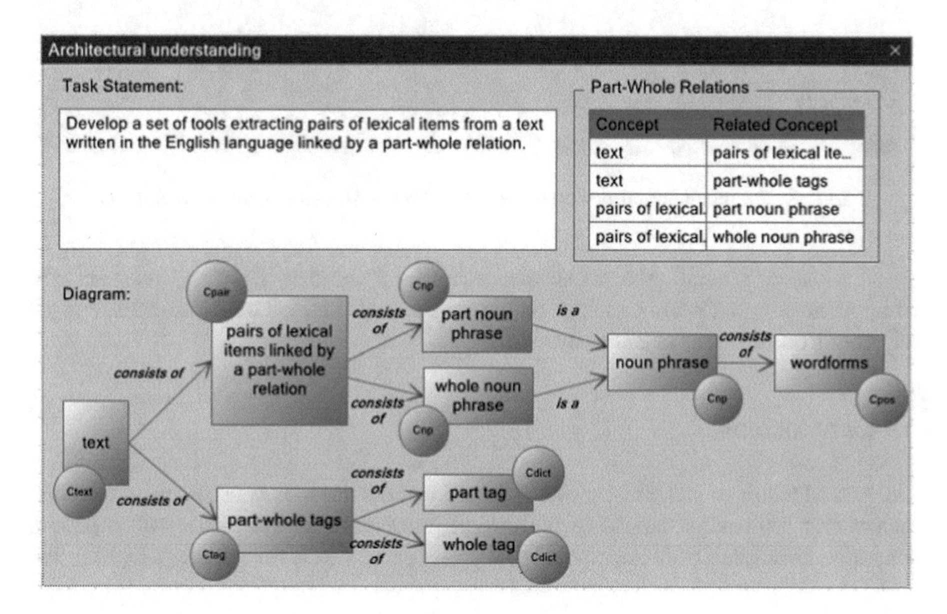

Fig. 4. An architectural prototype of the task from the viewpoint of data

The diagram demonstrates the main components of the tool (the ones in the circles) which use corresponding data types. This information was also extracted from the ontology developed at the define phase, i.e., by revealing the instrumental relations of each concept added to the scheme.

The workflow diagram (Fig. 5) was constructed not only based on the ontological data but also with the help of the previous two schemes. The concepts "POS-tagging," "Lexical category detection," "Noun phrase detection," "Part-whole tags detection" and "Getting pairs" means the **processes** taking place when our tool is being used. They are linked to each other by a sort of part-whole relation. However, the relations among these concepts are slightly different – they show that each process takes place right after another and, therefore, is a "part" of another one.

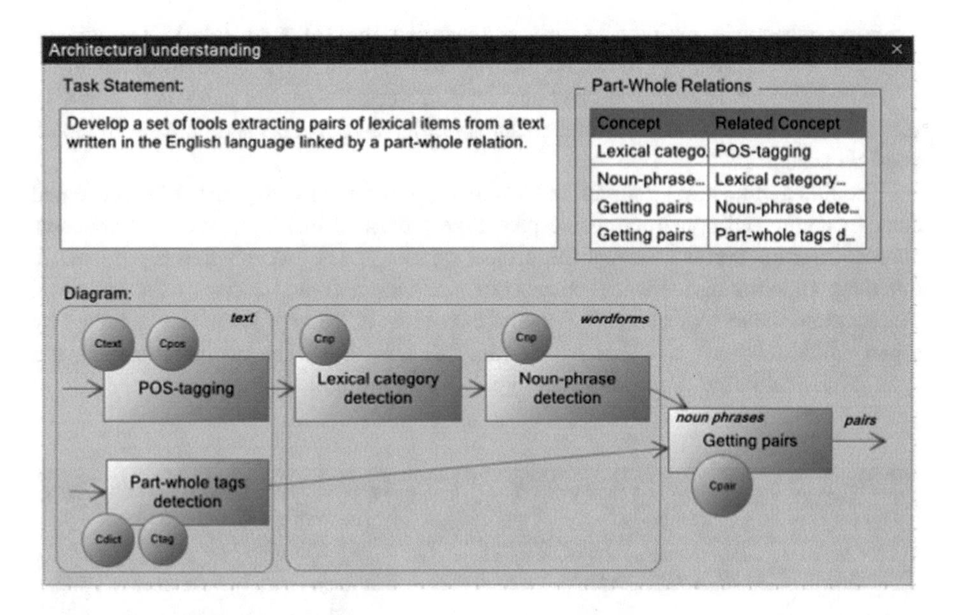

Fig. 5. An architectural prototype of the task from the viewpoint of workflows

The concepts mentioned above are grouped by the data type they use, and the components are performing each of the processes is also added to the scheme. We got this data from the project ontology.

6 Conclusions

The basic feature of any SIS-project is its uniqueness caused by numerous situational factors that influence a course of the design process. In its turn, such influence is a source of an unpredictable appearance of unexpected tasks, for solving of which the designers must apply the accessible experience. Usually, the first of these tasks is the root task of any innovative project. For reacting to new tasks, we suggest our version of the DT-approach that helps to designers to build architectural and cause-and-effect prototypes, the first kind of which is described above.

The described way is based on extracting the part-whole relations from textual units of used reasoning and operative notes, and registering discovered relations in the project ontology. For any new task, at the prototype phase of design thinking, the designers extract part-concepts of the investigated wholeness for automated drawing the diagram presented the wholeness as a system. This process can be interpreted as extracting the system assemblies of conceptual objects from the space CS(t). For providing of prototypes' quality, the designer should build them both for conceptual solving of corresponding tasks and increments of the project theory $Th^P(t)$ that also extracted from the space CS(t) in parallel with other actions of designing.

Designers must implement the suggested way in the instrumental environment WIQA supporting the conceptual designing of the SISs. Discovering and registering the indicated type of relations are functions that are embedded in the toolkit WIQA, the graphical subsystem of which provides building the architectural prototypes.

Acknowledgments. This work was supported by the Russian Fund for Basic Research (RFBR), Grants #18- 07-00989a, #18-47-730016p-a, #18-47-732012 p_мк and the State Contract №2.1534.2017/4.6.

References

1. Reports of Standish Group (2017). www.standishgroup.com/outline
2. Clarke, P., O'Connor, R.V.: The situational factors that affect the software development process: towards a comprehensive reference framework. J. Inf. Softw. Technol. **54**(5), 433–447 (2012)
3. Häger, F., Kowark T., Krüger, J., Vetterli Ch., Übernickel, F., Uflacker, M.: DT@Scrum: integrating design thinking with software development processes. In: Design Thinking. Understanding Innovation, pp. 263–289 (2014)
4. Sosnin, P.: Experience-Based Human-Computer Interactions: Emerging Research and Opportunities. IGI-Global, Hershey (2017)
5. Young, R.M.: Mental Space. Process Press (1994)
6. Heape, C.: The design space: the design process as the construction, exploration and expansion of a conceptual space. Ph.D. diss. The University of Southern Denmark, Sønderborg, Denmark (2007)
7. Dorst, K.: The nature of design thinking. In: DTRS8 Interpreting Design Thinking, Proceeding of Design Thinking Research Symposium, pp. 131–139 (2010)
8. Charmaz, K.: Constructing Grounded Theory, 2nd edn. Sage, London (2014)
9. Gärdenfors, P.: Semantics based on conceptual spaces. In: Banerjee, M., Seth, A. (eds.) Logic and Its Applications. LNCS (LNAI), vol. 6521, pp. 1–11. Springer, Heidelberg (2011)
10. Dove, G., Hansen, N.B., Halskov, K.: An argument for design space reflection. In: Proceedings of the 9th Nordic Conference on Human-Computer Interaction (2016)
11. Introduction to Design Thinking (2017). https://experience.sap.com/skillup/introduction-to-design-thinking/

12. Krysin, L.P.: Slovo v sovremennykh tekstakh i slovaryakh. Dostupno po adresu (2017). https://www.e-reading.club/chapter.php/137738/39/Krysin_-_Slovo_v_sovremennyh_tekstah_i_slovaryah.html
13. Rakhilina, Ye.V.: Kognitivnyy analiz predmetnykh imen: semantika i sochetayemost'. Russkiye slovari, 416 (2008). http://rakhilina.ru/files/01_Rah_single.pdf
14. Girju, R., Badulescu, A., Moldovan, D.: Automatic Discovery of Part-Whole Relations. http://www.hlt.utdallas.edu/~adriana/Publications/PartWhole_CL2006.pdf

Multi-level Ontological Model of Big Data Processing

Victoria V. Bova, Vladimir V. Kureichik(✉), Sergey N. Scheglov, and Liliya V. Kureichik

Southern Federal University, Rostov-on-Don, Russia
vvbova@yandex.ru, vkur@sfedu.ru, srg_sch@gmail.com

Abstract. The paper presents a possible solution to the problems of structuring data of a large volume, as well as their integrated storage in structures that ensure the integrity, consistency of their presentation, high speed and flexibility of processing unstructured information. To solve mentioned problems, the authors propose a method for developing a multi-level ontological structure that provides a solution to interrelated problems of identifying, structuring and processing big data sets that has primarily natural-linguistic forms of representation. This multi-level model is developed based on methods of semantic analysis and relative modeling. The model is suitable for the interpretation and effective integrated processing of unstructured data obtained from distributed sources of information. The multilevel representation of the big data determines the methods and mechanisms of the unified meta-description of the data elements at the logical level, the search for patterns and classification of the characteristic space at the semantic level, and the linguistic level of the procedures for identifying, consolidating and enriching data. The modification of this method consists in applying a scalable and computationally effective genetic algorithm for searching and generating weight coefficients that correspond to different similarity measures for the set of observed features used in the data-clustering model.

Keywords: Semantic similarity · Ontology · Unstructured information
Big data · Semantic analysis · Semantic meta-model · Genetic algorithms

1 Introduction

Nowadays, the research directions in the field of big data processing with complex structure and digital technologies has determined the transition to semantic technologies, which required the mandatory consideration of the design of formal ontologies of subject areas of modern multi-purpose information systems (MIS) represented in the form of semantic networks [1–3]. The relevance of the study is caused by the need to develop methods for analyzing complex distributed information for the development of models for structuring large-scale data based on the integration of ontological modeling methods and semantic analysis. The paper presents a possible solution to the problem of big-data structuring, as well as their integrated storage in structures that ensure the

A. Abraham et al. (Eds.): IITI 2018, AISC 874, pp. 171–181, 2019.
https://doi.org/10.1007/978-3-030-01818-4_17

integrity, consistency of their presentation, high speed, and flexibility of processing unstructured information.

The method of forming a multilevel ontological structure with the interpretation of metadata is proposed to solve the above problems. The method can be used to identify and consolidate data items from different subject areas (SA). The model and clustering algorithm developed based on this model will enable us to solve the problem of ensuring the structural and semantic interoperability of big data presented in various subject areas of MIS. Modification of this method consists in applying a scalable and computationally effective genetic algorithm (GA) for searching and generating weight coefficients that correspond to different similarity measures of the set of observed features used in the formation of the data-clustering model.

2 Problem Statement

Currently, the problem of big data processing is one of the most urgent problems in the field of information technology; moreover, it generates the most serious problems of an algorithmic nature associated with ensuring the accuracy and computational efficiency of the processing processes [4]. Analysis of the state of research and development on formulated problems shows that existing methods and algorithms for big data processing do not meet the expectations and needs of specialists in this field and determine the relevance of the work [2–5]. The paper suggests an approach to the integrated processing of unstructured data of large volume, based on methods of evolutionary modeling, semantic and cluster analysis of unstructured data. For the task of structuring data within the framework of one conceptual conceptualization system, an extended knowledge representation model is proposed-a formal ontology focused on processing natural language descriptions of data elements necessary for creating and filling an ontological knowledge base [2].

The BigData clustering occurs in many application areas related to the development of tools for organizing and structuring hypertext space (Semantic Web concept, Web site development and administration, web analytics, etc.). To solve it, many algorithms have been developed [2, 5–7], and in recent years, within the framework of the Big Data concept, special attention has been paid to the processing of information stored in either very large databases (VLDB) or for on-line processing in the form of a data stream [1]. To solve these problems, the mathematical apparatus of computational intelligence [4–6] and soft computing can be successfully used.

Modern approaches in the field of semantic analysis and processing of unstructured data are researched and analyzed. For example, the most widely used methods of clustering are algorithms based on k-means, principal component analysis (PCA), EM-algorithm [5–7]. Each method has its advantages, but it also has a number of limitations: the initial data must have a random nature and obey the normal distribution law; possibly "getting stuck" in the process of optimization in local extremes; computational complexity; the array of data to be clustered is predefined and does not change during the processing. These methods of computing intelligence should be significantly modified to process large amounts of information.

3 Method of Developing a Multilevel Ontological Model

Generally, we can state the research problem as follows. For a given five <*O*, *F*, *D*, *Term*, *C*>, where *O* is the SA ontology that is expanded by a linguistic level, *F* is a set of metadata models (schemes of facts), *D* is a text fragment of a meta description (hereinafter referred to as a document), *Term* is the terminological coverage of *D*, *C* is the segment coverage (clusters) of *D*. We need to find all semantic structures corresponding to ontology *O* that are cover *D*, and which can be obtained by applying rules from the *F* to *Term* taking *C* into account.

Choosing the subject area (ontology) is the first step in bringing unstructured data to a structured form [6]. The entire process of development an ontology is divided into several independent stages in the solution of a particular problem, the results of which serve as initial data for the task of the next more complex level. The algorithm contains the following sequence of actions: extraction from the documents in the natural language of candidate terms → splitting terms into groups (clustering) → assigning a general concept-concept to each group → defining relations between concepts → forming output rules (extending the concepts of the meta-description dictionary). In Fig. 1 presents, a logical scheme for constructing a formal ontology based on the semantic analysis of concepts and relationships in a variety of natural language descriptions of data.

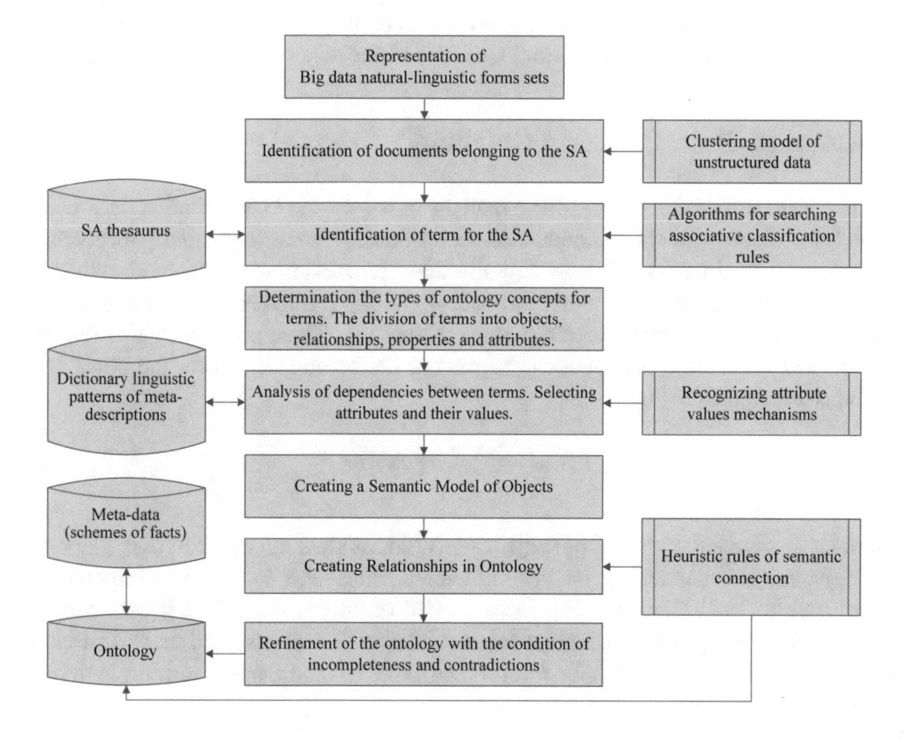

Fig. 1. Generalized scheme for development ontological model

It is possible to use the following tools for development a model of ontology as an extensible thesaurus: algorithms for the formation and replenishment of linguistic templates (dictionaries), mechanisms for recognizing attribute values, heuristic rules that allow us to identify the relationship between concepts in ontology and the relationship between objects of the schema of facts [7].

Thus, the abstract ontology O can be presented as follows

$$O = \langle W\{X, E_x, N_o\}, V\{I, S, D_s\}, R, A\rangle,$$

where W is a SA thesaurus, which has X is a set of ontology concepts, i.e. generalizing a class of concepts (terms) having the same properties and relations; E_x – *is* multiple instances of concepts, and N_o is a set of concept names for which a mapping as E_x: $X \rightarrow 2^{E_x}$; V is a dictionary of linguistic patterns of meta-descriptions, which includes I is a set of information inputs (language expressions, the values of which are presented in W); $S_{ij} = \{x_i, i_j\}$ is a set of relations of semantic connection between I and X; D_s is mapping of a set of schemes of facts of specified documents to information inputs and concepts of SA thesaurus; D_s: $(F, D) \rightarrow (I, X)$; R is a set of relations $R = \{R_1, R_2, \ldots, R_n\}$ between notions W and V, which are defined by heuristic search algorithms for connectivity analysis; A is axioms based on the properties of transitivity and inheritance. The ontological model includes meta-information about the data, their syntax and semantics that are represented by semantic dictionaries.

Their structure is formed from the hierarchy of concepts of the data ontology and the hierarchy of the corresponding formal concepts of the schema of facts (meta-data) [5].

4 Modified Data Clustering Method

Let us consider a modified clustering method based on the GA to find weight coefficients for evaluation of the semantic closeness of data that are sequentially received for processing in on-line. At the stage of preliminary processing and subject classification we consider the document as a "set of concepts" using numerical characteristics of the use of certain terms, regardless of the order of their use. Then, probability that the term $w \in W$, which occurs in the meta-description d, i.e. belongs to one or the other SA t, can be calculated as follows:

$$P(w|d) = \sum_{t \in T} P(w|t)(t|d) \tag{1}$$

where t is an element of a set T of subject areas. To estimate the maximum likelihood of model parameters depending on the hidden variables, the authors use Expectation-Maximization (EM) algorithm [6]. In it, instead of cluster centers, it is assumed that there is a probability density function for each cluster C. It is assumed that all variables are independent and all data have k joint distributions. The main algorithm is divided into two steps.

On the E-step we can calculate $P(t|w, d)^{(r)}$ – estimation of joint distribution parameters:

$$P(t|w,d)^r = \frac{P(w|t)^{(r-1)}P(t|d)^{(r-1)}}{\sum_{t' \in T} P(w|t')^{(r-1)}P(t'|d)^{(r-1)}} \tag{2}$$

On the M-step we can estimate the following pre-semantic analysis parameters $P(w|t)$ and $P(t|d)$:

$$P(w|t)^r = \frac{\sum_{d \in D} N(w,d)P(t|w,d)^r}{\sum_{w' \in W} \sum_{d \in D} N(w',d)P(t|w',d)^r} \quad \text{and} \quad P(t|d)^r = \frac{\sum_{w \in W} N(w,d)P(t|w,d)^r}{\sum_{t' \in T} \sum_{w \in W} N(w,d)P(t'|w,d)^r} \tag{3}$$

where $N(w, d)$ is quantity of occurrences of the thesaurus element w in the document d, r is quantity of iterations. The learning process is repeated until the parameters converge. However, the parameters often fall within the region of the local optimum. Efficiency does not improve as a result of training. We introduce the parameter $0 < \beta \leq 1$ for controlling the learning speed.

Thus, for the M-step we have:

$$P(t|w,d)^r = \frac{(P(W|t)^{(r-1)}P(t|d)^{(r-1)})^\beta}{\sum_{t' \in T} (P(w|t')^{(r-1)}P(t'|d)^{(r-1)})^\beta} \tag{4}$$

Define the total probabilities $W(w, t)$ and $D(d, t)$ as follows:

$$W(w,t)^r = \sum_{d \in D} N(w,d)P(t|w,d)^r \quad \text{and} \quad D(d,t)^r = \sum_{w \in W} N(w,d)P(t|w,d)^r \tag{5}$$

In (4) we get:

$$W(w|t)^r = \sum_{d \in D} \frac{N(w,d)(P(W|t)^{(r-1)}P(t|d)^{(r-1)})^\beta}{\sum_{t' \in T} (P(w|t')^{(r-1)}P(t'|d)^{(r-1)})^\beta} \tag{6}$$

$$D(d|t)^r = \sum_{w \in W} \frac{N(w,d)(P(W|t)^{(r-1)}P(t|d)^{(r-1)})^\beta}{\sum_{t' \in T} (P(w|t')^{(r-1)}P(t'|d)^{(r-1)})^\beta} \tag{7}$$

To form relations between the semantic network of the thesaurus W and the calculation of the semantic proximity $sim\ (a, b)$ between *Terms* a and b we introduce ϖ – the weight function that is defined over the set of semantic relations $R(a; b)$ and shows the degree of the semantic connection between a and b. Let us denote the frequency of occurrence in *Term* the pair $(a; b)$ as $f(r; a; b)$.

$$\varpi(R(a,b)) = \sum_{r_i \in R(a,b)} w_i \times f(r_i, a, b) \tag{8}$$

where w_i is weight associated with r_i. This approach has several disadvantages: the increase in the number of parameters with increasing complexity of the model, together with the assumption of the mutual independence of the parameters of the model. The algorithm for clustering lexical patterns (Fig. 2) is developed to overcome these obstacles and determine semantically related terms. The algorithm uses the following notation: P is the pairs frequency vector (a, b) belonging to the sets D and $W; f(a_i; b_i; p)$ accordingly in *Term* p; Θ is a similarity limit (specified by the expert). The calculation of similarity between p_i and centroid cluster c_j is carried by the cosine coefficient.

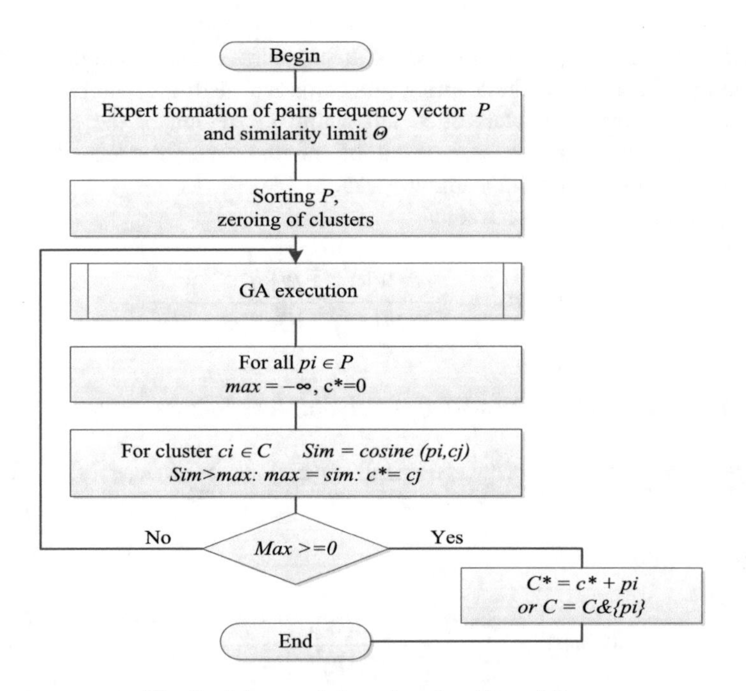

Fig. 2. Scheme of clustering algorithm of *Term*

To increase the efficiency of the calculation of the measure of semantic proximity and the definition of semantically related terms, it is proposed to use the GA (Fig. 3), which allows efficiently finding quasi-optimal solutions in polynomial time [8, 9]. To determine the optimum values of the coefficients, the objective function:

$$F = max\,(\varpi(R\,(a,\,b)) \tag{9}$$

In the first step of the GA work, the initial parameters of the proximity model elements (population size and chromosome length) are generated, and the weighting coefficients w_i and probabilities for the crossover operators and mutation. Next, the initial population of chromosomes is formed based on the available training data from the set $C = \{c_i\}$ for which we can see semantic relations r_i, where each element of the chromosome (gene) is a vector describing the terms a and b in the logical representation of the cluster c_i. To assess the fitness of each chromosome, the value objective function (4) is calculated. After, the elite selection is carried out [8]. The generation of a new set of individuals for each pair of selected parental chromosomes is performed using crossover operators and a mutation with a predetermined probability [10]. After an estimate by formula (9) the quality of each chromosome in the population and the choice of the best of them, a decision is taken to continue the evolutionary procedure for generating the next generation or to complete the training procedure. It is checked whether the stop criterion is reached. If this criterion is not met, the process is repeated iteratively. It is performed based on a recombination block that analyzes the current population of alternative solutions and manages the search process [1, 8]. Otherwise, the resulting quasi-optimal solutions. The time complexity of the developed algorithm is approximately $O(n^2)$.

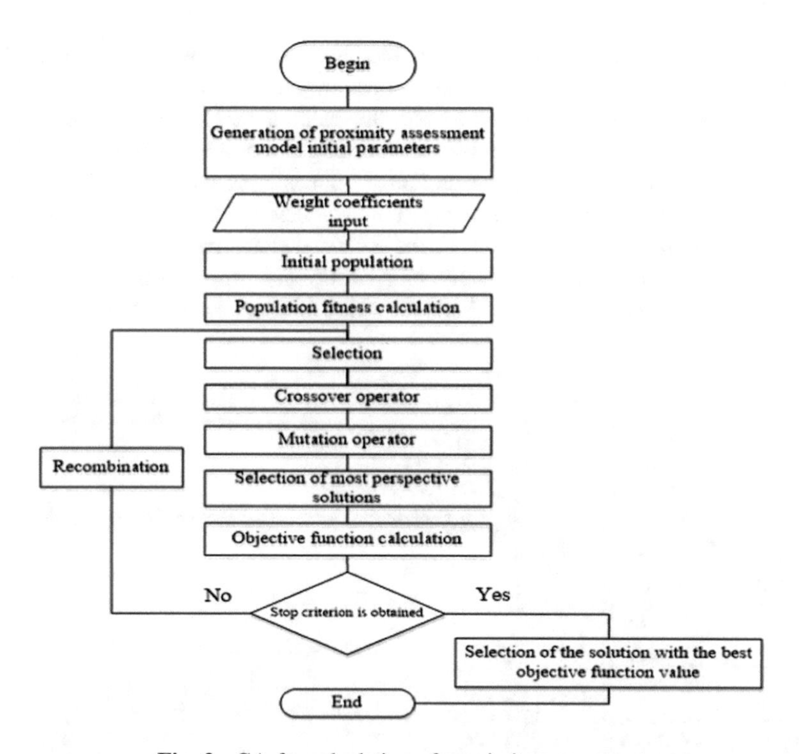

Fig. 3. GA for calculation of proximity assessment

For a comparative analysis of algorithms for semantic proximity estimating, the Miller-Charles dataset was used, which contains 30 sets of word pairs, estimated by the expert group from 0 (no likeness) to 4 (identity). The results of calculating the degree of correlation between expert estimates and various algorithms are given in Table 1. So, we can conclude that the developed method and algorithm are adequate and effective.

Table 1. Methods' correlation coefficients on Miller-Charles dataset

Method	Correlation on Miller-Charles dataset
Jaccard index	0.260
Dice's coefficient	0.267
Overlap coefficient	0.382
Method PMI (point-wise mutual information)	0.549
Normalized distance Google	0.205
Method SH (Sahami)	0.580
Method Chen (Chen, Lin, Wei 2006)	0.834
Developed algorithm	0.867

5 Experimental Results

Computational experiments were carried out to check the performance of the developed GA. As test data, the authors had used the collection (http://people.csail.mit.edu/jrennie/20Newsgroups), which is designed to test the method of clustering. The collection 20Newsgroups contains around 20 000 text messages that are divided among 20 thematic groups of new. Each group has around 1000 messages among 400 documents for testing and 600 documents for classifier training. Using this collection, experiments were performed to verify the accuracy of the proposed clustering method and the size of the intercluster correlation matrix of the joint occurrence. The dependence of the classifier accuracy on the algorithm used and the value of the distance z is shown in Fig. 4.

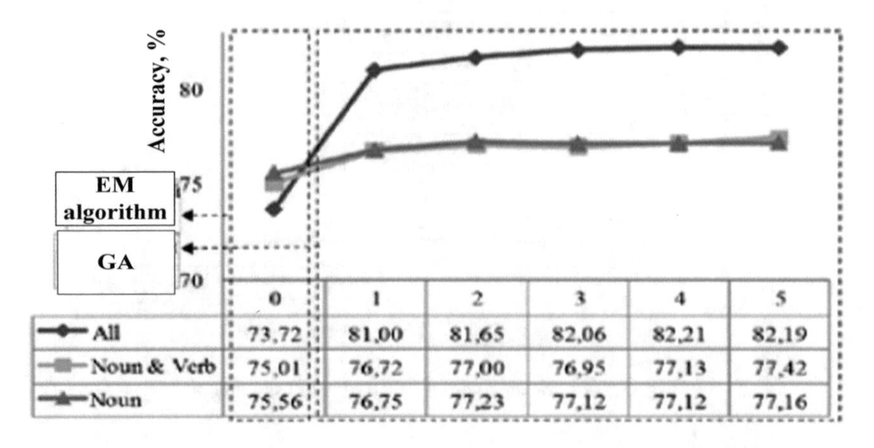

Fig. 4. Chart for estimating clustering accuracy by the EM algorithm and GA

The cosine similarity measure is used as an estimate of the proximity of the vectors. The horizontal axis represents the values of the coefficient z, and the vertical axis - the accuracy values of classification by two algorithms (EM and GA). It is believed that the term t_i is met with the term t_j, if the distance between them in the document does not exceed the limit z: $|i–j| \leq z$.

Three series of data correspond to the accuracy values obtained using the methods of clustering different terms: *All* – all terms are used; *Noun & Verb* – only nouns and verbs are used; *Noun* – only nouns are used. The dependence of the size of the intercluster correlation matrix of the joint occurrence of terms and the value of the distance z is presented in the graphs in Fig. 5.

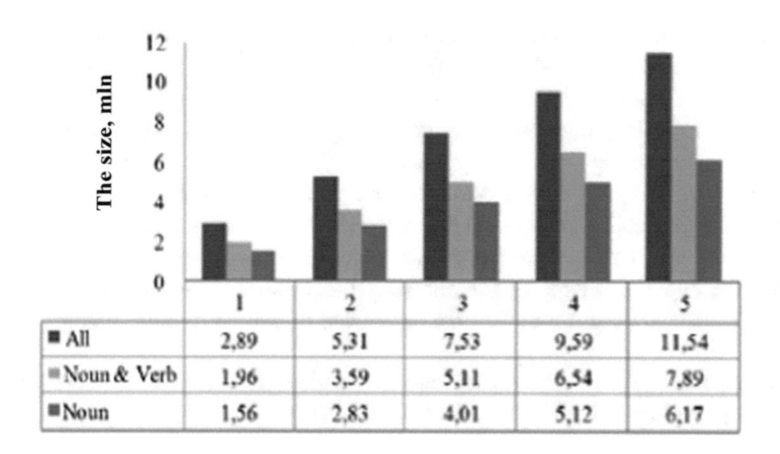

	1	2	3	4	5
All	2,89	5,31	7,53	9,59	11,54
Noun & Verb	1,96	3,59	5,11	6,54	7,89
Noun	1,56	2,83	4,01	5,12	6,17

Fig. 5. Chart of estimation matrix of joint occurrence

Based on the obtained data, it is evident that in the case of $z = 0$ (classical EM algorithm) the maximum accuracy (75.56%) is obtained when only nouns are saved in the collection of documents. However, in other cases when $z > 0$ (the estimation of the semantic proximity of the GA is used), the highest accuracy ($\approx$82%) is obtained using all terms.

The data obtained demonstrate that as a result of clustering on the basis of GA, estimates of the semantic proximity of terms significantly reduce the size of the co-occurrence matrix in comparison with the use of all terms: approximately 30% in the case of using only nouns and verbs and approximately 50% of nouns.

In this paper, the proposed method is compared with the known PCA and k-means clustering methods by the criterion of accuracy, search completeness and f-criterion (Table 2).

According to the results presented in Table 2, the efficiency of the proposed method is 0.8143 for the accuracy criterion (high). By other criteria, the proposed method also allowed to obtain higher results, which indicates its effectiveness.

Table 2. Results of evaluation and comparison the developed algorithm with clustering methods

Method	Accuracy	Completeness of search	f-criterion
Proposed method	0.8143	1.00	0.8976
K-means	0.7142	0.86	0.7803
PCA	0.3333	1.00	0.0073
EM	0.7854	0.76	0.7890

6 Conclusion

The paper proposes a method for the formation of a multilevel ontological structure with interpretation of metadata, for the tasks of identifying and consolidating data elements from various SA. Tools for searching, visualization, analysis, filtering, modeling, forecasting, highlighting patterns, highlighting emotional coloring, categorization, and extracting facts are applicable to the framed structures. The developed model and clustering algorithm can solve the problem of ensuring the structural and semantic interoperability of big data presented in different subject domains of MIS. The data obtained in a series of computational experiments have confirmed the theoretical significance and the prospects for applying the clustering method with the GA to estimate the semantic proximity of data elements presented in the ontology.

Acknowledgment. The study was performed by the grants from the RFBR (project № 18-07-00055 and project № 17-07-00446) in the Southern Federal University.

References

1. Bova, V.V., Kureichik, V.V., Leshchanov, D.V.: The model of semantic similarity estimation for the problems of big data search and structuring. In: 11th IEEE International Conference on Application of Information and Communication Technologies, AICT 2017, pp. 27–32 (2017)
2. Karpenko, A.P., Trudonoshin, V.A.: Multi-criteria estimation of the relevancy of documents in the enterprise ontological knowledge base using thematic clusterization. J. Sci. Educ. Bauman MSTU, 311–328 (2013)
3. Bova, V.V., Leshchanov, D.V.: The semantic search of knowledge in the environment of operation interdisciplinary information systems based on ontological approach. In: Proceedings of SFU. Technical Sciences, vol. 192, pp. 79–90. Publishing house TTI SFU, Taganrog (2017)
4. Rodzin, S., Rodzina, L.: Theory of bioinspired search for optimal solutions and its application for the processing of problem-oriented knowledge. In: 8th IEEE International Conference Application of Information and Communication Technologies, AICT 2014, pp. 142–147. IEEE Press, Astana (2014)
5. Aggarwal, C.C.: Data Clustering. In: Algorithms and Application, 648 p. CRC Press, Boca Raton (2014)

6. Cherezov, D.V., Tukachev, N.A.: Overview of the main methods of classification and clustering of data. In: Proceedings of VGU. System Analysis and Information Technologies, vol. 2, pp. 25–29 (2009)
7. Kravchenko, Y.A., Kuliev, E.V., Kursitys, I.O.: Information's semantic search, classification, structuring and integration objectives in the knowledge management context problems. In: 10th IEEE International Conference on Application of Information and Communication Technologies, AICT 2016, pp. 136–141 (2016)
8. Zaporozhets, D.Y., Zaruba, D.V., Kureichik, V.V.: Hybrid bionic algorithms for solving problems of parametric optimization. World Appl. Sci. J. **23**, 1032–1036 (2013)
9. Bova, V.V., Nuzhnov, E.V., Kureichik V.V.: The combined method of semantic similarity estimation of problem oriented knowledge on the basis of evolutionary procedures. In: Proceedings of the 6th Computer Science On-line Conference 2017 (CSOC 2017), Warsaw, Poland, vol. 1, pp. 74–83 (2017)
10. Bova, V.V., Kureichik, V.V., Leshchanov, D.V.: Model of semantic search in knowledge management systems based on genetic procedures. J. Inf. Technol. **12**, 876–883 (2017)

Ontological Modeling for Industrial Enterprise Engineering

Alena V. Fedotova[1(✉)], Vadim V. Tabakov[2], Michael V. Ovsyannikov[1], and Jens Bruening[3]

[1] Bauman Moscow State Technical University, 105005 2-ya Baumanskaya, 5, Moscow, Russia
afedotova.bmstu@gmail.com, mvo50@mail.ru

[2] Institute of World Economy and International Relations of the Russian Academy of Sciences, Moscow, Russia
vvtabakov@gmail.com, jensbruening@gmx.de

[3] Bremen University, 28359 Bremen, Germany

Abstract. The problems of creating enterprise ontologies for Enterprise Engineering are addressed in this paper. The main purpose of the research is development of a system of industrial enterprise ontologies. First of all, some definitions and viewpoints on Enterprise Engineering are discussed. A brief overview of the most well-known foreign projects of enterprise ontological engineering is given. The bottom-up and top-down ontology design approach are considered. A set of ontologies for a typical industrial enterprise has been built, which ensures the coordination of various enterprises activities. The main classes and subclasses of the enterprise ontology are given.

Keywords: Enterprise engineering · Ontological modeling

1 Introduction

A new stage in the development of computer-integrated production in industrial corporation, which is a heterogeneous network of various enterprises, is the strategy and technology of Enterprise Engineering (EE), which involves the extensive use of scientific methods and computer modeling tools for research and design of enterprises.

According to the standard ISO 14258, the EE is defined as an extensive scientific and technical area related to the implementation of theoretical and practical developments in the design, modification and reorganization of enterprises of any industry. It covers diverse knowledge, principles and practical recommendations related to the analysis, design, development and operation of enterprises [1].

In [2], the main aspects of the EE are identified: computer modeling and computer integration of enterprises. The modeling of the enterprise is aimed at:

1. Systematization of the enterprise knowledge, the analysis of the general state of the enterprise;
2. Support for the design and engineering of the enterprise;
3. Improving the efficiency of enterprise management;

A. Abraham et al. (Eds.): IITI 2018, AISC 874, pp. 182–189, 2019.
https://doi.org/10.1007/978-3-030-01818-4_18

4. Implementation of variants for intra-organizational and inter-organizational integration.

In the work of Telnov [3] proposed a three-tiered EE methodology, which includes strategic engineering, enterprise architecture development and ontological engineering.

Nowadays, an extremely broad multi-disciplinary area of Enterprise Engineering has been developed based on systems engineering, organization theory strategic management, advanced information and communication technologies. The objective of Enterprise Engineering is the design and creation of modern networked enterprise as an open sophisticated holistic system by modeling and integrating its products, processes, resources, organization structures, business operations, etc. In other, words, EE considers the formation of enterprise as a sort of engineering activities. Moreover, it tends to examine each aspect of the enterprise, including various resources, business processes, information flows, organizational structures.

In this article, the main attention is paid to the development of ontological engineering of enterprises in the interests of coordinating the work of enterprises-subcontractors and ensuring the effective operation of the entire corporation, ontological EE, constructing a system of ontologies of industrial enterprise.

2 Industrial Enterprise Engineering: An Overview of Ontological Enterprise Engineering Projects

Let us discuss some viewpoints on the essence and basic disciplines for EE.

Enterprise Engineering is defined in [4] as a body of knowledge, principles, and practices having to do with the analysis, design, development, implementation and operation of an enterprise. It means the shift from Data Systems Engineering and Information Systems Engineering to Enterprise Ontological Engineering [5]. In [6] three main goals of EE are mentioned: intelligent manageability, organizational concinnity, social devotion.

In his turn, Martin [2] focuses on seven disciplines of EE grouped around value framework: (1) strategic visioning viewed as ongoing cycle of value positioning; (2) enterprise redesign – discontinuous change in the value definition; (3) value stream reinvention – discontinuous change in the value offering; (4) procedure redesign – discontinuous reinvention of value creation; (5) total quality management – continuing change in value creation; (6) organizational and cultural development – continuous value innovation; (7) information technology progress (continuous value enablement).

According to Vernadat [7] EE is the art of understanding, defining, specifying, analyzing and implementing business processes for the enterprise entire life cycle, so that the enterprise can achieve its objectives, be cost-effective, and be more competitive in its market environment. Here two basic disciplines for EE are enterprise modeling and enterprise integration.

In [8] was noted that the two largest projects of EE are the projects of TOVE [9] and Enterprise Project [10], started back in the 1990s.

In the TOVE project (TOrontoVirtual Enterprise), dedicated to the formalization of knowledge in the field of enterprise engineering, a general methodology for ontological engineering was developed, now known as the Gruninger-Fox methodology [11]. The system of enterprise ontologies was also built in which top-level ontologies are subdivided into basic ones (ontology of activity, organization ontology, ontology of resources, product ontology) and special ones (cost ontology, ontology of quality, ontology of incentives, etc.). It should be noted that the ontologies developed within the TOVE project are heavyweight, i.e. they do not boil down to visual representations of concepts and connections between them, but realize the construction of ontology as a logical theory that provides an inference on ontology and automatic execution of requests to ontology.

The basic entities in the TOVE model are represented by objects with their properties and relations. Objects are combined in a taxonomy. Objects, attributes and relations are specified, where possible, in the predicate of the first order. It defines a set of axioms that fix constraints on objects and predicates in ontology. The constructed set of axioms forms a microtheory (in the sense of D. Lenat). In particular, the ontology of activity is based on Reiter's situational calculus [12], in which there are three types of objects: actions, situations and fluents (formulas that become true or false at certain time points). The ontology of time is based on the temporal relations of Allen's logic.

Both informal (textual) and formal ontology representation using the Ontolingua system were also used in the Edinburgh Enterprise Ontology Project [10]. It developed: the ontology "Activity - Plan - Capacity - Resources" (APCR), the ontology of the organization; ontology of strategies; marketing ontology; ontology of time. The activity is performed during a certain time interval and has such attributes as: (a)agent; (b) preconditions; (c) results; (d) decomposition into actions; (e) resources; (e) authority.

The key concept of enterprise ontology is an Organizational Unit, and composite concepts, such as an Organizational Structure, are formed by establishing various relationships between organizational units [13, 26]. The basic enterprise ontology includes the following organizational classes: corporations; shareholders; suppliers; partnership; individuals (as executors of roles in the enterprise). Next section describes a preliminary version of the ontology for the industrial enterprise.

3 Bottom-Up and Top-Down Design of Ontologies

In the theory of systems [14], a distinction is made between homogeneous elements and inhomogeneous units of the system. The element is the simplest indivisible part of the system, while the unit is the minimal part of the system that preserves its properties as a whole [15]. Accordingly, among the elements of ontologies, it is possible to specify classes, instances, attributes, relationships, axioms, roles, constraints, etc. [16, 17, 25]. At the same time, ontological units are the relationship "entity-relationship", "concepts-roles" (in descriptive logic), "Subject-predicate-object" (in RDF), "object-attribute-value", "object-attribute-value-confidence", etc.

Most often, the construction of a single, consistent and consistent ontology of the subject area is impossible. To simplify the processes of development, integration and reuse of ontologies, a modular approach is applied and an ontology hierarchy is formed. In this case, the bottom-up and top-down design of ontologies are distinguished [18, 19]. The bottom-up approach (Fig. 1) of constructing ontologies involves the preliminary determination of the most specific classes with their subsequent integration into more general classes and categories [14, 20]. Its practical implementation in semiautomatic mode is connected with the choice of some initial "pre-ontological" material (thesauri, conceptual data models, Excel tables, frame systems, etc.) and translating it into the appropriate description language, for example OWL. It is a question of constructing ontological structures on the basis of elementary concepts and relations between them. The bottom-up approach to the development of the ontology system, outlined in the pioneering work of Guarino [21], involves the creation of an ontology of the subject domain at the beginning, with a subsequent transition to ontologies of tasks and applications, and also to ontologies of the upper level.

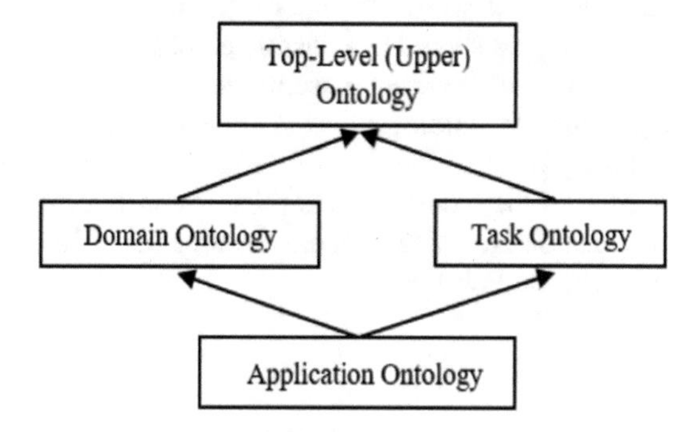

Fig. 1. Bottom-up hierarchical ontology system.

On the contrary, the top-down approach (Fig. 2) to the development of ontologies begins with the definition of the most general concepts that are invariant with respect to the subject domain, and then their sequential detailing is carried out. Here, in the first place, defines a metaontology that defines the properties of both the ontological system as a whole and its individual ontologies in particular. Namely, metaontology determines the choice of ontologies of the upper and lower levels.

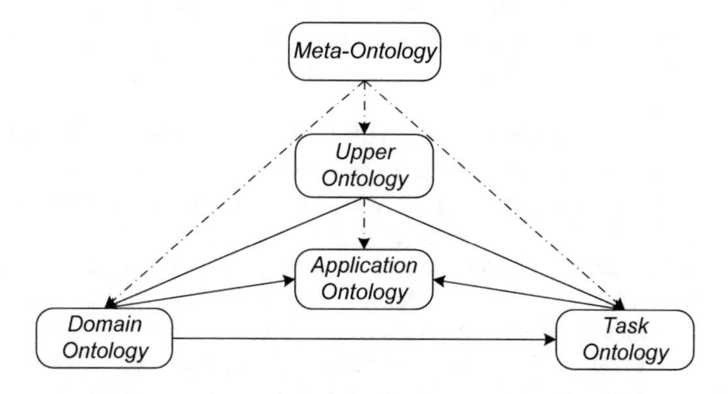

Fig. 2. Top-down ontological hierarchy representation.

In this paper, we use the bottom-up approach of designing the ontology of industrial enterprise.

4 Designing Industrial Enterprise Ontology

An ontology of a typical industrial (machine-building) enterprise is presented in the section (Fig. 3).

In this paper we use the bottom-up approach of construction of enterprise ontology according with Fig. 1.

Domain and tasks ontological modeling is necessary to increase the efficiency of joint work of various enterprises within the interoperability.

First of all simple ontologies should be constructed [22]. Then should be determine Variants of axioms, used Language of descriptive logic and creating system of ontologies in Protége (Simple ontologies → Variants of axioms → Language of descriptive logic → Protégé).

The general ontology of the industrial enterprise is constructed in the format.owl. Its more detailed visualization is available on the resource [23]. The owl file and related files can be obtained from the link [24].

4.1 Ontology Development

Industrial ontologies for management of corporate information about the enterprise and enterprise structures necessary to product lifecycle management. Here (Fig. 3), the ontology includes a universal reusable core that can be extended or specialized for use in a particular state. The basic concepts are: enterprise structure, product, lifecycle (LC), personnel (distributed human capital), process (distributed business processes). Each of the classes is divided into subclasses.

4.2 Main Classes and Subclasses of the Enterprise Ontology

The class “Lifecycle” contains the following subclasses: Product; Customer; Volume and Cost of Production; Technical Task; Technical Proposal; Conceptual Design; Preliminary Design; Technical Project; Working Project; Technological Equipment; Organizational and Technical Documentation; Prototype; Serial Copy; Use; Maintenance; Major Repairs; Recycle.

The class “Enterprise structure” is divided into the following subclasses: Archive, Accounting, Customer’s Directorate, Experienced Design Bureau, Department Of Chief Technologist, Contract Drafting Department, Technical Control Department, Economics and Labor Department, Legal Department, Trade Union Committee, Production and Technical Administration, Secretariat, Quality Service Department, Financial Management Department, Occupational Safety and Health Administration, Information Technology Center, Maintenance and Repair Department, Workshop.

The class “Personnel” includes subclasses: General Director, Deputy General Director, Chief Designer, Design Engineer, Process Engineer, Technician-Designer, Technician-Technologist, Department Head, Workshop Manager, Deputy Department

Fig. 3. A system of the enterprise ontologies.

Head, Deputy Workshop Manager, Brigade Chief, Controller, Programmer, System Administrator, Maintenance Personnel, Economist, Worker.

The class "Processes" is subdivided into subclasses: Details Manufacturing, Designing a Test Program, Designing a Technical Process, Software Setup, Training Employees, Tests Execution, Checking Design Documentation, Calculating the Strength, Calculating Material, Labor and Financial Costs, Repair Work, Assembly of Products, Creation of Notices, Coordination of Design Documentation, Support of Design Documentation, Drafting of a Report on the Work of the Team, Department, Drawing Up a Work Plan, Drafting the Technical Tasks, Design Documentation Account.

4.3 Examples of Rules Based on Simple Ontologies

In the form of simple ontologies the above classes and relations between them are written as follows: Details Manufacturing $\subseteq$ Processes, Designing a Test Program $\subseteq$ Processes, Designing a Technical Process $\subseteq$ Processes, Software setup $\subseteq$ Processes, Archive $\subseteq$ Enterprise structure, Accounting $\subseteq$ Enterprise structure, Enterprise structure $\subseteq$ Enterprise, LC $\subseteq$ Enterprise, Personnel $\subseteq$ Enterprise, Processes $\subseteq$ Enterprise, Designing a Technical Process $\cap$ Process Engineer $\neq \varnothing$, Designing a Technical Process $\cap$ Technologist $\neq \varnothing$, Details Manufacturing $\cap$ Worker $\varnothing$ etc. Further, based on the written rules constructed ontology: LC [24], Enterprise structure [24], Personnel [24] in the ontology modeling system Protégé.

Then the resulting simple ontologies are combined into a single ontology of the industrial enterprise with the help of rules and restrictions (Fig. 3). Ontology with all the rules and restrictions can be found on the resource [24].

5 Conclusion

A very important way of developing enterprise engineering paradigm consists in implementing enterprise knowledge management on the basis of ontological approach. The concept and main aspects of enterprise engineering are presented. The problem of enterprise ontological modeling is considered in the mainstream of the industrial enterprise engineering. The bottom-up and top-down approach for designing of ontologies are discussed. The paper considers an example of ontological enterprise engineering, which shows the possibility of automated analysis of the general state of the enterprise, as well as increasing the efficiency of life cycle management through communication and business systems in the form of interconnected ontologies systems. This work represents the first step towards the creation of a full-scale intelligent product lifecycle management system.

Acknowledgment. The reported study was funded by RFBR according to the research project № 18-07-01311, 17-07-01374.

References

1. ISO Concepts and Rules for Enterprise Models: International Standard 14258. www.mel.nist.gov/sc5wg1/std-dft.htm. Accessed 09 Mar 2017
2. Martin, J.: The Great Transition: Using the Seven Principles of Enterprise Engineering to Align People, Technology and Strategy. American Management Association, New York (1995)
3. Telnov, Y.F.: Evolution of the paradigm "enterprise engineering". In: Enterprise Engineering and Knowledge Management. Collection of Scientific Works of the XVIth Scientific and Practical Conference EE & KM-2013, Moscow, MESI, 25–26 April 2013, pp. 294–298. MESI, Moscow (2013)
4. Liles, D., Johnson, M.E., Meade, L.M., Ryan, D.: Enterprise engineering: a discipline? In: Society for Enterprise Engineering Conference Proceedings, vol. 6 (1195)
5. Dietz, J.: Enterprise Ontology – Theory and Methodology. Springer, Berlin (2006)
6. Dietz, J., Hoogervorst, J., et al.: The discipline of enterprise engineering. Int. J. Organ. Des. Eng. **3**(1), 86–114 (2013)
7. Vernadat, F.: Enterprise Modeling and Integration: Principles and Applications. Chapman and Hall, London (1996)
8. Guryanova, M.A., Efimenko, I.V., Khoroshevsky, V.F.: Ontological modeling of the enterprises economy and industries of modern Russia: Part 2. World research and development: an analytical review. Preprint WP7/2011/08 (Part 2). State University Higher School of Economics, Moscow (2011)
9. TOVE Ontology Project 2010. http://www.eil.utoronto.ca/enterprise-modelling/tove/
10. Uschold, M., King, M., Morales, S., Zorgios, Y.: The Enterprise ontology. Knowl. Eng. Rev. **13**(1), 31–89 (1998)

11. Fox, M.S., Barbuceanu, M., Gruninger, M.: An organization ontology for enterprise modeling: preliminary concepts for linking structure and behavior. Comput. Ind. **29**, 123–134 (1995)
12. Reiter, R.: The frame problem in the situation calculus: a simple solution (sometimes) and a completeness result for goal regression. In: Lifshitz, V. (ed.) Artificial Intelligence and Mathematical Theory of Computation, pp. 359–380. Academic Press Professional, San Diego Inc. (1991)
13. Tarassov, V.B.: Special session on intelligent agents and virtual organizations in enterprise. In: Binder, Z. (ed.) Proceedings of the 2nd IFAC/IFIP/IEEE Conference on Management and Control of Production and Logistics 2000 MCPL'2000, Grenoble, France, 5–8 July 2000, vol. 2, pp. 475–478. Elsevier Science Publishers, Amsterdam (2001)
14. Jamshidi, M. (ed.): System of Systems Engineering: Innovations for the Twenty-First Century. Wiley, New York (2008)
15. Evgenev, G.B.: Systemology of engineering knowledge. BMSTU, Moscow (2001)
16. Tarassov, V.B., Fedotova, A.V., Karabekov, B.S.: Granular meta-ontology, fuzzy and linguistic ontologies: enabling mutual understanding of cognitive agents. In: Proceedings of the 5th International Conference on Control, Automation and Artificial Intelligence CAAI2015, Phuket, Thailand, 23–24 August 2015, pp. 253–261. DEStech Publications Inc., Lancaster (2015)
17. Kovalev S.M., Tarassov V.B., Koroleva M.N., et al.: Towards intelligent measurement in railcar on-line diagnostics: from measurement ontologies to hybrid information granulation system. In: Abraham, A., Kovalev, S., Tarassov, V., et al. (eds.) Proceedings of the 2nd International Scientific Conference «Intelligent Information Technologies for Industry» IITI 2017, Varna, 14–16 September 2017 Advances in Intelligent Systems and Computing, vol. 679, pp. 169–181. Springer International Publishing, Cham (2018)
18. Tarassov, V.B., Fedotova, A.V., Stark, R., Karabekov, B.S.: Granular meta-ontology and extended allen's logic: some theoretical background and application to intelligent product lifecycle management systems. In: Proceedings of the 4th International Conference on Intelligent Systems and Applications INTELLI'2015, St. Julians, Malta, 11–16 October 2015, pp. 86–93. IARIA XPS Press, Copenhagen (2015)
19. Petrochenkov, A.B., Bochkarev, S.V., Ovsyannikov, M.V., Bukhanov, S.A.: Construction of an ontological model of the life cycle of electrotechnical equipment. Russ. Electr. Eng. **86** (6), 320–325 (2015)
20. Fedotova, A.V., Davydenko, I.T., Pförtner, A.: Design intelligent lifecycle management systems based on applying of semantic technologies. In: Proceedings of the First International Scientific Conference "Intelligent Information Technologies for Industry" IITI 2016, Sochi, Russia, 16–21May 2016, vol. 1, pp. 251–260. Springer International Publishing, Cham (2016)
21. Guarino, N.: Formal ontology and information systems. In: Guarino, N. (ed.) Proceedings of the 1st International Conference on Formal Ontologies in Information Systems, FOIS'98, Trento, Italy, 6–8 June 1998, pp. 3–15, IOS Press, Amsterdam (1998)
22. Valkman, Y.R., Tarasov, V.B.: From the ontology of designing to cognitive semiotics. Ontol. Des. **8**(1), 8–34 (2018)
23. Enterprise Ontology. http://app.ontodia.org/diagram?sharedDigram=d7h87avp22ehbcionnp9p61pcn. Accessed 07 Apr 2018
24. Enterprise Ontology. https://archive.org/details/EnterpriseOntology. Accessed 07 Apr 2018
25. Kuzenov, V.V., Ryzhkov, S.V.: Radiation-hydrodynamic modeling of the contact boundary of the plasma target placed in an external magnetic field. Appl. Phys. **3**, 26–30 (2014)
26. Kuzenov, V.V., Ryzhkov, S.V.: Numerical modeling of laser target compression in an external magnetic field. Math. Model. Comput. Simul. **10**, 255–264 (2018)

Designing the Knowledge Base for the Intelligent Inertial Regulator Based on Quasi-optimal Synthesis of Controls Using the Combined Maximum Principle

Andrey Kostoglotov[1(✉)], Sergey Lazarenko[1,2], Alexander Agapov[1], Zoya Lyaschenko[1], and Irina Pavlova[2]

[1] Rostov State Transport University, Rostov-on-Don, Russian Federation
{kostoglotov, lazarenkosv}@icloud.com,
aaalexander2794@gmail.com, lyashchenko.zoya@mail.ru

[2] Moscow State University of Technology and Management named after K.G. Razumovsky, The First Cossacs University, Moscow, Russian Federation
270300@mail.ru

Abstract. Knowledge engineering and the design of knowledge bases are now the most important sections of artificial intelligence. They require to develop a closed set of rules of logical inference based on effective control laws. The paper proposes a new algorithm for the synthesis the intelligent control systems, which mode of operation is determined by the closest proximity to the control law optimal for the chosen criterion and the physical realizability of the inertial regulator. The proposed approach allows us to determine the elements of the knowledge base based on the developed synthesis procedure in the problems to construct the set of product rules in the class of measurable piecewise-continuous and piecewise-constant controls.

Keywords: Analytical design of the regulator · Combined maximum principle Synthesis · Control

1 Introduction

A systematic approach to solving the problem of improving the control quality in complex systems involves the development of special mathematical apparatus for control, in which the central role belongs to models, approaches and methods for information processing that ensure the intellectual behavior of the control system [1–3]. Methods from sections of mathematics and modeling that allow to "work" with new types of information and, above all, with knowledge become particularly actual. Knowledge engineering and the design of knowledge bases are now the most important sections of artificial intelligence. They require to develop a closed set of rules of logical inference based on effective control laws.

The process of designing knowledge bases of control systems is closely related with the fundamental problem of synthesizing the laws of optimal or quasi-optimal control, which in essence are just "knowledge". At the same time, the synthesis

A. Abraham et al. (Eds.): IITI 2018, AISC 874, pp. 190–200, 2019.
https://doi.org/10.1007/978-3-030-01818-4_19

procedure is a complex mathematical task, and the designers of intelligent control systems must focus their efforts mainly on automating the procedures for constructing rules of logical inference, using as the basic elements known engineering solutions PID, the regulators that implement control laws on a set of linear structures. This is due to the fact that the solution of the problem of optimal control involves the boundary problem, which is very time-consuming procedure, and its complexity significantly exceeds the procedures for constructing production rules. The necessity of achieving the required accuracy of approximation of nonlinear control laws by means of linear constructions makes it necessary to increase the number of terms of linguistic variables. It makes many production rules to be complex, and this process is usually not mathematically formalized, but based on the experience of the researcher. At the same time, it should be noted that the construction of a set of solutions to the problems of control synthesis with respect to a certain criterion makes it possible to significantly reduce its cardinality and increase the efficiency of accumulating information to create a knowledge base. One of the actual problems of synthesizing intellectual systems under conditions of uncertainty with the linguistic variable associated with the terminal state is the problem to build a set of control laws that take into account the inertia of the controlled object and the regulator under restrictions on the domain of permissible controls. An effective approach to constructing such a set is the method of the combined maximum principle [4].

There are various types of mathematically substantiated methods of synthesis of control systems of inertial dynamical systems [5–7]. The first is to determine the reference mathematical model, then to compare it with the model of the invariable part of the system and to search for models of correcting elements that make it possible to approximate the synthesized system to the reference system [1]. Another type of synthesis is the construction of systems that are optimal with respect to a certain criterion. In this case, the application of the maximum principle of Pontryagin, method of dynamic programming of Bellman, and the combined maximum principle method [2–4] allow to synthesize systems that are optimal in terms of performance, or energy consumption, or accuracy [5]. The method of analytical construction of the Letov regulators allows to synthesize a control based on the condition of minimizing the integral of the quadratic form of variables [6, 7].

The paper develops a new algorithm for the synthesis the intelligent control systems, which mode of operation is determined by the closest proximity to the control law optimal for the chosen criterion and the physical realizability of the inertial regulator. It should be noted that the use of empirical laws does not allow to assess the effectiveness of the developed intellectual systems in comparison with the optimal systems. The proposed approach allows us to determine the elements of the knowledge base of the control systems based on the developed synthesis procedure in the problems to construct the set of product rules in the class of measurable piecewise-continuous and piecewise-constant controls.

2 Formulation of the Problem

The law of motion of the system can be written in the form of the Lagrange equations of the second kind [8]:

$$\frac{d}{dt}\frac{\partial T}{\partial \dot{q}_s} - \frac{\partial T}{\partial q_s} = y_s, \quad s = \overline{1,n}, \tag{1}$$

which follow from the Hamilton-Ostrogradskii stationarity principle [8]:

$$\delta' S = \int_0^{t_*} (\delta T + \delta' A) dt = 0. \tag{2}$$

Equation (2) determines the trajectory of motion, where the quantity $S = \int_0^{t_*} (T + A)dt$ is similar to the action integral. Here $q_s, \dot{q}_s$, $s = \overline{1,n}$ are phase variables (generalized coordinates) of the system; n - is the number of degrees of freedom; $T = \frac{1}{2}\sum_{s,k=1}^{n} a_{sk}\dot{q}_s\dot{q}_k$ is kinetic energy of the system; a_{sk}, $s,k = \overline{1,n}$ are the inertia coefficients; $\delta' A = \sum_{s=1}^{n} y_s \delta q_s$ is the elementary work of the generalized (control) forces $y_s = \overline{G}_u \subseteq R^n$, where the set R^n there is a convex polygon. As an admissible control we consider the measurable functions $y(q, \dot{q}) \in G_u$.

If the regulator has an inertia, then the law of variation of the control forces can be also written in the form of the Lagrange equation of the second kind:

$$\frac{d}{dt}\frac{\partial \theta}{\partial \dot{y}_s} - \frac{\partial \theta}{\partial y_s} = u_s, \quad s = \overline{1,n}, \tag{3}$$

which also follow from the Hamilton-Ostrogradskii stationarity principle [8]:

$$\delta' R = \int_0^{t_*} (\delta\theta + \delta' a) dt \tag{4}$$

for quantity $R = \int_0^{t_*} (\theta + a)\, dt$.

Here $y_s, \dot{y}_s$, $s = \overline{1,n}$ are the generalized variables of the control forces; $\theta = \frac{1}{2}\sum_{s,k=1}^{n} \alpha_{sk}\dot{y}_s\dot{y}_k$ is a quantity proportional to the kinetic energy, α_{sk} is the inertia coefficient of the regulator; $\delta' a = \sum_{s=1}^{n} u_s dy_s$ is the elementary work of signals affecting the regulator, $u_s \in \overline{G}_u$. As admissible functions we consider the measurable functions $u(y, \dot{y})$.

Let us define on the trajectories of the system (1) the functional

$$J_1 = \int_0^{t_*} F(q, \dot{q})dt \tag{5}$$

which is a measure of the quality of the controlled motion and also determines the optimal (quasi-optimal) control y^*. Let us define also on the movements of the regulator (3) the functional

$$I_1 = \int_0^{t_*} \frac{1}{2} \sum_{s=1}^{n} \left(y_s - y_s^*\right)^2 dt, \tag{6}$$

which characterizes the proximity of the control action of the inertial regulator and the optimal control.

The problem is to develop a procedure for synthesizing control u, which provides a minimum of the target functionals (5) and (6).

3 The Theorem of the Principle of the Generalized Power Maximum

Let us consider a reference (optimal or quasi-optimal) motion, for which it is required to find controls y_s providing a minimum to the functional (5).

To solve the problem, we need according the rule of the Lagrange multipliers to compose an extended functional [9–11]

$$J = \int_0^{t_*} [\lambda(T + A) + F]dt = \int_0^{t_*} Hdt, \tag{7}$$

where λ is the undefined Lagrange multiplier.

The theorem of the principle of the generalized power maximum [4, 9–11]. For optimal control $y(q, \dot{q}) \in \overline{G}_y$ and its corresponding trajectory $(q, \dot{q}) \in R^{2n}$ it is necessary that the function of generalized power

$$\Phi(q, \dot{q}, \lambda) = \max_{y \in \overline{G}_y} \sum_{s=1}^{n} (\lambda y_s + V_s)\dot{q}_s \tag{8}$$

where

$$V_s = \frac{\partial F}{\partial q_s} \tag{9}$$

reaches its maximum and $\lambda = \text{const} > 0$, and on the trajectory ends $t = 0,\ t = t_*$ we satisfy the transversality conditions for the Hamilton-Ostrogradskii function H

$$H = \lambda(T + A) + F = 0. \tag{10}$$

4 Synthesis of the Laws of Intelligent Control of the Inertial Regulator

We consider the problem to build an intelligent control system with two operation modes: "coarse" and "fine".

Let us define equations characterizing the dynamics of the system

$$\ddot{q} = y, |y| < Y, \tag{11}$$

$$\ddot{y} = u, |u| < U, \tag{12}$$

initial and final states

$$t = 0,\ q(0) = q_0,\ \dot{q}(0) = \dot{q}_0,\ t = t_*,\ q(t_*) = q_*,\ \dot{q}(t_*) = \dot{q}_*, \tag{13}$$

where q_* is fuzzy defined linguistic variable with the introduced terms "coarse" and "fine".

It is required to find the control laws that ensure the transition of the system from the initial state to the final state under the condition of a minimum of the functionals J, I given on the motions of the object (11)

$$J = \int_0^{t_*} \frac{1}{2}(q - q_*)^2 dt \tag{14}$$

and on the trajectory of the reference (optimal or quasi-optimal) control of the inertial controller (12)

$$I = \int_0^{t_*} \frac{1}{2}(y - y_*)^2 dt, \tag{15}$$

which characterizes the proximity of the control action of the inertial regulator y and the reference control y_*.

The function H of the reference motion in the case under consideration has the form

$$H = \lambda\left(\frac{\dot{q}^2}{2} + \int_{q(0)}^{q(t)} y dq\right) + \frac{1}{2}(q - q_*)^2 \leq 0. \tag{16}$$

The inertial control function H_p, respectively is:

$$H_p = \lambda\left(\frac{\dot{y}}{2} + \int\limits_{y(0)}^{y(t)} udy\right) + \frac{1}{2}(y - y_*)^2. \tag{17}$$

The order of solving the problem of synthesis of the inertial control system is determined by the theorem of the principle of maximum of the generalized power.

To do this, let us consider the procedure for synthesizing a reference mathematical model of the motion of the control object.

The condition (8) for the reference motion is:

$$\Phi(q, \dot{q}, u) = \max_{u \in \bar{G}_u}(\lambda y + (q - q*))\dot{q} \leq 0. \tag{18}$$

Let us consider the most typical practical case, when the control is chosen from the class of piecewise-constant functions, then condition (18) allows to obtain the expression

$$y_*(q, \dot{q}) = |y| \text{sign} \dot{q}, \tag{19}$$

where $|y|$ is the admissible control $(y_{\min} \leq y \leq y_{\max})$.

In this case, the transversality condition for the function H must be satisfied. Then, substituting u into Eq. (16), we can obtain after some transformations

$$\begin{aligned} H &= \lambda\left(\frac{\dot{q}^2}{2} + |y|\text{sign}\dot{q}(q - q_*) + \lambda^{-1}\frac{1}{2}(q - q_*)^2\right) \\ &= |y|\text{sign}\dot{q}\left[\frac{\left(\dot{q}^2 + \lambda^{-1}(q - q_*)^2\right)\text{sign}\dot{q}}{2|y|} + (q - q_*)\right] \leq 0. \end{aligned} \tag{20}$$

To hold the inequality, we must put

$$\text{sign}\dot{q} = \text{sign}\left[-\frac{\left(\dot{q}^2 + \lambda^{-1}(q - q_*)^2\right)\text{sign}\dot{q}}{2|y|} - (q - q_*)\right]. \tag{21}$$

The law of the reference control in the form of synthesis takes the form

$$y_* = |y|\text{sign}\left[-\frac{\left[\dot{q}^2 + \lambda^{-1}(q - q_*)^2\right]\text{sign}\dot{q}}{2|y|} - (q - q_*)\right]. \tag{22}$$

The dynamic model of the inertial regulator can be constructed as follows. Let the control be chosen from the class of piecewise continuous functions.

The condition (8) for the inertial regulator has the form

$$\Phi(y,\dot{y}) = \max_{v \in \bar{G}v} \left(\lambda_p u + (y - y_*)\right)\dot{y} \leq 0. \tag{23}$$

For the choice of piecewise-continuous controls it is assumed that

$$\lambda_p u + (y - y_*)\dot{y} = -\mu\dot{y}^2, \quad \mu = \text{const} > 0. \tag{24}$$

Then the control $u(y,\dot{y})$

$$u(y,\dot{y}) = \lambda_p^{-1}(-\mu\dot{y} - (y - y_*)). \tag{25}$$

The problem to choose a physically realizable regulator is reduced to the choice of coefficients λ, μ. An analysis of the behavior of invariants on the switching hypersurface using the operation of the Poisson brackets shows that to satisfy the transversality condition for function H_p it is necessary to establish a relation between the factors λ and μ in the form $\lambda^{-1}\mu = \sqrt{\lambda^{-1}}$.

Finally, the control law has the form

$$u(y,\dot{y}) = -\sqrt{\lambda_p^{-1}}\dot{y} - \lambda_p^{-1}(y - y_*). \tag{26}$$

5 Formation of the Knowledge Base Elements and the Mathematical Modeling

The mathematical model of the controlled motion of a dynamical system with an inertial regulator is determined by the model of the reference motion

$$\ddot{q} = y_*(q,\dot{q},q_*) = |y|\text{sign}\left[-\frac{(\dot{q}^2 + \lambda^{-1}(q - q_*))\,\text{sign}\,\dot{q}}{2|y|} - (q - q_*)\right] \tag{27}$$

and by the inertial regulator model

$$\ddot{y} = -\sqrt{\lambda_p^{-1}}\dot{y} - \lambda_p^{-1}\left(y - |y|\text{sign}\left[-\frac{(\dot{q}^2 + \lambda^{-1}(q - q_*))\,\text{sign}\,\dot{q}}{2|y|} - (q - q_*)\right]\right). \tag{28}$$

In accordance with the obtained results, we performed a numerical modeling to build the knowledge base elements of a fuzzy regulator with term sets for the mode “coarse” ($q^* = 1$) and “fine” ($q^* = 2$).

Figures 1, 2, 3 and 4 for the “coarse control” mode show the reference control and the control realized using the inertial regulator; the reference trajectory and the trajectory when taking into account the inertia of the regulator; the phase portraits; the

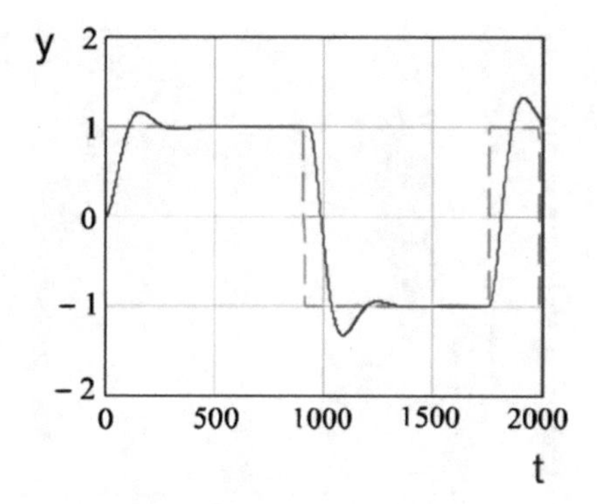

Fig. 1. The reference control and the control realized using the inertial regulator for the "coarse control" mode

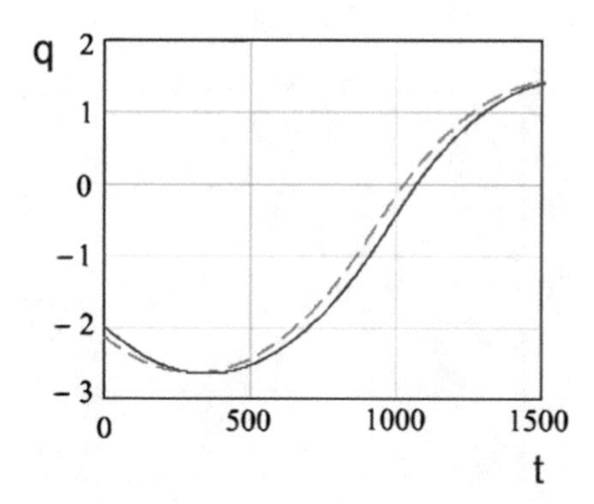

Fig. 2. The reference trajectory and the trajectory when taking into account the inertia of the regulator for the "coarse control" mode

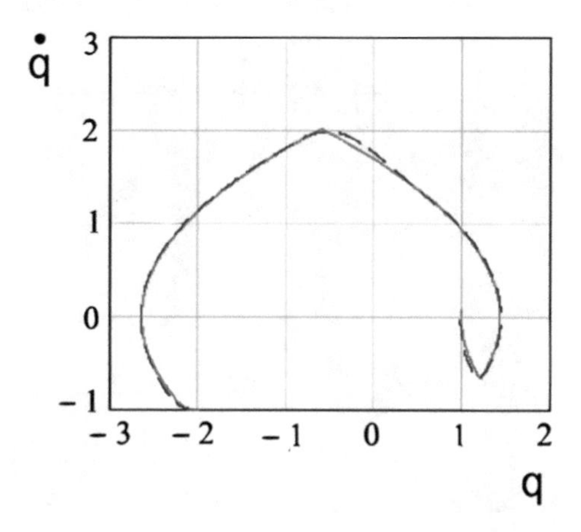

Fig. 3. The phase portraits for the "coarse control" mode

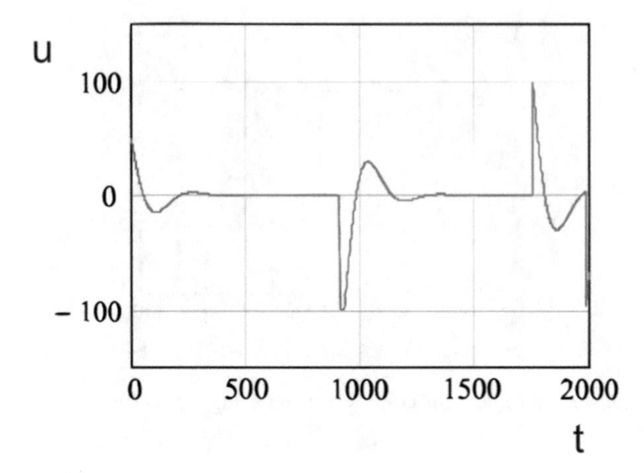

Fig. 4. The control at the input of the regulator for the "coarse control" mode

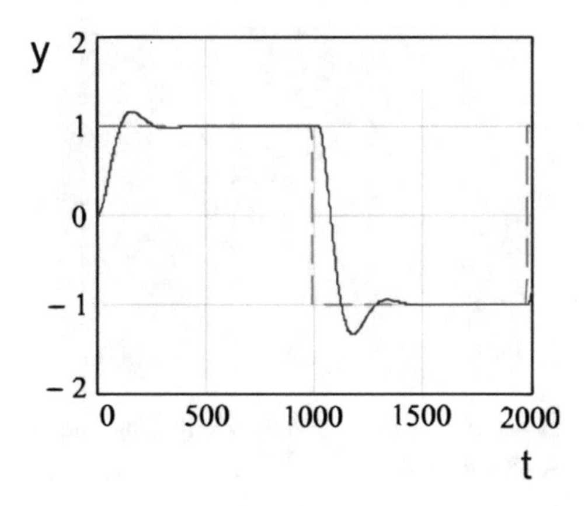

Fig. 5. The reference control and the control realized using the inertial regulator for the "fine control" mode

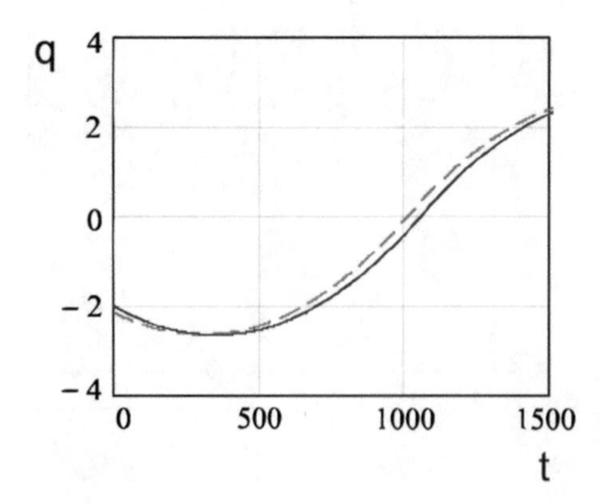

Fig. 6. The reference trajectory and the trajectory when taking into account the inertia of the regulator for the "fine control" mode

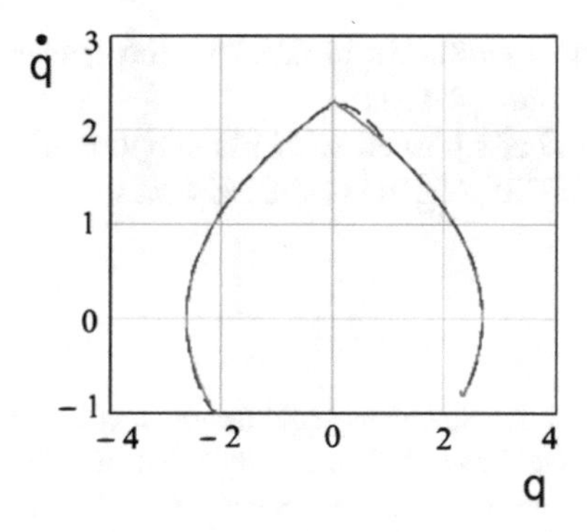

Fig. 7. The phase portraits for the “fine control” mode

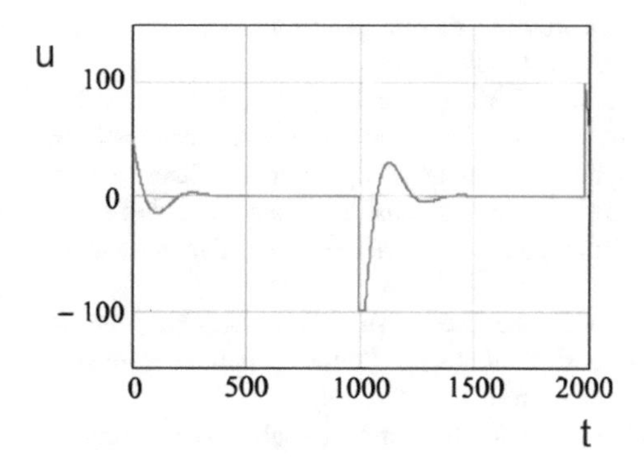

Fig. 8. The control for the “fine control” mode

control at the input of the regulator. Figures 5, 6, 7 and 8 show the same characteristics for the “fine control” mode. The results show that the elements of the knowledge base, i.e. controls at the input of the regulator (Figs. 4 and 6) for different term sets significantly differ in structure. Thus, to obtain effective solutions based on typical PID regulators are rather difficult due to the essential nonlinearity of the quasi-optimal control laws.

6 Conclusion

When designing the knowledge base of an intelligent control system with “coarse” and “fine” operating modes, we consider the variant to build the basic control laws explicitly associated with the specified elements of the term sets. The synthesis results allow to state that quasi-optimal control laws can be classified as nonlinear. To obtain the laws that are close to optimal using the linear combinations of the control laws of the PID regulators type is extremely difficult. Thus, the use of the method of the combined maximum principle allows to add "nonlinear elements" to the knowledge base of the intellectual control system.

The paper has been accomplished with the support of Russian Federal Property Fund grants No. 18-01-00385 A, № 18-08-01494 A.

References

1. Krut'ko, P.D., Palosh, V.E.: Stabilizing equilibrium states of double pendulum loaded by follower and conservative forces. J. Comput. Syst. Sci. Int. **48**, 165–178 (2009)
2. Pontryagin, L.S., Boltyanskii, V.G., Gamkrelidze, R.V., Mishchenko, E.F.: The Mathematical Theory of Optimal Processes. Wiley, New York (1986)
3. Bellman, R.: Dinamicheskoe programmirovanie (Dynamic Programming). Foreign Literature Publishing House, Moscow (1960)
4. Kostoglotov, A.A., Kostoglotov, A.I., Lazarenko, S.V.: The combined-maximum principle in problems of estimating the motion parameters of a maneuvering aircraft. J. Commun. Technol. Electron. **54**, 431–438 (2009)
5. Surkov, V.V., Sukhinin, B.V., Lovcenkov, V.I., Solovev, A.E.: Analiticheskoe konstruirovanie optimal'nyh regulyatorov po kriteriyam tochnosti, bystrodejstviya, ehnergosberezheniya (Analytical designing of optimum regulators on the criteria as well as the accuracy, speed, power-saving). Publishing House of Tula State University, Tula (2005)
6. Letov, A.M.: Dinamika poleta i upravlenie (Flight dynamics and control). Nauka, Moscow (1969)
7. Krasovskii, A.A.: Analiticheskoe konstruirovanie konturov upravleniya letatel'nymi apparatami (Analytical design of control loops letatel-governmental apparatus). Mechanical Engineering, Moscow (1969)
8. Lur'e, A.I., Analiticheskaya Mekhanika (Analytical mechanics). Gos. Izd. Fiz.-Matem. Liter., Moscow (1961)
9. Kostoglotov, A.A., Kostoglotov, A.I., Lazarenko, S.V.: Joint Maximum Principle in the Problem of Synthesizing an Optimal Control of Nonlinear Systems. Autom. Control Comput. Sci. **41**, 274–281 (2007)
10. Derabkin, I., Kostoglotov, A., Lazarenko, S., Lyashchenko, Z.: Intellectualization of industrial systems based on the synthesis of a robotic manipulator control using a combined-maximum principle method. Adv. Intell. Syst. Comput. **451**, 375–384 (2016)
11. Kostoglotov, A.A., Lazarenko, S.V., Kuznetcov, A.A., Lyashchenko, Z.V.: Method of quasi-optimal synthesis using invariants. In: MATEC Web of Conference, vol. 77, pp. 1–4 (2016)

Fuzzy Graphs, Fuzzy Networks and Fuzzy Inference for Planning and Cognitive Modelling

Assessing the Software Developer's Quality Using Fuzzy Estimates

Tatiana Afanasieva(✉) and Vlad Moiseev

Ulyanovsk State Technical University, Ulyanovsk, Russia
tv.afanasjeva@gmail.com

Abstract. The software developer's quality in the work has significant impact on the quality of a software, therefore the assessing developer's quality is actual. The results of that assessment need to be understandable and should described developer's quality in static and in dynamics. This article aimed on design of a methodology for obtaining these results. The proposed methodology for evaluating the quality of software developers is based on task metrics of a repository and uses fuzzy set theory for creating the linguistic estimates of developer's quality. The stages of proposed methodology are described in application to real project, obtained results meet the requirements and show usability in software developer's quality assessing.

Keywords: Software quality · Developer · Fuzzy assessment

1 Introduction

This article discusses issues related to the evaluation of the projects in the field of software development. To assess the quality of software development processes, many standards have been developed, in which the basic requirements for development processes, to the quality of results measured based on metrics, to quality assurance processes and so on are determined. In this direction, a set of works have been published, annual workshops are hold focused on the software development and methodologies in the field of improving the quality of software. All this allowed us to identify the actual problem associated with the need to assess the quality at all stages of software development.

The review of the works shows that there is a need for a systematic describing of the problem of quality assessment at various levels, for various entities of the software development process, and for the quality assessments both in statics (in the state of individual indicators) and in dynamics. Therefore, the aim of this article is to develop a methodology focused on solving this problem.

Software development are considered as a formal system, in which the following elements are combined: developers, projects and tasks. Tasks are the basic elements for evaluating and monitoring software quality. Metrics stored in task tracking systems can be used to quantify quality at different levels: for a task, for types of tasks, for a project, for a developer, and for development team. Characteristics and quality of software projects depends on quality of the developers.

A. Abraham et al. (Eds.): IITI 2018, AISC 874, pp. 203–212, 2019.
https://doi.org/10.1007/978-3-030-01818-4_20

In this study, software developer's quality is considered as the estimations of the static (quality level) and of the dynamics (type of quality level changing) quality based on the mining software repository. The results of implementing the proposed methodology are not only quantitative estimates, but also linguistic assessments of the quality that are understandable to human. Linguistic quality assessments improve the understanding of a software development and of developer's behavior. This information is useful in comparative analysis, as well as for motivating of the developers.

2 Related Works

Understanding the software development process is the key to improving the quality of software products. In this field, experience, results of research and publication of findings are accumulated. The focus is on data stored in the repositories of software support tools, such as issue&project tracking tools.

The author of the article [1] notes that "… the idea to use repositories of software support tools to explain and predict phenomena in software projects and to create tools that improve software productivity, quality, and lead times appears to be promising. Although the information in software support systems represents a vast amount of untapped resources for empirical work, it remains a challenge to create models of key problem…. Even though it appears that the use of software repositories should enable answering novel software engineering questions, most of these questions have yet to be identified".

In the article [2] an overview of more than 90 empirical studies in the understanding of the software development process using GitHub repositories was provided. The authors conclude that many studies have been published in this area, but not all of them are sufficiently systematic, sometimes these works describe poor sampling techniques and use a variety of methodologies.

The problem related to improving software quality and prevention faults is considered in paper [3]. In this paper, the authors discussed the quantitative metrics from the real projects based on standard software development processes. The data set static metrics (Lines of Code (LOC), McCabe [4], Halstead [5]) and change metrics (number of commits, number of errors, number of authors, etc.) for Matlab/Citylink modules. They focused on change metrics and bug statistics to extract useful information, in particular, about correlation between LOC and McCabe/Halstead metrics. Then they discussed fault prediction using this dependency.

Another approach to improving the quality of software development is discussed in the article [6], based on a study of the process of accomplishing tasks. It is based on the hypothesis that for this purpose it is necessary to more correctly determine the priorities of tasks and allocate resources accordingly. To improve the quality of solving these problems, the authors focused on research and prediction of the issue lifetime. Unlike predicting based on static features, in this work [6] it is suggested to combine static, dynamic and contextual features for which techniques of classification were used and investigated using data from more than 4000 GitHub projects.

In the study [7] the quality of software development is considered in relation with the quality of the development team, and the emphasis is on social characteristics. The effect of team diversity and contribution of diversity of software team's effectiveness is discussed. The authors argue that quantitative studies of large data sets are required to do the precise contribution of diversity on teams' effectiveness. They presented for the first time a large data set of social diversity attributes of programmers in GitHub teams, in particular, for studying the relationship between social diversity and technical activity in online teams based on distribution of some indexes. The authors note that the social and technical values of the diverse teams is more often recognized in task oriented settings.

However, there is no doubt that the quality of software development is determined not only by social characteristics, but much by the level of developer's quality, his ability to learn and improve his level of skills. The analysis of publications in the field of improving the quality of software development has shown that insufficient attention is paid to the problem of analyzing and evaluating the quality of developers. The authors hope that the proposed methodology for assessing the quality of software developers will help to eliminate this gap.

3 Methodology for Assessing the Quality of Software Developers

When developing a methodology for evaluating the quality of software developers, repository metrics were analyzed with focus on the key success factors, such as productivity, progress and code quality.

The main stages of the software quality assessment methodology include the following steps:

- Creation of a quality model in the form of the key quantitative indicators of the tasks.
- Summarization of developer's quality over the set of key indicators.
- Definition of fuzzy scales for each key indicator.
- Applying the fuzzy scales for linguistic evaluation of levels (quality in static) and changes (quality in dynamics) of the developer's quality for each key indicator and for the sum of all key indicators.
- Forming linguistic time series of quality in dynamics for each key indicator for extracting information about the progress of the software developer's quality in order to know who a leader is and who is an outsider.

3.1 Model of Software Developer's Quality

3.1.1 Key Indicators of the Software Developer's Quality

The data for each developer are stored in the form of a sequences of the tasks and the history of task statuses, which in the development process associated with commits in the version control system. The quality model is based on repository metrics, that show how software developers work on the task: (1) The developer to whom the task is assigned ("Assigned"); (2) Task status ("Status"); (3) Task type ("Type"); (4) Task start date ("Begin Date"); (5) Date, when task was executed ("Execution Date");

(6) Estimated time on a task ("Time Estimate"); (7) The time that the developer has spent on the solution of the problem ("Time Spent"). On the bases of these metrics we use below described indicators of developer's quality (R1, R2, R3, R4) for assessing their productivity, progress and code quality:

Indicator R1. Assessment of the programming style of the completed tasks (the tasks with the status "Done"). It is calculated as the ratio of the number of detected anomalies per LOC, which were found during code review of completed task. The smaller the indicator value, the higher quality of the programming style.
Indicator R2. The ratio of the time of the completed tasks to their estimated time. It is calculated as the average ratio of the "Time Spent" parameter to "Time Estimate". The smaller the indicator value, the higher quality of time planning of the tasks. It is assumed the "Time Estimate" is set in accordance with the internal standards for each type of the task.
Indicator R3. Return to the revision. It is calculated as the ratio of tasks that have changed their status to "revision", to all tasks. The smaller the indicator value, the higher the quality of the completed tasks.
Indicator R4. The ratio of the time that the developer has spent on bug fixing to the time of the completed tasks (the tasks with the status "Done"). The smaller the indicator value, the higher quality of execution or the smaller bugs in the code.

3.1.2 Formal Model of the Software Developer's Quality

Below, we consider the structure of the software developer's quality model which is presented at two levels: at the level of key indicators of quality and integrated quality. The proposed model is intended for assessing quality both in static and in dynamics and forms answers the following research questions.

For static quality analysis:

RQs 1. What is the quantitative evaluation of the developer's quality at a given point in time for a given quality indicator?
RQs 2. What is the qualitative (expressed in linguistics) level of the developer's quality at a given point in time based on a given quality indicator?
RQs 3. What is the integrated quantitative evaluation of the developer's quality at a given point in time?

For analysis of quality in dynamics:

RQd 1. How does the developer 's quality level on a key indicator change with respect to previous one?
RQd 2. How does the developer 's quality change with respect to all key indicator at a fixed time moment?
RQd 3. How does the developer 's quality change with respect to a key indicator over some period? What is a predictive trend of quality change?

The proposed formal model of the software developer's quality $Q_{developer}$ is defined by expressions given bellow:

$$Q_{developer} = \langle Q_{num}, Q_{ling} \rangle , \tag{1}$$

$$Q_{num} = \{q(i,t) \mid q(i,t) \in \mathbb{R}, \mathrm{i} = 1,2,\ldots,\mathrm{k}; \mathrm{t} = 1,2,\ldots,\mathrm{n}\}, \tag{2}$$

$$Q_{ling} = \{\tilde{q}(i,t), d\tilde{q}(i,t) \mid \tilde{q}(i,t) \in \mathrm{H}, d\tilde{q}(i,t) \in \Psi, \mathrm{i} = 1,2,\ldots,\mathrm{k}; \mathrm{t} = 1,2,\ldots,\mathrm{n}\}, \tag{3}$$

here Q_{num} and Q_{ling} denote numeric and linguistic components of a model; $q(i,t)$ and $\tilde{q}(i,t)$ are numeric and fuzzy linguistic estimates respectively of a developer's quality level at a given time t for a given quality indicator i; H = {Junior,Middle, Senior} denotes a set of linguistic terms (quality levels) used for static assessment of a developer's quality; $\mathrm{d}\tilde{q}(i,t)$ presents the fuzzy linguistic estimates of developer's quality changes, $\Psi = \{\text{Increase, Decrease, Stability}\}$ is a set of linguistic terms used for linguistic assessment of change of a developer's quality level; k is a quantity of key quality indicators (k = 4 in respect to Sect. 3.1.1); n is a quantity of observed time moments.

The proposed model (1) of the software developer's quality that combines static and dynamic assessments expressed both in quantitative and in qualitative values. When parameter $\mathrm{t} = \mathrm{const}$, model (1) describes the state of a developer's quality, that is the static quality in respect to a set of key indicators, $\mathrm{i} = 1,2,\ldots,\mathrm{k}$. In the case $\mathrm{i} = \mathrm{const}$ the model (1) presents dynamics of the developer's quality for a key indicator in the form of a time series.

3.2 Linguistic Assessment of Software Developer's Quality

As was mention in previous Section the model of Software Developer's Quality (1) combines numeric and linguistic component. Below we will consider the approach to generating qualitative estimates of the model of Software Developer's Quality (1), expressed in linguistic terms. In the proposed model, two types of linguistic terms are used. The first type is associated with a static assessment of the developer's quality at a given point in time for a given quality indicator and the second one is a dynamic assessment in the form of changes in static assessment over a monitoring interval.

Typically, quality scale and/or interval estimation for each numerical indicator are used to obtain static developer's quality levels. At the same time, with each numerical interval of the quality scale, a linguistic term could be associated, characterizing the qualitative level in the statics. Then to each value of a numerical indicator there corresponds a numeric interval of a scale and its linguistic evaluation. To combine numeric and linguistic values fuzzy sets and linguistic variable were introduced by Zadeh [8]. Their application to linguistic assessment of levels and changes was provided in the paper [9], where this tool was named ACL-scale. The purpose of ACL-scale is to form quality estimates for a given interval of numeric values using linguistic variable. ACL-scale is adaptive tool and requires setting follows parameters: a number and the names of the quality levels, as well as the left and right boundaries of the numeric interval. The parameters of ACL-scale for different numerical indicators should be set by software quality manager and are determined by internal standards of a company.

It is important to note that in ACL-scale the numerical intervals have the property of partial order. This property is inherited by the associated linguistic terms that characterize the quality level of the developer in static. The latter allows us to define the relations "more", "less" and "equivalence" on the set of static linguistic terms of quality, which are used to "determine" the dynamic linguistic terms "Increase", "Decrease", "Stability". Therefore, we used this ACL-scale [9] to transform the numerical time series of key quality indicators into linguistic ones described in the model (1):

$$q(i,t) \rightarrow \tilde{q}(i,t) \rightarrow d\tilde{q}(i,t).$$

4 Case Study

We applied the proposed methodology to assess Software Developer's Quality of RIAS HCS Laboratory in Russian company in Ulyanovsk.

Firstly, the values of key indicators of proposed model (1) $q(i,t)$ have been calculated using special tool. In case study i = 1, 2, 3, 4 and $q(1,t) = R1, q(2,t) = R2$, $q(3,t) = R3 \, and \, q(4,t) = R4$. The results of the static numeric assessing at a fixed time moment t on the key quality indicators, determined in Sect. 3.1.1, for ten developers are provided at the Table 1.

Table 1. This table show the quantitative evaluation of the developer's quality at a given point in time for quality indicators R1, R2, R3 and R4 (see Sect. 3.1.1).

Dev	R1	R2	R3	R4	Sum
P M	0.01	0.08	0.11	0.16	0.36
X E	0.04	0.94	0.16	0.21	1.35
M V	0.02	1.39	0.11	0.11	1.63
P S	0.03	1.41	0	0.29	1.73
Zh A	0.06	1.38	0.08	0.22	1.74
P E	0.04	1.39	0.17	0.19	1.79
F O	0.08	1.46	0.33	0.41	2.28
A R	0.08	1.63	0.35	0.31	2.37
A N	0.07	1.61	0.39	0.35	2.42
BA	0.09	1.74	0.26	0.42	2.51

Data at the Table 1 contain the answer the research question RQs1 about the quantitative assessing of the developer's quality at a given point in time for quality indicators R1, R2, R3 and R4 (see Sect. 3.1.1). The last column the answer the research question RQs 3.

To obtain the linguistic assessment $\tilde{q}(i,t)$ of numeric key indicators $q(i,t)$ the ACL-scales [9] were created for each indicator. In the Table 2 parameters of created ACL-scales for indicators R1, R2, R3 and R4 are presented. Table 3 summarizes the results of applying the ACL-scales with parameters shown in the Table 2 to the data in the Table 1 for two consecutive times t and (t + 1). Data at the Table 3 presents the static linguistic assessments of developer's quality and answer the research question the RQs 2. As can be seen from the left part of the Table 3 for time moment t four software developers (ZhA, AR, AN and BA) demonstrated the stable quality levels for all key indicators; the others had different linguistic assessments of quality levels.

Table 2. The parameters of ACL-scales for key indicators of software developer's quality.

ACL-scale for key indicators	Terms of quality levels		
	Senior	Middle	Junior
ACL-scale for R1	R1 $\leq$ 0.03	0.03 < R1 < 0.06	R1 $\geq$ 0.06
ACL-scale for R2	R2 $\leq$ 1	1 < R2 $\leq$ 1.5	R2 $\geq$ 1.5
ACL-scale for R3	R3 $\leq$ 0.1	0.1 < R3 < 0.2	R3 $\geq$ 0.2
ACL-scale for R4	R4 $\leq$ 0.1	0.1 < R4 < 0.35	R4 $\geq$ 0.35

At the right part of the Table 3 the linguistic assessments of the developers show that only two developers (XE, BA) had the same quality levels and others were assessed with different levels. This information is not obvious while analyzing the numeric data in the Table 1.

Table 3. Quality levels of software developers for two consecutive times t and (t + 1). Here the terms of developer's quality in static are used: S is Senior, M is Middle, J is Junior.

Time moment t					Time moment t + 1			
Dev	R1	R2	R3	R4	R1	R2	R3	R4
PM	S	S	M	M	S	S	M	M
XE	M	S	M	M	M	M	M	M
MV	S	M	M	M	S	M	S	M
PS	S	M	S	J	M	M	S	M
ZhA	M	M	S	M	M	J	S	M
PE	M	M	M	M	J	M	M	J
FO	J	M	J	J	M	M	J	M
AR	J	J	J	J	M	M	J	J
AN	J	J	J	J	J	M	J	J
BA	J	J	J	J	J	J	J	J

Besides that, data analysis of the Table 3 shows that estimating the quality levels (quality in static) is not enough to understanding the quality of the developers and it is necessary to assess their progress, that is, changes in their quality levels (quality in dynamics). In respect to data from the Table 3 and in accordance to the set $\Psi = \{\text{Increase}, \text{Decrease}, \text{Stability}\}$ of the quality model (1) (see Sect. 3.1.2) we formed the linguistic assessment of developer's quality in dynamics.

The data in the Table 4 are the linguistic estimates of developer's quality in dynamics for two consecutive times t and (t + 1). These estimates were obtained using the briefly described ACL-scale (see details in the paper [9]) in Sect. 3.2 and answer the question RQd 1. From the given data in the Table 4 it is possible to extract the following information (see column Summarization): compared to the previous time four developers have lowered, two developers have not changed, and four developers have improved their level of quality. Let us note these linguistic data contain the answer to the question RQd 2 in understandable form. Similarly, considering dynamic estimates over several periods, one can get a linguistic time series of dynamic quality estimates for each developer in respect to a key indicator. Table 5 contains 10 time series of linguistic estimates for each software developer for the period from October 2015 to February 2016. The last row summarized dynamic linguistic estimates for this period for each developer and these data answer the question RQd 3. To obtain these results we adopt the technique for linguistic description of a time series general tendency [10], this technique in the work [11] was named as GTI-algorithm for short. GTI-algorithm is aimed to assess the type of a general tendency of a time series in a linguistic term from the set {"Fall (Decrease)", "Stability", "Growth (Increase)", "Fluctuation"}.

Table 4. Quality changes of software developers for time moment (t + 1). Here Stab is "Stability", Inc is "Increase", Dec is "Decrease".

Time moment t + 1					
Dev	R1	R2	R3	R4	Summarization
P M	Stab	Stab	Stab	Stab	Stab
X E	Stab	Dec	Stab	Stab	Dec
M V	Stab	Stab	Dec	Stab	Inc
P S	Dec	Stab	Stab	Dec	Dec
Zh A	Stab	Dec	Stab	Stab	Dec
P E	Dec	Stab	Stab	Dec	Dec
F O	Inc	Stab	Stab	Inc	Stab
A R	Inc	Inc	Stab	Stab	Inc
A N	Stab	Inc	Stab	Stab	Inc
BA	Stab	Stab	Stab	Stab	Stab

Derived type of a general tendency not only describes quality changes in developer's skills, but is useful for prediction quality of developers, of the tasks and of a project.

Table 5. Dynamic assessments of developer's quality over five periods. Here the designations of developer's quality in dynamics are used: Stab – Stability, Inc – Increase, Dec – Decrease, Fluc – Fluctuation.

Periods	Dynamic assessments of developer's quality									
	AR	AN	BA	ZhA	MV	PC	PM	PE	FO	HE
Oct.15	Stab	Stab	Dec	Stab	Stab	Stab	Dec	Stab	Inc	Stab
Nov.15	Inc	Stab	Inc	Inc	Inc	Dec	Dec	Dec	Dec	Dec
Dec.15	Dec	Inc	Dec	Dec	Stab	Stab	Stab	Inc	Stab	Inc
Jan.16	Dec	Dec	Stab	Inc	Inc	Dec	Inc	Dec	Inc	Inc
Feb.16	Stab	Inc	Stab	Inc	Stab	Inc	Inc	Inc	Stab	Inc
Summarization	Dec	Inc	Dec	Inc	Inc	Dec	Fluc	Fluc	Inc	Inc

5 Discussion

In this work the methodology for assessing the quality of software developers is proposed and illustrated in case study. The goal of the methodology is mining information about software developer's quality from the data, stored in software repositories. The proposed methodology includes quality model with numeric and linguistic components and an approach to estimate linguistic components using fuzzy sets. At the other hand following the work [1] the proposed methodology is determined by six research questions and answers to them.

Unlike previously published works that analyzed the quality of the tasks, the dependence of faults on the attributes of the tasks for the purpose of prediction, the proposed methodology focuses on the evaluation of a quality.

Similarly, as was provided in the paper [4], the proposed methodology combines static and dynamic assessments, but generates linguistic understandable results, based on quality model. Like to the study [7], the proposed methodology focusses on the evaluating of effectiveness of the development team, but uses different indicators to assess the developer's quality in the work. Linguistic static and dynamic assessments that were formalized in the quality model determine the novelty of this work. Such assessments are more informative and understandable for a human.

To obtain linguistic assessments of the developer's quality, it has been proposed to use the theory of fuzzy sets [8], which makes it possible to integrate numeric and linguistic quality indicators and proven means. The application of this methodology in a real project to assessing the quality of developers in static and in dynamics has shown its effectiveness. Further research will be aimed at applying the quality assessment model developed within this methodology to solving the tasks of clustering developers, searching for abnormal values and changes in quality indicators, predicting the quality of tasks depending on the quality of the developers.

Acknowledgments. The authors acknowledge that this paper was partially supported by the Russian Foundation of Basic Research, projects № 16-07-00535 and № 16-47-730715.

References

1. Mockus, A.: Software support tools and experimental work. In: Basili, V., et al. (eds.): Empirical Software Engineering Issues: Critical Assessments and Future Directions. LNCS, vol. 4336, pp. 91–99. Springer, Heidelberg (2007)
2. Cosentino, V., Luis, J., Cabot J.: Findings from GitHub: methods, datasets and limitations. In: Proceedings of 2016 IEEE/ACM 13th Working Conference on Mining Software Repositories, MSR 2016, 14–15 May 2016, Austin, TX, USA (2016). https://doi.org/10.1145/2901739.2901776
3. Altinger, H., Siegl, S., Dajsurent, Y., Wotawa, F.: A novel industry grade dataset for fault prediction based on model-driven developed automotive embedded software. In: Proceedings of 12th Working Conference on Mining Software Repositories 2015, At Florence, Italy, MSR 2015 (2015). https://doi.org/10.1109/msr.2015.72
4. McCabe, T.J.: A complexity measure. IEEE Trans. Softw. Eng. **4**, 308–320 (1976)
5. Halstead, M.H.: Elements of Software Science (Operating and Programming Systems Series). Elsevier Science Inc., New York (1977)
6. Kikas, R., Dumas, M., Pfahl, D.: Using dynamic and contextual features to predict issue lifetime in GitHub projects. In: Proceedings of 2016 IEEE/ACM 13th Working Conference on Mining Software Repositories, MSR 2016, 14–15 May 2016, Austin, TX, USA (2016). http://dx.doi.org/10.1145/2901739.2901751
7. Vasilescu, B., Serebrenik, A., Filkov, V.: A data set for social diversity studies of GitHub teams. In: Proceedings of 2015 IEEE/ACM 12th Working Conference on Mining Software Repositories (RSS) (2015). https://doi.org/10.1109/msr.2015.77
8. Zadeh, L.A.: The concept of a linguistic variable and its application to approximate reasoning. Memorandum ERL-M 411 Berkeley, October 1973
9. Afanasieva, T., Yarushkina, N., Gyskov, G.: ACL-scale as a tool for preprocessing of many-valued contexts. In: Proceedings of Second International Workshop on Soft Computing Applications and Knowledge Discovery (SCAD 2016), Moscow, Russia, pp. 2–11, 18–22 July 2016
10. Afanasieva, T., Sapunkov, A.: Selection of time series forecasting model using a combination of linguistic and numerical criteria. In: Proceedings of 2016 IEEE 10th International Conference on Application of Information and Communication Technologies (AICT-2016), 12–14 October, Baku, Azerbaijan, pp. 341–345 (2016). https://doi.org/10.1109/icaict.2016.7991715
11. Afanasieva, T., Yarushkina, N., Sibirev, I.: Time series clustering using numerical and fuzzy representations. In: Proceedings of Joint 17th World Congress of International Fuzzy Systems Association and 9th International Conference on Soft Computing and Intelligent Systems (IFSA-SCIS 2017), Otsu, Shiga, Japan, 27–30 June 2017. https://doi.org/10.1109/ifsa-scis.2017.8023356

A Fuzzy Control Method for Priority Driven Embedded Device

Karolina Janosova(✉), Michal Prauzek, Jaromir Konecny, Monika Borova, and Martin Stankus

Department of Cybernetics and Biomedical Engineering, VSB-Technical University of Ostrava, Ostrava-Poruba 70833, Czech Republic
karolina.janosova@vsb.cz
http://www.vsb.cz

Abstract. The current goal is to minimize an embedded system consumption while maintaining all its features. Well-tuned power management becomes even more important in systems without any possibility of energy harvesting that rely solely on the battery power. The aim is a significant increase in the battery life without a desirable degradation in the quality of the service.

A fuzzy rule-based classifier is intended to secure a simulation of the priority of the operation and the battery voltage. The energy consumption is a critical concern in the battery operated embedded devices. Power management is one of the most important consider aspects in low-power embedded systems design.

This contribution studies the low power system in terms of the battery voltage level, depending on system priorities. The goal is to achieve the best possible efficiency of a whole device. The article offers a list of arrangements and applied methods having the most significant impact on the extension of the systems operating time.

Keywords: Low-power embedded system · Fuzzy expert system
Battery management system

1 Introduction

Portable devices, like cellular phones, tablets, GPS trackers, watches, laptops etc. can be found in the all everyday technical areas. The practical examples are described in [5,13,14]. The energy harvesting system is introduced in [9].

A large amount of methods based on the hardware or the software solution were designed for the prolonging operating time of portable devices. The aim of this study is to come up with one universal method for reducing system's power consumption using the fuzzy expert system. Many of software-based methods are characterized by advanced data analytics from acquired measurements, e.g. The neural network or fuzzy expert system methods. Furthermore, the study proposes a set of arrangements which could be applicable in many simple low-power embedded devices followed by the simulation of usable operating modes

A. Abraham et al. (Eds.): IITI 2018, AISC 874, pp. 213–222, 2019.
https://doi.org/10.1007/978-3-030-01818-4_21

using the fuzzy method. As described in this paper, one of the topics is to focus on the hardware (re)configuration and finding sources of the most significant power demands. The aim of this study is to propose one universal method for reducing system's power consumption using the fuzzy expert system.

This paper describes method how to simulate output modes with different settings of the operating parameters and the resulting consumption of the embedded low-power monitoring system. The paper is organized into five section. Section Background briefly discusses general characteristics of battery management system and fuzzy methods. Related work provides an overview of alternative approaches used to minimize portable system's energy consumption. Parts of the designed measurement apparatus along with methods used for the Fuzzy control system are described in the section System Structure and Methods. Testing and Results section solves the individual improvements of the apparatus', along with the proposed fuzzy' expert system and the impact of these measures on a power consumption. Finally, major findings and outlines of our future work are summarized in the Conclusion.

2 Background

The issue of the energy consumption is one of the biggest concerns of current embedded systems. One of the general aims of the power management for portable devices, described in [20], is to balance power consumption to maximize device's operating time while maintaining required functionalities that are needed for the respective application under given conditions. Other way of the solution for the power management is a battery management system (BMS), see [4].

2.1 Fuzzy Model

Artificial intelligence tools, such as expert system, fuzzy logic, and neural network are some of methods used in power electronics for the energy battery management system. The fuzzy expert system is a software that emulate the reasoning process of a human expert or provide in an expert manner in domain for which no human expert exists [3].

The linguistic model is an expression that formalizes the relationship between input and output linguistic variables of the system. It is used to describe the non-numeric behavior of complex systems. It defines the linguistic model as an expression. In this expression occur linguistic variables, their names, the logical clutches and truth values. The standard form of the expression of this relationship is IF-THEN rules. This rule is called the Mamdani type rule and this is one of the most effective form of expressing human knowledge. Fuzzy methods enable to deduce conclusions based on the imprecise description of the given situation thanks linguistically formulated fuzzy rules:

$$IF\ X1\ is\ A1\ AND\ X2\ is\ A2\ THEN\ Y\ is\ B \tag{1}$$

where A1, A2 and B are certain predicates characterizing the variables X1, X2 and Y. They are specified linguistically. The software enables to work with specific kind of linguistic expressions, or the user may specify his own ones. There exist approximate reasoning methods (type of inference) for explication fuzzy IF-THEN rules. Logical deduction, Fuzzy approximation with conjunctions or Fuzzy approximation with implications [11] can be included among these methods. For this system the Fuzzy approximation with conjunction was chosen. The aim of this method is to approximate the function hidden inside the fuzzy relation. The Center of Gravity method (CoG) is one of the way how to achieve the functional fuzzy approximation for defuzzification.

The fuzzy model Mamdani is an approximation of the indefinite function. To derive the size of the output linguistic variable, when considering the specific values of the input variables, the compositional rule is used. Its variants are given by a variant of the interpretation of the fuzzy logical coupling of the fuzzy session of the composition [3,11].

2.2 Defuzzification

This process specifies the method of defuzzification of the fuzzy set, i.e. the result of approximate deduction. There are several methods for the interference of the fuzzy system. In this article the Center of Gravity method was chosen. By using CoG, the linguistic version is convert by the defuzzification into the form of a common number.

3 Related Work

The following text elaborates options for the battery management system. High-level industrial low-power design methodology and evaluation of a wide spectrum of the hardware implementation are described in [17]. A hybrid verification approach for a system design in an abstract mode, see [8]. The following contribution [7] introduces the Dynamic Equalization for Lithium Battery Management System (LBMS). The paper [6] proposes an intelligent Li-ion battery charging management system, followed by the implementation of the battery monitoring system and the balance platform (BMBP) with a recharging.

A general precondition is that monitoring systems need to operate independently of the external sources of energy. To achieve a long-term sustainability, monitoring systems often rely on an energy obtain from the ambient environment, e.g. through solar harvesting, see [18] and [12]. Measurement of environment conditions and related harvesting energy system are described in [15,16]. Here, the energy management is based on a fuzzy rule system and methods that allow adaptation to the environmental energy profile. The following contribution proposes a low-power system for location tracking systems. The designed method found the solution for the power problem by limiting a power consumption of a tag through a power-saving mode and a concept of duty cycle [19].

Following article [1] presents a fuzzy rule-based classifier which is intended to secure an equilibrium between accuracy and the energy consumption, which is critical in battery operated embedded devices. Next paper describes the design and the control of such an on-board hybrid energy storage system and designed based on a fuzzy-logic controller [2]. In following paper [10] the fuzzy model is introduce for the low energy adaptive clustering hierarchy. Research article paper [21] presents a novel fuzzy logic technique, to optimize the real-time system feasibility and the energy consumption in the multicore architecture.

4 System Structure and Methods

4.1 Measuring Apparatus

The measuring apparatus includes the board Freescale FRDM-KL25Z (MCU). The MCU is connected via I/O pins to multi-sensor board FRDM FXS MULT2-B. The accelerometer from FXS MULT2-B board was used as electronic load during a system testing. The system further comprises a "gas gauge" battery monitoring circuit LTC 2941, connected to the MCU via I^2C communication.

The designed measuring system comprises one rechargeable Li-ion battery 6SP652535 as a power supply having capacity 400 mAh. The battery maximum charge voltage is 4.18 V and discharge cut-off voltage is 3 V.

As for the communication process, a sensor first transmits measured values to the MCU. Thereafter, the data is to send via J-Link Segger debug probe using J-Scope visualization software to the PC.

4.2 Proposal of the Fuzzy Expert System

The fuzzy expert system works with genuine linguistically defined rules forming a linguistic description of the given process, decision or classification situation. Fuzzy inference engine (Mamdani rule) is used to choose the most efficient mode for the improving the lifetime of the system. In this approach, suitable fuzzy descriptors are chosen: (1) the battery voltage and (2) the priority of the current desired action.

The aim of this proposal is simulate the lifetime of the designed device. The proposal is realized in the LFLC2000 software and presumes two input linguistic variables with five linguistic terms and one output variable with five linguistic terms. Based on created input and output variables follow by creation of 25 decision rules of the fuzzy expert system. This section defines input and output linguistic variables and their parameters of the linguistic values. Input linguistic variables include battery Voltage (Table 1) and Priority (Table 2). The input variable Priority specifies the priority claim for the required operation or the process. This means that it determines the order of operations required by the system according to their importance (e.g. data processing or data transfer). Determining the Priority is therefore dependent on the amount of the battery voltage (its viability). An example may be when the battery voltage level is at a very low level e.g. 25% and Priority is high e.g. 75%. In this case, a less energy-intensive mode will be determined (e.g. mode with reduced frequency of data transmission or MCU frequency reduced, etc.), to achieve longer system operation. The output linguistic variable is the Mode of the entire device and has five linguistic values (Table 3). The experiments will be realized after proposing the system.

Table 1. Input variable Voltage. The following variable includes five input linguistic terms.

Linguistic terms	Points of triangular function
Lowest	(3, 3, 3.25)
Low	(3.15, 13.33, 3.55)
Average	(3.42, 3.59, 3.77)
High	(3.7, 3.85, 4.05)
Highest	(3.9, 4.18, 4.18)

Table 2. Input variable Priority. The following variable includes five input linguistic terms.

Linguistic terms	Points of triangular function
Lowest	(0, 0, 25)
Low	(0, 25, 50)
Middle	(25, 50, 75)
High	(50, 75, 100)
Highest	(70, 100, 100)

5 Testing and Results

The basic principle of the measuring cycles was to measure the voltage supply and the accumulated enegy by the "gas gauge" circuit.

The measurement of the accelerometer parameters was performed using the variable frequency within the range of the accelerometer frequency spectrum, defined in the device datasheet. The following energy-related parameters of the connected devices were measured – the overall consumption and the battery voltage.

The continuous measurement of the battery voltage was provided and the register work as a coulomb counter to count the accumulated charge and voltage. The gas gauge unit indicates voltage and charge which shows the status of the battery. Settings of MCU frequency and connected peripheries was set according to the voltage of the battery and the operating priority. This allows longer operating time of the system.

Table 3. Output variable Mode. The following variable includes five output linguistic terms.

Linguistic terms	Points of triangular function
Mode 1	(0, 0, 1.25)
Mode 2	(0, 1.25, 2.5)
Mode 3	(1.25, 2.5, 3.75)
Mode 4	(2.5, 3.75, 5)
Mode 5	(3.75, 5, 5)

5.1 System Consumption Testing

The following parameters had substantial influence on the overall consumption of the system: Configuration of peripheries (LEDs, Bluetooth), Clock frequency of the microcontroller, connection of the OpenSDA interface or Sampling rate of the accelerometer. The settings of the individual mode is in the (Table 4).

Table 4. Arrangements for the individual modes of the the system's energy consumption. In the case of the individual mode was gradually reduces the MCU frequency and disconnected the unnecessary peripheries according to the system needs and priority of the user.

Mode	Settings of the mode
Mode 1	Default
Mode 2	48 MHZ, disconnect periph. and OpenSDA
Mode 3	8 MHz, disconnect periph. and OpenSDA, lower sampl. freq.
Mode 4	4 MHz, low sampling freq.
Mode 5	4 MHz, lower sampling freq., no processing

The default overall consumption of the testing apparatus was 35.56 mA. One of the arrangement for decreasing power demands was to switch off all the peripheries. Also following parameters had an influence on the energy consumption profile: the frequency of the measurement (regulated based on the state of battery voltage) and the data processing (e.g. convolution filter). The reduction was achieved by decreasing the overall consumption to less then 6 mA, which is more than 83% of the default consumption. The aim of the design of the fuzzy system is to simulate a way to find the best combination of modes for prolong the operating time and the efficiency of the system.

5.2 Fuzzy Expert System Design Testing

A simulation function verification was performed to verify the functionality of the proposed system. Based on the model, values were randomly selected for

Table 5. Testing values of linguistic variables

	Test1	Test2
Voltage	3.80	3.94
Priority	87.11	64.43
Mode	4	2

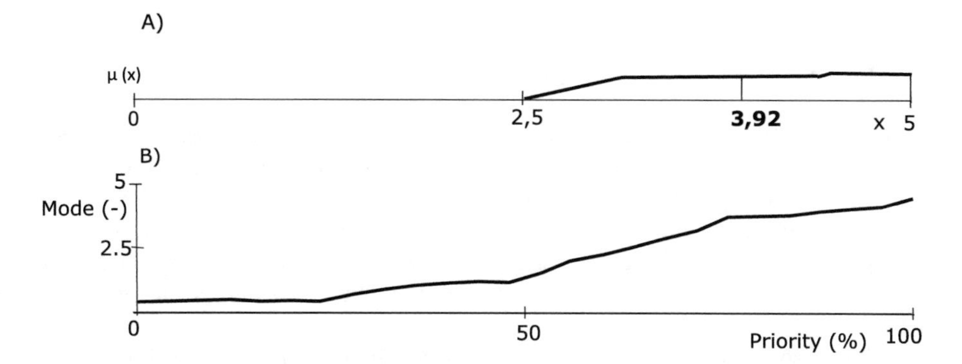

Fig. 1. The graphs illustrate the result from the Test 1 in (Table 5). In the section (A) is possible to see the result after defuzzification and in the section (B) Representation of variable Priority.

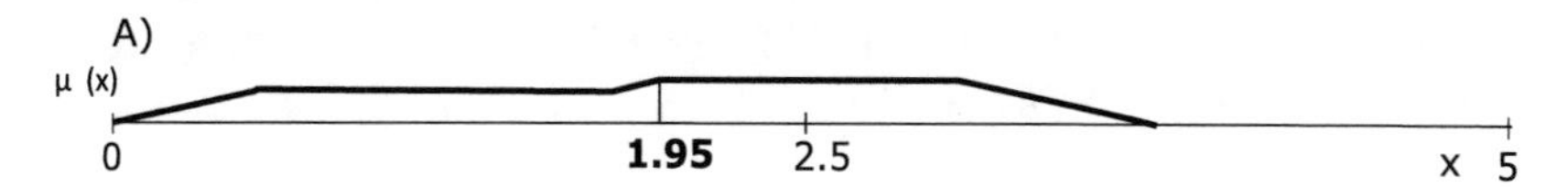

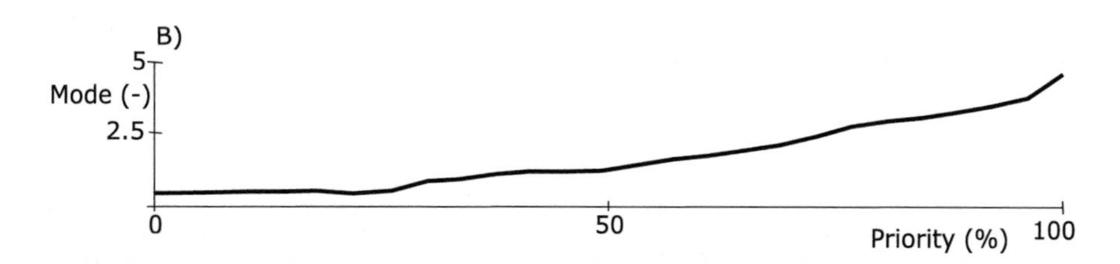

Fig. 2. The graphs illustrate the result from the Test 2 in (Table 5). In the section (A) is possible to see the result after defuzzification and in the section (B) Representation of variable Priority.

individual input variables (Table 5). The fuzzy approximation with conjunctions was chosen as an interference method in the LFLC 2000 program. Thereafter the Center of gravity method was selected for defuzzification.

The simulation tests were performed for input linguistic variables and its parameters of linguistic competence functions to test and display the result for following values. The first test was performed for battery voltage 3.80 V and Priority 87.11%. The winner of this test is Mode 4 and is possible to see in the (Fig. 1). The second test was realized for higher battery voltage 3.94 V and lower Priority 64.43%. Best result of this test is Mode 2 and is possible to see in the (Fig. 2).

6 Conclusion

The aim of this research article was to propose the functional fuzzy rule-based model for the evaluation of the operating mode, to enable the longest lifetime of the system. The system is composed from two input linguistic variables: Priority of the operations and Voltage of the battery. The output linguistic variable was the Mode. The Mamdani's Fuzzy Approach model was used for the design. Thereafter, as the method of the defuzzification, the Center of gravity was chosen.

The reason of our approach was to create a functional model of the device. The device operates in a low energy mode and can be used to power a wide range of portable devices.

Our future work will be directed to the investigation of other ways leading to lower consumption, so as to make the battery management more efficient. Enhanced equipment can then be used for real-life testing, environmental testing, medical sensors or portable devices.

Acknowledgement. This work was supported by the European Regional Development Fund in the Research Centre of Advanced Mechatronic Systems project, project number CZ.02.1.01/0.0/0.0/16_019/0000867 within the Operational Program Research, Development and Education s the project SP2018/160, "Development of algorithms and systems for control, measurement and safety applications IV" of Student Grant System, VSB-TU Ostrava.

References

1. Cocaña-Fernández, A., Ranilla, J., Gil-Pita, R., Sánchez, L.: Energy-conscious fuzzy rule-based classifiers for battery operated embedded devices. In: 2017 IEEE International Conference on Fuzzy Systems (FUZZ-IEEE), pp. 1–6. IEEE (2017)
2. Gao, C., Zhao, J., Wu, J., Hao, X.: Optimal fuzzy logic based energy management strategy of battery/supercapacitor hybrid energy storage system for electric vehicles. In: 2016 12th World Congress on Intelligent Control and Automation (WCICA), pp. 98–102. IEEE (2016)
3. Kandel, A.: Fuzzy Expert Systems. CRC Press, Boca Raton (1991)
4. Kansal, A., Hsu, J., Zahedi, S., Srivastava, M.B.: Power management in energy harvesting sensor networks. ACM Trans. Embed. Comput. Syst. **6**(4), 1–38 (2007)

5. Kromer, P., Prauzek, M., Musilek, P.: Harvesting-aware control of wireless sensor nodes using fuzzy logic and differential evolution. In: 2014 11th Annual IEEE International Conference on Sensing, Communication, and Networking Workshops, SECON Workshops 2014, pp. 51–56 (2014)
6. Lan, C., Lin, S., Syue, S., Hsu, H., Huang, T., Tan, K.: Development of an intelligent lithium-ion battery-charging management system for electric vehicle. In: Proceedings of the 2017 IEEE International Conference on Applied System Innovation: Applied System Innovation for Modern Technology, ICASI 2017, pp. 1744–1746 (2017)
7. Linlin, L., Xu, Z., Zhujinsheng, Jing, X., Shuntao, X.: Research on dynamic equalization for lithium battery management system. In: Proceedings of the 29th Chinese Control and Decision Conference, CCDC 2017, pp. 6884–6888 (2017)
8. MacKo, D., Jelemenska, K., Cicak, P.: Early-stage verification of power-management specification in low-power systems design. In: Formal Proceedings of the 2016 IEEE 19th International Symposium on Design and Diagnostics of Electronic Circuits and Systems, DDECS 2016 (2016)
9. Musilek, P., Prauzek, M., Kromer, P., Rodway, J., Barton, T.: Intelligent energy management for environmental monitoring systems, December 2017
10. Nayak, P., Devulapalli, A.: A fuzzy logic-based clustering algorithm for wsn to extend the network lifetime. IEEE Sens. J. **16**(1), 137–144 (2016)
11. Novak, V., Perfilieva, I., Mockor, J.: Mathematical Principles of Fuzzy Logic, vol. 517. Springer Science & Business Media, Berlin (2012)
12. Pimentel, D., Musilek, P.: Power management with energy harvesting devices. In: 2010 23rd Canadian Conference on Electrical and Computer Engineering (CCECE), pp. 1–4, May 2010
13. Prabhakar, T., Devasenapathy, S., Jamadagni, H., Prasad, R.: Smart applications for energy harvested WSNs. In: 2010 2nd International Conference on COMmunication Systems and NETworks, COMSNETS 2010 (2010)
14. Prauzek, M., Konecny, J., Hamel, A., Hlavica, J.: Fuzzy energy management of autonomous weather station. IFAC-PapersOnLine **28**(4), 226–229 (2015)
15. Prauzek, M., Musilek, P., Watts, A.G.: Fuzzy algorithm for intelligent wireless sensors with solar harvesting. In: IEEE SSCI 2014 - 2014 IEEE Symposium Series on Computational Intelligence - IES 2014: 2014 IEEE Symposium on Intelligent Embedded Systems, Proceedings, pp. 1–7 (2014)
16. Prauzek, M., Watts, A.G., Musilek, P., Wyard-Scott, L., Koziorek, J.: Simulation of adaptive duty cycling in solar powered environmental monitoring systems. In: 2014 IEEE 27th Canadian Conference on Electrical and Computer Engineering (CCECE), pp. 1–6, May 2014
17. Pursley, D., Yeh, T.: High-level low-power system design optimization. In: 2017 International Symposium on VLSI Design, Automation and Test, VLSI-DAT 2017 (2017)
18. Rodway, J., Musilek, P., Lozowski, E., Prauzek, M., Heckenbergerova, J.: Pressure-based prediction of harvestable energy for powering environmental monitoring systems. In: 2015 IEEE 15th International Conference on Environment and Electrical Engineering, EEEIC 2015 - Conference Proceedings, pp. 725–730 (2015)
19. Son, S., Jeon, Y., Baek, Y.: Design and implementation of low-power location tracking system based on IEEE 802.11. In: Proceedings - 16th IEEE International Conference on High Performance Computing and Communications, HPCC 2014, 11th IEEE International Conference on Embedded Software and Systems, ICESS 2014 and 6th International Symposium on Cyberspace Safety and Security, CSS 2014, pp. 562–565 (2014)

20. Stankovic, J.A., He, T.: Energy management in sensor networks. Philos Trans. A Math. Phys. Eng. Sci. **370**(1958), 52–67 (2012)
21. Yousra, N., Samir, B.A.: Fuzzy multiprocessor architecture reconfiguration based on dynamic frequency scaling. In: 2017 12th International Conference on Intelligent Systems and Knowledge Engineering (ISKE), pp. 1–6. IEEE (2017)

Interpretability of Fuzzy Temporal Models

Alexander N. Shabelnikov(✉), Sergey M. Kovalev, and Andrey V. Sukhanov

Rostov State Transport University, Rostov-on-Don, Russia
drewnia@rambler.ru

Abstract. The paper presents a new approach to assessment of interpretability of fuzzy models. The approach differs from conventional ones, which consider interpretability from the point of structural complexity of both fuzzy model and its elements. In terms of developed approach, the interpretability means the ability of fuzzy model to reflect the same information presented in different forms to different users. Different forms of fuzzy model are given by use of specific inference system, which provides equivalent transformations of fuzzy rules from knowledge base on the linguistic level.

In our work, the inference system providing the equivalent transformations of fuzzy rules is developed for the specific class of fuzzy-temporal models. The necessary and sufficient conditions for properties of fuzzy rules are found. Such conditions provide semantic equivalence for equations obtained during fuzzy inference.

The formalized criterion is presented for interpretability of fuzzy model. The criterion is based on ability of model to keep information semantics on the fuzzy sets level when it is changed on the linguistic level.

Keywords: Assesment of fuzzy models · Cointension
Fuzzy interpretation of subjective information

1 Introduction

One of the main advantages of fuzzy logic is ability of knowledge linguistic presentation as well as ability of meaningful interpretation. The models developed in terms of fuzzy logic are "transparent" and clear for a common user. That condition is the main advantage of these models over other ones, especially in such areas, where meaningful interpretation of the results is less important than accuracy of modeling [1–4].

In practice, fuzzy models are frequently used as universal approximators, which allow to detect complex dependences of input-output type in the experimental data and present them via user-clear language. However, these models have less accuracy than conventional numerical ones. New classes of adaptive neuro-fuzzy models and fuzzy-genetic models were born during development of

A. Abraham et al. (Eds.): IITI 2018, AISC 874, pp. 223–234, 2019.
https://doi.org/10.1007/978-3-030-01818-4_22

hybrid technologies. The accuracy of such models can be significantly increased due to specific training. During this training, adjusting parameters and structure of fuzzy model can be significantly changed so resulting model lose its interpretational properties. As a result, the pursuit of accuracy leads to the loss of important competitive advantage of fuzzy model that is interpretability. Because of this, development of fuzzy models providing the acceptable accuracy and, simultaneously, satisfying the criterion of interpretability is an actual problem.

Interpretability is not common notion in mathematical modelling, so the different researches consider it in different aspects and estimate from the different points. The bigger part of researchers connects this property with the set of conditions and restrictions imposed on description of a task. They do this according to the common sense and intuitive notions about interpretability [5–9]. Most often, interpretability is connected with structural specifications of fuzzy model and its complexity caused by such factors as number of fuzzy rules included into knowledge base, total number of linguistic variables and fuzzy terms, features of fuzzy partition of base set, etc. [10–13]. However, despite on possibility of such simple formalization of this approach to interpretability assessment, nowadays, there is no uniform decision for choice of one or the another feature from rule base to assess the complexity of fuzzy model and the conditions, which could be a comprehensive feature of interpretability of fuzzy rules.

Present paper develops a fundamentally new approach to assessment of interpretability, which is based on notion of "cointension", introduced by Zadeh [5]. According to [5], cointension is relation between similar notions and objects. Paper [14] proposes to use this notion to assess interpretability of fuzzy classifiers. There, the assessment is based on cointension between external semantics (semantics of fuzzy knowledge base) and internal semantics of user. Interpretability of fuzzy models is considered from the point of cointension between these semantics, i.e. as estimation of the operation efficiency of two rule bases. The first rule base is related to initial model and the second one is obtained as initial rule base minimization using logical transformations according to Quine algorithm. When results of both models are similar for test samples, the semantics are considered as cointensive and initial model is considered as interpretable.

Present paper elaborates the idea of cointension use for assessment of interpretability of fuzzy models and proposes the common approach, which is based on matching of different forms of fuzzy model representation at linguistic and semantic levels and on its similarity assessment. Proposed approach is developed for fuzzy temporal models, which are oriented on temporal data mining. Section 2 presents the idea inspiring the current research, Sect. 3 presents formations of fuzzy temporal rules used in the experiments, Sect. 4 describes equivalent transformations of fuzzy rules, Sect. 5 shows the proposed notion of interpretability, Sect. 6 presents the conditions, which allows to assess interpretability of fuzzy model and the last section provides conclusions and future work.

2 Common Approach for Interpretability Assessment

Our representation of interpretability of fuzzy model means the ability of different groups of people to perceive equally well the same information contained in fuzzy model. Commonly, the fuzzy rules of knowledge base are formulated in such a manner, that they look like sentences of natural language. These sentences form the classes of cognitive structures for user, inside of which rules have the different forms of representation, but they are equally well interpreted at the semantical level by different groups of people.

Proposed approach for interpretability assessment is based on the assumption that different users consider the same rules of the knowledge base of the fuzzy model from the different points of view depending on their preferences in perception of one or the another logical-linguistic construction. For example, during interpretation of fuzzy sentence, which contains expression NOT LONG AGO and NOT JUST NOW, some users prefer to use its inversion RECENTLY. Naturally to assume, that a set of all fuzzy rules obtained from knowledge base of fuzzy model using formally equivalent transformations of its rules fully conforms to the possible representations of these rules by different users of this model. Herein, all rules obtained using logical transformations of initial rule must keep the uniform semantics because the same model is considered. In other words, formal transformations of fuzzy rules based on logical operators, which are defined in propositional logic, must be kept in the implicit semantics, which is defined by hidden cognitive structures of thinking. Formalized representation of such implicit semantics is fuzzy sets, which describe elementary linguistic sentences in form of fuzzy rules, fuzzy relations and inference operators. Thus, naturally to propose, if all the variants of equivalent transformations for the fuzzy rules from the same knowledge base lead to the same or similar results on the set of real data, then the model is the interpretable in terms of proposed approach.

3 Presentation of Fuzzy Temporal Model

In fuzzy logic applications, fuzzy linguistic terms included in production rules of the fuzzy model contain the information about linguistic variables (LV). Linguistic values are presented on the numerical scale in form of ordered fuzzy subsets and have the mentions of the following sentences: SMALL, BIG, MEDIUM, etc. The preconditions of fuzzy IF-THEN rules forming knowledge base of fuzzy model are formed on the basis of combinations of fuzzy terms using fuzzy logical operations. In [15], the interpretability of fuzzy model is proposed to be considered from the point of the assessment of the interpretability of separated LV, included in the fuzzy rules.

Analogically, fuzzy terms of temporal LV included into description of production rules of fuzzy temporal model (FTM) contain information about the time of an event connected with the appearance of one or another value of features in relation to the current time t^c. Fuzzy terms of temporal LV are presented by ordered fuzzy subsets at the relative time scale and have the following meaning:

"PRECEEDED BY", "RECENTLY", "SOME TIME AGO", "LONG TIME AGO", etc. Fuzzy-temporal formulas included into descriptions of IF-THEN preconditions of FTM rules are constructed on the basis of the features of fuzzy numbers and fuzzy temporal terms using logical and temporal operators.

The following structure of fuzzy temporal formulas are common for FTM and oriented on the processing of temporal data presented in form of time series. Time series (TS) is ordered sequence of the elements of arbitrary semantics, which are related to fixed time values $t_i = i$ at the discrete time scale T:

$$X = \big(x(t_1), x(t_2), \dots, x(t_i), \dots\big),$$

where $x(t_i) = x(i)$ is the TS at i-th time point.

Let the model of elementary temporal knowledge in FTM be in form of fuzzy production rule (fuzzy-temporal production) IF-THEN defined in relation to current time t_c:

$$\text{IF } G\big(t^c, t|(t \le t^c)\big) \text{ THEN } r(t)|(t > t^c), \tag{1}$$

where $G\big(t^c, t|(t \le t^c)\big)$ is the FTM described temporal scenario of events on a preceding temporal interval; $r(t)$ is the predicted event in the following time $t > t^c$.

Fuzzy production (1) reflects the causality between following events obtained at $t > t^c$ and preceding ones obtained at $t < t^c$.

Depending on the FTM type, numerical values $x(t)$, class labels of temporal patterns λ_n ($n \in \mathbb{N}$), function of preceded values $F\big(x(t^c), x(t^c-1), \dots, x(t^c-k)\big)$ ($k \in \mathbb{N}$) or linguistic values (fuzzy terms) α_n ($n \in \mathbb{N}$) can be used as predicted events $r(t)$ presenting consequents from (1).

Formula $G\big(t^c, t|(t \le t^c)\big)$ describing the antecedent of fuzzy production (1) is given in one of the following forms:

$$G(t^c, t) := \exists t \le t^c \bigvee_{j=1}^{m} \Big(\bigwedge_{i=1}^{n_j} P_{i_j}\big(x(t)\big) \wedge AT_{i_j}(t, t^c)\Big) \tag{2}$$

or

$$G(t^c, t) := \forall t \le t^c \bigvee_{j=1}^{m} \Big(\bigwedge_{i=1}^{n_j} P_{i_j}\big(x(t)\big) \wedge AT_{i_j}(t, t^c)\Big), \tag{3}$$

where $j \in [1, \dots m]$ is the pattern id, m is the total number of assumed patterns, i is the time point id inside j-th pattern, n_j is the total number of time points inside j-th pattern, $P_{i_j}\ (x(t))$ is the fuzzy predicate describing temporal event, which is occurred at time t (for example, $P_{i_j}\big(x(t)\big) = \big(x(t) \ge 0.9\big)$ or $P_{i_j}\big(x(t)\big) = \big(x(t)$ IS "VERY SMALL"$)\big)$, $AT_{i_j}(t, t^c)$ is the fuzzy predicate presented as follows:

$$AT_{i_j}(t, t^c) = \big(t^c - t = \tau_{i_j}\big),$$

where $\tau_{i_j} \in \mathrm{T}$ is the term of temporal LV "PERIOD" T.

To make the accent on operation $P \wedge AT$ in the following descriptions, let $P \wedge AT$ and $P \bullet AT$ be identical.

It should be noted that time variable t from $G\big(t^c, t|(t \le t^c)\big) = G(t^c, t)$ is associated with one of two quantifiers ($\exists$ or $\forall$). Because of this, formula $G\big(t^c, t|(t \le t^c)\big) = G(t^c, t)$ depends only on t^c. Therefore, the following designation is permissible $G(t^c) := G\big(t^c, t|(t \le t^c)\big)$.

Temporal LV "PERIOD" is defined on the discrete scale in relation to time Δ, the values of which are presented as difference between current time t^c and considered one $t < t^c$. An example of temporal LV T with 3 fuzzy temporal terms (τ_1(long time ago), τ_2(some time ago), τ_3 (recently)) is presented in Fig. 1.

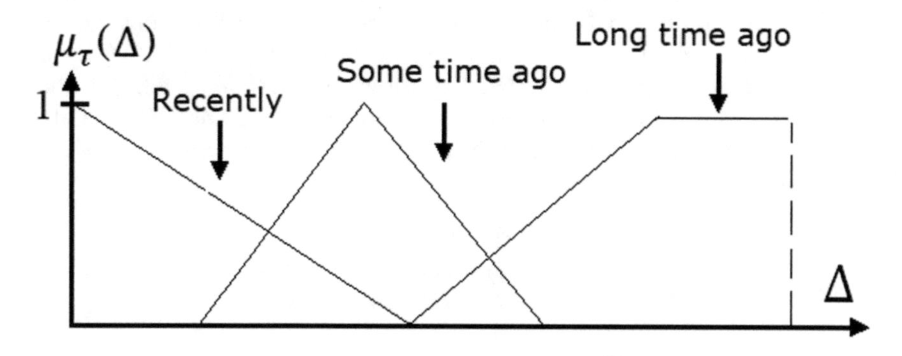

Fig. 1. Membership functions for 3 fuzzy terms of temporal LV "PERIOD"

For fuzzy term τ, the value of membership function (MF) $\mu_\tau(\Delta)$ characterizes the truth of proposition "Time $t^c - \Delta$ corresponds to PERIOD τ_i". For example, $\mu_{\text{RECENTLY}}(3) = 0.8$ means that event at the time $t^c - 3$ is recent in relation to t^c with the confidence 0.8.

Fuzzy proposition $t^c - t = \tau_i$ in conjunct $P_{i_j}(x(t)) \bullet AT_{i_j}(t, t^c)$ becomes assigned at certain current time $t^c = t^{c*}$ and gets the truth value $\mu_{\tau_i}(t^{c*} - t)$ for each t.

In the common case, temporal LV T has N time ordered fuzzy terms $\tau_1, \tau_2, \ldots, \tau_N$, for which the condition of the ordering is the following:

$$\forall \tau_i \in \mathrm{T}, \exists t_{trig} \in \Delta,$$
$$\big(\forall t < t_{trig},\ \mu_{\tau_{i-1}}(t^c - t) > \mu_{\tau_i}(t)\big) \& \big(\forall t \ge t_{trig},\ \mu_{\tau_{i-1}}(t) < \mu_{\tau_i}(t^c - t)\big).$$

Fuzzy-temporal formulas (2) and (3) describe some temporal scenario occurred at the TS by the current time t^c, which is significant for decision making. Significance of t^c means that, for example, it may be the precedent before some key event in the future or it may set causal relationship between preceding and subsequent temporal patterns.

As example, fuzzy temporal formula describing the antecedent of the rule from fuzzy model of the monitoring of railway car negative dynamics can be shown as follows [16]:

$$\forall t \le t^c,$$
$$\Big((x(t) = \mathrm{L}) \bullet (t^c - t = \mathrm{ST})\Big) \vee \Big((x(t) = \mathrm{S}) \bullet (t^c - t = \mathrm{NL\ RCT})\Big), \tag{4}$$

where $x(t)$ is the value of TS, which characterizes the deviation of car center from its normal state at time t; SMALL and LARGE are the fuzzy terms of LV "DEVIATION"; RCT (recently) and ST (some time ago) are the fuzzy terms of LV "PERIOD"; NL (not later than) is the temporal modifier.

Formula (4) describes fuzzy-temporal scenario of increase of car vibration amplitude, which is preceded to abnormal emergence situation. Parenthetical expressions in the formula are atoms describing fuzzy periods of event occurring in relation to the current time t^c.

Representation grammar for fuzzy-temporal formulas of form (2) and (3) in Backus-Naur form can be defined more strictly as the following. Let T = $\{\tau_i\}$ $(i = 1, 2, \ldots, N)$ be a time-ordered term set of temporal LV. To increase expressiveness of the grammar, ANT linguistic value (IN ANY TIME) can be added in term set T. As well, two temporal modifiers NL (not later) and NE (not earlier) applied to fuzzy terms from T can be added. Semantics of ANT, NL and NE is given as follows:

$$\begin{aligned} &\forall t < t^c,\ \mu_{\mathrm{ANT}}(t^c - t) = 1 \\ &\forall t < t^c,\ \mu_{\mathrm{NL}\ \tau_i}(t^c - t) = \sup\left\{\mu_{\tau_i}(t^c - t^*)|t^* \le t\right\} \\ &\forall t < t^c,\ \mu_{\mathrm{NE}\ \tau_i}(t^c - t) = \sup\left\{\mu_{\tau_i}(t^c - t^*)|t^* \ge t\right\}. \end{aligned}$$

Figure 2 illustrates application of temporal modifiers on MF $\mu_{\tau_i}(t)$ of fuzzy temporal term τ_i.

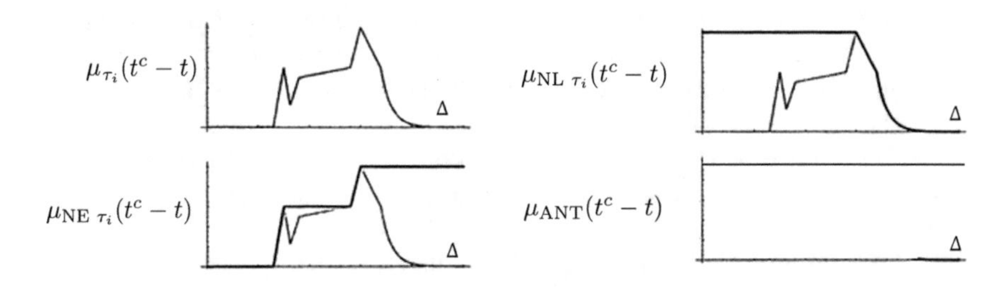

Fig. 2. Illustration of temporal modifiers ANT, NL, NE

Grammar Γ for representation of fuzzy temporal formulas describing antecedents of rules from knowledge base of FTM is given as follows:

$$\begin{aligned} \langle\text{Start}\rangle &\models \langle\exists t \le t^c\rangle,\langle\text{formula}\rangle \mid \langle\forall t \le t^c\rangle,\langle\text{formula}\rangle \\ \langle\text{formula}\rangle &\models \langle\text{atom}\rangle \mid \langle\text{atom}\rangle\vee\langle\text{atom}\rangle \\ \langle\text{atom}\rangle &\models \langle P\rangle\bullet\langle\text{fuztemp}\rangle \mid \langle P\rangle\bullet\langle\text{fuztemp}\rangle\wedge\langle\text{atom}\rangle \\ \langle\text{fuztemp}\rangle &\models \langle t^c - t = \tau_i\rangle \mid \langle t^c - t = \text{adverb } \tau_i\rangle \\ \langle\text{adverb}\rangle &\models \langle\text{NL}\rangle \mid \langle\text{NE}\rangle \mid \langle\text{ANT}\rangle \end{aligned}$$

Therefore, knowledge base of FTM includes the system of fuzzy production rules of type "IF-THEN" (1), where antecedents are defined by grammar Γ and consequents are elements of set $\mathfrak{R}$. As elements of set $\mathfrak{R}$, numerical or symbol value of $x(t)$, class labels of temporal patterns λ_n $(n \in \mathbb{N})$, function of preceding values $F\big(x(t^c), x(t^c-1), \ldots, x(t^c-k)\big)$ $(k \in \mathbb{N})$ or fuzzy terms of fuzzy numerical variables α_i can be used.

4 Logical Transformation of Knowledge Base

Logical transformations of knowledge base is performed on the basis of the system of inference rules $\mathfrak{R} = \{R_i\}$ providing equivalent transformation of internal fuzzy production rules to get different forms of representation for the same knowledge and transform the knowledge to different user groups in different manner. As well, equivalent transformations are useful for knowledge base minimization due to more compact description with less number of included variables in production rules. Inference rules are applied to antecedents of fuzzy production rules included into knowledge base of FTM and presented by logical equivalences of form $R_i : G \Leftrightarrow F$, where G and F are the temporal formulas of form (2) or (3).

Inference rule system $\mathfrak{R}$ for knowledge base transformation includes two groups of logical equivalences. The first one reflects the common laws of commutativity, associativity and distributivity, which take place for propositional logic and presented as follows:

$$\begin{array}{r}
P\big(x(t)\big) \bullet \big(AT_i \wedge AT_j\big) \Leftrightarrow P\big(x(t)\big) \bullet \big(AT_j \wedge AT_i\big) \\
P\big(x(t)\big) \bullet \big(AT_i \vee AT_j\big) \Leftrightarrow P\big(x(t)\big) \bullet \big(AT_j \vee AT_i\big) \\
P\big(x(t)\big) \bullet \big((AT_i \vee AT_j) \vee AT_l\big) \Leftrightarrow P\big(x(t)\big) \bullet \big(AT_i \vee (AT_j \vee AT_l)\big) \\
P\big(x(t)\big) \bullet \big((AT_i \wedge AT_j) \wedge AT_l\big) \Leftrightarrow P\big(x(t)\big) \bullet \big(AT_i \wedge (AT_j \wedge AT_l)\big),
\end{array} \tag{5}$$

where $AT_k = (t^c - t = \tau_k)$ $k \in \mathbb{N}$.

The second group is presented by individual inference rules defined automatically based on intuitive imaginations of human user about temporal relationships between time events and temporal modifiers.

Let $\mathrm{T} = \{\tau_i\}$ $(i = 1, 2, \ldots, N)$ be a set of linguistic values of temporal LV T, which are ordered in time in relation to current time t^c. Individual inference rules are presented by the following system of equivalent transformations:

$$\begin{array}{l}
P\big(x(t)\big) \bullet \big(AT_1 \vee \ \ldots \ \vee AT_j\big) \Leftrightarrow P\big(x(t)\big) \bullet \big(t^c - t = \mathrm{NL}\ \tau_j\big) \\
P\big(x(t)\big) \bullet \big(AT_j \vee \ \ldots \ \vee AT_n\big) \Leftrightarrow P\big(x(t)\big) \bullet \big(t^c - t = \mathrm{NE}\ \tau_j\big) \\
P\big(x(t)\big) \bullet \big(AT_1 \vee \ \ldots \ \vee AT_n\big) \Leftrightarrow P\big(x(t)\big) \bullet \big(t^c - t = \mathrm{ANT}\big).
\end{array} \tag{6}$$

It should be noted that above presented inference rules take place in propositional logic, where the internal structure of elementary propositions is not considered. There, the way of combination of these elementary propositions just based on logical operations and temporal modifiers. If inference rules (5) and (6) are considered as formulas $G \Leftrightarrow F$ taking into account truth values of included elementary fuzzy propositions, then these formulas may not be equivalences in general case, because their truth values, which depend on the properties of propositions (such as current time t^c, truth value of fuzzy predicate $P\big(x(t)\big)$, membership functions of temporal terms $\mu_\mathcal{T}(t^c - t)$), may differ for left and right side of formula.

In case of their equivalence on the semantic level, individual inference rules are the powerful tool for knowledge base simplification due to use of temporal modifiers, which allows to transform formulas into more compact expressions.

5 Definition of Interpretability of Fuzzy Temporal Model

As previously noted, the proposed approach to assessment of interpretability of fuzzy model is based on operation results matching at different forms of representation of knowledge base obtained by equivalent transformation of production rules included into knowledge base. Equivalent transformations are related to formulas describing antecedents of fuzzy rules from FTM. The following conceptual definition of interpretability refers to this approach:

Definition 1. *Fuzzy-temporal model is interpretable if fuzzy sets of decisions, which are based on any fuzzy rules representation obtained as a result of equivalent transformations* (5) *and* (6), *are equivalent.*

The following set of definitions are related with the above one and allow to more strictly represent it.

Definition 2. *In terms of fuzzy propositions* $(t^c - t) = \tau$, *assignment for current time* t^c *as substitution* $t^c = t^*$ *is called temporal interpretation of fuzzy-temporal formula* $G = G(t^c)$.

Based on association of t with quantifiers $\forall$ or $\exists$ in (2) and (3), assignment of current time variable $t^c = t^*$ allows to compute the truth value of (2) and (3) at current time $t^c = t^*$ on the basis of one of the following expressions:

$$G(t^c, t) := \forall t \le t^c\ T\big(G(t^*)\big) = \sup_t \bigvee_{j=1}^{m} \left(\bigwedge_{i=1}^{n_j} T\Big(P_{i_j}\big(x(t)\big)\Big) \bullet \mu_{\mathcal{T}_{i_j}}(t^* - t) \right)$$

or

$$G(t^c, t) := \exists t \le t^c\ T\big(G(t^*)\big) = \sup_t \bigvee_{j=1}^{m} \left(\bigwedge_{i=1}^{n_j} T\Big(P_{i_j}\big(x(t)\big)\Big) \bullet \mu_{\mathcal{T}_{i_j}}(t^* - t) \right),$$

where $T\Big(P_{i_j}\big(x(t)\big)\Big)$ is the truth value of predicate P_{i_j} at time t; $\mu_{\mathcal{T}_{i_j}}(t^* - t)$ is the value of MF $\mu_{\mathcal{T}_{i_j}}(\Delta)$ for $\Delta = t^* - t$.

Definition 3. *Let knowledge base* Ξ *of fuzzy model be represented by the following system of fuzzy temporal rules:*

$$\Xi := G_i(t^c) \to r_i(t)\ (i \in \mathbb{N}),$$

where $G_i(t^c)$ *is the fuzzy temporal formula of form* (2) *or* (3); $r_i(t) \in R$ *is the predicted event.*

Fuzzy inference in knowledge base Ξ is mapping of decisions into fuzzy scale or computing of its truth value based on one of the following formula depending on type of consequent $r_i(t) \in \mathfrak{R}$:

$$x(t^c+1) = \frac{\sum_i T\big(G_i(t^c)\big) \cdot F_i\big(x(t^c), \ldots, x(t^c-k)\big)}{\sum_i T\big(G_i(t^c)\big)},$$

where $T(\cdot)$ is the truth value.

Based on above mentioned definition of general fuzzy inference scheme, the unambiguous relation between inferential decisions and truth values of preconditions from fuzzy rules included into knowledge base of FTM is followed. It allows to form the condition of interpretability in a stricter way.

Definition 4. *Formulas G and F are called as logically equivalent based on inference rules $G \overset{\mathfrak{R}}{\Leftrightarrow} F$ if the sequence of inference rules $R_1 \in \mathfrak{R}, R_2 \in \mathfrak{R}, \ldots, R_k \in \mathfrak{R}$, which transform one formula to another, exists, i.e.:*

$$G \overset{\mathfrak{R}}{\Leftrightarrow} F \rightarrow \exists R_1, \ldots, R_k \in \mathfrak{R}, \big(G \overset{R_1}{\Leftrightarrow} F_1'\big) \& \ldots \& \big(F_{k-1}' \overset{R_k}{\Leftrightarrow} F\big). \tag{7}$$

Definition 5. *Formulas G and F are called as semantically equivalent if their truth values are equivalent in every interpretation $t^c = t^*$, i.e.:*

$$\forall t^c = t^*,\ T\big(G(t^*)\big) = T\big(F(t^*)\big). \tag{8}$$

FTM is interpretable if formulas describing antecedents of all its fuzzy rules are interpretable.

FTM interpretability conditions. The following definition characterizes the inference rule system property, which is important in definition of FTM interpretability conditions.

Definition 6. *Inference rule $R_i : G_i(t) \Leftrightarrow F_i(t)$ is called applicable if $T\big(G_i(t^*)\big) = T\big(F_i(t^*)\big)$ takes place in every interpretation $t^c = t^*$.*

The following proposition can be given.

Proposition 1. *Let G and F be logically equivalent in relation to inference rule system $\mathfrak{R}$. Formulas G and F are semantically equivalent if and only if inference rules $R_i \in \mathfrak{R}$ are applicable.*

Therefore, interpretability of fuzzy rules from FTM knowledge base depends on conditions, for which inference rules are applicable. Applicability conditions for inference rules (5) and (6) depend on basis of fuzzy operations. As well, they depend on MF character of fuzzy temporal terms and fuzzy partitions generated by these fuzzy temporal terms in relation to time scale Δ.

Applicability of inference rule group (5) is provided for the corresponding choice of fuzzy operation basis (particularly, Lukasiewicz basis):

$$a \vee b = min\{a+b, 1\}; \qquad a \wedge b = max\{a+b-1, 0\}. \tag{9}$$

The conditions of applicability for individual inference rule group (6) is presented as follows. It should be noted that it is enough to set equivalence for expressions with quantifiers $\forall$ and $\exists$ at every t^c and t to assess the applicability of inference rules $R := G \Leftrightarrow F$, where fuzzy formulas G and F are presented by (2) or (3). Inference rules (6) can be rewritten via truth values as follows:

$$T_i(t) \bullet \left(\mu_{\tau_1}(t_c - t) \vee \cdots \vee \mu_{\tau_j}(t_c - t)\right) \Leftrightarrow T_i(t) \bullet \mu_{\mathrm{NL}\ \tau_j}(t_c - t)$$

$$T_i(t) \bullet \left(\mu_{\tau_j}(t_c - t) \vee \cdots \vee \mu_{\tau_N}(t_c - t)\right) \Leftrightarrow T_i(t) \bullet \mu_{\mathrm{NE}\ \tau_j}(t_c - t)$$

$$T_i(t) \bullet \left(\mu_{\tau_1}(t_c - t) \vee \cdots \vee \mu_{\tau_N}(t_c - t)\right) \Leftrightarrow T_i(t) \bullet \mu_{\mathrm{ANT}}(t_c - t).$$

$T_i(t)$ can be reduced:

$$\left(\mu_{\tau_1}(t_c - t) \vee \cdots \vee \mu_{\tau_j}(t_c - t)\right) \Leftrightarrow \mu_{\mathrm{NL}\ \tau_j}(t_c - t) \tag{10}$$

$$\left(\mu_{\tau_j}(t_c - t) \vee \cdots \vee \mu_{\tau_n}(t_c - t)\right) \Leftrightarrow \mu_{\mathrm{NE}\ \tau_j}(t_c - t) \tag{11}$$

$$\left(\mu_{\tau_1}(t_c - t) \vee \cdots \vee \mu_{\tau_n}(t_c - t)\right) \Leftrightarrow \mu_{\mathrm{ANT}}(t_c - t). \tag{12}$$

Applicability of above presented formulas for chosen basis of fuzzy operations (9) on MF $\mu_{\tau_i}(d)$, which conforms to normality and concavity conditions, depends on mutual position of MF on time scale Δ. Applicability of formula (12) is provided based on definition of MF $\mu_{\mathrm{ANT}}(t_c - t)$ and formalization way of fuzzy disjunction if and only if:

$$\forall t < t_c \quad \mu_{\tau_1}(t_c - 1) + \cdots + \mu_{\tau_c}(t_c - t) = 1. \tag{13}$$

Condition (13) presents well-known Ruspini restriction for fuzzy partition of numerical universe [17].

The following proposition takes place.

Proposition 2. *Let* $T = \{\tau_1, \tau_2, ..., \tau_N\}$ *be a time ordered set of fuzzy terms included into antecedent descriptions of fuzzy rules of form* (1). *Formulas* (5), (6) *r fuzzy temporal formulas represented in form* (2), (3). *Then, inference rule system* (5), (6) *is applicable for fuzzy logical operation basis* (9) *if and only if MF of fuzzy terms satisfy the requirements of normality and concavity and fuzzy partition, which is induced by the terms, satisfy the condition* (13) *in relation to time scale* Δ.

Proposition 2 defines necessary and sufficient conditions for MF of fuzzy temporal terms, for which the system is applicable and FTM is interpretable according to criterion of Definition 2.

6 Conclusion

The paper is dedicated to development of a new approach to interpretability assessment in fuzzy systems. The approach totally differs from conventional ones, where interpretability is assessed through complexity of fuzzy model and its elements description. In terms of proposed approach, interpretability is considered as ability of fuzzy model to reflect the same knowledge in different forms of the representation to different groups of users without semantics loss. This ability is assessed by using matching of operation results for different implementation of the same fuzzy model obtained via formal-logical transformation of initial model.

The interpretability criterion is proposed for fuzzy models and necessary and sufficient conditions of interpretability are found for special class of fuzzy models with fuzzy temporal variables. Obtained conditions are represented as restrictions over normality and concavity of membership functions, which is related to well-known Ruspini requirement. These conditions conform to many known heuristic criteria of interpretability. However, they have strict rationale in form of corresponding statement.

Developed approach to interpretability assessment can be applied to any type of fuzzy models and used for development of interactive human-computer systems, which able to describe the decision making process to wide range of users.

Acknowledgement. The work was supported by RFBR (Grants No. 17-20-01040 ofi_m_RZD, No. 16-07-00032-a and No. 16-07-00086-a).

References

1. Zadeh, L.: From computing with numbers to computing with words - from manipulation of measurements to manipulation of perceptions. In: IEEE Transactions on Circuits and Systems - I: Fundamental Theory and Applications, pp. 81–117. Springer, Heidelberg (2002)
2. Mencar, C., Fanelli, A.M.: Interpretability constraints for fuzzy information granulation. Inf. Sci. **178**(24), 4585–4618 (2008)
3. Delgado, M.R., Zube, F.V.: Interpretability issues in fuzzy modelingstudies in fuzziness and soft computing. In: Hierarchical Genetic Fuzzy Systems: Accuracy, Interpretability and Design Autonomy, pp. 379–405. Physica-Verlag, New York (2003)
4. Bargiela, A., Pedrycz, W.: Granular computing. In: Handbook on Computational Intelligence: Fuzzy Logic, Systems, Artificial Neural Networks, and Learning Systems, vol. 1, pp. 43–66 (2016)
5. Zadeh, L.A.: Is there a need for fuzzy logic? Inf. Sci. **178**(13), 2751–2779 (2008)
6. Zhou, S.M., Gan, J.Q.: Extracting Takagi-Sugeno fuzzy rules with interpretable submodels via regularization of linguistic modifiers. IEEE Trans. Knowl. Data Eng. **21**(8), 1191–1204 (2009)
7. Alonso, J.M., Magdalena, L.: Combining user's preferences and quality criteria into a new index for guiding the design of fuzzy systems with a good interpretability-accuracy trade-off. In: 2010 IEEE International Conference on Fuzzy Systems (FUZZ), pp. 961–968 (2010)
8. Mencar, C., Castellano, G., Fanelli, A.M.: On the role of interpretability in fuzzy data mining. Int. J. Uncertain., Fuzziness Knowl.-Based Syst. **15**(05), 521–537 (2007)

9. Gacto, M.J., Alcala, R., Herrera, F.: Integration of an index to preserve the semantic interpretability in the multiobjective evolutionary rule selection and tuning of linguistic fuzzy systems. IEEE Trans. Fuzzy Syst. **18**(3), 515–531 (2010)
10. Ishibuchi, H., Nojima, Y.: Analysis of interpretability-accuracy tradeoff of fuzzy systems by multiobjective fuzzy genetics-based machine learning. Int. J. Approx. Reason. **44**(1), 4–31 (2007)
11. Marquez, A.A., Marquez, F.A., Peregrin, A.: A multi-objective evolutionary algorithm with an interpretability improvement mechanism for linguistic fuzzy systems with adaptive defuzzification. In: IEEE International Conference on Fuzzy Systems (FUZZ), pp. 277–283 (2010)
12. Alonso, J.M., Magdalena, L., Guillaume, S.: HILK: a new methodology for designing highly interpretable linguistic knowledge bases using the fuzzy logic formalism. Int. J. Intell. Syst. **23**(7), 761–794 (2008)
13. Riid, A., Rustern, E.: Interpretability improvement of fuzzy systems: reducing the number of unique singletons in zeroth order Takagi-Sugeno systems. In: IEEE International Conference on Fuzzy Systems (FUZZ), pp. 2013–2018 (2010)
14. Mencar, C., Castiello, C., Cannone, R., Fanelli, A.M.: Interpretability assessment of fuzzy knowledge bases: a cointension based approach. Int. J. Approx. Reason. **52**(4), 501–518 (2011)
15. Bodenhofer, U., Bauer, P.: A formal model of interpretability of linguistic variables. In: Interpretability Issues in Fuzzy Modeling, pp. 524–545. Springer, Heidelberg (2003)
16. Kovalev, S.M., Tarassov, V.B., Dolgiy, A.I., Dolgiy, I.D., Koroleva, M.N., Khatlamadzhiyan, A.E.: Towards intelligent measurement in railcar on-line monitoring: from measurement ontologies to hybrid information granulation system. In: International Conference on Intelligent Information Technologies for Industry, pp. 169–181. Springer, Cham (2017)
17. Ruspini, E.H.: A new approach to clustering. Inf. Control. **15**(1), 22–32 (1969)

Control of the Cognitive Process in Hard Real-Time Environment in the Context of the Extended Stepping Theories of Active Logic

Michael Vinkov[1], Igor Fominykh[2](✉), and Sergey Romanchuk[2]

[1] Bauman Moscow State Technical University, Moscow, Russia
vinkovmm@mail.ru
[2] Moscow Power Engineering Institute, Moscow, Russia
igborfomin@mail.ru, theerror133@gmail.com

Abstract. Issues related to ensuring stability to unpredictable situations that arise when solving tasks by a cognitive agent in the hard real time environment. The underlying basis for maintaining stability is the control of the cognitive process, during which these situations are identified, and the process is adapted taking into account time resources available to the agent. The work provides the definition of the basic principles of control execution and suggests an approach for their implementation in the context of the extended logical programs of Active Logic.

Keywords: Active Logic · Hard real time · Stepping theories
Cognition process · Meta-reasoning · Argumentative semantics

1 Introduction

The problem addressed by this report is the control of reasoning process of a cognitive agent (hereinafter, the agent) capable of reasoning on the basis of its knowledge and observations of the external environment, task solving under hard real-time conditions. Operating under these conditions is characterized by a critical time constraint (deadline), where time-out exposes the agent to severe consequences, which can be catastrophic, and for the agent this is unacceptable. It is clear that agents working in such time pressure conditions must build their behavior differently from "soft" real-time situations with no strict deadline, and the delay in problem solving only leads to a decrease in the quality of the agent's functioning, approximately in proportion to the size of the time delay.

Below is the description of a reasoning process (hereinafter, the cognitive process) carried out in a hard real-time environment and the review of its basic principles.

The problem addressed by this report is the control of reasoning process of a cognitive agent (hereinafter, the agent) capable of reasoning on the basis of its

This work was supported by the Russian Foundation for Basic Research. Grants No. 17-97-00696, 15-07-02320.

A. Abraham et al. (Eds.): IITI 2018, AISC 874, pp. 235–243, 2019.
https://doi.org/10.1007/978-3-030-01818-4_23

knowledge and observations of the external environment, task solving under hard real-time conditions. Operating under these conditions is characterized by a critical time constraint (deadline), where time-out exposes the agent to severe consequences, which can be catastrophic, and for the agent this is unacceptable. It is clear that agents working in such time pressure conditions must build their behavior differently from "soft" real-time situations with no strict deadline, and the delay in problem solving only leads to a decrease in the quality of the agent's functioning, approximately in proportion to the size of the time delay.

Below is the description of a reasoning process (hereinafter, the cognitive process) carried out in a hard real-time environment and the review of its basic principles.

2 Reasoning in Hard Real-Time Environment

Since a cognitive process in a hard real-time environment has strict time limits which exceeding is unacceptable, the agent must closely monitor this process, aiming to identify situations fraught with unexpected difficulties (hereinafter, anomalies) and the process comes to a "standstill" without leading to expected results. Obviously, the agent needs meta-level information, which characterizes the cognitive process as a whole. In these circumstances, at any point in time while pursuing to solve the problem, it must be capable of doing at least the following interrelated tasks:

1. Assess the available time resource.
2. Monitor the intermediate results and the time when they should have been achieved.
3. Correct the cognitive process, changing its strategy as needed.

Having these capabilities is a necessary condition providing the possibility for reasoning process control to tackle the threats posed by anomalies that arise during task solving under strict time constraints. Below we will consider issues related to the implementation of these capabilities in the context of the advanced stepping theories of Active Logic [1]. Active Logic [2–4] is a conceptual system, which includes a number of formalisms applied for modelling of meta-reasoning, which is focused on specific features of task solving in a hard real-time environment. Among other formalisms of Active Logic, extended stepping theories occupy roughly the same niche as the formalism of extended logic programs has in relation to other formalisms of non-monotonic reasoning.

3 Extended Stepping Theories of Active Logic (General Issues)

One of the fields of Active Logics introduced in [5] deals with various systems of the so-called stepping theories. The semantics of these systems is described in terms of literal sets, which in some ways makes them relevant to logic programming systems of "traditional" logic. Below we will expand on the most common system of extended stepping theories (with two types of negation) [1]. An extended stepping theory includes a set of rules expressed in the following way:

$$N: a_1 \wedge a_2 \ldots \wedge a_m \wedge not^t c_1 \wedge not^t c_2 \ldots \wedge not^t c_n \Rightarrow b, \quad (1)$$

where N – is a string of symbols designating the name of the rule, b is a propositional literal, $a_1 \ldots a_m$ are propositional literals or first order logic literals of now (j) kind, later (j) or ¬ later (j) kind, where j is a natural number, $c_1, \ldots, c_n$ are propositional literals, not^t is a subjective negation operator. Informally, the not^t *q* expression in the antecedent of the stepping theory rule means that the agent has failed to infer the literal q by the current point of time.

Hereafter, any propositional literal of a_i kind, which is not preceded by a not^t subjective negation operator will be called an objective literal. Any literal of not^t c_j kind will be hereafter called a subjective literal.

Here it should be noted that any further reasoning will remain valid even when the first order logic literals without functional symbols, i.e. with the finite Herbrand universe, are used instead of the propositional literals.

The rules reflect the principle of negative introspection in the following interpretation: if function $a_1 \wedge a_2 \ldots \wedge a_m$ is executed and at this inference step it is unknown whether formula – b (which is a literal) is plausible or not, then it can be assumed that formula b is true (it should be reminded that –b means the (always objective) literal complementing to a contrary pair of b). Wherever it is convenient, the antecedents of rules are regarded as sets of literals. If such a set is empty, then the rules are regarded as being equivalent to rules with the same consequents, but whose antecedents only have a single literal later now (0).

An extended stepping theory is a pair T = (R, Ck), where R is a finite set of rules of the form (1), Ck is a so-called clock of a stepping theory, which is a finite subsequence of the sequence of natural numbers, for example, Ck = (0, 1, 2, 4, 7, 11). Members of this subsequence (representing natural numbers or 0) are the moments of completion of successively performed deductive cycles that determine a cognitive process of reasoning in all systems of Active Logic. Let's take *t* as a point of time on the Ck timeline. The next sequential point of time in the Ck sequence is designated as *next t*. A sequence number of time point *t* in the given sequence is the value of a function *rank (t)*. If *n* is a sequence number of the time point, then this point of time itself is the value of function *clock (n).* A detailed review of this issue is given in [6].

For any extended stepping theory of the form T = (R, Ck) let R [q] refer to the set of all rules whose consequent is q. The set of literals forming the antecedent of rule r will be designated as A (r). Let Lit_T be the set of all literals occurring in the rules of stepping theory T. *The belief set* of stepping theory T = (R, >) is a set of the form $\{now (t)\} \cup L_T^t$, where t is a natural number or 0 representing a point in time on the clock Ck of this stepping theory, $L_T^t \subset Lit_T$. Let's consider operator ϑ_T that transforms the belief sets into other belief sets in such a way that if B is a belief set such that now (t) $\in$ B, then now (next t) $\in \vartheta_T(B)$. The sufficient conditions under which the marked literals will belong to belief set $\vartheta_T(B)$ will be different for different stepping theories, see, for example, [5, 7]. In the next section, we will look at these conditions as they apply to a system of step theories with two kinds of negation in the context of their argumentation semantics.

Now let B be a belief set of theory T, such that literal now (t) $\in$ B. Then B *is a quasi-fixed point* of operator ϑ_T iff $B \backslash \{now (t)\} = \vartheta_T(B) \backslash \{now (next\ t)\}$ for any t. *The history* in stepping theory T is a finite sequence of belief sets $\boldsymbol{B}$. $\boldsymbol{B}$(i) is the i-th member in the history, $\boldsymbol{B}(0) = \{\text{now }(0)\}$, for any t $\boldsymbol{B}(\text{rank }(t) + 1) = \vartheta_T(\boldsymbol{B}(\text{rank }(t)))$. The last element in the history is a belief set designated as B_{fin}, (*final*). It is the smallest quasi-fixed point of operator ϑ_T in the meaning indicated above. *An inference step* in stepping theory T = (R, Ck) is any pair of the form ($\boldsymbol{B}$(i), $\boldsymbol{B}$(i + 1)), and *the inference step number* is the number equal to (i + 1). The consequent (t-*consequent*) of stepping theory T is a literal belonging to belief set B_{fin} ($\boldsymbol{B}$(rank (t)), t ∈ Ck*), where Ck* is the set of all members of sequence Ck.

4 Argumentation Semantics for Extended Stepping Theories

The argumentation theory [8] has proved to be quite prolific in presenting of non-monotonic reasoning [9]. In [7] argumentation semantics for formalisms of active logic stepping theories was suggested. As basic elements, reasoning systems usually have a certain *logical language* and definitions for *argument, conflict between arguments* and *argument status.* A reasoning system, consisting of the above-specified elements constructed considering the specifics of active logic stepping theories, is presented below.

Definition 1. We assume that T= (R, Ck) is an extended stepping theory. Let an argument for T be defined as follows:

(1) any literal (of the first order logic) of now (t) kind, later (t) or ¬ later (t), where $t \geq 0$, for which there exists rule r ∈ R, so that l ∈ A (r);
(2) sequence of rules Arg = [r_1, …, r_n], where r_1, …, r_n ∈ R, so that for any $1 \geq i \geq n$, if p ∈ A (r_i), where p is the objective propositional literal, then there might be such j < i that r_j ∈ R [p];
(3) any subjective literal of not^t q kind.

For this stepping theory T = (R, >, Ck), *a set of all its arguments* is designated $Args_T$. If an argument is the first order logic literal of kind (1), then we will call such reason *limiting* (the function of the other reasons after and before in time).

Argument of kind (2) is called *supporting argument.* Propositional literal b t. and t. t. is called a *conclusive inference* of supporting argument Arg = [r_1, …, r_n]., when r_n ∈ R [b].

Argument of not^t q kind is called a *subjective argument.*

Any subsequence of [r_1, …, r_n] sequence, meeting definition 1 is called *supporting subargument* of argument Arg = [r_1, …, r_n]. Limiting, or subjective argument is a *subargument* of the argument Arg = [r_1, …, r_n] if a corresponding first-order logic literal, or a subjective literal is included in the antecedent of any of the rules r_1, …, r_n.

Any supporting subargument of the argument Arg = [r_1, …, r_n] is called its *maximum subargument* if a literal being a conclusive inference of the said subargument is included in the antecedent of r_n. rule. Limiting, or subjective argument is a maximum subargument of argument Arg = [r_1, …, r_n] if a corresponding literal is included in the antecedent of r_n rule.

Further, for the purpose of simplicity, an inference step with number i will be referred to as step i.

Definition 2. Putting arguments of extended stepping theories in action is performed according to the following rules:

1. Any limiting argument of *now (t)* kind is put into action on step i = rank (t) (at time point t).
2. Any limiting argument of *later (t), t* > 0 is put into action on step i, so that clock (i – 1) $< t \leq$ clock (i). Any limiting argument of $\neg$ later (t) kind is activated on step 1 (at time point 0).
3. If the supporting argument does not have any subarguments, then it will be activated on step 1.
4. Any supporting argument is activated on step i, i > 1, iff all of its subarguments are put into action on the previous steps and there is a maximum subargument activated on step (i − 1).
5. Any subjective argument is put into action on step 1(at time point 0).

Definition 3. Withdrawal of limiting arguments from action is performed by the following rule:

1. Any limiting argument of now (t) kind is withdrawn from action on step i = (rank (t) + 1).
2. Any limiting argument of $\neg$ later (t) kind is withdrawn from action on step i, where i is such that clock (i – 1) $\leq t <$ clock (i).

All the other arguments, including limiting arguments of $\neg$ later (t) after their introduction into subsequent steps of inference, have the status *active*.

Definition 4. Attacking arguments

1. Arg_1 supporting argument *attacks* the other supporting arguments with conclusion of q, or subjective argument $Arg_2 = not^t - q$ on step i iff the following conditions are satisfied:
   (1) conclusion of Arg $_1$ is the literal − q;
   (2) Arg_1 and all of its subarguments are active on step i;
   (3) none of the Arg_1 subarguments is attacked on step i by none of the other supporting arguments.
2. The subjective argument $Arg_1 = not^t$ q *attacks the* subjective argument Arg $_2 = not^t - q$ at step i, and t. t., when it is not attacked at step i by any supporting argument with the conclusion q.

Any set of beliefs in the extended stepping theory, in addition to a literal of now (i) kind, consists of objective literals, which are conclusions of supporting arguments. The definition given below establishes the necessary and sufficient condition of attributing the objective literal to a belief set in the extended stepping theory.

Definition 5. Let ***B*** (i) be a belief set of a certain extended stepping theory. An objective literal is q ∈ ***B***(i) iff

(1) there is a supporting argument Arg_1 of an extended stepping theory, which conclusion is literal q;
(2) Arg_1 and all of its subarguments (excluding limiting subarguments of ¬ later (t) and now (t) kind, provided that q ∈ ***B*** (i − 1)) are active on step *i* and Arg_1 is put into action not later than on step (i – 1);
(3) none of Arg_1 subarguments is attacked on step *i* by no other supporting arguments.

It should be noted that the supporting argument obtains an ability to attack other arguments one step earlier than it obtains an ability to support an objective literal which is its conclusion. Thus, any contradiction in belief sets is prevented.

5 Evaluation of Available Time Resource

The majority of logic systems developed to simulate the behavior of a cognitive agent using the modal approach are based on the assumption that the results of reasoning performed by the agent capable to do this are made available to it immediately when necessary. There are situations in which such an assumption is considered to be justified. However, in situations where the agent is operating under strict time constraints, such assumption proves to be too far and exorbitant. The unrealistic assumption leads to the so-called logical omniscience problem, which is very well known in epistemic logic. Consequently, agents that are subject to this problem are known as omniscient or ideal agents [10]. It is clear that an ideal agent does not need to care about the amount of time resource it has because it obtains all reasoning results instantaneously. However, if we cannot assume omniscience for our agent, then a way must be found to deal with the logical omniscience problem. Solutions to this problem have been proposed both within the modal approach (for example, refer to [11]), and outside it [2]. As was shown in [12] the solutions found within the modal approach share one principal drawback, specifically, the fundamental impossibility to obtain logical results interpreted as follows: the agent is capable (or incapable) to derive formula F within time limit t. It is noteworthy that the results of this kind are extremely important in situations when there are hard constraints on the time available to the agent. Outside the modal approach this drawback is successfully dealt with in logical systems that are relevant to the concept of 'Active Logic' or similar systems [13], in which the agent's reasoning is interpreted not as a sequence of formulas (statements) regarded as a whole and existing outside of time, but rather, as a process that has a certain duration in time. Below, we talk about logic systems that fall within the concept of Active Logic.

It seems obvious that for agents acting under hard real time constraints, control of this time resource cannot be performed without correlating the results obtained during the cognitive process with the time points when these results were obtained. In accordance with the concept of Active Logic, the cognitive process implies sequential performance of deductive cycles referred to as inference steps (see Sect. 2 of this article). In the above-mentioned system of extended stepping theories (hereinafter, theories), keeping track of time is performed using a special literal *now* (.). Its arguments are time points of completion of deductive cycles (this information is always reflected in the clock of Ck theory). In accordance with the argumentation semantics of extended stepping theories described in the previous section, each set of beliefs in the theory contains exactly one literal now (t), where t is a time point on clock Ck, indicating when the given set of beliefs was formed.

This information is sufficient to monitor the decrease in the time resource as the cognitive process continues. Specifically, using a literal later (t) makes it possible to estimate the current time resource at the upper level, whereas using literal $\neg$ later (t) enables estimating the lower limit of the current time resource.

Example 1. Let us consider theory $T_1 = (R_1, Ck_1)$, where R_1 = {N1: $\Rightarrow$ a, N2: a $\Rightarrow$ b, N3: b $\Rightarrow$ c, N4: $\neg$ later (7)) $\wedge$ c $\Rightarrow$ set_a_subgoal_A, N5: later (7) $\wedge$ not^t set_a_subgoal_A $\Rightarrow$ set_a_subgoal_B}, Ck_1 = (1, 3, 8, 10). The history of this theory T_1 is expressed as follows: ***B*** (0) = {now(0)}, ***B*** (1) = {now(1), a}, ***B*** (2) = {now(3), a, b,}, ***B*** (3) = {now(8), a, b, c}, ***B*** (4) = B_{fin} = {now(10), a, b, c, set_a_subgoal_B}. Regarding information content, in the course of the cognitive process formalized using the theory T_1, an estimate of the time resource on step ***B*** (3) was made for possible setting a local sub-goal A with the help of rule N 4. The resource was insufficient and a local sub-goal B was set (the rules used to achieve it are not shown in this example). It should be noted that in theory T_2 = (R1, Ck_{12}), which differs from T_1 only with Ck_{12} = (1, 3, 6, 8). B (4) = Bfin = {now (8), a, b, c, set_a_subgoal_A}. Here, step 3 has a shorter duration than in the previous case, and the available time resource at the end of step 3 was sufficient to set a sub-goal A.

6 Monitoring Interim Results

Under conditions of strict timing constraints, it is extremely important to continuously monitor the reasoning process, identifying the arising anomalies. In a hard real-time environment, anomalies emerge primarily due to a delay in obtaining the expected results. It is necessary that the agent be able to *be aware* not only of what it knows at a given moment of time, but also what it *does not know* at that moment. In the formalism of extended stepping theories of Active Logic, this ability (which can be called self-awareness) is realized through the use of the subjective negation operator not^t. It is noteworthy that in other existing formalisms of stepping theories, where a subjective negation operator is absent, self-awareness ability also does not exist.

Example 2. Consider theory $T_3 = (R_3, Ck_3)$, where R_3 = {N1: $\Rightarrow$ a, N2: a $\Rightarrow$ b, N3: b $\Rightarrow$ c, N4: later (6) $\wedge$ not^t c $\Rightarrow$ anomaly}, Ck_3 = (1, 7, 9). The history of the theory T_3 looks like this: ***B*** (0) = {now (0)}, ***B*** (1) = {now (1), a}, ***B*** (2) = {now (7), a, b,}, ***B*** (3) = B_{fin} = {now (9), a, b, c, anomaly}. In this case, in step 2, when the specified time fence was reached, but the literal did not appear in the set of beliefs ***B*** (2) contrary to expectations, rule N4 kicked in, which fixed the anomaly.

7 Cognitive Process Correction

The analysis of identified anomalies under strict time constraints in the face of possible catastrophic consequences if the deadline is not met does not imply exhaustive completeness. The main task to be solved in the course of such analysis is to evaluate the degree of threat of the deadline crossing in the detected anomaly, as well as the size of the available time resource that can be used to avoid this threat. It seems that in such cases special "emergency" strategies should be provided, when the quality of the decision is sacrificed for the speed of its finding. Below is an example where, when an anomaly is detected, rules specifically tailored for such an event are activated. Of course, in practice, situations associated with anomalies of different types are possible, and then each of these types of rules are provided, but this does not change the essence of what is considered in this example.

Example 3. Consider theory $T_3 = (R_3, Ck_3)$ wherein R_3 = {N1: ⇒ a, N2: not^t anomaly ∧ a ⇒ b, N3: not^t anomaly ∧ b ⇒ c, N4: not^t anomaly ∧ c ⇒ d, N5: later (6) ∧ not^t c ⇒ anomaly, N6: anomaly ∧ a ⇒ e, N7: anomaly ∧ e ⇒ f}, Ck_3 = (1, 7, 9, 11, 13). The history of T_3 will be: ***B*** (0) = {now (0)}, ***B*** (1) = {now (1), a}, ***B*** (2) = {now (7), a, b,}, ***B*** (3) = {now (9), a, anomaly},}, ***B*** (4) = {now (11), a, anomaly, e},}, ***B*** (5) = B_{fin} = {now (13), a, anomaly, f}. In this case, the rule N5 in step 3, because of the absence of the literal c, resulted in an objective literal anomaly, after which the cognitive process passed under "crisis management": only rules that have an anomaly literal in the antecedent now apply, rules that have in the antecedent subjective literal not^t anomaly steel are not applicable.

8 Conclusion

The studies of the above principles of control of the agent's cognitive process, implemented in hard real-time environment, cannot be considered to have been performed in full. It also appears that scientific literature provides only fragmentary information when addressing these issues. It does not deal with the specifics of hard real-time multiagent intelligent systems [14], emphasizing the cooperation of agents and their joint actions within a MAS. At the same time, the above principles seem to be relevant even in the circumstances when the agent is one of the components of a more complex multi-agent system.

References

1. Vinkov, M.M., Fominykh, I.B.: Stepping theories of active logic with two kinds of negation. In: Advances in Electrical and Electronic Engineering, vol. 15, no. 1, pp. 84–92 (2017). Web of Science
2. Elgot-Drapkin, J.: Step logic: reasoning situated in time. Ph.D. thesis. Department of computer science, University of Maryland, Colledge-Park, Maryland (1988)

3. Purang, K., Purushothaman, D., Traum, D., Andersen, C., Traum, D., Perlis, D.: practical reasoning and plan executing with active logic. In: 1999 Proceedings of the IJCAI 1999 Workshop on Practical Reasoning and Rationality (1999)
4. Perlis, D., Purang, K., Purushothaman, D., Andersen, C., Traum, D.: Modeling time and metareasoning in dialog via active logic. In: Working Notes of AAAI Fall Symposium on Psychological Models of Communication (2005)
5. Vinkov, M.M.: Active logic in the context of well-founded semantics of logical programs with priorities. In: Proceedings of the 9th National Conference on with International Participation "KII – 2004", vol. 1, pp. 86–94. Fizmatlit Publ., Moscow (2004). Sb. nauch. tr. IX nats. konf. s mezhdunar. uchastiem "KII – 2004". (in Russian)
6. Vinkov, M.M.: Time as an external entity in the modeling of reasoning of a rational agent with limited resources. In: Proceedings of the 11th National Conference on Artificial Intelligence with International Participation KII-2008. Fizmatlit, Moscow (2008). (in Russian)
7. Vinkov, M.M., Fominykh, I.B.: Argumentation semantics for active logic step theories with granulation time. Iskusstvenny intellekt i prinyatie resheniy Artificial intelligence and decision-making, vol. 3, pp. 3–9. URSS Publ., Moscow (2015). (in Russian)
8. Dung, P.M.: Negation as hypothesis: an abduction foundation for logic programming. In: Proceedings of the 8th International on Logic Programming. MIT Press, Paris (1991)
9. Vagin, V.N., Zagoryanskaya, A.A.: Argumentatsiya v pravdopodobnom vyvode. In: Proceedings of the 9th National Conference on with Int. Participation "KII – 2000". Fizmatlit Publ., Moscow, vol. 1, pp. 28–34 (2000). Sb. nauch. tr. VII nats. konf. s mezhdunar. uchastiem "KII – 2000". (in Russian)
10. Bezhenitshvili, M.N.: The logic of the modalities of knowledge and opinion (2007). Preface. VC. Finna. M. ComBook. (in Russian)
11. Fagin, R., Halpern, J.Y.: Belief, awareness and limited reasoning. Artif. Intell. **34**, 39–76 (1988)
12. Vinkov, M.M., Fominykh, I.B.: Discourse on knowledge and the problem of logical omniscience. Part I: modal approach . In: Artificial Intelligence and Decision Making, pp. 4–12 in Russian. ISSN 2071
13. Alechina, N., Logan, B., Whitsey, M.: A complete and decidable logic for resource-bounded agents. In: Proceedings of Third International Joint Conference on Autonomous Agents and Multi-Agent Systems (AAMAS 2004) (2004)
14. Emelyanov, V.V.: Multiagent model of decentralized management of the flow of productive resources. In: Proceedings of the International Conference "Intellectual Management: New Intellectual Technologies in Control Problems" ICIT 1999, Pereslavl-Zalessky, 6–9 December 1999, pp. 121–126. Science. Fizmatlit, Moscow (1999). (in Russian)

Method of the Maximum Dynamic Flow Finding in the Fuzzy Graph with Gains

Victor Kureichik and Evgeniya Gerasimenko(✉)

Southern Federal University, Taganrog, Russia
{vmkureychik, egerasimenko}@sfedu.ru

Abstract. The paper illustrates the method of the maximum flow value finding in a fuzzy weighted directed graph, which presented as a generalized network. The interest to such type of networks is explained by their wide practical implementation: they can deal with water distribution, money conversion, transportation of perishable goods or goods that can increase their value during transportation, like plants. At the same time the values of arc capacities of the considered networks can vary depending on the flow departure time, therefore, we turn to the dynamic networks. Network's parameters are presented in a fuzzy form due to the impact of environment factors and human activity. Considered types of networks can be implemented in real roads during the process of transportation. The numerical example is given that operated data from geoinformation system "ObjectLand" that contains information about railway system of Russian Federation.

Keywords: Fuzzy network · Fuzzy dynamic generalized graphs
Fuzzy dynamic graph with gains and losses

1 Introduction

Conventional flow tasks have different variations and practical implementations and assume transmission the flow from the source to the sink of the network to find the maximum flow value or to calculate the minimum transshipment cost of the given flow value. In all these problem statements the initial data is capacitated network and the goal is to find the extreme flow value without exceeding arc capacities of the network [1, 2].

Another similarity of conventional flow problems is flow conservation condition, particularly, the assumption that the flow is conserved on every arc. However, various types of practical applications violate this assumption. For example, while transporting of special kinds of cargo the last can be damaged or spoiled; one the one hand transportation of perishable goods leads to decreasing of the initial transported value, and on the other hand some kinds of goods can increase their value during transportation (for example, plants).

Therefore, in such problem statements we deal with generalized networks (networks with losses or gains) [3]. In these networks, there is a positive multiplier assigned to each arc of the network (flow efficiency factor), which represents the part of the flow during the process of transportation along the corresponding arc.

A. Abraham et al. (Eds.): IITI 2018, AISC 874, pp. 244–253, 2019.
https://doi.org/10.1007/978-3-030-01818-4_24

Described networks were firstly considered by Dantzig and Jewell [3] and different types of modified algorithms [4–8] to solving such tasks were proposed. However, the bulk of such publications is devoted to improving the running time of the modified approaches that don't operate with other network peculiarities.

In our algorithm, we offer to consider uncertainty peculiar to all kinds of networks and such parameters, as arc capacities, costs of transportation. These parameters are influenced by weather conditions, human activity (repairs on the roads, errors in measurement), and lack of relevant information. Therefore, the mentioned network parameters should be represented in a fuzzy form.

Another significant remark is that solving flow problems it is important to take into account dynamic nature of arc capacities, as they can change their values in time and be influenced by the flow departure time. These modifications lead to necessity of dealing with fuzzy dynamic networks and considering generalized flow tasks at such types of networks [9].

In summing up, the present paper considers the method of the maximum flow finding in a fuzzy generalized dynamic network, assuming losses and gains.

Present paper has the following structure: Sect. 2 presents proposed method of the maximum dynamic flow finding in generalized networks. Practical realization and numerical example are given in Sect. 3. Section 4 is conclusion and future work.

2 Proposed Method of the Minimum Cost Flow Finding in the Fuzzy Network, Considering Vitality Degree

As was already mentioned above, generalized flow problem is a generalization of the conventional maximum flow problem that violates assumption of the flow conservation [3]. Similar to the conventional flow networks capacitated network is considered, where each arc has a capacity $\tilde{u}$ that limits the amount of the flow passed into his arc. Moreover, the generalized network has additional parameter – flow gain factor or flow efficiency. This parameter ρ is a positive multiplier and indicates that if $\tilde{\xi}$ flow units enter the arc, then $\tilde{\xi} \times \rho$ flow units exit, as presented in Fig. 1.

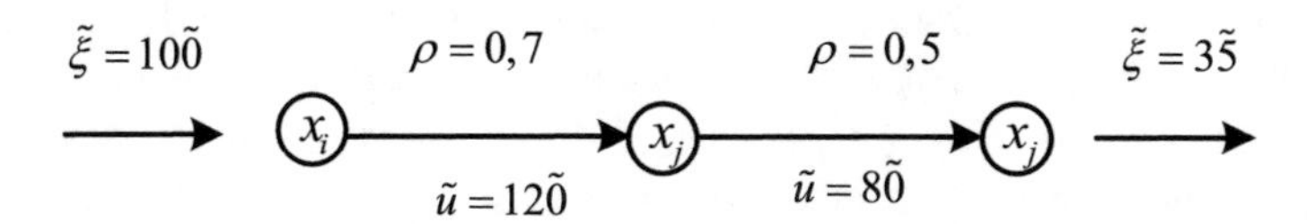

Fig. 1. Efficiency product factor

Let's pose a problem statement of the maximum dynamic flow finding in a fuzzy generalized network:

$$Maximize\ \tilde{v}(p), \tag{1}$$

$$\sum_{\theta=0}^{p}\left(\sum_{x_j\in\Gamma(x_i)}\tilde{\xi}_{ij}(\theta)-\sum_{x_j\in\Gamma^{-1}(x_i)}\tilde{\xi}_{ji}(\theta-\tau_{ji}(\theta))\right)=\tilde{v}(p), x_i=s, \quad (2)$$

$$\sum_{\theta=0}^{p}\left(\sum_{x_j\in\Gamma(x_i)}\tilde{\xi}_{ij}(\theta)-\sum_{x_j\in\Gamma^{-1}(x_i)}\tilde{\xi}_{ji}(\theta-\tau_{ji}(\theta))\right)=\tilde{0},\ x_i\neq s,\ t;\ \theta\in T, \quad (3)$$

$$\sum_{\theta=0}^{p}\left(\sum_{x_j\in\Gamma(x_i)}\tilde{\xi}_{ij}(\theta)-\sum_{x_j\in\Gamma^{-1}(x_i)}\tilde{\xi}_{ji}(\theta-\tau_{ji}(\theta))\right)=-\tilde{v}(p),\ x_i=t, \quad (4)$$

$$\tilde{\xi}_{\substack{ij\\ x_i\in\Gamma^{-1}(x_j)}}(\theta+\tau_{ij})=\rho_{ij}\times\tilde{\xi}_{\substack{ij\\ x_j\in\Gamma(x_i)}}(\theta), \quad (5)$$

$$\tilde{0}\leq\tilde{\xi}_{\substack{ij\\ x_j\in\Gamma(x_i)}}(\theta)\leq\tilde{u}_{ij}(\theta),\ \forall\,(x_i,\,x_j)\in\tilde{A},\ \theta\in T. \quad (6)$$

Equation (1) reflects the maximization of the flow $\tilde{v}(p)$. The Eq. (2) is the flow conservation condition for the source $\sum_{\theta=0}^{p}\sum_{x_j\in\Gamma(x_i)}\tilde{\xi}_{ij}(\theta),\ x_i=s$. The Eq. (4) is the flow conservation condition for the sink $\sum_{\theta=0}^{p}\sum_{x_j\in\Gamma^{-1}(x_i)}\tilde{\xi}_{ji}(\theta-\tau_{ji}(\theta)),\ x_i=t$. The Eq. (3) is the flow conservation condition for the intermediate nodes. The equality (5) indicates the dependence between entering flow, efficiency factor and leaving flow. The inequality (6) indicates that the flows $\tilde{\xi}_{ij}(\theta)$ for all time periods should be less than arc capacities of the corresponding arcs $\tilde{u}_{ij}(\theta)$ in the same time periods.

To represent the proposed algorithm let us introduce the rules of transition to the time-expanded variant of the initial dynamic graph and transition to the fuzzy residual network of the "time-expanded" graph.

Rule 1 of transition to the time-expanded variant of the initial dynamic graph

The goal is to transit from the initial version of dynamic network, presented as a graph to the time-expanded version via creating a distinct copy of each node $x_i\in X$ at each time period $\theta\in T$. $\tilde{G}_p=(X_p,\,\tilde{A}_p)$ can be denoted as fuzzy time-expanded version of the given dynamic fuzzy graph. The set of nodes X_p can be determined as $X_p=\{(x_i,\,\theta):(x_i,\,\theta)\in X\times T\}$. The set of arcs $\tilde{A}_p$ includes arcs from the node-time pair $(x_i,\,\theta)\in X_p$ to every node-time pair $(x_j,\,\vartheta=\theta+\tau_{ij}(\theta))$, where $x_j\in\Gamma(x_i)$ and $\theta+\tau_{ij}(\theta)\leq p$. Values of fuzzy arc capacities $\tilde{u}(x_i,x_j,\,\theta,\,\vartheta)$ joining $(x_i,\,\theta)$ with $(x_j,\,\vartheta)$ are equal to $\tilde{u}_{ij}(\theta)$. Flow efficiency factors $\tilde{p}(x_i,x_j,\,\theta,\,\vartheta)$ joining $(x_i,\,\theta)$ with $(x_j,\,\vartheta)$ are equal to $\tilde{p}_{ij}(\theta)$. Transit times $\tau(x_i,\,x_j,\,\theta,\,\vartheta)$ joining $(x_i,\,\theta)$ with $(x_j,\,\vartheta)$ are equal t
o $\tau_{ij}(\theta)$.

<u>*Rule 2 of transition to the fuzzy residual network of the "time-expanded" graph*</u>

The construction of the fuzzy residual network $\tilde{G}^{\mu}_{p} = (X^{\mu}_{p}, \tilde{A}^{\mu}_{p})$ is implemented according to the "time-expanded" graph $\tilde{G}_p = (X_p, \tilde{A}_p)$ and the flow values $\tilde{\xi}(x_i, x_j, \theta, \vartheta = \theta + \tau_{ij}(\theta))$ issuing from the nodes (x^{μ}_{i}, θ) and entering the nodes (x^{μ}_{j}, ϑ): each arc in the residual network $\tilde{G}^{\mu}_{p}$, which head-node is (x^{μ}_{i}, θ) and tail-node is (x^{μ}_{j}, ϑ) with the flow $\tilde{\xi}(x_i, x_j, \theta, \vartheta)$ issuing from the node x_i at the time $\theta \in T$ has residual fuzzy arc capacity $\tilde{u}^{\mu}(x_i, x_j, \theta, \vartheta) = \tilde{u}(x_i, x_j, \theta, \vartheta) - \tilde{\xi}(x_i^{+}, x_j^{-}, \theta^{+})$ with the transit time $\tau^{\mu}(x_i, x_j, \theta, \vartheta) = \tau(x_i, x_j, \theta, \vartheta)$ and efficiency factor $\rho^{\mu}(x_i, x_j, \theta, \vartheta) = \rho(x_i, x_j, \theta, \vartheta)$ and reverse arc, with the head-node (x^{μ}_{j}, ϑ) and the tail-node (x^{μ}_{i}, θ) with fuzzy arc capacity $\tilde{u}^{\mu}(x_j, x_i, \vartheta, \theta) = \tilde{\xi}(x_j^{-}, x_i^{+}, \theta + \tau_{ij}^{-})$, transit time $\tau^{\mu}(x_j, x_i, \vartheta, \theta) = -\tau(x_i, x_j, \theta, \vartheta)$ and efficiency factor $\rho^{\mu}(x_i, x_j, \vartheta, \theta) = 1/\rho(x_i, x_j, \theta, \vartheta)$, where $\tilde{\xi}(x_i^{+}, x_j^{-}, \theta^{+})$ – the flow, issuing from the node x_i at time period θ towards the node x_j. $\tilde{\xi}(x_j^{-}, x_i^{+}, \theta + \tau_{ij}^{-})$ is the flow, entering the node x_j at time period $\theta + \tau_{ij}$ from the node x_i.

Therefore, considering the rules described above, we can introduce the proposed algorithm of the maximum dynamic flow finding in the generalized network.

Algorithm of the Maximum Dynamic Flow Finding in the Generalized Network

Step 1. Transit to the time-expanded version $\tilde{G}_p$ of the given directed graph $\tilde{G}$ by creating a distinct copy of every node $x_i \in X$ at every time $\theta \in T$. Due to the *rule 1* the goal is to determine the maximum flow in the dynamic generalized fuzzy network passing from the set of sources, expanded for p periods and entering the set of sinks expanded for p periods no later than p. We should add artificial source s' and sink t' and connect with each given source and sink, respectively. Artificial arcs issuing from and entering artificial nodes have infinite capacities.

Step 2. Create a fuzzy residual network $\tilde{G}^{\mu}_{p}$ taking into account flow values travelling the arcs of the graph $\tilde{G}_p$. Fuzzy residual network $\tilde{G}^{\mu}_{p} = (X^{\mu}_{p}, \tilde{A}^{\mu}_{p})$ is based on the "time-expanded" graph $\tilde{G}_p = (X_p, \tilde{A}_p)$ and flow values, obtained by the *rule 2*.

Step 3. Find the path $\tilde{P}^{\mu}_{p}$ with the highest efficiency factor product $r = \prod_{(x_i, x_j, \vartheta, \theta) \in A_f} \rho^{\mu}(x_i, x_j, \vartheta, \theta) \times \prod_{(x_i, x_j, \vartheta, \theta) \in A_b} 1/\rho^{\mu}(x_i, x_j, \theta, \vartheta)$.

3.1. Go to the **step 4** if the path $\tilde{P}^{\mu}_{p}$ is found.

3.2. If there is no such a path, the maximum flow value $\tilde{\xi}(x_i, x_j, \theta, \vartheta) + \tilde{\delta}^{\mu}_{p} \times \tilde{P}^{\mu}_{p} = \tilde{v}(p)$ is found in the time-expanded fuzzy static graph, then go to the **step 6.**

Step 4. Transmit the maximum flow value depending on the arc with minimal fuzzy residual capacity of the issuing flow and considering efficiency factors as follows: $\tilde{\delta}^{\mu}_{p} = \min[\tilde{u}(\tilde{P}^{\mu}_{p})]$, $\tilde{u}(\tilde{P}^{\mu}_{p}) = \min[\tilde{u}^{\mu}(x_i, x_j, \theta, \vartheta) \times \rho_{ij}]$, $(x_i, \theta), (x_j, \vartheta) \in \tilde{P}^{\mu}_{p}$, taking into account efficiency factors in the head and tail of the arc. Therefore, received flow value can be fined as the product of the sent flow value $\tilde{\xi}(x_i^+, x_j, \theta^+)$ and *r:*

$$\tilde{\xi}(x_j^-, x_i^+, \theta + \tau_{ij}^-) = \tilde{\xi}(x_i^+, x_j^-, \theta^+) \times \prod_{(x_i, x_j, \vartheta, \theta) \in A_f} \rho^\mu(x_i, x_j, \vartheta, \theta) \times \\ \times \prod_{(x_i, x_j, \vartheta, \theta) \in A_b} 1/\rho^\mu(x_i, x_j, \theta, \vartheta).$$

Step 5. Update the fuzzy flow values in the graph $\tilde{G}_p$: replace the fuzzy flow $\tilde{\xi}(x_j^-, x_i^+, \theta + \tau_{ij}^-)$ by $\tilde{\xi}(x_j^-, x_i^+, \theta + \tau_{ij}^-) - \tilde{\delta}_p^\mu$ for arcs connecting node-time pair (x_i^μ, ϑ) with (x_j^μ, θ), such as $((x_i^\mu, \vartheta), (x_j^\mu, \theta)) \notin \tilde{A}_p$, $((x_i^\mu, \vartheta), (x_j^\mu, \theta)) \in \tilde{A}_p^\mu$ in $\tilde{G}_p^\mu$ along the corresponding arcs going from (x_j, θ) to (x_i, ϑ) from $\tilde{G}_p$ and replace the fuzzy flow $\tilde{\xi}(x_i^+, x_j^-, \theta^+)$ by $\tilde{\xi}(x_i^+, x_j^-, \theta^+) + \tilde{\delta}_p^\mu$ for arcs connecting node-time pair (x_i^μ, θ) with (x_j^μ, ϑ), such as $((x_i^\mu, \vartheta), (x_j^\mu, \theta)) \in \tilde{A}_p^\mu$ in $\tilde{G}_p^\mu$ along the arcs going from (x_i, θ) to (x_j, ϑ) from $\tilde{G}_p$. Replace the fuzzy flow value in $\tilde{G}_p$: $\tilde{\xi}(x_i, x_j, \theta, \vartheta) \to \tilde{\xi}(x_i, x_j, \theta, \vartheta) + \tilde{\delta}_p^\mu \times \tilde{P}_p^\mu$ and go to the **step 2,** starting with the updated flow values.

Step 6. Transit to the given dynamic fuzzy generalized graph $\tilde{G}$ if the maximum flow $\tilde{\xi}(x_i, x_j, \theta, \vartheta) + \tilde{\delta}_p^\mu \times \tilde{P}_p^\mu = \tilde{v}(p)$ is found in $\tilde{G}_p$. Therefore, the maximum flow of the value $\tilde{v}(p)$ is obtained in the original graph, which is equivalent to the flow from the sources (initial node, expanded on the *p* periods) to the sinks (terminal node, expanded on the *p* periods) in $\tilde{G}_p$ after cancelling artificial nodes. Each path connecting the node-time pairs (s, ς) and $(t, \psi = \varsigma + \tau_{st}(\varsigma))$, $\psi \in T$ which the flow $\tilde{\xi}(s, t, \theta, \varsigma)$ goes along is equal to the flow $\tilde{\xi}_{st}(\vartheta)$.

3 Numerical Example

In the following section, we perform a numerical example, reflecting the proposed method. The required data for the task are taken from the geoinformation system «ObjectLand» 2.6 [10]. The system contains information about railway system of Russian Federation. User can select a part of the map for processing and obtain information about stations, paths etc. The problem statement is to find the maximum flow value in the initial dynamic generalized network in the form of the. We'll obtain two values of the maximum flow: for submitted flow and for received flow as well as the generalized network is considered.

Fuzzy arc capacities are set by experts and incorporated in the GIS in the forms of tables either as crisp initial data, such as time parameters and efficiency factors. Consider the solution of the maximum flow finding task in a dynamic generalized network. Represent selected area as fuzzy generalized graph, as shown in Fig. 2. Tables 1 and 2 give us parameters assigned to arcs of the network.

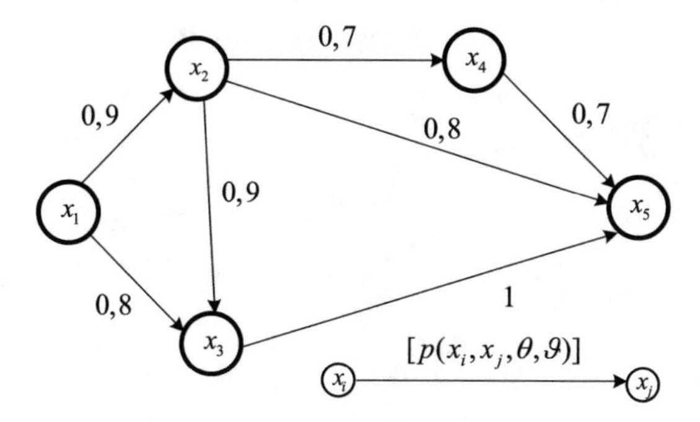

Fig. 2. Initial fuzzy generalized graph

Table 1. Fuzzy arc capacities with the efficiency factors $\tilde{u}_{ij}$, depending on the flow departure time θ

Arcs of the graph	Fuzzy arc capacities $\tilde{u}_{ij}$ at the time periods θ, time units			
	0	1	2	3
(x_1, x_2)	$\tilde{40}$	$\tilde{20}$	$\tilde{25}$	$\tilde{40}$
(x_1, x_3)	$\tilde{16}$	$\tilde{12}$	$\tilde{55}$	$\tilde{37}$
(x_2, x_3)	$\tilde{18}$	$\tilde{25}$	$\tilde{55}$	$\tilde{35}$
(x_2, x_4)	$\tilde{15}$	$\tilde{30}$	$\tilde{35}$	$\tilde{18}$
(x_2, x_5)	$\tilde{42}$	$\tilde{10}$	$\tilde{38}$	$\tilde{16}$
(x_3, x_5)	$\tilde{36}$	$\tilde{45}$	$\tilde{18}$	$\tilde{55}$
(x_4, x_5)	$\tilde{20}$	$\tilde{22}$	$\tilde{19}$	$\tilde{26}$

Turn to the time-expanded static graph according to the rule 1 (Fig. 3).

Find a path with the maximum flow efficiency factor product and pass the minimum flow value along it according to arc capacities and efficiency factors (Fig. 4).

Table 2. Time parameters τ_{ij} depending on the flow departure time θ

Arcs of the graph	Time parameters τ_{ij} at time periods θ, time units			
	0	1	2	3
(x_1, x_2)	1	1	1	2
(x_1, x_3)	1	1	3	2
(x_2, x_3)	5	1	1	1
(x_2, x_4)	4	1	3	1
(x_2, x_5)	1	4	2	2
(x_3, x_5)	1	3	1	1
(x_4, x_5)	5	1	1	3

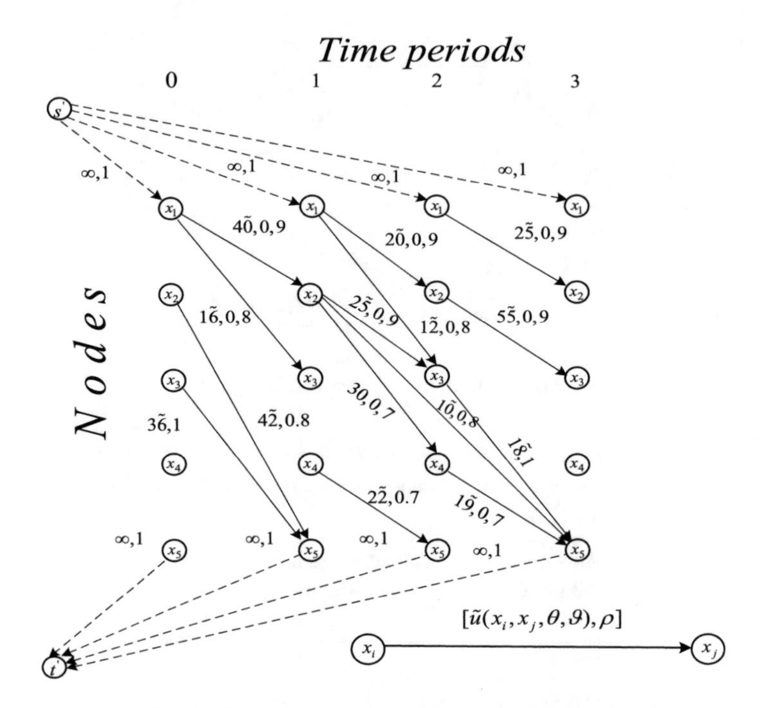

Fig. 3. Fuzzy time-expanded generalized network

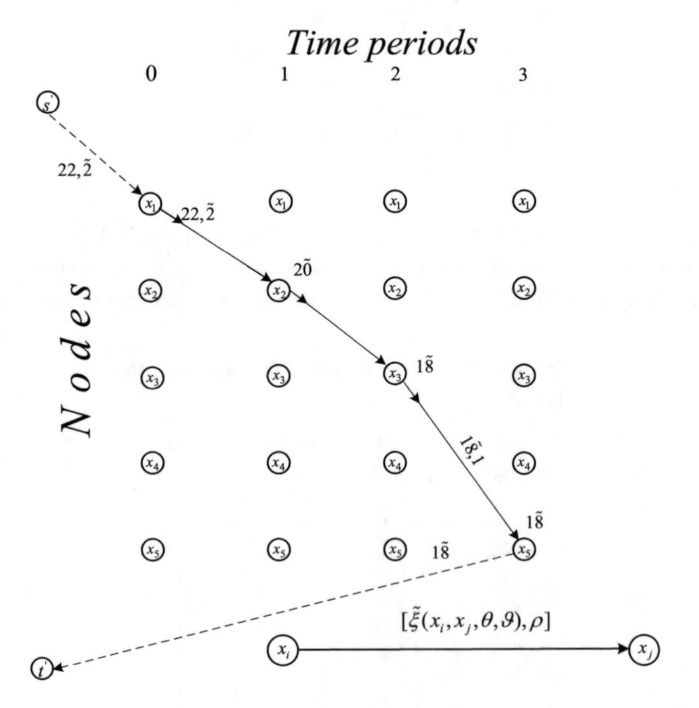

Fig. 4. Generalized graph with the path of the maximum efficiency factor and the flow along it

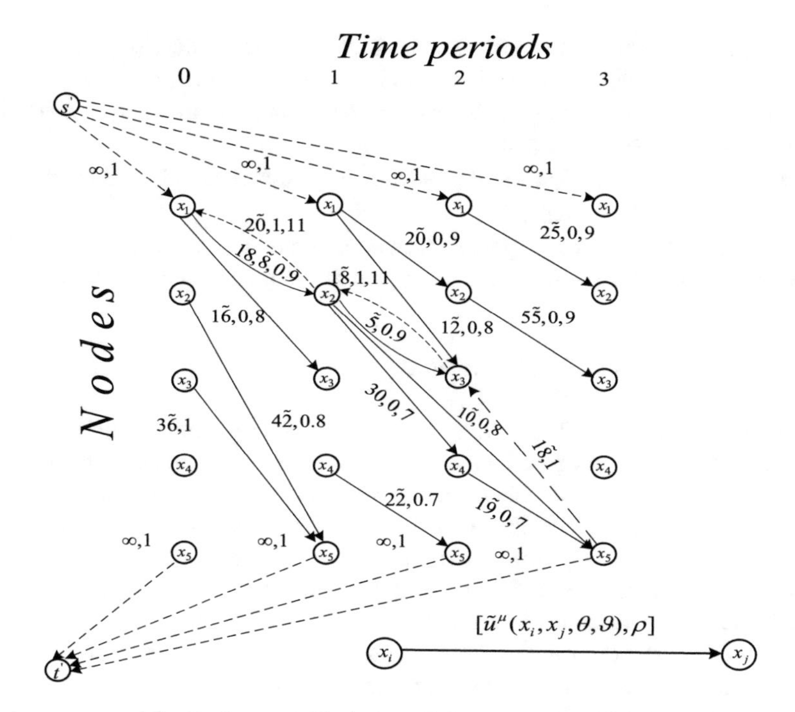

Fig. 5. Fuzzy residual network for the graph in Fig. 4

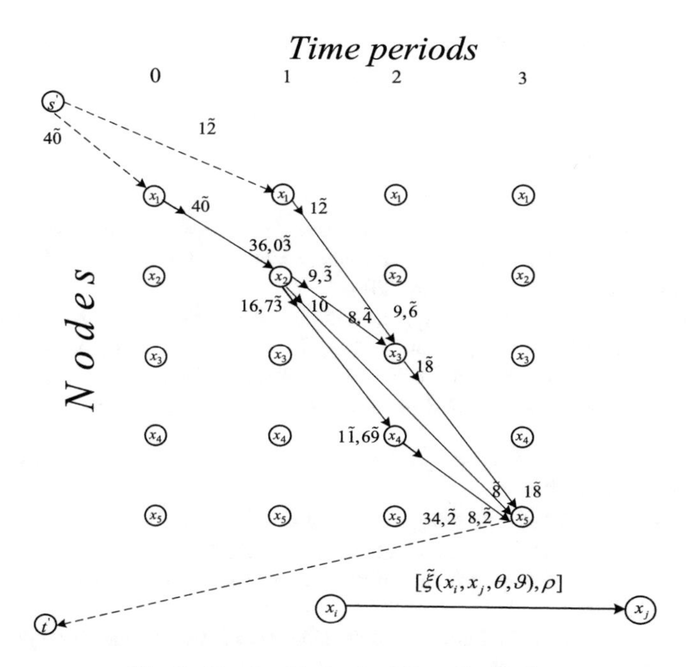

Fig. 6. Graph with the final flow distribution

Construct a residual network according to already passed flow (Fig. 5).

Repeating the following procedure iteratively we obtain the final flow distribution in a generalized network (Fig. 6), because the following construction of residual networks shows no augmenting paths.

Finally, the maximum flow value sent from the source is $5\tilde{2}$ units and the flow value entering the sink is $34,\tilde{2}$ units.

The proposed algorithm is optimal, because the saturated path is found at each step of the method and there is no augmenting path at the final step of the method.

4 Conclusion and Future Work

Method of the maximum flow finding in fuzzy dynamic generalized network is considered, based on the proposed rules of residual and time-expanded network construction. Proposed method takes into account efficiency parameters assigned to network arcs, fuzziness, peculiar to real types of networks due to various factors, connected with human activities, environmental conditions, accidents etc. Dynamic nature of the arc capacities and their dependence on the flow departure time is considered. Numerical example is presented based on GIS«ObjectLand». Considered method has significant practical value, because researchers can solve real optimization tasks, particularly, tasks of the maximum cargo transportation on real railway roads. Moreover, the quantity of sent and received cargo will vary. In the future works methods of minimum cost dynamic flow finding in generalized networks will be proposed.

Acknowledgments. This work has been supported by the Russian Foundation for Basic Research, Projects 18-0700050, № 16-01-00090 a.

References

1. Bozhenyuk, A.V., Gerasimenko, E.M., Kacprzyk, J., Rozenberg, I.N.: Flows in Networks Under Fuzzy Conditions. Studies in Fuzziness and Soft Computing, vol. 346. Springer International Publishing Switzerland (2017)
2. Kureichik, V., Gerasimenko, E.: Approach to the minimum cost flow determining in fuzzy terms considering vitality degree. In: Silhavy, R., Senkerik, R., Kominkova Oplatkova, Z., Prokopova, Z., Silhavy, P. (eds.). CSOC 2017, Artificial Intelligence Trends in Intelligent Systems, Advances in Intelligent Systems and Computing, vol. 573, pp, 200–210. Springer, Cham (2017)
3. Wayne, K.D.: Generalized maximum flow algorithm. Ph.D. dissertation, Cornell University, New York, January 1999
4. Murray, S.M.: An interior point approach to the generalized flow problem with costs and related problems. Ph.D. thesis, Stanford University (1993)
5. Radzik, T.: Faster algorithms for the generalized network flow problem. Math. Oper. Res. **23** (1), 69–100 (1998)
6. Wayne, K.D., Fleischer, L.: Faster approximation algorithms for generalized flow. In: ACM Transactions on Algorithms (1998)

7. Eguchi, A., Fujishige, S., Takabatake, T.: A polynomial-time algorithm for the generalized independent-flow problem. J. Oper. Res. **47**(1), 1–17 (2004)
8. Krumke, S.O., Zeck, C.: Generalized max flow in series–parallel graphs. Discret. Optim. **10** (2), 155–162 (2013)
9. Groß, M., Skutella, M.: Generalized maximum flows over time. In: Solis-Oba, R., Persiano, G. (eds.) Approximation and Online Algorithms. WAOA 2011. Lecture Notes in Computer Science, vol. 7164. Springer, Heidelberg (2012)
10. ObjectLand/Geoinformation system. http://www.objectland.ru/

Hybrid Bioinspired Algorithm of 1.5 Dimensional Bin-Packing

Boris K. Lebedev, Oleg B. Lebedev(✉), and Ekaterina O. Lebedeva

Southern Federal University, Rostov-on-Don, Russia
lebedev.b.k@gmail.com, {lebedev.ob,lbedevakate}@mail.ru

Abstract. The paper deals with the problem of 1.5 dimensional bin packing. As a data structure carrying information about packaging, a sequence of numbers of rectangles is used, representing the order of their packing. An essential role in obtaining the solution is played by a decoder, which performs the laying of rectangles according to the rules laid down in it. New methods for solving the packing problem are proposed, using mathematical methods in which the principles of natural decision-making mechanisms are laid. Unlike the canonical paradigm ant algorithm to find solutions to the graph $G = (X, U)$ is constructed with a partition on the route of the formation and on the tops within each part, subgraphs whose edges are delayed pheromone. The structure of the solution search graph, the procedure for finding solutions on the graph, the methods of deposition and evaporation of pheromone are described. The time complexity of the algorithm, experimentally obtained, practically coincides with the theoretical studies and for the considered test problems is $O(n^2)$. In comparison with existing algorithms, the improvement of results is achieved by 2–3%.

Keywords: Swarm intelligence · Ant colony · Adaptive behavior
1.5 dimensional bin packing · Optimization

1 Introduction

Bin Packing tasks occupy an important place in modern combinatorial optimization and attract the attention of many scientists, both in Russia and abroad [1, 2]. There are many varieties of this task: 2-Dimensional Bin Packing, 2-Dimensional Strip Packing (2 DSP or 1.5 DBP), linear packing, packing by weight, packing by cost and others. Packaging tasks can be applied in the field of optimal filling of containers, loading trucks with weight restrictions, creating backup copies on removable media, etc.

Thus, finding a solution to such problems has attracted many scholars. Literature has proposed a variety of algorithms for solving the rectangle packing problem in two-dimensions; some examples include the greedy algorithm, genetic algorithm, particle swarm optimization, heuristic algorithm and hybrid algorithm [3–10]. The paper deals with the problem of packing a semibounded strip. A set of rectangles is specified. Given

This work was supported by the grant from the Russian Foundation for Basic Research the project № 17-07-00997.

A. Abraham et al. (Eds.): IITI 2018, AISC 874, pp. 254–262, 2019.
https://doi.org/10.1007/978-3-030-01818-4_25

one large object (called a strip) whose width W is given, and the height of H is the desired value of the variable. The goal is to minimize the height H of the strip containing the rectangles placed in the strip without overlapping them. Nesting-packing problems belong to the class of *NP*-difficult combinatorial optimization problems. In this regard, the development and research of iterative methods for solving nesting-packing problems, including methods of local search that have proved themselves, is of great importance. As a structure of data carrying information about packaging, the sequence of numbers of rectangles, representing the order of their packing, is most often used. An essential role in the general process of finding a solution is played by a decoder, which performs the laying of rectangles according to the rules laid down in it. An important characteristic of the decoder is its ability to obtain optimal packaging according to a pre-known optimal code (priority list), i.e. the ability to correctly decode. Both the process of decoder operation and the process of sequence formation use algorithms with a heuristic and meta-heuristic - solution method for obtaining quasi-optimal results. This representation is convenient for its use in various meta-heuristics (genetic algorithms, ant algorithms), since they just work with the sequences encoding the package. However, the quality of the solution largely depends on the methods of forming the components of meta-heuristics. Proceeding from the foregoing, it is of interest to develop and apply new search algorithms to solve the problem of cutting-and-packing rectangular objects on the basis of effective meta-heuristics – the principles of encoding packages. The result of the ongoing search for the most effective packaging methods has been the use of bionic methods and algorithms. One of the new directions of such methods is multi-agent methods of intellectual optimization based on the modeling of collective intelligence [11–13]. To such methods it is possible to carry, and ant algorithms (Ant Colony Optimization – ACO) [14, 15]. The basis of the behavior of the ant colony is self-organization, which ensures achievement of the common goals of the colony on the basis of low-level interaction. The paper presents a method for solving the packing problem, based on modeling the adaptive behavior of the ant colony.

2 Formulation of the Problem

The problem of rectangular packing in the strip (1.5 Dimensional Bin Packing, 1.5 DBP) is as follows. The initial information is given by the following data set: H – width of the semi-infinite strip; $R = \{r_i | i = 1, 2, ..., m\}$ is a set of rectangles, where m is the number of rectangles; $W = \{w_i | i = 1, 2, ..., m\}$ is the vector of the widths of the rectangles; $L = \{l_i | i = 1, 2, ..., m\}$ is the vector of the lengths of the rectangles. We introduce a rectangular coordinate system *XOY*, in which the axes *OX* and *OY* coincide, respectively, with the lower unbounded and lateral sides of the strip. The solution of the problem is represented as a set of elements $<X, Y>$, where $X = \{x_i | i = 1, 2, ..., m)$, $Y = \{y_i | i = 1, 2, ..., m)$ are the coordinate vectors of the rectangles. (x_i, y_i) are the coordinates of the lower-left corner of the rectangle, respectively, along the axes *OX* and *OY*. A set of elements is called an admissible rectangular packing if the following conditions are satisfied:

1. The sides of the rectangles are parallel to the faces of the strip.
2. Rectangles do not overlap.
3. Rectangles do not go beyond the borders of the strip.

Since the problem of packing is *NP*-difficult, the research conducted on this problem is focused mainly on the development of approximate solution algorithms. Approximate algorithms find a solution with a certain accuracy, but do not guarantee the optimal packing for any data set. Most of the decoders developed to date based on heuristic rules are divided into three groups: *level, shelf, plane*.

Flat algorithms place the rectangles strictly close to each other.

Shelfs examine several shelves at once, and distribute rectangles along them.

The approach of all level algorithms is that the strip is divided into levels, based on the height of the rectangles available at this stage. Rectangles are placed along the bottom of the current level from left to right as long as possible. A rectangle that does not fit is packed to the next level. The height of each level is determined by the highest rectangle in it. Rectangles on one level form a row. Here are three options for packaging:

Next Fit Level - when the next level opens, the previous one "closes" and is no longer considered;

First Fit Level - at each step of the algorithm, each level is viewed from the lowest level, and the rectangle is packed into the first suitable one, which has enough space;

Best Fit Level - at each step of the algorithm, all levels are viewed, and the most suitable one is selected, the one on which the minimum space remains after packing.

The modification of the *Best Fit Level* algorithm is the *Floor Ceiling No Rotation* algorithm. Packaging is carried out by blocks. Each block is limited to two levels. The lower level of the block is "*floor*" and the upper level of the block – "*ceiling*". While there is a possibility, rectangles are packed on the "*floor*" from left to right (Fig. 1). When the place ends, an attempt is made to pack on the "*ceiling*" from right to left; if there is no room on the ceiling, then only the formation of a new block begins. In accordance with the *BFDH* heuristic, at every step all levels are looked at - first "*floor*", then "*ceiling*" - for the presence of the most suitable place. The method successfully packs the smallest rectangles on the "*ceilings*".

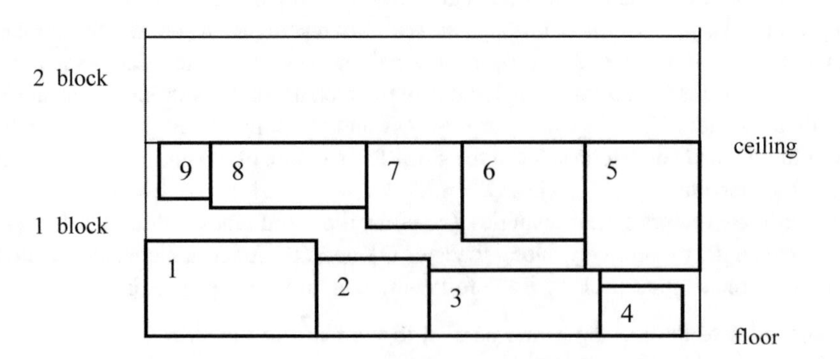

Fig. 1. Packaging in accordance with the *FCNR* heuristic

In the work, the *Floor Ceiling No Rotation* heuristic *(FCNR)* was chosen as the base structure of the decoder. To build the decoder and code sequence, heuristic modifications *(FCNR)* and meta-heuristics based on modeling the adaptive behavior of the ant colony.

3 Packaging Mechanisms Based on Ants Colony

Usually the levels are filled in accordance with the sequence formed by sorting the entire list by not increasing the height of the rectangles.

The paper uses the following approach to packaging the semi-infinite strip. In the first stage, the set of rectangles $R = \{r_i | i = 1, 2, ..., m\}$ is divided into two subsets $R1$ and $R2$, $R_1 \cup R_2 = R$ so that the total width $H1$ of the rectangles entering $R1$ and the total width $H2$ rectangles, entering $R2$, would be approximately (practically) equal, $H1 \approx H2$, and the total width $H = H1 + H2$.

In the second step, the rectangles of each subset R_j are sorted by non-increasing height. In the third stage, the blocks are sequentially filled, beginning with the first. When filling the block, first fill the lower level, and then fill the upper level. The levels are filled in accordance with the sequences obtained in the second stage. The lower level is filled with rectangles from the R_1 list, and the top level is filled with rectangles from the R_2 list. The process of level filling is completed, if there is no place for selecting the next element r_i from the list. The next level is filled in from the r_i element. A more dense packing level can be achieved by scanning the list to the end, placing the next selected item, if possible in the free zone of the level.

The filled levels in each block approach to the minimum allowable distance, excluding the overlapping of rectangles. At the final stage, compression is performed along the OX axis, and the size of the filled part of the strip – L.

In this paper, the set of rectangles $R = \{r_i | i = 1, 2, ..., m\}$ is divided into two subsets of R_j, by the method of an ant colony.

Most of the meta-heuristic models of adaptive behavior of ants colony are based on the search paradigm of the shortest way ants on the graph. However, for a number of tasks it is very difficult or simply impossible to use this paradigm.

The main focus of the work is the modification of the canonical paradigm of the ant algorithm to extend the scope of its application. Let there be some solution of the problem of laying m rectangles on c levels, given by some sequence R. Let m_i be the number of rectangles stacked at the *i-th* level.

The search for solutions is carried out on a complete graph of solutions $G = (X, U)$ [13]. The vertices of the $X = \{x_i | i = 1, 2, ..., m\}$ correspond to the rectangles r_i. In the general case, the search for the solution of the packing problem is carried out by the collective of ants $A = \{a_\kappa | \kappa = 1, 2, ..., m\}$. At each iteration of the ant algorithm, each ant a_κ creates a specific solution - a graph divided into two parts. Splits are built by ants in turn one by one. In other words, ants, independently of one another, construct m partitions using the data accumulated in the previous iterations. The vertices of each part of the graph correspond to rectangles stacked on levels of the same type (lower or upper). To uniformly distribute the ants and create equal starting conditions as the initial vertices for the ants formed by an anomaly into two subsets, the vertices of X are used.

Simulation of the behavior of ants in the packing problem is associated with the distribution of pheromone on the edges of the graph. At the initial stage, the same (small) amount of pheromone Q is deposited on all the edges of graph G. The parameter Q is given a priori. The process of finding solutions is iterative. Each iteration s includes three steps.

In the first stage, each ant finds a solution (partitioning the set R into 2 subsets of R_1 and R_2). Then, using the decoder, the rectangles are stacked in layers and the solution estimate L_k is determined. In the second stage, the ants deposit pheromone, in the third stage, the total evaporation of the pheromone. The work uses the ant-cycle method of ant systems. In this case, pheromone is deposited by each agent on the edges of the graph after complete formation of solutions by all ants. At each iteration, each ant a_{κ} forms its subgraph G_{k1}. The process of finding solutions is iterative. Each subgraph $G_{k1} \in G$ includes vertices X_{k1} corresponding to the set of rectangles R_1 placed in blocks on the lower levels. The process of forming the subgraph $G_{k1} \in G$ is step-by-step. At each step t, the agent a_k applies a probability rule for selecting the next vertex to include it in the generated subgraph $G_{k1} \in G$. Let $G_{k1}(t - 1)$ be the subgraph formed after $(t - 1)$ steps. At the step t, the set of vertices $X^o_{k1}(t) \in X$, is selected such that if $x_i \in X^o_{k1}(t)$, then the total width H_1 of the elements of the set $(x_i \cup X_{k1}(t - 1))$ is less than or equal to half the total width H of the elements of the set X. This rule is due to the fact that the graph G splits into two subgraphs so that the total width of the elements in both subgraphs is approximately the same. So if $x_i \in X^o_{k1}(t)$, then x_i can be included in the generated subgraph $G_{k1} \in G$. The agent looks through all the vertices $x_i \in X^o_{k1}(t)$, and for each vertex the parameter f_{ik} is calculated - the total level of pheromone on the edges of the graph G that connect x_i to the vertices of the set $X_{k1}(t - 1)$. The probability P_{ik} of the inclusion of the vertex x_i in G_{k1} is determined by the following relation:

$$P_{ik} = f_{ik} / \sum_i f_{ik}. \tag{1}$$

Agent a_k with probability P_{ik} chooses one of the vertices $x_i \in X^o_{k1}(t)$, which is included in G_{k1}. After this, the inclusion of the rectangle r_i, corresponding to x_i in the subset R_1 is fixed. If $X^o_{k1}(t)$ turns out to be empty, then the formation process of the ant of a_k of the subgraph G_k ceases.

The decompositions of the subset R_1 and R_2 are sorted by not increasing the height of the rectangles. With the help of the decoder, the rectangles are stacked over the levels and the estimate of the solution $Lk(s)$ obtained by the a_k at the iteration s is determined.

In the second stage of the iteration, each ant a_k lays the pheromone on the edges from the full subgraphs, each of which is constructed on the set of vertices that make up one of the parts $M_{kj}(t)$ of the route $M_k(t)$. The amount of pheromone $\Delta\tau_k(s)$, deposited by the ant a_k on each edge of the subgraphs built on the s-th iteration is determined as follows:

$$\Delta\tau_k(s) = \Phi / L_k(s), \tag{2}$$

where s is the iteration number, Φ is the total amount of pheromone deposited by the ant on the edges of the full subgraphs G_{k1}, $L_k(s)$ is the target function for the solution obtained by the a_k at the s-th iteration. The smaller $L_k(s)$ – the more pheromone is deposited on the edges of the full subgraphs and, the more is the probability of selecting these edges when constructing the routes at the next iteration.

After each agent has formed a solution and postponed the pheromone, in the third stage, the total evaporation of the pheromone occurs on the edges of the complete graph G in accordance with the formula (3).

$$f_{ik} = f_{ik}(1 - \rho), \tag{3}$$

where ρ is the update coefficient.

After performing all the actions on the iteration, there is an agent with the best solution that is remembered. Next, go to the next iteration.

The time complexity of this algorithm depends on the lifetime of the colony l (the number of iterations), the number of vertices of the graph n, and the number of ants m, and is defined as $O(l \cdot n^2 \cdot m)$.

The algorithm of one-dimensional packaging based on the method of the ant colony is formulated as follows.

1. In accordance with the initial data, a complete graph of search solutions $G = (X, U)$ is formed.
2. The number of agents is specified and they are placed into the initial vertices of $G = (X, U)$.
3. The value of the parameter Φ and the number of iterations N_s is set.
4. On all edges of the graph G the initial quantity pheromone is deposited. $s = 1$.
5. At the first stage of the s-th iteration by each agent a_k forms on the G its subgraph G_{k1} and splits the set R into two subsets of R_1 and R_2.
6. For each decision received by the ant produced sorting R_1 and R_2 by not increasing the height of the rectangles.
7. Using the decoder for each R_1 and R_2, a stacking rectangles in blocks by levels.
8. For each solution of the packing problem, the value of the criterion $L_k(s)$ is determined.
9. In the graph G on the edges of the complete subgraphs G_{k1} and G_{k2} constructed each pile is deposited pheromone. The amount of pheromone, which is postponed by each agent, is proportional to $L_k(l)$.
10. The procedure for evaporation of pheromone on the edges of graph G.
11. The best solution received throughout executed iterations.
12. If all the iterations are fulfilled, then the end of the algorithm work, in otherwise, go to step 4 to perform the next iteration.

4 Experimental Studies

The packaging algorithm was implemented in C++ in a Windows environment. Experimental studies were carried out on IBM computers. For these purposes, the procedure for synthesizing control examples with a known optimum was used. To draw up reliable conclusions, not one, but a series of experiments-experiments was carried out.

When investigating the convergence of the algorithm, a generation number was memorized for each experiment, after which no improvement in the estimate was observed. In each series of 50 tests, the minimum and maximum generation numbers were determined. In addition, the average value of the number of generations was calculated, after which there was no improvement in the estimate. For each series of tests, the best solution was determined, which in fact was optimal. As a result of the experiments it was established that with the population size $M = 100$, the number of iterations at which the algorithm found the best solution lies in the range 110–130. The algorithm converges on average by 120 iterations (Fig. 2). The initial amount of pheromone Q should be 14 times greater than the average quantity of the ferromone $\tau_k(l)$ value deposited by the agents at each iteration. The update coefficient is $\rho = 0.95$.

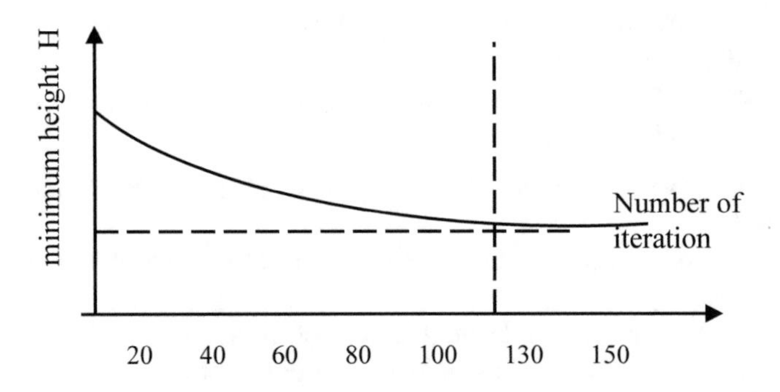

Fig. 2. The dependence of the quality of the hybrid bioinspired algorithm on the number of iterations

A set of data is randomly assigned from OR-Benchmark, The tasks on which the developed algorithm was tested are available in the OR-objects library (OR-Library Beasley).

The hybrid algorithm explained in study [6] combines a constructive method and a genetic algorithm CAGA, so it has high computing efficiency. The BHA (Bricklaying Heuristic Algorithm) algorithm of study [7] is based on the experience of human transformation of nature, which has higher computational efficiency. The Heuristic Dynamic Decomposition Algorithm (HDDA) algorithm explained in study [6]. The SCPL1 ~ SCPL9 data of the OR-Benchmark are used to test the above three algorithms, and the calculation results are shown in Table 1.

In Table 1, $H*$ is the Optimum height, which can be calculated by the following formula:

$$H* = \left(\sum_{i=1}^{n} w_i h_i\right) \Big/ W$$

H is the real strip height, which varies according to different test environments.

N - set of rectangles.

Table 1. Comparison results of three algorithms

Inst	N	Opt	HDDA	CAGA	BHA	HBA
		H*	H	H	H	H
SCPL1	425	110	213	148.08	183.5	180
SCPL2	127	120	148	150.16	463.5	106
SCPL3	225	84	158	155.84	321	152
SCPL4	365	102	136	158.263	684.5	136
SCPL5	165	102	117	128.89	480	117
SCPL6	657	126	172	152.4	234	170
SCPL7	357	198	606	219	242	218
SCPL8	475	156	955	163.419	900	160
SCPL9	175	117	185	189.96	507	182

As Table 1 shows, the results of the hybrid bioinspired algorithm (HBA) are significantly better than the previous three algorithms HDDA, GAGA, BHA, and some results have reached the limit height.

Compared with the existing algorithms [4, 6, 9], the results are improved by 2–3%. The time complexity of the algorithm, obtained experimentally, practically coincides with the theoretical studies and for the considered test problems is $\approx O(n^2)$.

5 Conclusion

New mechanisms for solving the packing problem are proposed, using mathematical methods in which the principles of natural decision-making mechanisms are laid. In contrast to the canonical paradigm of the ant algorithm, an ant on the search graph $G = (X, U)$ constructs a route with a division into two parts and the formation of subgraphs on the vertices that make up each part, on the edges of which pheromone is deposited. The agent on the decision graph $G(X, U)$ does not move from the vertex to the vertex, but from the vertex group to the vertex. Experimental studies were conducted on the IBM PC. The temporal complexity of the algorithm, obtained experimentally, practically coincides with the theoretical studies and for the considered test problems is $O(n^{2)}$.

To conduct objective experiments, we used well-known test problems presented in the literature and the Internet. The tasks on which the developed algorithm was tested are available in the *OR*-objects library (OR-Library Beasley). To draw up reliable conclusions, not one, but a series of experiments-experiments was carried out. In comparison with existing algorithms, the improvement of results is achieved by 2–3%.

Acknowledgements. This research is supported by grants of the Russian Foundation for Basic Research of the Russian Federation, the project № 17-07-00997.

References

1. Lodi, A., Martello, S., Monaci, M.: Two-dimensional packing problems: a survey. Eur. J. Oper. Res. **141**, 241–252 (2002)
2. Leung, S.C.H., Zhang, D., Sim, K.M.: A two-stage intelligent search algorithm for the two-dimensional strip packing problem. Eur. J. Oper. Res. **215**, 57–69 (2011)
3. Martello, S., Monaci, M., Vigo, D.: An exact approach to the strip-packing problem. INFORMS J. Comput. **15**, 310–319 (2003)
4. Duanbing, C., Wenqi, H.: Greedy algorithm for rectangle-packing problem. Comput. Eng. **33**, 160–162 (2007)
5. Timofeeva, O.P., Sokolova, E.S., Milov, K.V.: Genetic algorithm in optimization of container packaging. In: Proceedings of NSTU, Informatics and Management Systems, vol. 4, no. 101, pp. 167–172 (2013)
6. Hifi, M., M'Hallah, R.: A hybrid algorithm for the two-dimensional layout problem: the cases of regular and irregular shapes. Int. Trans. Oper. Res. **10**, 195–216 (2003)
7. De Zhang, F., ShuiHua, H., WeiGuo, Y.: A bricklaying heuristic algorithm for the orthogonal rectangular packing problem. Chin. J. Comput. **33**, 509–515 (2008)
8. Leung, C.H., Zhang, D., Zhou, C., et al.: A hybrid simulated annealing metaheuristic algorithm for the two-dimensional knapsack packing problem. Comput. Oper. Res. **39**, 64–73 (2012)
9. Shi, W.: Solving rectangle packing problem based on heuristic dynamic decomposition algorithm. In: 2nd International Conference on Electrical and Electronics: Techniques and Applications, pp. 187–196 (2017)
10. Lebedev, O.B., Zorin, V.Yu.: Packaging on the basis of the method of the ant colony. Izvestiya Southern Federal University. Publishing House of TTI SFedU, no. 12, pp. 25–30 (2010)
11. Lebedev, V.B., Lebedev, O.B.: Swarm Intelligence on the basis of integration of models of adaptive behavior of ants and bee colony. Izvestiya Southern Federal University. Publishing House of TTI SFedU, no. 7, pp. 41–47 (2013)
12. Lebedev, B.K., Lebedev, V.B.: Optimization by the method of crystallization of placers of alternatives. Izvestiya Southern Federal University. Publishing House of TTI SfedU, no. 7, pp. 11–17 (2013)
13. Lebedev, B.K., Lebedev, O.B.: Modeling of the adaptive behavior of an ant colony in the search for solutions interpreted by trees. Izvestiya Southern Federal University. Publishing House of TTI SfedU, no. 7, pp. 27–35 (2012)
14. Dorigo, M., Stützle, T.: Ant Colony Optimization, p. 244. MIT Press, Cambridge (2004)
15. Lebedev, O.B.: Models of adaptive behavior of ant colony in the task of designing. Southern Federal University. Publishing House of SfedU, p. 199 (2013)

Community Detection in Online Social Network Using Graph Embedding and Hierarchical Clustering

Vang Le[1(✉)] and Vaclav Snasel[2]

[1] Faculty of Information Technology, Ton Duc Thang University, Ho Chi Minh City, Vietnam
levanvang@tdtu.edu.vn

[2] Department of Computer Science, VSB-Technical University of Ostrava, Ostrava, Czech Republic
vaclav.snasel@vsb.cz

Abstract. The community detection plays an important role in social network analysis. It can be used to find users that behave in a similar manner, detect groups of interests, cluster users in e-commerce application such as their taste or shopping habits, etc. In this paper, we proposed an algorithm to detect the community in online social networks. Our algorithm represents the nodes and the relationships in the social networks using a vector, agglomerative clustering (the most famous clustering algorithm) will cluster those vectors to figure out the communities. The experimental results show that our algorithm performs better traditional agglomerative clustering because of the ability to detect the community which has better modularity value.

Keywords: Community detection · Graph embedding · Hierarchical clustering Social network analysis

1 Introduction

Today, the online social networks such as Facebook, Twitter becoming a popular channel for transmission of information such as news, brochures, and marketing, … Social network analysis is an extremely important area of research. In recent years, a number of studies in the field of complex networks have attracted considerable amounts of attention from the scientific community.

One of the most important research areas in this area is community detection. Community detection is the most widely studied structural features of complex networks. Communities in a network are dense groups of vertices, which are tightly coupled to vertices inside the the communities and loosely coupled to the rest of the vertices outside. Community detection plays an important role in understanding the functions of complex networks. How to determine efficiently the community in social networks which maximize the modularity is a major challenge up to the present.

A. Abraham et al. (Eds.): IITI 2018, AISC 874, pp. 263–272, 2019.
https://doi.org/10.1007/978-3-030-01818-4_26

Kernighan and Lin [18] proposed The Kernighan-Lin algorithm. It divides the graph into g clusters of predefined size, such that the number of links in a cluster is denser than the number of edges between the clusters.

In general, for a social data set we know very little about the internal clustering structure. Therefore, it is often difficult to pre-define the number of partitions for the graph partitions algorithm. In such cases, clustering procedures such as graph partitioning can not be applied because we have to make some reasonable assumptions about the number and size of clusters, which usually is fixed and non-adjustable. Hastie et al. [33] proposed a hierarchical clustering technique which reveals the multi-level structure of the graph. Hierarchical clusters are very popular in social network analytics, biology, engineering, marketing, etc.

In this paper, we further propose a hydrid method which combines the hierarchical clustering algorithm and graph embedding algorithm. The experiments on the datasetshow that our approach can considerably improve the modularity of the original hierarchical clustering algorithm. Our approach is based on the idea:

(1) Finds the way to turn the vertices of a social network into independent vectors but preserves network information including the links between the vertices of the network.
(2) Apply the common clustering algorithms on the list of vectors collected in the previous step to discover social network community optimally in the value of modularity.

The main contribution of our research is the improvement of the accuracy, our social network community obtained from our approach is better than the result obtained from traditional hierarchical clustering algorithms. The remaining parts of this paper are organized as follows. We summary the related work in Sect. 2. In Sect. 3, we introduce our proposed clustering method. Experiments are given in Sect. 4. Finally, we conclude the paper in Sect. 5.

2 Related Works

Identifying the most influential spreaders in a network is critical for ensuring efficient diffusion of information. For instance, a social media campaign can be optimized by targeting influential individuals who can trigger large cascades of further adoptions. This section presents briefly some related works that illustrate the various possible ways to measure the influence of individuals in the online social network.

2.1 Clustering Algorithm

In data mining and statistics, hierarchical clustering is a method of cluster analysis that seeks to build a hierarchy of clusters. Classical clustering strategies are divided into categories as below:

Graph partitioning approach, the problem of graph partitioning consists in dividing the vertices in g groups of predefined size, such that the number of edges lying between the groups is minimal [17], Kernighan et al. [18] propose a method to present a heuristic method for partitioning arbitrary graphs which is both effective in finding optimal partitions, and fast enough to be practical in solving large problems. Barnes et al. [19] propose a heuristic algorithm for partitioning the nodes of a graph into a given number of subsets in such a way that the number of edges connecting the various subsets is a minimum. Shahriar et al. [20] propose EGAGP algorithm, an enhanced genetic algorithm for producing efficient graph partitions. Reijnders et al. [21] propose a new pre-processing heuristic for partitioning road networks using multilevel graph partitioning (MGP) implementations.

Hierarchical clustering approach, is a clustering algorithm based on similarity of the input data points. Based on the data matrix on the similarity of the data points we will construct a hierarchical structure commonly referred as the dendrogram. Then we will use a cutting technique on this dendrogram tree to create the desired clusters. Depend on the technique of building hierarchical dendrogram tree we can divide the hierarchical clustering into two groups: Agglomerative algorithms, in which clusters are iteratively merged if their similarity is sufficiently high; Divisive algorithms, in which clusters are iteratively split by removing edges connecting vertices with low similarity. Sharma et al. [29] propose a method DRAGON, a computationally efficient divisive hierarchical clustering method which was verified on mutation and microarray data.

Partitional clustering approach, an algorithm that cluster data points into the number of clusters (for example k) that is preassigned. The points are embedded in a metric space and we define a function to measure the dissimilarity distance between pairs of points in the space. The goal is to separate the points in k clusters such to maximize/minimize a given cost function. MacQueen et al. [22] describe a process which is called 'k-means' for partitioning an N-dimensional population into k sets. Hlaoui et al. [23] propose an extension of k-means clustering algorithem. The key elements of the new approach are an efficient graph matching algorithm for computing the similarity measurement. Bezdek et al. [24] propose a method to improve k-means clustering which is called fuzzy k-means clustering.

Spectral clustering approach, an algorithm that cluster data points using eigenvectors of matrices derived from the data. This algorithm obtains data representation in the low-dimensional space that can be easily clustered, this approach is very useful in case of hard non-convex clustering problems. Donath et al. [25] is the first contribution on spectral clustering, which applied the eigenvectors of the adjacency matrix for graph partitions. Li et al. [26] propose a method that substantially reduces the computational requirements of grouping algorithms based on spectral partitioning based on a technique for the numerical solution of eigenfunction - the Nyström method. Bhattacharyya et al. [27] propose spectral clustering methods for identifying communities in dynamic or multiple networks. Li et al. [28] propose the effectiveness of spectral methods in clustering multi-scale data, which can be adaptive in many cases of various sizes and densities.

Divisive clustering approach, the basic concept of divisive clustering algorithm is to find the edge acting as the connection between the community and removing them from the graph and to obtain the separate communities. Murata et al. [30] propose a divisive clustering method which removes edges in a network based on low-similarity to separate communities from each other. Girvan et al. [1] propose Girvan-Newman algorithm which removes edges iteratively based on edge-betweenness measurement. Madani et al. [31] propose a divisive method which removes edges iteratively based on the edge clustering coefficient measurement.

2.2 Graph Embedding Algorithm

Graph analysis gives us more comprehensive views behind the data and therefore can benefit many useful applications like node classification, node recommendation, link prediction, and more. Graph embedding is an effective way to solve graph analysis problems. It converts graph information into a low-dimensional space in which the structure information and graph properties are maximally preserved. There are many different graphing embedding models:

Matrix Factorization approach, obviously, factorize a matrix, i.e. to find out two (or more) matrices such that when you multiply them you will get back the original matrix. This approach represents the original graph property in the form of a matrix then find the way to factorize this matrix to get node embedding. In the direction of matrix factorization, there are several research such as Hofmann et al. [2] proposes a novel efficient framework to perform feature selection for graph embedding, Yan et al. [4] proposes a new supervised dimensionality reduction algorithm called marginal Fisher analysis, Chen et al. [3] introduces a method name Locality Preserving Projections (LPP) can be seen an alternative to Principal Component Analysis (PCA) – a classical linear technique that projects the data along the directions of maximal variance.

Edge Reconstruction based Optimization approach, Zhou et al. [5] propose an asymmetric proximity preserving (APP) graph embedding method via random walk with restart, which captures both asymmetric and high-order similarities between node pairs, Feng et al. [6] proposes a method name GAKE, which formulates knowledge base as a directed graph, and learns representations for any vertices or edges by leveraging the graph's structural information, Tang et al. [7] propose a novel network embedding method called the LINE. This method optimizes a carefully designed objective function that preserves both the local and global network structures. An edge-sampling algorithm is proposed that addresses the limitation of the classical stochastic gradient descent and improves both the effectiveness and the efficiency of the inference. Xie et al. [8] propose a novel method named Type-embodied Knowledge Representation Learning (TKRL) to take advantages of hierarchical entity types to encode both entities and relations into a continuous low-dimensional vector space.

Generative Model approach, Alharbi et al. [9] designs a model to embed a location-based social network (LBSN), this method borrows the idea from LDA algorithm where documents are embedded in a "topic" space where documents with similar words have similar topic vector representations graph

Deep Learning approach, Deep learning is increasingly being used in a wide variety of research areas such as computer vision, natural language processing and graph embedding is also not an exception. By transforming graph data to reuse existing deep learning models or build a completely new neuron network, we can apply a deep learning model to the study of graph embedding. Deep modeling used in graph embedding can be divided into two main groups: deep learning combined with random walk and deep learning without random walk.

In the direction of deep learning combined with random walk, there are several research directions using skip-gram model combined with hierarchical softmax or negative sampling techniques. Perozzi et al. [10] proposed a method name DeepWalk, a novel approach for learning latent representations of vertices in a network which encode social relations in a continuous vector space, Yang et al. [11] propose a method name GenVector, a multi-modal Bayesian embedding model, to learn social knowledge graphs. GenVector uses latent discrete topic variables to generate continuous word embeddings and network-based user embeddings, Li et al. [12] propose a method name Discriminative Deep Random Walk (DDRW), a novel method for relational network classification. This method extends DeepWalk by jointly optimizing the classification objective and the objective of embedding entities in a latent space that maintains the topological structure, Pan et al. [13] propose a method name TriDNR, a tri-party deep network representation model, this method borrows the idea from coupled deep natural language module; Dong et al. [14] propose method name metapath2vec which formalizes the random walks based on meta-path to construct the heterogeneous neighborhood of a node and then apply a heterogeneous skip-gram model to perform node embeddings

In the direction of deep learning combined without random walk, there are several research directions using autoencoder model. The encoder maps input data to a representation space and the decoder maps the representation space to a original space. The transformation of data from the new space to the original space will ensure that data loss, deviation is inevitable. The autoencoder algorithm seeks to minimize the loss of data. Wang et al. [15] propose a method name Structural Deep Network Embedding (SDNE) - a semi-supervised deep learning model, which has multiple layers of non-linear functions. As a popular deep learning model, Convolutional Neural Network (CNN) and its variants have been widely adopted in graph embedding neural network model such as convolutional neural network model (CNN), graph convolutional network model (GNN). Niepert et al. [16] propose a framework for learning representations for classes of directed and undirected graphs builds on concepts from convolutional neural networks for images and extends them to arbitrary graphs.

3 Proposed Model

The first step in our approach is graph embedding, a process to convert the vertices of the networks into vectors but still maintain the relationship of the connections between them. As mentioned in the review, we have a lot of graph embedding algorithms to perform this conversion. In the example below (Fig. 1), we illustrate the steps of conversion process using the node2vec algorithm of proposed by Grover et al. [32]. Table 1 represents the random walk process which random generates walk steps from a node in graph and Table 2 represents the output of node2vec algorithm.

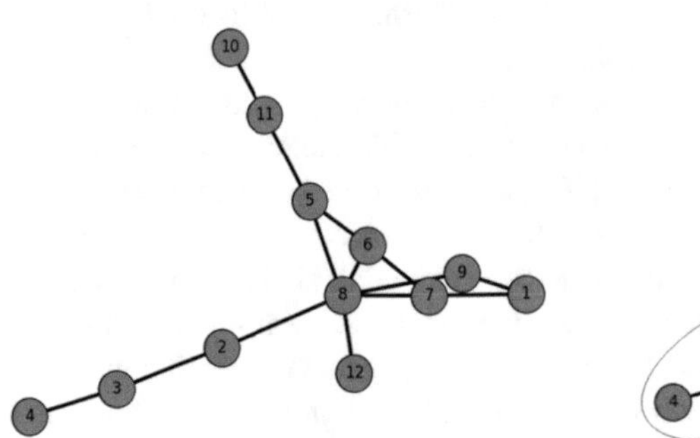

Fig. 1. The original example social network for community detection

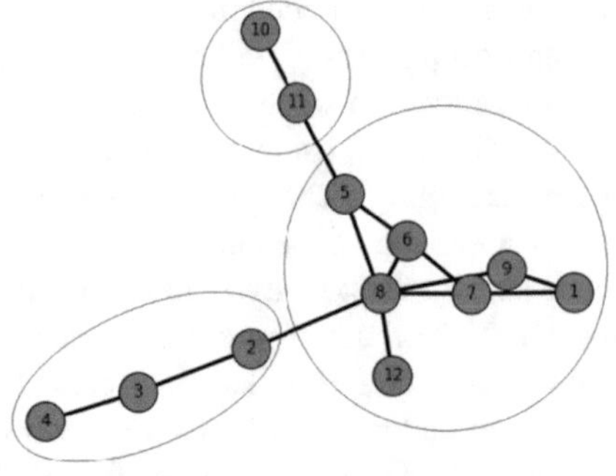

Fig. 2. The communities detected

In the next step we will use a clustering algorithm to separate the vectors obtained in the previous step into different clusters and each cluster will act as a community in the network. As mentioned in the review, there are many different algorithms for clustering on vectors such as k-means, hierarchical clustering, Girvan-Newman clustering, etc. In Fig. 2, we illustrate the agglorative hierarchical clustering, algorithm details are presented in Algorithm 1.

Table 1. Random walks simulation

Walk	Random walk steps
Walk 1st	[2, 3, 2, 3, 2, 3, 4, 3, 4, 3, 4, 3, 2, 8, 9, 1, 7, 6, 8, 2, 8, 5, 11, 5, 8, 2, 8, 6, 8, 2, 8, 12, 8, 9, 1, 7, 6, 8, 6, 7, 6, 7, 1, 7, 8, 9, 8, 5, 8, 9, 8, 9, 1, 7, 6, 7, 8, 9, 1, 7, 1, 7, 8, 5, 8, 5, 6, 7, 8, 6, 7, 6, 7, 6, 5, 8, 7, 6, 7, 8]
Walk 2nd	[6, 5, 8, 2, 3, 4, 3, 4, 3, 4, 3, 2, 8, 7, 8, 7, 1, 9, 8, 9, 1, 9, 8, 2, 8, 2, 3, 4, 3, 4, 3, 4, 3, 4, 3, 2, 3, 2, 3, 4, 3, 4, 3, 4, 3, 2, 3, 2, 3, 2, 3, 4, 3, 2, 8, 9, 1, 7, 1, 7, 8, 7, 1, 9, 1, 9, 1, 7, 6, 5, 11, 10, 11, 5, 6, 5, 11, 5, 8, 2]
	[10, 11, 10, 11, 10, 11, 5, 11, 5, 6, 8, 12, 8, 5, 8, 2, 8, 5, 6, 5, 8, 7, 1, 7, 8, 5, 11, 5, 11, 5, 8, 7, 6, 5, 11, 10, 11, 5, 6, 8, 7, 8, 5, 6, 8, 2, 8, 7, 1, 7, 1, 7, 1, 7, 1, 7, 6, 5, 8, 12, 8, 2, 8, 7, 8, 5, 8, 12, 8, 6, 5, 11, 10, 11, 10, 11, 5, 11, 10, 11]
…	…
Walk nth	[12, 8, 2, 8, 2, 3, 4, 3, 4, 3, 4, 3, 2, 8, 7, 8, 5, 8, 6, 8, 6, 7, 6, 5, 8, 9, 8, 7, 8, 9, 8, 6, 5, 6, 5, 11, 10, 11, 5, 8, 7, 8, 5, 6, 8, 2, 8, 5, 6, 7, 6, 5, 8, 9, 8, 9, 8, 5, 11, 10, 11, 5, 6, 8, 6, 5, 8, 2, 8, 9, 8, 5, 11, 10, 11, 10, 11, 10, 11, 5]

Table 2. Vector representation of vertices in the network

Node	Vector representation
Node 1st	[0.47701791 0.8146677 0.19374575 −0.34769022 −0.76072752 0.85946429 0.33675942 −0.00882156]
Node 2nd	[0.72927964 0.10661288 0.64459532 0.48989671 0.61123133 0.25582007 1.01145279 0.18808879]
…	…
Node nth	0.67689693 0.21316138 −0.37620094 −0.27378342 0.40693155 1.05078983 −0.10819849 0.07817915]

Algorithm 1 Agglomerative Clustering

1: **function** AGGLOMERATIVE CLUSTERING(D, L, m) ▷ Where D - The distance matrix, L - Clusters by level, m - current level
2: If all objects are in one cluster
3: return
4: $k_{u,v} \leftarrow min(d_{i,j})$ ▷ Where r is value of matrix D at row i, column j
5: $m \leftarrow$m+1
6: $t \leftarrow$(u,v)
7: $L(m) = t$
8: Update D by:
9: remove $D_{(u,\,)}$, $D_{(v,\,)}$, $D_{(\,,u)}$, $D_{(\,,v)}$ from D
10: add row, column D_t
11: **for** $i \in$V **do**
12: $d_{i,t} \leftarrow min(d_{ui}, d_{vi})$
13: **end for**
14: AGGLOMERATIVE CLUSTERING(D, L, m)
15: **end function**

Algorithm 2 describes the general definition of community in the network. The input of the algorithm is the G network, the vertices of the network V, and the edges of network E. Line 2 starts with d which is the number of dimensions we want to embed, in this case we will represent the vertices of G with a 300 dimensions vector. In line 3, we will use a graph embedding algorithm to convert the networks to discrete vectors with dimension d = 300 as described above. In line 4, we construct a matrix D that represents the distance between the vectors obtained above (each vector equals a node/vertice in the network). In lines 5, 6, 7 we use the agglorative clustering algorithm to construct the dendrogram tree. In line 8 we perform a dendrogram cut to obtain clusters (equivalent to the communities in the network).

Algorithm 2 Community Detection

1: **function** COMMUNITY DETECTION(G, V, E) ▷ Where G - The social network, V - The vertex, E - The edge connections in G
2: $d \leftarrow 300$
3: $M \leftarrow$ NODE2VEC(G, V, E, d)
4: $D \leftarrow$ CREATECONNECTIONMATRIX(M, G, V, E)
5: Initialize L
6: $m \leftarrow 0$
7: AGGLOMERATIVE CLUSTERING(D, L, m)
8: CLUSTERING CUT(L)
9: return L
10: **end function**

4 Experiments

To validate the effectiveness of our algorithm, we run the experiments on the Zachary's karate club dataset. To evaluate the experimental results and compare with other approaches we used the modularity measure proposed by Gervan Newman as presented in Eq. 1. In this equation, L is the total number of the edges within the network, l_k is the total number of edges within the kth community, d_k is the degree of the nodes in the kth community, s is the total number of communities obtained from algorithm 2. With the above calculation, the greater the modularity value, the greater the density of intra-community connections than the density of connections between communities, this also show that the higher the value of modularity, the better the communities we detected (Table 3).

$$Modularity = \sum_{k=1}^{s} \left[\frac{l_k}{L} - \left(\frac{d_k}{2L} \right)^2 \right] \quad (1)$$

Table 3. The statistic of dataset in the experiments

Algorithm	Communities	Modularity value
Girvan Newman	[1, 2, 4, 5, 6, 7, 8, 11, 12, 13, 14, 17, 18, 20, 22], [3, 9, 15, 16, 19, 21, 23, 24, 25, 26, 27, 28, 29, 30, 31, 32, 33, 34], [10]	0.348783695
Our approach	[3, 10, 15, 16, 19, 21, 23, 24, 25, 26, 27, 28, 29, 30, 31, 32, 33, 34], [1, 2, 4, 8, 9, 12, 13, 14, 18, 20, 22], [5, 6, 7, 11, 17]	0.350345168

5 Conclusion

Our approach is the hybrid method which combines the graph embedding algorithms and hierarchical clustering algorithms. Experiments on the dataset show that our approach can considerably improve the modularity of social network community compare with traditional clustering algorithm.

Hierarchical clustering is an efficient method, however, it can only be applied in case of vectorization problems. In this paper, we further improve the hierarchical clustering algorithm by apply the graph embedding which is a tool to convert nodes of graph into independent vectors.

References

1. Girvan, M., Newman, M.E.: Community structure in social and biological networks. Proc. Natl. Acad. Sci. **99**(12), 7821–7826 (2002)
2. Hofmann, T., Buhmann, J.M.: Multidimensional scaling and data clustering. In: NIPS, pp. 459–466 (1994)

3. Chen, M., Tsang, I.W., Tan, M., Jen, C.T.: A unified feature selection framework for graph embedding on high dimensional data. IEEE Trans. Knowl. Data Eng. **27**(6), 1465–1477 (2015)
4. Yan, S., Xu, D., Zhang, B., Zhang, H., Yang, Q., Lin, S.: Graph embedding and extensions: a general framework for dimensionality reduction. IEEE Trans. Pattern Anal. Mach. Intell. **29**(1), 40–51 (2007)
5. Zhou, C., Liu, Y., Liu, X., Liu, Z., Gao, J.: Scalable graph embedding for asymmetric proximity. In: AAAI, pp. 2942–2948 (2017)
6. Feng, J., Huang, M., Yang, Y., Zhu, X.: GAKE: graph aware knowledge embedding. In: COLING, pp. 641–651 (2016)
7. Tang, J., Qu, M., Wang, M., Zhang, M., Yan, J., Mei, Q.: Line: large-scale information network embedding. In: WWW, pp. 1067–1077 (2015)
8. Xie, R., Liu, Z., Sun, M.: Representation learning of knowledge graphs with hierarchical types. In: IJCAI, pp. 2965–2971 (2016)
9. Alharbi, B., Zhang, X.: Learning from your network of friends: a trajectory representation learning model based on online social ties. In: ICDM, pp. 781–786 (2016)
10. Perozzi, B., Al-Rfou, R., Skiena, S.: Deepwalk: online learning of social representations. In: KDD, pp. 701–710 (2014)
11. Yang, Z., Tang, J., Cohen, W.: Multi-modal Bayesian embeddings for learning social knowledge graphs. In: IJCAI, pp. 2287–2293 (2016)
12. Li, J., Zhu, J., Zhang, B.: Discriminative deep random walk for network classification. In: ACL (2016)
13. Pan, S., Wu, J., Zhu, X., Zhang, C., Wang, Y.: Tri-party deep network representation. In: IJCAI, pp. 1895–1901 (2016)
14. Dong, Y., Chawla, N.V., Swami, A.: metapath2vec: Scalable representation learning for heterogeneous networks. In: KDD, pp. 135–144 (2017)
15. Wang, D., Cui, P., Zhu, W.: Structural deep network embedding. In: KDD, pp. 1225–1234 (2016)
16. Niepert, M., Ahmed, M., Kutzkov, K.: Learning convolutional neural networks for graphs. In: International Conference on Machine Learning, pp. 2014–2023, 11 June 2016
17. Fortunato, S.: Community detection in graphs. Phys. Rep. **486**(3–5), 75–174 (2010)
18. Kernighan, B.W., Lin, S.: An efficient heuristic procedure for partitioning graphs. Bell Syst. Tech. J. **49**(2), 291–307 (1970)
19. Barnes, E.R.: An algorithm for partitioning the nodes of a graph. SIAM J. Algebr. Discret. Methods **3**(4), 541–550 (1982)
20. Shahriar, F., Nesar, A.B., Mahbub, N.M., Shatabda, S.: EGAGP: an enhanced genetic algorithm for producing efficient graph partitions. In: 2017 4th International Conference on Networking, Systems and Security (NSysS), pp. 1–9. IEEE, 18 December 2017
21. Reijnders, B.J.: Pre-processing Road Networks for Graph Partitioning Using Edge-Betweenness Centrality. Bachelor's thesis
22. MacQueen, J.: Some methods for classification and analysis of multivariate observations. In: proceedings of the fifth Berkeley symposium on mathematical statistics and probability, vol. 1, no. 14, pp. 281–297, 21 June 1967
23. Hlaoui, A., Wang, S.: A direct approach to graph clustering. Neural Netw. Comput. Intell. **4** (8), 158–163 (2004)
24. Bezdek, J.C.: Objective Function Clustering. In: Pattern Recognition with Fuzzy Objective Function Algorithms, pp. 43–93. Springer, Boston (1981)
25. Donath, W.E., Hoffman, A.J.: Lower bounds for the partitioning of graphs. IBM J. Res. Dev. **17**(5), 420–425 (1973)

26. Li, L., Wang, S., Xu, S., Yang, Y.: Constrained spectral clustering using nyström method. Procedia Comput. Sci. **31**(129), 9–15 (2018)
27. Bhattacharyya, S., Chatterjee, S.: Spectral clustering for multiple sparse networks: I. arXiv preprint arXiv:1805.10594, 27 May 2018
28. Li, X., Kao, B., Luo, S., Ester, M.: ROSC: robust spectral clustering on multi-scale data. In: Proceedings of the 2018 World Wide Web Conference on World Wide Web, pp. 157–166. International World Wide Web Conferences Steering Committee, 10 April 2018
29. Sharma, A., López, Y., Tsunoda, T.: Divisive hierarchical maximum likelihood clustering. BMC Bioinform. **18**(16), 546 (2017)
30. Murata, T.: Detecting communities in social networks. In: Handbook of Social Network Technologies and Applications, pp. 269–280. Springer, Boston (2010)
31. Madani, F.: 'Technology Mining' bibliometrics analysis: applying network analysis and cluster analysis. Scientometrics **105**(1), 323–335 (2015)
32. Grover, A., Leskovec, J.: node2vec: scalable feature learning for networks. In: Proceedings of the 22nd ACM SIGKDD International Conference on Knowledge Discovery and Data Mining, pp. 855–864. ACM, 13 August 2016
33. Hastie, T., Tibshirani, R., Friedman, J.: The Elements of Statistical Learning, vol. 1. Np. Springer, New York (2001)

Cognitive Visualization of Carbon Nanotubes Structures

Vadim A. Shakhnov, Lyudmila A. Zinchenko,
Vadim V. Kazakov(✉), Andrei A. Glushko, Vladimir V. Makarchuk,
and Elena V. Rezchikova

Bauman Moscow State Technical University, Moscow, Russia
shakhnov@mail.ru, kazakov.VADIM.2012@Yandex.ru

Abstract. An overview of carbon nanotubes structures visualization techniques is given in this paper. The methods based on cognitive technologies have been applied. A new version of the NanoTube Analytics tool for visual analytics of carbon nanotubes is presented. The software testing on the users has shown that the perception of the visualized information is easier and without additional explanations. Used approaches help to work easily and faster. Approaches presented in this paper can be applied for visualization of complicated virtual objects and multidimensional data.

Keywords: Cognitive technology · Visualization · Carbon nanotubes Multidimensional data

1 Introduction

Carbon is widely used in many industrial applications. One of promising applications is carbon computers [1]. In the case, carbon can be used in the form of carbon nanotubes (CNT). The carbon computers would be significantly faster. In addition, electrical power consumption will be less. Silicon transistors in computers will be replaced by transistors, based on carbon nanotubes. However, carbon nanotube transistors applications are inhibited by material imperfections inherent to carbon nanotubes. Process variations specific to carbon nanotubes result in diameter normal distribution with $\mu = 1.2$ nm and $\sigma = 0.3$ nm [2]. In addition, variations in electrical properties of the carbon nanotubes are observed. They can be either metallic or semiconductive type [3].

It is important to analyze the structure of the CNTs. However, standard tools do not show nanotubes structures in a form suitable for visual analytics. Cognitive approaches have to be used for visualization of the CNTs structures.

The main problem in analyzing CNTs is the complexity of visualizing their structure and parameters. It is not easy to display everything in an accessible and understandable form for the researcher. The basic principles of interaction with the outside world and the perception of visual information by a person were taken into account, while cognitive visualizations in the software were developing. It is necessary to simplify the work with information for the user without losing its content. Cognitive methods of perception help people in the right way to submit information.

A. Abraham et al. (Eds.): IITI 2018, AISC 874, pp. 273–279, 2019.
https://doi.org/10.1007/978-3-030-01818-4_27

In this paper, problems of CNTs structure visualization are discussed. The related works are reviewed. Then the proposed new cognitive approach is discussed to simplify the work with information. Features of the tool are outlined. Finally, conclusions are derived.

2 Related Works

The simplest representation of a carbon nanotube is a line [4]. Figure 1 shows an example of a random CNT network representation. Each line represents a CNT.

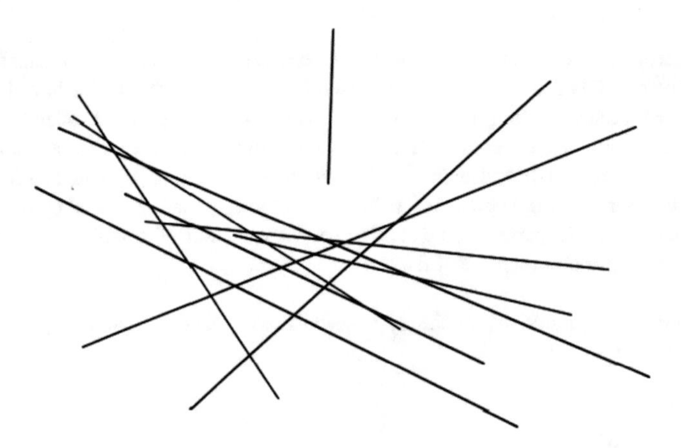

Fig. 1. A random CNT network.

However, application of the CNT representation can result in wrong conclusions, e.g. about a number of nanotubes intersection contacts.

In [5], the circle is used as another form of representation of nanotubes. However, a role of chirality indices is not obvious for this representation.

Some authors use a cylinder as a graphical representation of nanotube [6]. Disadvantage of the approach is incorrect representation of a CNT structure.

Computer methods of nanotube structure generation include several approaches. In [7], coordinates are generated for specified chirality indices. Both single walled and multi walled nanotubes are supported. In [8], coordinates of heterojunction between two arbitrary CNTs are calculated. However, CNTs visual images are generated only.

In [3], application of visual analytics to single-walled CNT properties investigation has been proposed. The first version of our tool NanoTube Analytics with cognitive approaches in visualization has been introduced. The user interactively defines chirality indices n, m. Two CNTs with the given chirality indices n, m are generated and visualized. A user can compare their properties including diameter D, thermal conductivity G etc. Figure 2 illustrates the approach.

In [9], this approach has been expanded for a single-walled CNTs family that can be produced for the given technological variations. A user defines two values: minimal diameter D_{min} and maximal diameter D_{max} and then the corresponding family generates. A CNT from the family can be chosen by a user and then its representation generates.

Shakhnov et al. [10] has applied the cognitive approach for visualization of thermal and electrical single-walled CNTs properties simultaneously. It is important that a summary about the generated CNTs family is given as well. Figure 3 shows an example of this representation.

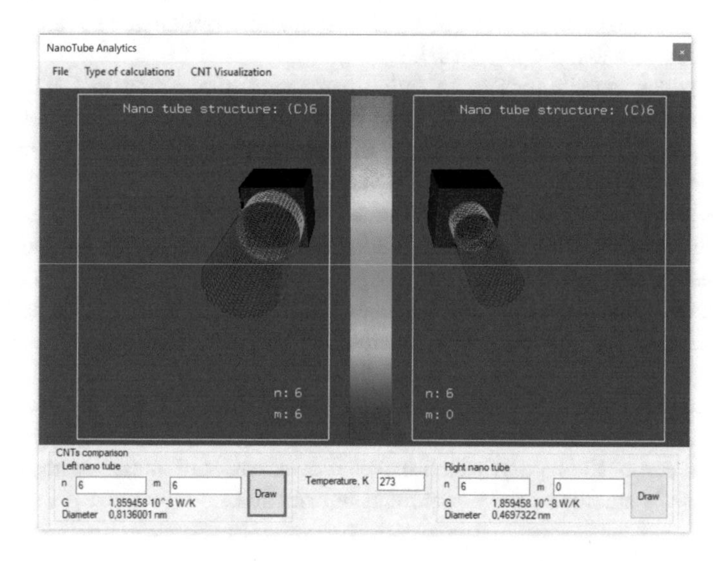

Fig. 2. An example of two single-walled CNTs.

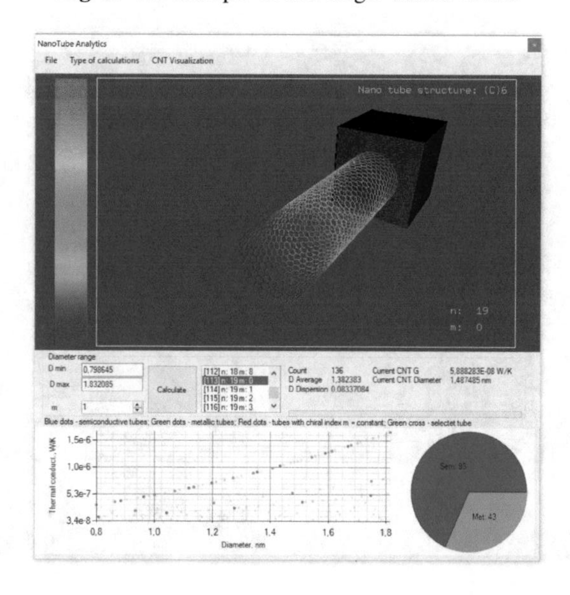

Fig. 3. Thermal and electrical properties of a single-walled CNT.

However, too much information is provided for a user in [9, 10]. Therefore, cognitive technologies have to be used to overcome these deficiencies.

3 Application of Cognitive Technologies in Visual Analytics of CNTs Structures

A CNT is not visible for a human eye. Therefore, a user is unable to recall stored visual memories. Therefore, a problem of CNT visualization requires new approaches of visualizing objects that are not visual.

In the paper, possibilities of selective perception application in CNT properties investigation are discussed. A tool has to highlight important parameters and images and to do not show all available data in different applications of CNTs. In addition, relations between parameters have to be visualized as well.

In the previous version of a tool NanoTube Analytics, a CNT was shown as a hollow cylinder (Fig. 3). New functionalities were added and in the current version a user can use a scope and increase an image to observe the correspondent atoms and bonds. Figure 4 illustrates this new functionality. Simple geometrical calculations for a six-member cycle are performed to generate the correspondent hexagons and then a hexagonal lattice. It should be noted that the bond length of carbon atoms d_0 is fixed and equal to 0.142 nm [7]. It's used the class *MoveController* (see more details in [3]), that has been used in the 3D data visualization mode. Also, interactions can occur via the touch screen of the laptop, which makes it even easier to interact with objects and focus the user's attention on the important information.

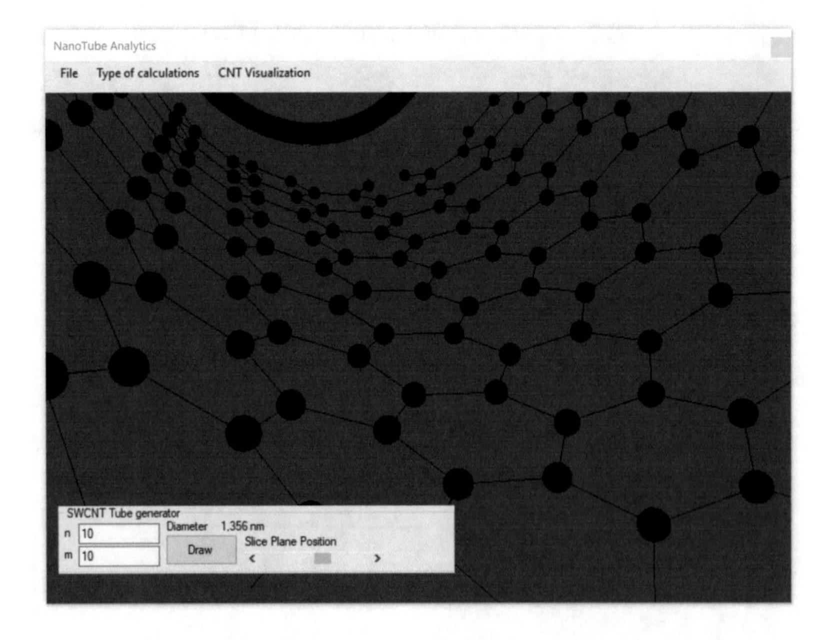

Fig. 4. A single-walled CNT structure.

It is important that user can set a chirality index to be constant and analyze relations between atoms and bonds (see Fig. 4).

Visualization of multi-walled CNTs is more difficult problem. In order to overcome too much information, it is necessary to show layers and a sketch of the tube with the corresponding sizes. Figure 5 shows the proposed representation. The user interactively defines the chirality indices *n*, *m* of the inner tube and a number of layers *L*. Then the family of CNTs is generated. A diameter D_i is calculated as follows:

$$D_i = D_{in} + 2 \times i \times d_l, \tag{1}$$

where

D_{in} is the inner diameter *Inner D*;
d_l is the distance between graphite planes and d_l = 0.335 nm [7];
i is a number of a layer.

The inner diameter is calculated as follows:

$$D_{in} = \sqrt{m^2 + n^2 + m * n}\frac{\sqrt{3} * d_0}{\pi}. \tag{2}$$

In current version of the tool, the new class *MWCNT* containing the generated nanotubes has added. This is the only list of standard nanotubes. The specific function for visualization of multi-walled CNTs has added in order to visualize. A user can scale, move and rotate the multi-walled CNT. Again, the class *MoveController* is exploited [3], which is used in 3D data visualization mode.

In this mode, a problem was observed with standard visualization techniques on the example of tubes including more than 10 layers. Many atoms and bonds became unrecognizable. Cognitive approaches to the construction of visualizations made it possible to understand which form of representation of the object is most suitable for solving this problem. It was solved by the use of the color differentiation of the layers and by using the ability of a layer selection. The lines showing layers, rings showing minimum and maximum diameters and the plane of the tube cut were visualized.

Since a person reacts to a bright color that stands out against the background, the selected CNT layer is clearly visible even between the other layers, but drawn in black. The human brain itself draws the missing pieces of information hidden behind other layers, and thus it is necessary to see a separate layer distinctly. For the analysis of each layer, the possibility of interactive selection was added. The mode of visualization of each layer by a separate color, focusing on an individual color was added too. So, the user can analyze each layer (Fig. 5). In this case, the person himself mentally chooses which layer to see. The ability to interact with the object allows people to select a convenient view. Without this approach, it is almost impossible to distinguish layers (Fig. 6).

In order to visualize each object, user needs to choose contrasting colors or try to ensure that similar colors do not overlap, otherwise it may influence badly on perception. Each visualization is unique. Person can come up with a trick understanding the basic principles of cognitive perception. This trick allows to use the inherent capabilities of peoples brain, to solve problems of presenting complex information.

It is not possible to build a model of visualization in advance without knowledge in the field of cognitive perception. In this case, many different options would have to be implemented.

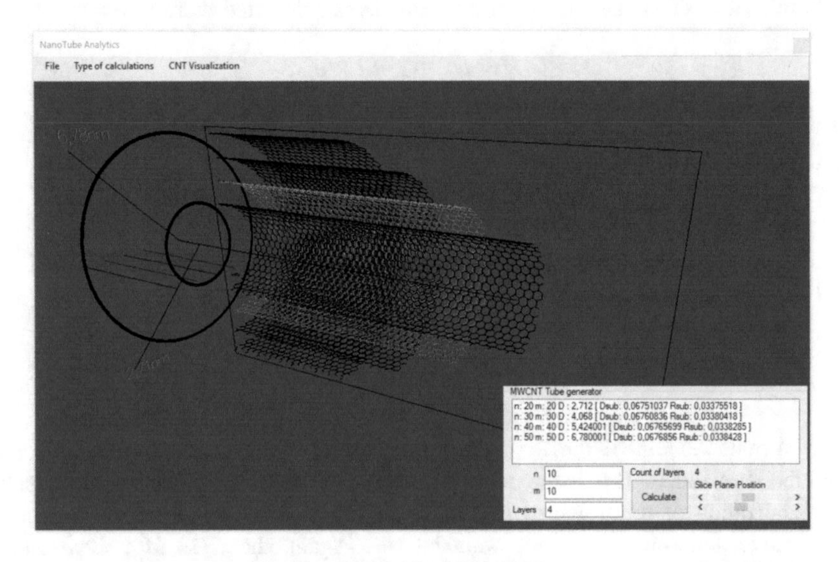

Fig. 5. Multi-walled CNT.

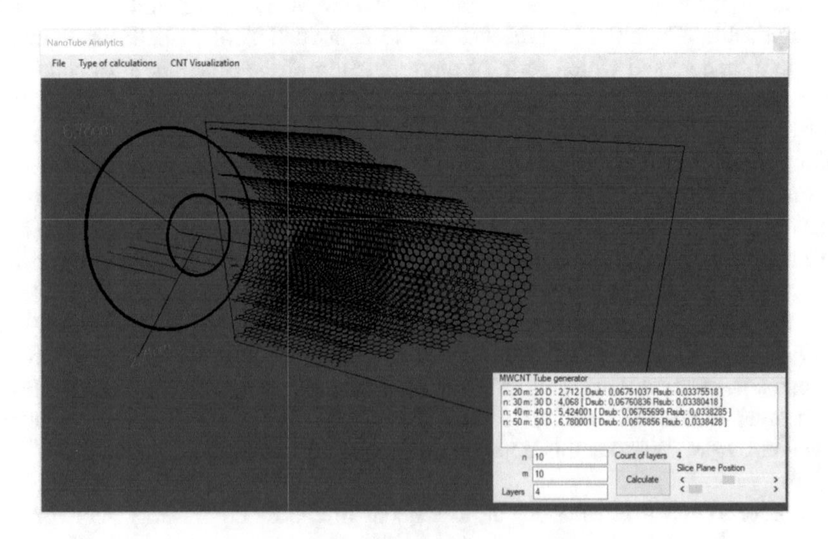

Fig. 6. Multi-walled CNT without coloring.

4 Conclusion

Preliminary conducted tests show efficiency of the approach that is based on cognitive technologies. A new version of the tool NanoTube Analytics is implemented to help an user to analyze several CNTs properties simultaneously. Important parameters and their relations are shown, so user can analyze CNT properties in different modes. It is obviously that selective perception is a powerful method of attracting user's attention to important features of CNTs. The approaches are proposed to show single-walled CNTs and multi-walled CNTs and implementation of them.

This approach based on cognitive technologies and selective perception. The current model of visualization can be applied in other areas, where several objects interfere with the survey of each other. Also when it is necessary to focus on a particular object or its area. Such approaches allow to work not only with the presented information, but also with the user and his perception. Many mechanisms are already laid by nature and their use will help to simplify the development of visualizations, and to increase its effectiveness.

The tool can be used by manufacturing engineering during technological process development and equipment calibrations.

This work was partially supported by grant RFBR 15-29-01115 ofi-m and 16-06-00404 a.

References

1. Franklin, A.D., Luisier, M., Han, S.-J., et al.: Sub-10 nm carbon nanotube transistor. Nano Lett. **12**(2), 758–762 (2012)
2. Patil, N., et al.: Wafer-scale growth and transfer of aligned single-walled carbon nanotubes. IEEE Trans. Nanotechnol. **8**(4), 498–504 (2009)
3. Kazakov, V., Verstov, V., Zinchenko, L., Makarchuk, V.: Visual analytics support for carbon nanotube design automation. In: Samsonovich, A., Klimov, V., Rybina, G. (eds.) Advances in Intelligent Systems and Computing, vol. 449, pp. 71–78 (2016)
4. Bush, S., Li, Y.: Graph spectra of carbon nanotube networks: molecular communication. In: MRS Proceedings, p. 951 (2006). 0951-E04-06
5. Dumlich, H., Gegg, M., Hennrich, F., Reich, S.: Bundle and chirality influences on properties of carbon nanotubes studied with van der Waals density functional theory. Phys. Status Solidi B **248**, 2589–2592 (2011)
6. Tanachutiwat, S., Wang, W.: Carbon Nanotubes Interconnect Analyzer (CNIA) (2015). https://nanohub.org/resources/cnia. Accessed 12 Mar 2018
7. Maruyama, S.: Nanotube coordinate generator with a viewer for windows. http://www.photon.t.u-tokyo.ac.jp/~maruyama/wrapping3/wrapping.html. Accessed 12 Mar 2018
8. Melchor, S., Dobado, J.: CoNTube: an algorithm for connecting two arbitrary carbon nanotubes. J. Chem. Inf. Comput. Sci. **44**, 1639–1646 (2004)
9. Shakhnov, V.A., Zinchenko, L.A., Makarchuk, V.V., Rezchikova, E.V., Kazakov, V.V.: Visual analytics in investigation of chirality-dependent thermal properties of carbon nanotubes. J. Phys.: Conf. Ser. **829**(1), 012008 (2017)
10. Shakhnov, V., Kazakov, V., Zinchenko, L., Makarchuk, V.: Cognitive data visualization of chirality-dependent carbon nanotubes thermal and electrical properties. In: Samsonovich, A., Klimov, V. (eds.) Advances in Intelligent Systems and Computing, vol. 636, pp. 302–307 (2017)

Decision Making Model for Outsourcing by Analysis of Hierarchies of T. Saaty Under Fuzzy Environment

Natalia Egorova and Yana Sorokina(✉)

Central Economics and Mathematics Institute (CEMI), Russian Academy of Sciences (RAS), Moscow, Russia
yana_325@mail.ru

Abstract. The article describes disadvantages of existing one-criterion methodologies: only the monetary effect of outsourced process is considered, the analysis of all possible positive and negative consequences of outsourcing is not taken into account. It outlines the main difficulties of multi-criteria approaches for decision making about outsourcing, related to the fact that the improvement of one important criterion of the enterprise's activity can lead to the deterioration of another criterion.

It outlines the fact that there is no unified algorithm for making decisions about production outsourcing. The existing multicriteria decision-making methodology was improved, the Saati method was proposed for ranking the criteria taking into account their importance, as well as a pairwise comparison of the parameters. It explains necessity of applying the proposed decision-making method for outsourcing at an industrial enterprise. It is shown that, along with the economic effect, the mentioned parameters are important for making decision about outsourcing. A detailed analysis of these parameters is carried out on example of production enterprise.

Keywords: Outsourcing analysis · Evaluation · Decision-making

1 Introduction

The efficiency of enterprises in the industrial sector directly depends on the degree of organization of internal and external processes. In conditions of unstable market conjuncture and economy as a whole, outsourcing is one of the most successful methods to increase efficiency of business processes.

There are several definitions of «outsourcing»:

- Outsourcing (from "out", outside, and "source" - source) is transferring of the internal division or divisions of the enterprise and all related assets to the organization which is service provider offering to provide a certain service for a certain time according to the agreed price [3].
- Outsourcing is transferring of non-core (secondary) business processes (production, service, information, financial, management, etc.) to the organization (outsourcer) for their implementation and functioning [1].

A. Abraham et al. (Eds.): IITI 2018, AISC 874, pp. 280–289, 2019.
https://doi.org/10.1007/978-3-030-01818-4_28

The concept of outsourcing is rather widely discussed in the works of Russian and foreign scientists, but there is no unified algorithm for its rational application.

Obviously, the transfer of a business process to a third-party organization in the long term will allow:

- Significantly reduce transaction costs: outsourcing will allow the company to focus on the core activities, while none-core activities will be delegated to a third-party organization;
- provide access to new skills and knowledge on an international scale;
- distribute risks.

It is the problem of scientific research that there is no effective decision-making method about outsourcing or insourcing. Among the approaches to making decisions about outsourcing, there are 2 types:

1. One-criterion models use one criterion evaluation when making a decision about outsourcing, most often the production costs are compared with outsourcing costs. Companies believe that using outsourcing, the costs of transferred business processes will be significantly less than using their own production. To evaluate the effectiveness of outsourcing, it is necessary to take into consideration all costs connected with its use. According to the report of Deloitte, 2010, hidden costs of outsourcing were presented (Table 1): Obviously, lower costs of business processes does not mean lower costs for outsourcing. Lower costs of business processes can be used as the primary information for decision-making about outsourcing [4].

Table 1. Hidden costs of outsourcing (in % of the total budget for outsourcing)

Hidden costs	% of the total budget for outsourcing
Conclusion of the contract	1–2
Transfer of know-how from client to outsourcing company	5–7
Transport and travel expenses	2–3
Transferring of the business process	1–2
Communications	1–3
Change, adaptation of the business process	0–30
Total expenses:	10–47

2. Multicriteria models – based on evaluation of outsourcing influence on various factors of the enterprise activities. The criteria used in the multicriteria method the most fully describe the enterprise, as well as its interaction with the external environment. Multicriteria models include measuring and comparing all the entered criteria and determine the weights of each of the criteria.

2 Background and Related Works

From the Industrial Revolution to the early 1980s, the manufacturer's strategy was based on establishing processes and requirements related to all products or ordered products inside the organization depends on existing resources and labor. However, faced with many difficulties, many organizations began to focus on core activities, outside divisions of labor and planning of issues. In this way, it was possible to obtain a competitive advantage by improving the main qualities. Obviously, specialization and certain areas of activity will become possible using outsourcing. In fact, outsourcing is handing over some of the primary or non-primary chores based on the rational decision-making process; therefore, competitive advantages will be reached through outsourcing. This will lead to a decrease in the vertical integration rate of the system.

In general, outsourcing is used to reduce production costs, obtain new higher technology and skills, effectively use the organization's available time and limited resources, prevent activities confusion, and ultimately prevent the organization's unlimited increase and related costs.

Probert [5] presented a strategic methodology for making decisions about production or purchase, which was based on a thorough analysis of all the different aspects of production technology.

McIvor et al. [2], in his work outlined the establishment of relations with the selected supplier, tried to present a conceptual model for the production or purchase of strategic goods. One of the applications of this model is for organizations in which so much strategic attention should be paid to making decisions in the production or purchase.

Padilo and Dibey [6] first tried to use multicriteria model. They proposed a method of analysis and decision-making, and evaluated the strategy of production or purchase in seven stages. This method includes a comparison model with four goals at the same time: to maximize competitive strategic performance, to maximize management performance, and to minimize risk and maximize financial performance. In this model, various methods such as complex programming and AHP are used.

Lancedale [7] proposed a conceptual framework for effective management of outsourcing risks, emphasizing the consideration of organizational competitive advantage. Comann and Ronan [8] presented a model that indicates the situation in which demand is greater than supply, and management must decide how much of this product to produce and what quantity to buy from the contractors. They showed outsourcing as a problem of linear programming based on financial and production capacity parameters.

Vals-pierre and Clain-Hans [9] developed a series of laws and regulations based on the criteria of purchasing or production decisions, which can be used for strategic decisions.

Aktan et al. [10] developed a financial model to evaluate the impact of outsourcing options. In fact, this model provides a comprehensive framework for evaluation the full expected cost of outsourcing with the supplier network when purchasing is faced with unknown exchange rate. For this purpose, the Monte Carlo simulation method has been used.

Tills and Dreary [11] developed a model that supports purchasing or manufacturing decision-making based on the study of strategic goods and investments.

Our research builds on experience from previous above mentioned researches and we offer a new hybrid multi-criteria decision-making model. The scientific novelty is that we tried to offer model for making decisions about outsourcing and insourcing of production activities in condition of variety of qualitative and quantitative criteria. This model based on a combination of ANP and DEMATEL methods in a fuzzy environment can clarify decision-makers' verbal evaluations and overcome one of the difficulties of the ANP, which can use a large number of pairs to achieve the importance of weights of these criteria, having an internal connection with each other.

3 The Proposed Method for Outsourcing Decision-Making

In the real world, so many decisions include ambiguity in the definition of goals, limitations, and possible actions that are not clearly defined the roots of this ambiguity are inestimable data, incomplete data, and unavailable information. In order to solve this problem, Zade [12] proposed a fuzzy set theory as a mathematical method for ambiguity in decision-making. This theory is used when decision-making faces unclear and ambiguous human language discourse. The decision makers often tend to evaluate everything based on their experience and knowledge, and often use ambiguous linguists' words to make estimates. In order to unify the experience, beliefs, and ideas of decision makers, it is best to convert language estimates into fuzzy numbers. Therefore, the decision in the real world makes it necessary to use fuzzy numbers [13]. Decision-making in the case of outsourcing also can use this principle.

When making decisions on issues related to production activities outsourcing, a variety of qualitative and quantitative factors need to be reasonably taking into consideration. Therefore, the decision-making of these issues is a multi-criteria one and requires to use systematic and highly-dependent methods.

Most MCDM common methods are based on the assumption of element independence. But a criterion cannot always be independent. In order to solve the problem of element-oriented interaction, Saati proposed ANP as a new method of MCDM. ANP can consider various dependencies in a systematic way. This method has been successfully applied to so many cases. However, ANP's function in processing dependencies is not perfect [14]. From another perspective, the DEMATEL method can not only transform causality into a clear structural model, but it can also be used as a good method for facing internal dependencies within a set of criteri. In fact, DEMATEL can provide decision makers with more valuable information than ANP method [14, 15]. Considering that the ANP and DEMATEL methods are used in decision making, it is necessary to use of the opinions of experts. These opinions usually exist in the form of linguistic expressions, including ambiguous and vague concepts. This two methods cannot clearly explain and use the decision makers language. To integrate and eliminate indefinite points, it is useful to turn fuzzy linguistic expressions into fuzzy numbers, and in fact these two methods will be used in a fuzzy environment.

With respect to the mentioned advantages of ANP and DEMATEL and the possibility of using them in uncertainty circumstances, a new effective method based on the combination of ANP and DEMATEL methods in a fuzzy environment is proposed to help in making decisions for outsourcing and outsourcing production activities.

The implementation phase of the proposed model is based on the eight steps of the common ANP method. The difference between this method and the ordinary method is that the internal criteria of each group or the secondary criteria have mutual influences with each other, technically speaking, have an internal dependence. To determine the degree of this mutual effect and their dependence, instead of using the ordinary method of comparisons in ANP, we use the results obtained by the matrix of the general relationship used in the fuzzy DEMATEL.

The execution phase of this model is shown in Fig. 1. Thus, not only the amount of calculations is reduced, but also the linguistic analysis of decision-makers is used in a more efficient and convenient way.

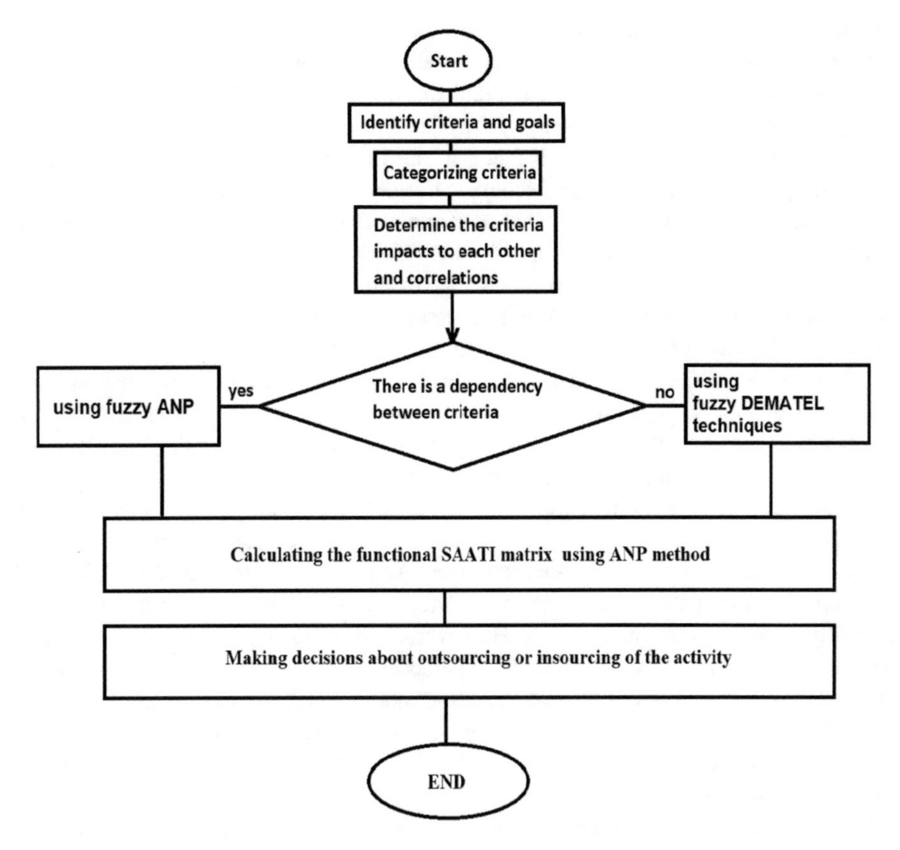

Fig. 1. Outsourcing decision-making model

4 Case Study

In order to show the applicability of the proposed model, we tried to apply this decision-making model about outsourcing on the example of «Turbomash-Service» industrial enterprise for the production of power equipment. For a comparison of the resources cost, we made calculations, according to which the cost of own production and outsourcing costs are reflected in Table 2 [4].

Since, as described above, the outsourcing process contains hidden costs, which are difficult to estimate in money due to their dependence on production volumes and a number of other factors. Besides, the evaluation of the two alternatives will be quantitative and qualitative.

Table 2. Cost of own production/outsourcing of spare parts for CJSC "Turbomash-service"

Name of turbine spare part	Price of «in house» production (RUB)	Price of outsourced production (RUB)	Required quantity
Traverse	203 768	170 000	20
Valve seat	193 400	160 000	34
Oil separator	1 050 000	927 000	21
Ring	108 000	95 000	21
Transportation cost	-	Included	-

To calculate the savings, apply the following formula:

$$E = [C - (P + T)] \tag{1}$$

where:

E - the resulting economic effect from the outsourcing;
C - cost of own production;
P - the price of manufacturing the product according to the drawing by a outsourced company;
T - transportation costs;

Then,

$$E = 34968960 - 30302000 = 4,666,960\,\text{Rubles}. \tag{2}$$

However, to make a decision to transfer the process to outsourced organization, it is not enough to consider only the positive effect, it is necessary to take into account the terms of supply, as well as the one-time costs and other hidden costs connected with outsourcing.

$$E = \sum_{i=1}^{n} \frac{Si - Pi}{\frac{(1+d)^i}{100}\%} - C_0 + D_0 \tag{3}$$

where:

E - the economic outsourcing effect;
N - duration of the period during which outsourcing is expected to be used;
Si - expected costs for the own production process in the i-th year;

Pi - the total cost of the process when it is performed by the outsourcer in the i-th year (includes the cost of outsourcer services, transportation costs, customer costs for interaction with the outsourcer, etc.);
D - is the discount rate, %.
C_0 - one-time costs connected with outsourcing (include, for example, the amount of compensation benefits paid to employees);
D_0 - one-time income connected with outsourcing (for example, income from the sale of machine tools).

This formula has a significant lack: only the monetary effect of outsourcing process is considered, and the analysis of all possible positive and negative consequences of outsourcing is not taken into account [4].

To take into account several qualitative criteria when deciding on outsourcing, it is necessary to consider the application of the ANP Saati method. The method is based on criteria analysis. The final rating is calculated using a special algorithm discussed in Sect. 3.

Because of so many ambiguities facing humanity when evaluating, we ignored the ordinary Saati model and instead of the one we use Fuzzy language comparison.

The first step: determine the ideals of decision making and establish a committee to collect ideas. To reflect the results of pair comparisons of criteria and alternatives, the following scale is used:

1 - the criteria are equivalent; 3 - one criterion is moderately better than other; 5 - one criterion is significantly better than other; 7 - one criterion greatly exceeds the other; 9 - one criterion is the best.

The next step of Fuzzy method is to determine evaluation criteria. The proposed method for decision-making about outsourcing, is based on a number of criteria, which include [4]:

- Quality of products before and after outsourcing;
- Level of critical knowledge protection;
- Organization, communication and coordination of processes.

Than using the evaluations made by decision makers we obtaining the normal direct-relation fuzzy matrix. The result of comparison of alternatives for the listed criteria is presented in the Table 3.

Table 3. Comparison of criteria and alternatives

Criteria	Alternatives	Outsourcing	Insourcing	Amount	The normalized value
Quality of products before and after outsourcing	Outsourcing	1	5/1	6	0.83
	Insourcing	1/5	1	1.2	0.17
Level of protection of critical knowledge	Outsourcing	1	1/4	1.25	0.2
	Insourcing	4	1	5	0.8
Organization, communication and coordination of processes	Outsourcing	1	1/2	1.5	0.22
	Insourcing	2/1	1	3	0.78

Communication and coordination of processes in outsourcing will be slightly worse than for own production due to differences in the standards of the customer's company and outsourcer, due to external communications of personnel, as well as due to the movement of production.

In our case, JSC "Turbomash-service" does not focus on improving the quality of products, because it is semi-finished products for further processing. At the same time, the company aim to increase the level of organization of production processes after their transfer to the outsourcer, therefore this criterion is no less important for making a decision about outsourcing.

Taking into account the above, the results of the evaluation of the criteria are presented in Table 4:

Table 4. Evaluation results

Criteria	Quality of products	Level of protection of critical knowledge	Organization, communication and coordination of processes	Total	Weight	The weight of the criterion for 2 alternatives	
						Outsourcing	Insourcing
Product quality	1	1/7	3/1	4.14	0.19	0.83	0.17
Level of protection of critical knowledge	7	1	9	17	0.75	0.2	0.8
Organized, communication and coordination processes	1/3	1/9	1	1.45	0.06	0.22	0.78

The applied method provides a pairwise comparison of the criteria themselves according to their significance. Obviously, for a company using production outsourcing to reduce costs, the level of protection of critical knowledge is more important criteria than organization, communication and coordination of processes. Being an exclusive developer and holder of design documentation, transferring it to competitors, the enterprise risks losing part of its profits.

Applying linear convolution to obtain integral estimates of alternatives, the values obtained for outsourcing and insourcing are, respectively, 0.58 and 1.25 (Table 5).

Table 5. Integral estimates of alternatives

	Cost of the process, rub.	Normalized cost	Integral evaluation	Attitude
Outsourcing	30 302 000,00	0.46	0.3209	0.58
Insourcing	34 968 960,00	0.54	0.6791	1.25

The obtained data show that, despite the positive economic effect from outsourcing, the most preferable option is the production of spare parts «in-house».

5 Conclusions and Future Work

5.1 Conclusions

Thus, all the current methods of making decisions about outsourcing can be conditionally divided into 2 groups: one-criteria and multi-criteria. The main disadvantage of multicriteria approaches is that the deterioration of one criterion is compensated by the improvement of another criterion. Although the proposed method is multi-criteria, but it lacks this disadvantage, because our algorithm use comparing the importance of the criteria themselves.

5.2 Future Work

It would be interesting to see the results of a study that would be conducted at more companies, in order to compare them to the results of the present study. In addition, another way of forming the data samples could indicate different results, which is worth looking into. Moreover, future work in the field of decision-making about outsourcing will be connected with suppler evaluation model. We hope that future paper will make a huge contribution to the development of decision-making theory and will actively use in practice. The proposed evaluation methodology allows synthesizing the problem and making informed decisions at the initial stage of the outsourcing project, and can also be used to manage the efficiency in the project implementation process.

References

1. Konik, N.A.: Outsourcing. Knorus, Moscow (2008)
2. Gertl, G., Geidos, M., Potkani, M.: Evaluation of the effectiveness of outsourcing through the costs of its use. Procedia Econ. Financ. **26**, 1080–1085 (2015). [trans. from English]
3. Heywood, J.B.: Outsourcing: in search of competitive advantages: [trans. from English], p. 40. Izd. House "Williams", Moscow (2004)
4. Sorokina, Y.S.: Features of the multicriteria methodology for decision-making on outsourcing at an industrial enterprise. Netw. Electron. J. KubSU **128**, 710–720 (2017)
5. Probert, D.R.: The practical development of a make or buy strategy: the issue of process positioning. Integr. Manuf. Syst. **7**(2), 44–51 (1996)
6. Padillo, J.M., Diaby, M.: A multiple-criteria decision methodology for the make-or-buy problem. Int. J. Prod. Res. **37**, 3203–3229 (1999)
7. Lonsdale, C.: Effectively managing vertical supply relationships: a risk management model for outsourcing. Int. J. Supply Chain Manag. **4**, 176–183 (1999)
8. Coman, A., Ronen, B.: Production outsourcing: a linear programming model for the theory-of- constraints. Int. J. Prod. Res. **38**, 1631–1639 (2000)
9. Vallespir, B., Kleinhans, S.: Positioning a company in enterprise collaborations: vertical integration and decisions. Prod. Plan. Control **12**, 478–487 (2001)

10. Aktan, M., Nembhard, H.B., Shi, L.: A real options design for product outsourcing. In: Proceedings of the 2001 Winter Simulation Conference, pp. 548–552 (2001)
11. Tayles, M., Drury, C.: Moving from make/buy to strategic sourcing: the outsource decision process. Long Range Plan. **34**, 605–622 (2001)
12. Zadeh, L.A.: Fuzzy sets. Inf. Control **8**(2), 338–353 (1965)
13. Kaufmann, A., Gupta, M.M.: Introduction to Fuzzy Arithmetic: Theory and Applications. Thomson Computer Press, NewYork (1991)
14. Wu, W.W.: Choosing knowledge management strategies by using a combined ANP and DEMATEL approach. Expert Syst. Appl. **35**(3), 828–835 (2007)
15. Tsai, W.H., Chou, W.C.: Selecting management systems for sustainable development in SMEs: a novel hybrid model based on DEMATEL, ANP, and ZOGP. Expert Syst. Appl. **36**, 1444–1458 (2009)

Evolutionary Modeling, Bionic Algorithms and Computational Intelligence

Forecasting of Results of Dynamic Interaction Between Space Debris and Spacecrafts on the Basis of Soft Computing Methods

Boris V. Paliukh, Valeriy K. Kemaykin, Yuliya G. Kozlova(✉), and I. V. Kozhukhin

Tver State Technical University, Tver, Russia
pboris@tvstu.ru, {vk-kem, jul_kozl, kozhukhin}@mail.ru

Abstract. Space debris (SD) is a real danger for spacecraft (S) in-orbit operation. Taking this danger into account is a S flight safety requirement. SD particles are detected by the S on-board equipment. The integrated intelligent information system forecasts, within its execution time, the results of the impact caused by these particles. Such forecasting enables one to evaluate potential damage from the collision and to take sufficient measures to ensure the S safety. The article presents an approach to forecasting the results of dynamic interaction between SD objects and a S on the basis of fuzzy logic rules and the mechanism of knowledge base training, carried out by generative adversarial network (GAN).

Keywords: Space debris · Knowledge database · Fuzzy systems
Damage risk assessment

1 Introduction

The problem of outer space exploration is substantially determined by orbital spacecraft (S) durability, which, in turn, is related to their ability to endure space debris (SD) impact. Large space debris (SD) is monitored and cataloged by space monitoring (SM) resources. At present S safety management comes to a collision avoidance manoeuvre in case the near collision criterion is fulfilled and (or) the calculated value of a probable collision is exceeded. Average and small SD is not monitored by ground facilities since SM resources can discover SD over 10 cm in size. At the same time SD particles over 1 cm in size deliver damage influencing on functioning S and its active existence fact in case high-velocity interactions. Forecasting SD particle impact provides to evaluate damage expectancy of collision consequence and to assume the adequate measures to S safety management. It is difficult to study the dynamic nature of interaction between SD and S structural elements because of quiet short time of interaction, space velocities and high kinetic energy of collisions leading to forecast ambiguity of the structural element destruction or S loss.

A. Abraham et al. (Eds.): IITI 2018, AISC 874, pp. 293–302, 2019.
https://doi.org/10.1007/978-3-030-01818-4_29

2 The Modern Approach

The main parameter indicating S persistence against mechanical impact caused by high-velocity particles of natural and technogenic origin is its undamage probability during service life (UDP) [7]. It is probability that none of dangerous collisions of a high-velocity particle with a vulnerable surface of S (n = 0) will occur during time τ. UDP is calculated according to the formula:

$$UDP = e^{-\overline{N}} \tag{1}$$

where $\overline{N}$ is an average number of collisions.

Given that it is convenient to use UDP of one element of a vulnerable surface for operational estimates, the value is defined as:

$$\overline{N} = N_{ex} S_{el} \tau \tag{2}$$

where N_{ex} is the value of integrated average flux of high-velocity particles featuring a critical mass of m_{ex}, m_{ex} is extreme minimum mass of a single SD particle which damaged the considered element of S surface (according to the adopted criterion), $m^{-2}\cdot year^{-1}$; S_{el} is the considered area of a vulnerable element of the S surface, m^2; τ is residence time of S in the SD environment (in-service time of S), year. It is assumed that the density of SD material is known. The task is to determine the value of d_{ex} which will allow us to estimate the number of S wall through damages in the flow of SD high-velocity particles. There is an approach to calculating d_{ex} using ballistic limit equations (BLE) [2]. BLE-based analysis of the S protection implies of a number of requirements: BLE must take into account all relevant factors of the possible impact caused by SD object; the range of variables change in the BLE must reflect the real-life processes of a possible impact. The disadvantage of the proposed approach is the impossibility of constructing a universal equation since a significant part of the information on the physics of the collision process between a particle and a wall is contained in numerical coefficients demanding constant accounting. Therefore the models must be constantly improved, and their calibration must be regularly updated with newly-obtained measurements. Thus it is possible to state the following main problems which are inherent to the problem of forecasting the results of interaction between SD and S: the data on the interaction between SD and S is so limited and statistically unstable that it puts into question the credibility of the model developed by the methods of probability theory and classical mathematical statistics; sufficient reliability can't be ensured in case the forecasting results are presented in the form of deterministic damage functions since many influencing factors are fuzzy and difficult to formalize using traditional modeling methods.

3 The Main Results

The main cause of the pragmatic problem is the uncertainty of the characteristics and the parameters of the relative motion of S and SD objects at the contact point. This problem can be solved by using soft computing methods. Soft computing (SC) is a consortium of computational methodologies that collectively provide the basis for understanding, designing and developing intelligent systems. The SC main components in this consortium are fuzzy logic (FL), neurocomputing (NC), genetic computing (GC), probabilistic computing (PC), evidential reasoning, belief networks, chaotic systems, branches of the machine learning theory [1, 10]. The fuzzy inference engine is chosen to evaluate the S predicted damage under the influence of SD objects in conditions of uncertainty [3, 9]. The starting point for evaluating the possible impact of a SD object on a S is the statement: "If the collision is possible, then the damage caused by the SD object depends on the sizes and velocities of the S and the SD, on the composition of their materials, on the angle at which the SD strikes the S and also on the vulnerability and the importance of the S area (the S component) where the impact occurred." From this point of view, forecasting the results of the impact caused by SD objects on a S can be considered as an identification problem having the following features:

1. In order to evaluate the results, it is necessary to establish the relationship between the input and output variables.
2. The input variables are associated with the characteristics and parameters of motion in relation to each other at the contact point as well as their mass-dimensional and design features.
3. The output variable is associated with the state of the S after the impact of the SD object and allows forecasting the expected damage.

Collision with a S single dense wall, having a sufficient thickness, can create a perforation, a crater on its surface or a hole in the wall (in case higher velocity and a thinner shell). When a thin shell is breached, the particle collapses. After that the particle fragments (the largest fragment is called the "leading element") and the shell fragments fly inside the instrument container. Collision with a porous shell creates either a thimble or end-to-end channel in the wall. At the same time the particle or the leading element of the previously breached barrier does not change its trajectory but it slows down and ablates. If the S has modules with multilayer shells and the layers can be both dense and porous, then the analysis task is to evaluate the wall breakdown resistance, that is, to determine whether the particle will break the whole packet or will be delayed on one of the layers. The vulnerability (damage risk) of the S in the SD flow can be defined as a combination of UDP of elements when the importance (criticality) of each functionally important element of S is taken into account.

4. The input and output variables can have quantitative and qualitative estimations.
5. The structure of the relationship between the output and input variables is described by the logical dependencies IF <inputs>, THEN <output> using qualitative estimates of variables and representing a fuzzy knowledge base.

By $X = (x_1, x_2, \ldots, x_n)$ let us denote the parameters that affect character and magnitude of possible damage to S (its individual element). Then, according to cybernetics standpoint, the task resolves itself into constructing a forecasting model which defines a representation $X \rightarrow D = \{d_1, d_2, \ldots, d_m\}$. Influencing factors X and the output parameter d_j, including the ranges of values and the term-sets used for description, are presented in Table 1.

Table 1. The characteristics of formed term-set

	Symbol	Variable range (universal set) and dimension	Term-sets of the linguistic variable value
High-velocity collision parameters			
Interaction velocity of SD and S at contact point, v_n	*X1*	0…20 km·s^{-1}	Low (L), under medium (UM), medium (M), above medium (AM), high (H)
The angle between the particle velocity vector and a normal to the surface, θ	*X2*	0… ± 180°	Small (S), medium (M), normal (N),big (B),huge (H)
Parameters of SD			
Particle diameter, d_p	*X3*	0.1…1000 cm	Small (S), medium (M), big (B)
Particle velocity, v_p	*X4*	3…8 km·s^{-1}	Low (L), medium (M), high (H)
Particle material density, ρ_p	*X5*	2.5-7 g/cm^3	Aluminum alloy (Aa_p), titanium alloy (Ta_p), stainless steel (Ss_p)
Particle mass, m_p	*X6*	10^{-5}…10^3 t	Small (S), medium (M), big (B)
S wall parameters			
Wall thickness, l_w	*X7*	0.13–0.4 cm	Small (S), medium (M), big (B)
Wall material density, ρ_w	*X8*	2.5–7 g/cm^3	Aluminum alloy (Aa), titanium alloy (Ta), stainless steel (Ss)
Wall material hardness by Brinell, HB	*X9*	60–150 MPa	Small (S), medium (M), big (B)
Bumper parameters			
Bumper thickness, l_b	*X10*	0.03–0.055 cm	Small (S), medium (M), big (B)
Bumper material density, ρ_b	*X11*	0.01–2.5 g/cm3	Small (S), medium (M), big (B)
The gap between a bumper and the S back wall, G	*X12*	0…5 cm	Small (S), medium (M), big (B)
Yield stress of the back wall, σ_w	*X13*	130-540 MPa	Low (L), medium (M), high (H)
S velocity, v_s	*X14*	3…8 km·s^{-1}	Low (L), medium (M), high (H)
Wall damage criterion, *D*	d_j	d_1 is perforation, dent (roughness), d_2 is a crater (a thimble channel) $\left(d_p/l_w < 1\right)$, d_3 is a hole with a diameter of *d* (an end-to-end channel) $\left(d_p/l_w > 1\right)$, d_4 is element damage for the case: $d_3 \wedge (d_p \geq S_e)$.	

The information provided by digitized X-ray images processing results, which were obtained during bench tests, is represented by a set of *N* pairs of input value vectors and their corresponding d_j, defined as the values of the linguistic variable:

$$N = k_1 + k_2 + \ldots + k_j + \ldots k_m \tag{3}$$

where k_j is number of experimental data units corresponding to d_j, $j = 1 \ldots m$, m is the number of terms for the linguistic variable of damage S and in the general case $k_1 \neq k_2 \neq k_3 \neq k_m$. It is clear that $N < l_1, l_2 \ldots l_n$, that is, number of selected experimental data units is less then complete enumeration of different combinations of levels (l_i, i = 1… n) of input variables changes. A table formed according to the rules is called a knowledge matrix [3]:

1. the dimension of this matrix equals (n + 1)N, where (n + 1) is number of columns while $N = k_1 + k_2 + \ldots k_m$, m is number of rows;
2. the first n columns of the matrix correspond to input variables x_i, $i = 1 \ldots n$, and (n + 1)th column correspond to the values of $d_j (j = 1 \ldots m)$;
3. each row of the matrix represents a combination of input variables values, this combination classified by an expert as corresponding to one of the possible values of the output variable. The first k_1 rows correspond to output variable value d_1, the second k_2 rows correspond to value d_2, the last k_n rows correspond to value d_m;
4. The element a_i^{jp}, standing at the intersection of the column i and row j, correspond to linguistic estimation of the parameter x_i in j_p row of the fuzzy knowledge base. The linguistic estimate a_i^{jp} is chosen from the term-set of the variable x_i, that is, $a_i^{jp} \in A_i, i = 1 \ldots n, j = 1 \ldots m, p = 1 \ldots k_j$;
5. w_{jp} is a number within the range [0,1] that characterizes an expert's judgemental degree of confidence in part of the statement having the number p = k_j.

The operations $\cup$(OR) and $\cap$(AND) can help to rewrite the system of logical statements in a more compact way:

$$\cup_{p=1}^{k_j} \left[w^{jp} \cap_{i=1}^{n} \left(x_i = a_i^{jp} \right) \right] \rightarrow d_j, \quad j = 1 \ldots m \tag{4}$$

Forecasting results of influence SD objects on S implies solving FL equations. These equations are created basing on a knowledge matrix or logical statements system which is isomorphic to this matrix. They also enable one to calculate the values of membership functions of the S damage for fixed values of the input variables. FL equations are derived by replacing linguistic terms a_i^{jp} and d_j with corresponding membership functions and the operations $\cup$ and $\cap$ with $\vee$ and $\wedge$ accordingly.

The system of logical equations can be written in the following compact way:

$$\mu^{d_j}(x_1, x_2, \ldots x_n) = \bigvee_{p=1}^{k_j} \left[w^{jp} \bigwedge_{i=1}^{n} \mu^{a_i^{jp}}(x_i) \right], \quad j = \overline{1, m} \tag{5}$$

When calculating, one should replace the logical operations AND ($\vee$) and OR ($\wedge$) over the membership functions with min and max operations accordingly:

$$\mu^{d_j}(x_1, x_2, \ldots x_n) = \max_{p=\overline{1,k_j}} \left\{ w^{jp} \quad \min_{i=\overline{1,n}} \left(\mu^{a_i^{jp}}(x_i) \right) \right\}, j = \overline{1, m} \tag{6}$$

The fuzzy knowledge base corresponds to the forecasting model in the form of design ratios (6), where we use membership functions for x_i represented as a double-sided Gaussian curve.

$$\mu(x_i) = \begin{cases} \text{gmf}(x, b_1, c_1), & \text{if } x_i < b_1 \\ 1, & \text{if } x_i \in [b_1, b_2] \\ \text{gmf}(x, b_2, c_2), & \text{if } x_i > b_2 \end{cases} \tag{7}$$

where gmf is the Gaussian membership function:

$$\mu(y) = \exp\left(-\frac{(y-b)^2}{2c^2}\right) \tag{8}$$

where b and c are the membership function parameters, the former being the maximum coordinate, the latter being the minimum coordinate. Then in general form:

$$D = F(X, W, B, C) \tag{9}$$

where $X = (x_1, x_2, \ldots x_n)$ is the input vector, $W = (w_1, w_2, .., w_N)$ is the weight vector of rules (rows in fuzzy knowledge base), $B = (b_1, b_2, \ldots, b_q)$ and $C = (c_1, c_2, \ldots, c_q)$ are vectors of adaptation parameters of fuzzy terms membership functions, N is the total number of rules-rows, q is a general number of terms, F is communication statement <inputs-output> corresponding to the usage of the relation (7). A fuzzy knowledge base must meet the requirement of adequacy, completeness and consistency as well as accuracy and interpretability of the acquired results. The mechanisms of its construction must provide for a possibility of entering new information about the subject area and use this information for using this information for keeping the knowledge base up-to-date, including its timely correction (training).

The issue of knowledge base training exists as the task of construction a matrix. This matrix, meeting the limitation of the parameter range (W, B, C) and of the number of rows, provides for:

$$\sum_{p=1}^{M} \left[\mu^{d_j}(X_p, W, B, C) - \mu_p^{d_j}\right]^2 = \min_{W,B,C} \tag{10}$$

In this respect, simultaneous application of neural networks and FL is well-grounded since it totally gives new quality. The fuzzy model, derived as the result of this combination and immersed into a neural network (a neuro-fuzzy network), has two most important human (intelligent) properties as linguisticness, that is, using knowledge in natural language and trainability in real time [3]. The network structure is presented in Fig. 1 and the node content is shown in Fig. 2. The neuro-fuzzy network structure has four layers: layer 1 is identification object inputs; layer 2 is fuzzy terms used in the knowledge base; layer 3 includes rows which are conjunctions of fuzzy knowledge base; layer 4 includes the rules combined into classes d_j, $j = 1 \ldots m$. The number of neuro-fuzzy network nodes is defined as follows: the total number of layer 1 neuro-fuzzy network nodes equals to the number of identification objects inputs; that of layer 2 equals to the number of fuzzy terms in the knowledge base; that of layer 3 equals to the number of conjunction rows in the knowledge base; layer 4 equals to the number of predicted result of cooperation.

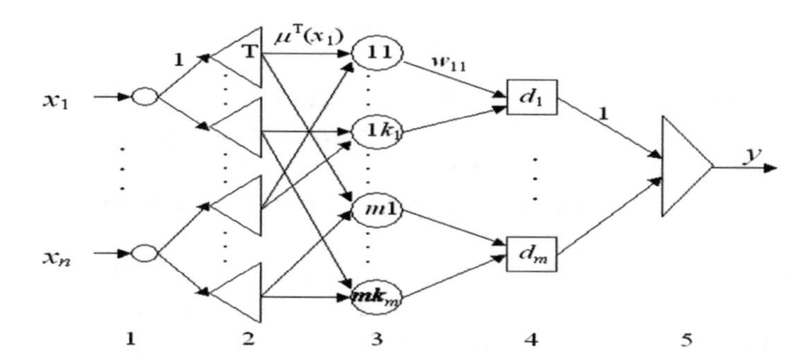

Fig. 1. The neuro-fuzzy network structure

Node	Name	Function
u ○ v	An input	$v=u$
u T v	A fuzzy term	$v=\mu^T(u)$
u_1 … u_l ○ v	A fuzzy rule	$v=\prod_{i=1}^{l} u_i$
u_1 … u_l □ v	A rules class	$v=\sum_{i=1}^{l} u_i$

Fig. 2. The elements of neuro-fuzzy network

The arcs of the graph are weighed as follows: 1 (one) is used for the arcs between layer 1 and layer 2; membership functions of input to fuzzy term are used for the arcs between layer 2 and layer 3; rules weights are applied for the arcs between layer 3 and layer 4. When defining the element of the fuzzy rules and the rules class included into the table, we substitute the FL operations of minimum and maximum with multiplication and addition operations. The possibility of such a substitution is proved in [5]. This allows us to obtain analytical expressions that are convenient for differentiation. We suggest solving the problem of fuzzy knowledge base training by using a generative adversarial network (GAN). GAN architecture networks consist of two neural networks. The first network is a generating one; it implements the function $yG = G(X_{gen})$. It takes a set of generated values *Xgen* at the input and the value *yG* at the output. The second network is a discriminating one; it implements the function $yD = D(X_{tr})$, that is, it takes a set of training values X_{tr} at the input and the value *yD* at the output [4]. The generating network must learn to generate such states yG that a discriminating network D cannot discriminate between them and existing reference states. Vice versa, the discriminating network must learn to discriminate between these generated states and real ones. Training the former implies maximization of the functionality:

$$D(G(X_{gen})) \rightarrow \max \tag{11}$$

The network tends not to maximize its own result, but the result of the discriminating one. The generating network must learn to generate a particular value at the output for every value given at the input. This particular output value, given to the input of the discriminating network, must provide for the maximum value at its output. If there is a sigmoid at the output of the discriminating network, it is possible to say that the sigmoid returns the probability that a correct value is given at the network input. Thus the generating network tends to maximize the probability that the discriminating one will not be able to discriminate between the result of generating network operation and reference patterns (that means the generating network generates states which are close to reference states). In order to take a step in the generating network training, it is necessary to calculate not only the result of the generating network operation, but also the operation result of the discriminating one. The generating network will learn upon the gradient of the operation result of the discriminating network. Training the discriminating network implies maximization of the functionality:

$$D(X_{tr})(1 - D(G(X_{gen}))) \rightarrow \max. \tag{12}$$

The network must generate "ones" for reference states and "naughts" for examples which are generated by the generating network. And the network doesn't know if the input data are a reference or imitation. Only the functionality knows that. At the same time the network is learning towards a gradient of this functional. The structure of GAN is shown in Fig. 3.

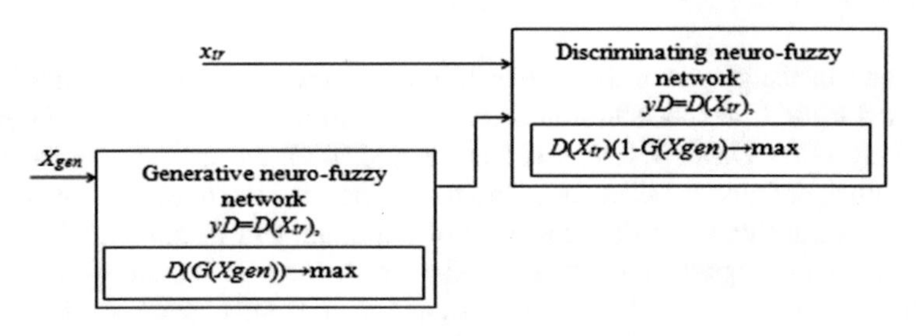

Fig. 3. The structure of GAN

At each step of the discriminating network we compute the operation result of the generating network once, and then do the same twice for the operation result of the discriminating network. First, the discriminating network receives a reference sample at the input. Secondly, it receives the result of the generating network. Both networks are united within the joint process of mutual training. The training is carried out using all available experimental data for $D = d_j \in [d_1, d_2, \ldots, d_n]$. At each step of the training the generating network receives at the input some values of parameters X from the stated interval of allowed values x_i. At each step of the generating network receives at the input a state with known effect and the current operation result of the generating network. The main result of GAN operation is a trained fuzzy knowledge base with tuned values $F(X, W, B, C)$. This base takes into account possible combinations of input parameters X during modelling of the functional dependence to forecasting a value μ_i^j. The training task with using GAN can be regarded as the inverse problem of identification [8]. It is ascertained for a set of predictable damages, $D = d_j \epsilon [d_1, d_2, \ldots, d_n]$ which parameters at the input are more characteristic for a given d_j, and which are not. Having the given result and the result provided by the network, one can calculate the value of the error function gradient for the input parameters of the generating network using the known technology of the loaded dual networks method [6]. In accordance with the values of the gradient, the values at the input of the network change so as to reduce the error. This iteratively allows obtaining an input parameters vector generating the required result. At the same time the structure of the generating network remains unchanged. Thus, the values established at the input of the generating network at the end of the training procedure allow us to refine the intervals of values for the applied term sets of input parameters. These intervals correspond to the predicted damages d_j.

One of the crucial aspects of creating a complex information intelligent system is the initial data preparation based on the results of modeling and experimental research of superhigh-velocity particle interaction including the superhigh-velocity radiography method. The consequences of superhigh-velocity impact of SD on the shell elements of a S is modelled using modern software ANSYS, LS-DYNA, AUTODYN, ABAQUS, FlowVision and others. Some results of ANSYS-developed numeric evaluation of stress state of a descent module structural member suffering from the impact of a debris particle is shown in Fig. 4a.

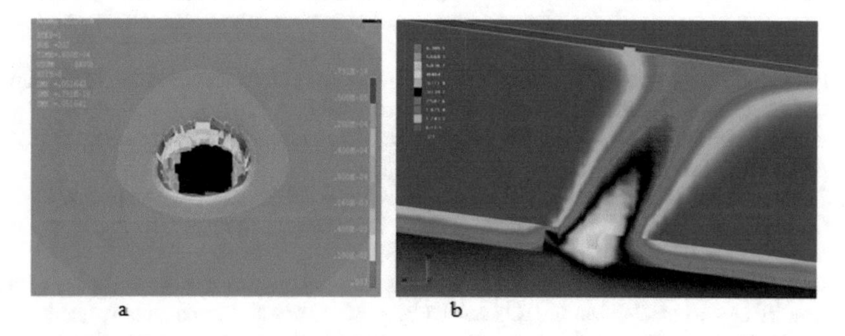

Fig. 4. Stress-state (a) of S structural element made of an aluminum alloy under the impact of the debris particle having 7 mm diameter and velocity 4 km/s and the temperature field (b) in the damage zone

The generated hole represents a potential hazard not only on the orbit but also in the dense layers of the atmosphere during the module descent. Modeling in FlowVision has showed that the generated high-temperature gas flow can significantly extend the hole and break into the module (Fig. 4b) in a few milliseconds. It poses a real threat for the life support systems and endangers for the module operation. Verification of models by the experiment results will improve their reliability.

4 Conclusions

Forecasting of results of possible impacts caused by SD on S is carried out during the operation cycle of onboard computer. Thus we increase S autonomy and create the basis for taking timely and justified measures for protecting the S against SD. Forecast accuracy is ensured by the hybrid knowledge base of the intelligent system. This base contains prior (fuzzy) and adaptive (neuro-fuzzy) models. Training of these adaptive (neuro-fuzzy) models is conducted with the help of GAN. Mutual training of generative and discriminating neural networks is conducted during the interaction process. The training results define which parameters of S, SD and conditions of their relative motion are more typical for each predictable damages of the S. The range for the applied term sets of input parameters $X = \{x_i\}$, corresponding predictable damage (d_j) are also defined more exactly.

Acknowledgements. The research was done within the government task of the Ministry of Education and Science of the Russian Federation. The number for the publication is 2.1777.2017/4.6.

References

1. Zade, L.A.: The role of soft computing and fuzzy logic in understanding, designing and developing information/intelligent systems. Fizmat, Moscow (2001)
2. Mironov, V.V., Tolkach, M.A.: The ballistic limit equations for optimization of system of protection of spacecrafts against micrometeoroid and space debris. Space Tech. Technol. **3** (14), 26 (2016)
3. Mityushkin, Yu.I., Mokin B.I., Rotshteyn A.P.: Soft computing: identification of regularities indistinct knowledge bases. Universum, Vinnitsa (2002)
4. GAN- generative adversarial network. http://robocraft.ru/blog/machinelearning/3693.html. Accessed 13 May 2018
5. Zimmermann, H.J.: Fuzzy Set Theory - and Its Applications. Kluwer, Dordrecht (1991)
6. Gorban, A.N., Rossiev, D.A.: Neural networks on a personal computer. Nauka, Novosibirsk (1996)
7. OST 134-1031-2003: Space technique products. General requirements to aerospace systems security against mechanical impacts caused by particles of natural and anthropogenic origin
8. Li, Y., Jiang, Y., Huang, C.: Shape design of lifting body based on genetic algorithm. Int. J. Intell. Syst. Appl. (IJISA) **1**, 37–43 (2010)
9. Bodyanskiy, Y.V., Tyshchenko, O.K., Kopaliani, D.S.: An extended neo-fuzzy neuron and its adaptive learning algorithm. Int. J. Intell. Syst. Appl. **02**, 21–26 (2015)
10. Rotshteyn, A.P.: Intellectual technologies of identification: fuzzy logic, genetic algorithms, neural networks. Universum, Vinnitsa (1999)

A Fuzzy Evaluation of Quality for Color Vision Disorders Diagnostic

Monika Borova(✉), Jaromir Konecny, Michal Prauzek, and Karolina Janosova

Department of Cybernetics and Biomedical Engineering,
VSB-Technical University of Ostrava, 70833 Ostrava-Poruba, Czech Republic
monika.borova@vsb.cz
http://www.vsb.cz

Abstract. The quality of testing represents one of the most important aspect in color vision testing, because the testing software can be as good as possible, but if the examination is not done under the right conditions, the results will not be accurate. This paper proposes a fuzzy system, which is able to determine the quality of software based testing alike the manual testing. Proposed fuzzy system determines the quality based on visual disorder, error score and condition of testing. System is based on CCD model (conjunction-conjunction-disjunction) and the defuzzification uses method of Center of Gravity. Data for testing come from testing of reliability comparison of software based and standard based method of Farnsworth-Munsell 100 Hue Test.

Keywords: Farnsworth-Munsell · Color vision disorders
Fuzzy expert system · Quality of testing

1 Introduction

As well as diagnostics of color vision disorders, it is also important to determine the quality of the performed examination. The vision is one of the most important sense, and early diagnosis of any of its disorders can help prevent serious illness such as diabetes, cataract, glaucoma or Parkinsons disease.

The software that deals with this article is able to provide testing and subsequent evaluation of color vision disorders based on the same principle as standard test forms of Farnsworth-Munsell 100 Hue Test and Farnsworth D15 test. In this article the proposed fuzzy expert system evaluates the quality of testing software based and standard based method, with input variables being the patient's visual disorder, error score, and conditions of testing.

This paper is organized as follows. Section 2 includes detailed description of color vision standard testing methods that are currently in use in medical applications and the basic knowledge about making fuzzy systems. Section 3 on related work provides an overview of alternative solutions. Section 4 expands on structure of the system developed within the scope of this research, and ways of its utilization and proposal of Fuzzy Expert System. Testing methods along with obtained results are covered in Sect. 5. Finally, major conclusions are summarized in Sect. 6.

A. Abraham et al. (Eds.): IITI 2018, AISC 874, pp. 303–312, 2019.
https://doi.org/10.1007/978-3-030-01818-4_30

2 Background

The main symptoms of illnesses such as diabetes, cataract, glaucoma or Parkinsons disease, see [6] can be early diagnosed through examination of color vision disorders. During routine medical examination the most commons examination methods are the pseudoisochromatic plate and the Farnsworth D15 Test, described by [15]. The most reliable type of examination is The Farnsworth-Munsell 100 Hue Test, but this test is very time-consuming [3].

2.1 Pseudoisochromatic Plate

The pseudoisochromatic plate is one of the options of color perception testing consisting of circle of dots which have different sizes and colors. The people with normal trichromatic vision can recognize the shape, which is created by dots. However, people with color vision disorders are incapable to recognizing these shapes, due to their orientation according to dots of same brightness, which does not create meaningful shape. The pseudoisochromatic plate provides only screening examination [10].

2.2 The Farnsworth-Munsell 100 Hue Test

The Farnsworth-Munsell 100 Hue Test is consider to be international standard in ophthalmology, because it provides detailed diagnostics of inborn or acquired color vision disorders. This test consists of 93 caps divided into four boxes, where first and last cap are fixed. The task of patient is arranged the caps between them with the smallest difference in hue of adjacent caps. Each cap has specific number written on the back, which serve for evaluation of test, see [11,18]. Figure 1 shows score graph of patient with particular color vision disorder.

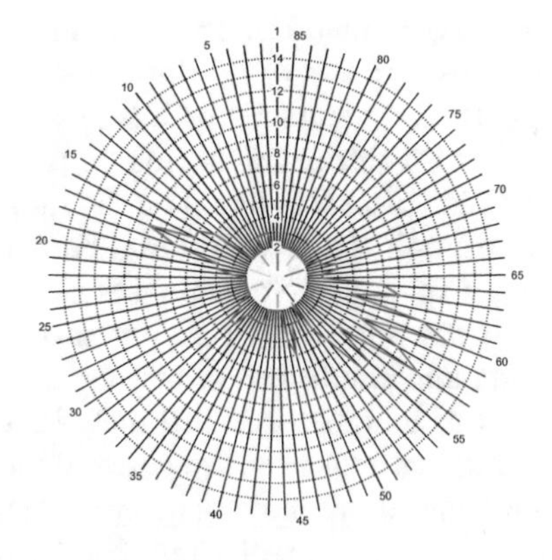

Fig. 1. Farnsworth-Munsell 100 Hue Test result: score graph of patient with particular color vision disorder

2.3 Farnsworth D15 Test

This test is not time-consuming as Farnsworth-Munsell 100 Hue Test and therefore it is more used in clinical practice described in [15]. The Farnsworth D15 Test is used to diagnose medium and strong defect of color vision. It consists of 16 caps, 15 for testing and one cap as reference. The caps with special hue have to be arranged in the correct order. The resulting graph is created by sequence of the patients arrange, see [15]. Figure 2 shows results graph of D15 test of patient with protanopia.

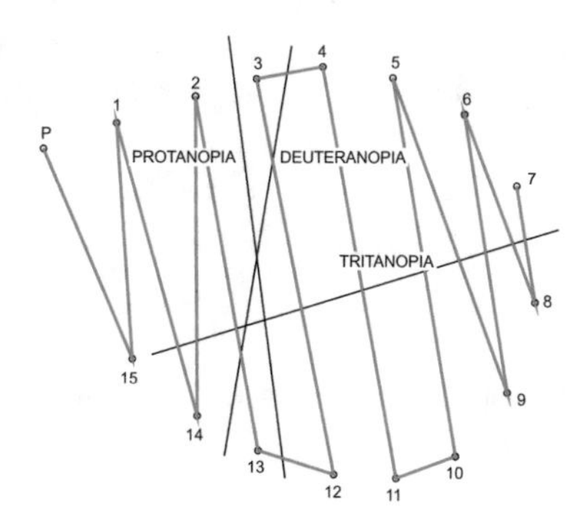

Fig. 2. Farnsworth D15 test: the representation of color vision defect (protanopia).

2.4 Fuzzy Logic

The origins of system modeling and control based on using of fuzzy set theory as a mathematical background can be traced back to 1975, when the first rule-based controller powered by a fuzzy inference mechanism was introduced by Mamdani and Assillian [2].

Thanks to the fuzzy expert system, it is possible to emulate the reasoning process of a human expert or provide in an expert manner in domain for which no human expert exists [7]. Due to the linguistically formulated fuzzy IF-THEN rules (1), conclusions on the basis of imprecise description can be deduce:

$$IF\ X1\ is\ A1\ AND\ X2\ is\ A2\ THEN\ Y\ is\ B \tag{1}$$

where A1, A2 and B are certain predicates characterizing the variables X1, X2 and Y. All types of variables are specified linguistically. Each variable is characterized by name, upper and lower bound and discretization of the universe. Upper and lower bound specify the interval in which fuzzy sets interpreting the linguistic expressions. Discretization of the universe specifies number of points on the universe in which the membership functions are compute [12].

For the explication of fuzzy IF-THEN rules, it can be used several types of approximate reasoning method such as logical deduction, fuzzy approximation with conjunctions or fuzzy approximation with implications. Method applied for proposal system is fuzzy approximation with conjunctions, known as Mamdani-Assilian method. The body of knowledge of expert system is composed of Mumdani rules connected by a fuzzy logic operator OR, fuzzy logic operators between partial fuzzy statements of antecedents of rules AND and fuzzy implications THEN between antecedents and consequents (CCD model - conjunction-conjunction-disjunction) [7,12].

As well as fuzzification, also duzzification works with more available methods. For example Mean of Maxima Method, Deffuzification of Evaluating Expressions or Center of Gravity Method [17]. The last of these methods was applied as a method of defuzzification of the proposal system. It consists of finding the centroid of the area bounded by the controller output membership function and it is expressed by the crisp controlling value, according (2):

$$COG[C(z)] = \frac{\sum_{i=1}^{q} z_i C(z_i)}{\sum_{i=1}^{q} C(z_i)}, \tag{2}$$

where q is the number of sample values of the output and z_i is the value of the control output at the sample value.

3 Related Work

Many researchers tried to develop computer programs to plotting of score graphs for evaluation of the Farnsworth-Munsell 100 Hue Test, because it is the most time-consuming methods in the area of color vision testing.

One of older application described by [1,9] is written in Applesoft BASIC. It provides plotting of score graph, correct placement of patient order do not have to be scored, only false order. The program also computes the axes of confusion, on the basis of which it is determined dominant wavelength using in testing. Printed output provides a histogram of distribution of error scores by sample.

Many software shown in [3,4] use a bar code scanner, reading the data about patient order. The scanner scans number written on the back of each cap and loads number sequence. It is more reliably than written by hands, because it makes some errors. System reduces time to generate report from 60 min to 4 min, together system checks the user inputs numbers to make sure that the values are correct.

The other method described by [5], using scanning system for the Farnsworth-Munsell 100 hue test, consist of a light pen, omni-directional bar-codes attached on the back of each cap and small micro-computer. This program provides printing of patients data, statistical analysis of the total error score, but it doesnt provide standard score graph.

The authors of [8] presents a novel five-layer fuzzy ontology which models the domain knowledge with uncertainty and extend the fuzzy ontology to the diabetes domain. The semantic fuzzy decision making mechanism simulates the semantic description of medical staff for diabetes-related application.

4 Methods and Applications

4.1 Hospital Information System Architecture

The software can be used in hospital as a part of diagnostics color vision disorders, where fulfils the task of staff, because allows automatic plotting of score graph. But also this software can be used as a part of HomeCare system, enabling testing in the comfort of home. This kind of testing is very important for prevention and for following treatment.

4.2 Hue Test Testing Application

The software for color vision testing is based on classic methods of testing color blindness, mainly the Farnsworth-Munsell 100 Hue Test and the Farnsworth D15 Test. Compared to the classic methods, this software is able to plot graphs automatically from obtained results and save this results into database.

The software is composed of modules for user edition, for adding patient into database and the last one for color vision testing. In user module it is possible to create and edit users, who have access rights to the software. The modul for adding patient serves to add some information about the patient to the database (name and date of birth, usually contains address, personal identification number, insurance company or contact information).

The modul for color vision testing creates the main part of the software. It provides testing by the Farnsworth-Munsell 100 Hue Test, the Farnsworth D15 Test and manual mode. The manual mode is particularly useful in case that the patient is tested by the standard methods and doctor or other medical staff want to show score graph and save results into database. The main advantages of this software are speed and reliability of evaluation, as the medical staff can make mistakes during manual plotting.

The software version of Farnsworth-Munsell 100 Hue Test is based on the same principle as standard method. The colors, which are used in this application come from Munsell Color system [16]. For testing, it is absolutely necessary to keep the same brightness all of color, because people with color vision disorders orientate according to them.

4.3 Proposal of Fuzzy Expert System

The aim of the proposal of fuzzy expert system is to determine the quality of color vision testing using a standardized method with respect to the patient's *visual disorders*, their *error score* when they are tested by the software based or standard based method and *examination conditions*. The proposed fuzzy system consists of three input variables with three linguistic variables (Tables 1, 2 and 3) and one output linguistic variable with five output variables (Table 4). Based on input and output variables, 27 decision rules were created for the fuzzy system. The proposed system was developed in the LFLC 2000 program. The input variables have been chosen to reflect as much as possible about the quality of color vision testing. The input variable *visual disorder* refers to the visual difficulties the patient suffers from (short or long-sighted, astigmatism, etc.).

Table 1. Input variable: visual disorder

Linguistic variables	Points of triangular function
Healthy	(0, 0, 0.5)
With mild difficulties	(0, 0.5, 1)
With serious difficulties	(0.5, 1, 1)

The *error score* is based on data that can be measured. Patients without defect have error score at very low levels.

Table 2. Input variable: error score

Linguistic variables	Points of triangular function
Without defect	(0, 0, 10)
Middle defect	(0, 10, 30)
Serious defect	(30, 100, 100)

Testing conditions may include, for example, the quality of the illumination or whether the patient performs the examination according to a standardized procedure, which is often not observed in practice.

Table 3. Input variable: conditions of testing

Linguistic variables	Points of triangular function
Ideal	(0, 0, 0.5)
Sufficient	(0, 0.5, 1)
Inconvenient	(0.5, 1, 1)

Table 4. Output variable: quality of color vision testing

Linguistic variables	Points of triangular function
Unacceptable	(0, 0, 0.25)
Unfavourable	(0, 0.25, 0.5)
Average	(0.25, 0.5, 0.75)
Sufficient	(0.5, 0.75, 1)
Excellent	(0.75, 1, 1)

5 Testing and Results

5.1 Testing Group

As to compare the proposal software reliability with the standard manual method, it had to be tested in the Eye clinic at University Hospital Ostrava. The testing group was consisted of 30 people at the age of 12–70 and with various disorders (myopia, hyperopia, cataract, tumor, diabetes).

5.2 Testing Methodology

First, a patient was tested by the standard Farnsworth-Munsell 100 Hue Test and after that, patient used the software based method. Important think was a ten minute break for relaxation of eyes between these two tests to get more reliable results.

The results of standard method of the Farnsworth-Munsell 100 Hue Test was saved in database using manual mode of this software. The database saves information about the caps ordered for both tests. According to Eq. (3), representing calculation of index of statistical file, two statistical files were created – one for results of the standard method, the other for software testing results.

The equation for calculation of index:

$$e = 85 - \sum_{i=1}^{21-1} (o_i - o_{i+1}), \qquad (3)$$

where e is error index of statistical file, o_i is patient cap order. The index of statistical file was determined as a difference between numbers of two adjacent caps. Ideally, the index is 0, meaning that patient has normal color vision, so the caps were in the correct order. Although, a person with normal trichromatic vision (i.e. without any disorder) switch caps order in small scale, and thus small deviation is considered as normal. On average the index of statistical file is 11 for software testing and 16 for manual testing. Standard deviation of this file is 10 for software testing and 19 for manual testing. It can be said that testing by software shows significantly better results than manual testing.

5.3 Testing of Proposal Fuzzy Expert System

For testing of quality model values were randomly selected for individual input variables (Table 5). In the LFLC 2000 program, the fuzzy approximation with conjunctions was selected as an interference method and center of gravity was selected as the defuzzification method.

The graphs below show results. Graphs with (a) represent the resulting fuzzy set for a given combination of input variables, and the graphs with the label (b) represent the resulting values for the input variable *Visual disorder*. The combination (Fig. 3) of values of 0.34 for visual disorder which means mild difficulties,

Table 5. Testing values of linguistic variables

	Visual disorder	Error score	Conditions of testing	Value of center of gravity
Test 1	0.34	21.65	0.21	0.63
Test 2	0.68	14.95	0.87	0.31

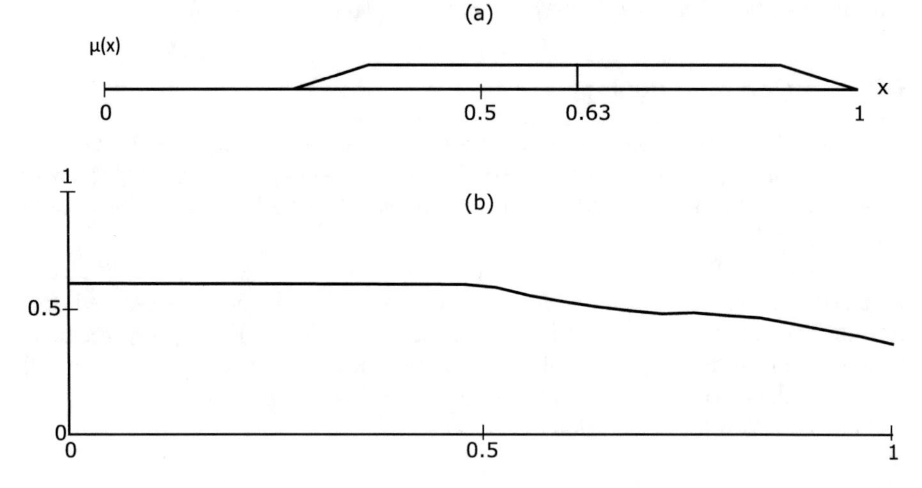

Fig. 3. (a) Test 1 - fuzzy set with 0.63 value of center of gravity (b) representation of variable *visual disorder*

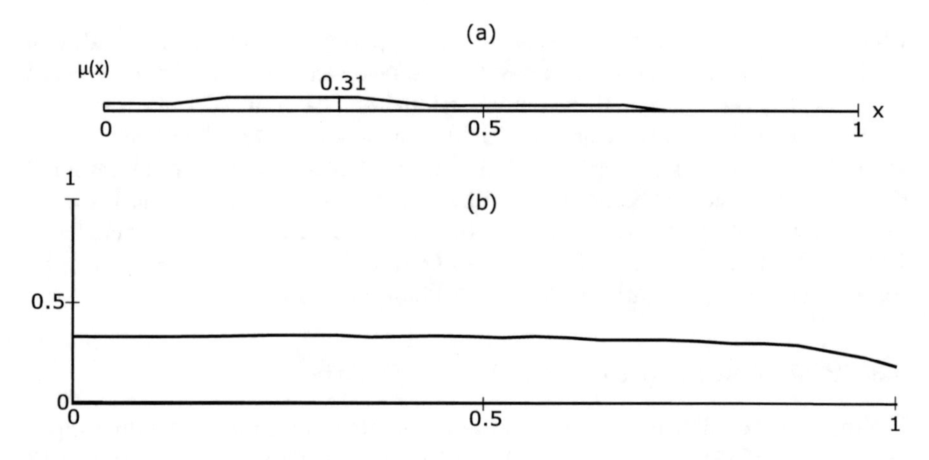

Fig. 4. (a) Test 2 - fuzzy set with 0.31 value of center of gravity (b) representation of variable *visual disorder*

21.65 for error score which means middle defect and ideal conditions with value 0.21 the fuzzy system evaluates as average to sufficient (value of center of gravity 0.63).

The second test (Fig. 4) for value 0.68 of mild difficulties, 14.95 for error score and 0.87 for conditions the fuzzy system evaluates as average to unfavorable with value 0.31 for center of gravity.

6 Conclusion and Future Work

This paper proposes a fuzzy evaluation of quality of testing color vision disorders. The principle of this evaluation je classic version of Mamdani model. Fuzzy system is composed of three input variables with three linguistic variables. The quality of testing depends on value of visual disorders, error score and on conditions of testing. The linguistic output value is deffuzificate with method of Center of Gravity.

It's important to consider the quality of results, because some visual disorders can influence the results alike the conditions of testing. In case of using software based testing, the conditions are always ideal, because the monitor is calibrated, but in case of manual testing, the quality of lighting can be poor. It has huge influence on quality testing. The software we are talking about in the article is based on standard tests such as the Farnsworth-Munsell 100 Hue Test and the Farnsworth D15 Test. This software allows quick automatic plotting of score graph and reliabe evaluation.

In the future, the software for testing color vision will be extended to be able to diagnose the particular color blindness, using neural networks [13,14].

Acknowledgement. This work was supported by the European Regional Development Fund in the Research Centre of Advanced Mechatronic Systems project, project number CZ.02.1.01/0.0/0.0/16_019/0000867 within the Operational Program Research, Development and Education s the project SP2018/160, "Development of algorithms and systems for control, measurement and safety applications IV" of Student Grant System, VSB-TU Ostrava.

References

1. Coren, S.: The Farnsworth-Munsell 100-hue test: a BASIC program for scoring and analysis. Behav. Res. Methods Instrum. Comput. **18**(3), 323–324 (1986)
2. Duţu, L.C., Mauris, G., Bolon, P.: A fast and accurate rule-base generation method for mamdani fuzzy systems. IEEE Trans. Fuzzy Syst. **26**(2), 715–733 (2018)
3. Hidajat, R.R., Hidayat, J.R., McLay, J.L., Elder, M.J., Goode, D.H., Pointon, R.C.: A fast system for reporting the Farnsworth-Munsell 100-hue colour vision test. Doc. Ophthalmol. **109**(2), 109–114 (2004)
4. Hidayat, R.: Generating fast automated reports for the Farnsworth-Munsell 100-hue colour vision test. In: Proceedings of New Zealand Computer Science Research Student Conference, NZCSRSC 2008, pp. 37–40 (2008)
5. Hill, A.R., Reeves, B.C., Burgess, A.: Huematican automated scorer for the Farnsworth-Munsell 100 hue test. Eye (Basingstoke) **2**(1), 80–86 (1988)
6. Hlavica, J., Prauzek, M., Peterek, T., Musilek, P.: Assessment of Parkinson's disease progression using neural network and anfis models. Neural Netw. World **26**(2), 111–128 (2016)

7. Kandel, A.: Fuzzy Expert Systems. CRC Press, Boca Raton (1991)
8. Lee, C.S., Wang, M.H.: A fuzzy expert system for diabetes decision support application. IEEE Trans. Syst. Man Cybern. Part B (Cybern.) **41**(1), 139–153 (2011)
9. Lugo, M., Tiedeman, J.S.: Computerized scoring and graphing of the Farnsworth-Munsell 100-hue color vision test. Am. J. Ophthalmol. **101**(4), 469–474 (1986)
10. Mercer, M.E., Drodge, S.C., Courage, M.L., Adams, R.J.: A pseudoisochromatic test of color vision for human infants. Vis. Res. **100**, 72–77 (2014)
11. Mntyjrvi, M.: Normal test scores in the Farnsworth-Munsell 100 hue test. Doc. Ophthalmol. **102**(1), 73–80 (2001)
12. Novak, V., Perfilieva, I., Mockor, J.: Mathematical Principles of Fuzzy Logic, vol. 517. Springer, New York (2012)
13. Prauzek, M., Hlavica, J., Michalikova, M.: Automatic fuzzy classification system for metabolic types detection. Advances in Intelligent Systems and Computing, vol. 370 (2015)
14. Prauzek, M., Hlavica, J., Michalikova, M., Jirka, J.: Fuzzy clustering method for large metabolic data set by statistical approach. In: Proceedings of the 7th Cairo International Biomedical Engineering Conference, CIBEC 2014, pp. 87–90 (2015)
15. Roy, M., Gunkel, R., Rodgers, G., Schechter, A.: Lanthony desaturated panel D15 test in sickle cell patients. Graefe's Arch. Clin. Exp. Ophthalmol. **226**(4), 326–329 (1988)
16. Roy Choudhury, A.K.: Colorimetric study of scotdic colour specifier. Color. Technol. **124**(5), 273–284 (2008)
17. Saade, J.J., Diab, H.B.: Defuzzification methods and new techniques for fuzzy controllers (2004)
18. Taylor, W.O.: Clinical experience of electronic calculation and automatic plotting of Farnsworth's 100-hue test. Mod. Probl. Ophthalmol. **19**, 150–154 (1978)

Improving the Efficiency of Solution Search Systems Based on Precedents

Alexander Eremeev, Pavel Varshavskiy, and Roman Alekhin(✉)

Applied Mathematics Department of the National Research University "MPEI", Krasnokazarmennaya str., 14, Moscow 111250, Russia
eremeev@appmat.ru, VarshavskyPR@mpei.ru, r.alekhin@gmail.com

Abstract. In this paper, actual issues of improving the efficiency of solution search systems based on precedents – Case-Based Reasoning Systems (CBR systems) are considered. To improve the efficiency of CBR systems and accelerate the search for solutions, it is proposed to use a modified CBR cycle, which allows to create a base of successful and unsuccessful precedents and reducing the number of precedents in the database of successful and unsuccessful precedents through the use of classification and clustering methods.

Keywords: Precedent · Case-Based Reasoning
Intelligent decision support system · Classification · Clustering

1 Introduction

Currently, the actual task in the field of artificial intelligence (AI) is the development of methods for data mining (DM) and related software [1]. DM methods are actively used in intelligent systems (IS), database management systems (DBMS) and knowledge management systems (KMS), business applications, machine learning systems, document automation systems, etc. In the DM, statistical and inductive methods, genetic algorithms, artificial neural networks (ANN), cluster analysis, CBR (Case-Based Reasoning) methods are used to extract new knowledge from available data.

It should be noted that in modern DM tools for DBMS and KMS, including those oriented to the platform of Business Intelligence (BI) [2], there are practically no developed tools for data analysis based on precedents. For this reason, the problem of developing effective CBR tools for empowerment the DM tools on the BI platform for modern DBMS and KMS is actual.

2 Case-Based Reasoning

The CBR approach is based on the concept of a precedent, defined as a case that occurred earlier and serves as an example or justification for the following cases of this kind, and a fairly simple principle that similar problems have a similar solution.

This work was supported by RFBR (projects №18-01-00459, №17-07-00553, №18-51-00007, №18-29-03088)

A. Abraham et al. (Eds.): IITI 2018, AISC 874, pp. 313–320, 2019.
https://doi.org/10.1007/978-3-030-01818-4_31

In general, the case representation model includes a description of the situation, a solution for the given situation (e.g. diagnosis and recommendations), and the result of applying the solution, which may include a list of completed actions, additional comments and references to other precedents, and in some cases, the justification of the choice of this solution and possible alternatives: *CASE* = (*Situation, Solution, Result*). In most cases, a simple parametric representation model is used [3].

But in some cases, parametric representation is not enough, because there are limitations associated with the expressive capabilities of the parametric model, it is difficult to account for the relationship between the precedents parameters (for example, time dependencies or cause-and-effect relationships).

One possible way to solve this problem is to represent precedents using the ontological approach [4]. The choice of ontology for case representation is due to a number of important advantages that distinguish it from other models of knowledge representation. Using ontology to represent cases allows specifying a complex case structure that includes data of different types.

The ontology contains domain knowledge that is used to support the CBR cycle, and also defines the case structure and ensures its storage.

Domain knowledge and a case model are described as a hierarchy of ontology concepts, and every case from the case library (CL) as a hierarchy of concept instances related using ontology description language relations for the Semantic Web (OWL) [5].

As a rule, CBR methods are based on the so-called CBR cycle [6], which includes four main stages:

- retrieve the most similar case(s) from the CL;
- reuse the retrieved case(s) to attempt to solve the current problem;
- revise the proposed solution in accordance with the current problem if necessary;
- retain the new solution as a part of a new case.

3 Method of Case Retrieval

There are different methods can be used to retrieve cases from the CL [7], such as:

- Nearest Neighbor method (NN) and its modifications (for example, the method of k nearest neighbors (k-NN));
- Induction method of case retrieval;
- Method of case retrieval on the basis of knowledge;
- Method of case retrieval, taking into account the applicability of cases, etc.

To determine the similarity degree of case represented in the form of complex structures using the ontology, it is suggested to use the retrieve method based on the Structure-Mapping Theory (SMT) [8].

According to SMT, it is assumed that the analogy is a mapping of knowledge of one domain (base) into another domain (target), based on the system of relations that exist between the base objects and the target domain objects, and also that the decision making person (DMP) prefers to operate with some integral system of interconnected relations, and not with a simple set of superficial and weakly connected facts.

The process of finding a solution based on analogies according to SMT includes the following main steps:

1. Identification of potential analogues. Having a target situation (target), determine another situation (base) that is similar to the target;
2. Mapping and inference. Construct a mapping consisting of the correspondences between the base and the target. This mapping may include additional knowledge (facts) about the base that can be mapped to the target.
3. Mapping quality estimation. Estimate the correspondence obtained by using structural criteria such as the number of similarities and differences, the degree of structural conformity.

It is proposed to retrieve and determine the similarity of the case and the current situation in two stages [9]:

- determination of the similarity of the case and the current situation on the basis of the domain ontology and the formation of paired matches using the algorithm based on SMT;
- determination of the similarity of the case and the current situation using the NN algorithm, taking into account obtained pair matches.

As a result, we get a set of cases, each of which is compared with two estimates of similarity, which can be expressed as a percentage:

- estimate on the basis of domain ontology: $S_{struct} = \sum_{i=1}^{k} LS_i/SES_{MAX}$, k – number of correspondences, LS_i – plausibility estimation of i correspondence, SES_{MAX} – estimation for the case where a base is selected as the target;
- estimate by nearest neighbor method: $Sim(Q, C) = 1 - d_{QC}/d_{MAX}$, d_{QC} – distance between the current situation Q and case C, d_{MAX} – maximum distance in the selected metric.

4 Modified CBR Cycle

The modified CBR cycle (Fig. 1) is proposed in which test samples with examples can be used to verify the correctness of the solution found.

In the presence of expert knowledge (test samples), before retain in the CBR cycle, you should check the correctness of the solution on the test sets. If the solution is tested and accepted by the DPM, it is stored in the CL as a new case. If the test of the correctness of the solution on the test sets fails, the case is stored in the unsuccessful case library (UCL).

A case that does not worsen the quality of the CBR system will be called successful and case that worsens the quality of the CBR system will be called unsuccessful. Thus, it is proposed to use the test (expert) sample at the last stage of the CBR cycle for the formation CL and UCL.

Modified CBR cycle was considered on the example of the solution of the data classification problem using data set from UCI Machine Learning Repository, including a test sample and a database with information on students' knowledge status about the subject of Electrical DC Machines, which contains 258 entries [10]. The initial CL was formed on the basis of the first 20 records from the database, and the remaining 238 entries were included in the training set.

The results of estimation of the classification quality by the CBR system are shown in Fig. 2. This estimation is defined as the percentage of correctly classified examples from the entire training set. The use of a modified CBR cycle and the formation of CL based on successful precedents allowed improving the classification quality by CBR system up to 87%.

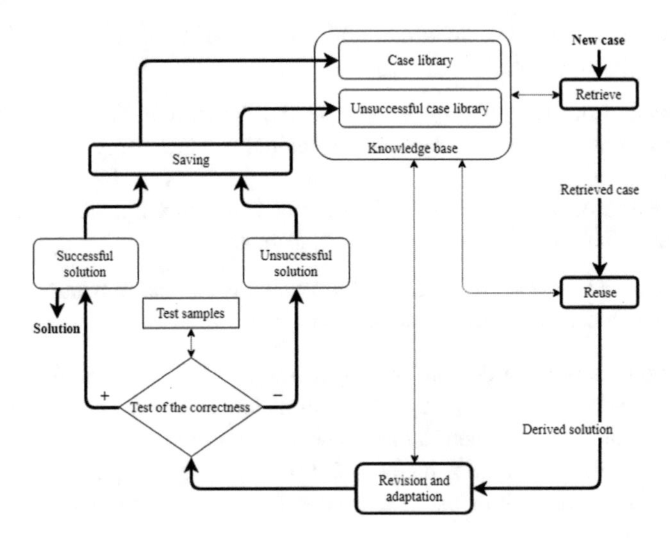

Fig. 1. Modified CBR cycle

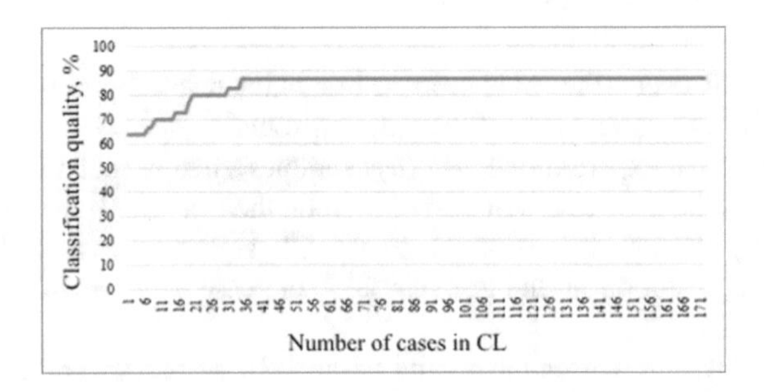

Fig. 2. Estimation of the classification quality formed on the basis of a modified CBR cycle

5 Improving the Efficiency of CBR System

In the life cycle of CBR systems, the number of cases in CL increases, that is, CBR systems accumulate experience and solve the tasks assigned to them, but at the same time, the time and storage costs to maintain the CBR cycle are significantly increased. For this reason the actual task is to increase the efficiency of CBR system operation by increasing of their performance and reducing the amount of memory required to store cases in the CL.

5.1 Reducing the Number of Cases in the CL Using the Classification Method

One way to reduce the number of precedents in a CL is to classify (for example, using the k-NN algorithm), and then delete all cases that are defined as closest to the selected. This allows you to represent the CL in a compact form while preserving the distribution by classes, reducing the amount of memory required to store use cases, and significantly improve the performance of the CBR system, but it is possible a significant reduction in the quality of the solution of problems by the CBR system.

Algorithm 1. Algorithm for reducing the number of precedents in CL based on the classification.

Input data: ***CL*** – non-empty set of cases; ***n*** – number of parameters in cases; ***m*** – number of cases in *CL*; $S(C_i, C_j)$ – specified metric to calculate the distance between the cases C_i and C_j; ***H*** – threshold of similarity.
Output data: ***CL*** – set of cases after reduction of *CL*.
Intermediate data: ***SC*** – set of closest precedents to the selected; ***i, j*** – cycle parameters.
Step 1. $j = 1$, $SC = \varnothing$, go to the next step.
Step 2. If $j \leq m$, then choose case C_j from *CL* ($C_j \in CL$); go to the step 3, otherwise all the cases from the *CL* are considered, go to the step 5.
Step 3. Define *SC* – set of closest cases C_i ($C_i \in CB$, $i \neq j$) to the case C_j in accordance with the given metric $S(C_i, C_j)$, go to the next step.
Step 4. If $SC \neq \varnothing$, then delete from *CL* all cases in the set *SC*. Then $m = m - |SC|$, $SC = \varnothing$, $j = j + 1$, go to the step 2.
Step 5. End (completion of the algorithm).

5.2 Reducing the Number of Cases in the CL Based on Cluster Methods

Clustering involves the allocation of compact, remote from each other groups of objects, characterized by internal homogeneity and external isolation. Regarding to CBR systems, clustering algorithms allow you to get sets of cases – "centers" of clusters, in relation to which other cases in the cluster are close. That is, a set of "centers" of clusters in the CL can be used instead of the CL itself, since situations close to any case from the cluster will be close to the center of this cluster.

The paper proposes an approach to reducing the number of cases in the CL based on clustering (the k-means algorithm), with subsequent removal of all cases in each cluster except its central representative. Thus, clustering allows us to represent the CL in a more compact form and shorten the search time for the solution and the amount of memory required to store use cases in the CL. The disadvantage of this method is the loss of accuracy at the boundaries of clusters.

Algorithm 2. The k-means algorithm for reducing the number of cases in the CL.

Input data: ***CL*** – case library; ***n*** – number of parameters in cases; ***m*** – number of cases in *CL*; ***k*** – number of clusters; $S(C_i, C_j)$ – specified metric to calculate the distance between the cases C_i and C_j.
Output data: ***CL*** – set of cases after reduction of *CL*.
Intermediate data: CL_i – received clusters; **μ** – centers of mass of clusters; ***V*** – the total quadratic deviation of objects (cases) in clusters from their centers; ***i, j*** – cycle parameters.
Step 1. At the first stage, select *k* random cluster centers (cases) from *CL*.
Step 2. For each cluster center μ_i ($i = 1, \ldots, k$), define a set of cases for which the corresponding center is the nearest CL_i those the distance $S(C_j, \mu_i)$ from the precedent C_j ($C_j \in CL$) to the center of the cluster μ_i is minimal.
Step 3. If for one of the centers, the set of nearest cases corresponding to it is empty, instead of it we select a new random center (case).
Step 4. In each set of immediate precedents (cluster CL_i) the new center of mass of the clusters is determined, which is used at the next iteration.
Step 5. The total distance of cases in the *CL* to the centers is estimated $V = \sum_{i=1}^{k} \sum_{C_j \in CL_i} (C_j - \mu_i)^2$, if the total distance has not decreased, then to step 6, otherwise return to step 2.
Step 6. All precedents are deleted from *CL* except the received cluster centers.
Step 7. End (completion of the algorithm).

The Table 1 shows the results of computational experiments evaluating the effectiveness of the proposed algorithms to reduce the number of precedents in the CL by the example of the classification problem considered above.

As can be seen from the Table 1 with a reduction in the number of cases in the accumulated CL (from 171 to 25 cases) using the classification algorithm 1, the quality of the classification on the test set is reduced by 15%. Reducing the number of cases in the accumulated CL (from 171 to 4 cases) using the cluster algorithm 2, the quality of the classification on the test set is reduced by only 7%. These results are permissible with a significant increase in the performance of the CBR system and a decrease in the amount of memory necessary to store use cases in the CL.

Table 1. Research results.

CL	Number of cases	Classification quality, %
Initial	20	64
Accumulated	171	87
Result of algorithm 1	25	72
Result of algorithm 2	4	80

6 Architecture of the CBR System Prototype

The architecture of the CBR system prototype consists of the following main components (Fig. 3):

- *user interface* – interface to interact with an expert or DMP and display the results;
- *case retrieval unit* – case retrieve based on the proposed two-stage algorithm;
- *knowledge base* – contains a successful case library and unsuccessful case library (respectively CL and UCL);
- *set of test samples* – expert information containing data (examples) with correct decisions, for its consideration when extracting precedents and the formation of CL and UCL;
- *CL optimization module* – to reduce the number of cases in the CL using proposed algorithms.

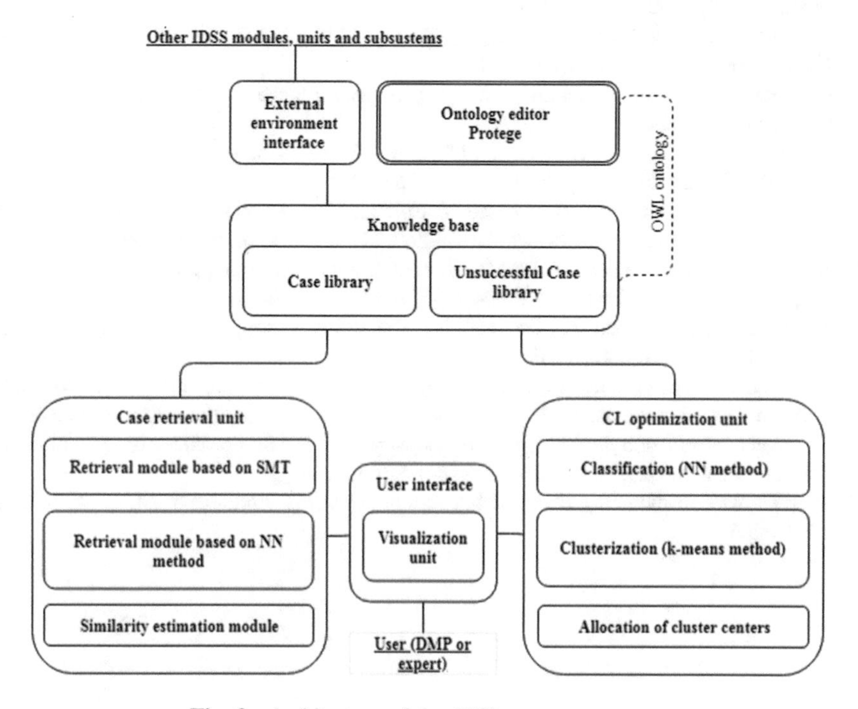

Fig. 3. Architecture of the CBR system prototype

7 Conclusion

The article discusses the perspective possibility of empowerment the existing DM tools for modern DBMS and KMS by CBR tools, and also consider current issues of improving the efficiency of CBR systems. The opportunity to represent cases based on the ontological model is implemented. A method for finding a solution based on cases using the domain ontology, the structure-mapping theory, and the method of the nearest neighbor is proposed. A modified CBR cycle using expert information (test datasets) to verify the correctness of the solution and the formation of CL and UCL is proposed. To improve the efficiency of CBR systems, algorithms to reduce the number of cases in the CL based on classification and clustering methods are proposed. Using the developed CBR system prototype, computational experiments were implemented on the data set from the UCI Machine Learning Repository.

References

1. Russell, S., Norvig, P.: Artificial Intelligence: A Modern Approach 3rd edn. Prentice Hall, 1152 p. (2009)
2. Rausch P., Sheta A. F., Ayesh A. Business Intelligence and Performance Management: Theory, Systems and Industrial Applications. Advanced Information and Knowledge Processing, Springer, 269 p. (2013)
3. Wilson, D.C., Leakes, D.B.: Maintaining case-based reasoners: dimensions and directions. Comput. Intell. **17**(2), 196–213 (2001)
4. Gangemi, A., Presutti, V.: Ontology design patterns. In: Staab, S., et al. (eds.) Handbook on Ontologies 2nd edn. Springer (2009)
5. McGuinness, D.L., van Harmelen, F.: OWL web ontology language. W3C Recommendation, 10 February 2004. http://www.w3.org/TR/owl-features/
6. Aamodt, A., Plaza, E.: Case-based reasoning: foundational issues, methodological variations, and system approaches. Artif. Intell. Commun. **7**(1), 39–59 (1994)
7. Varshavskii, P.R., Eremeev, A.P.: Modeling of case-based reasoning in intelligent decision support systems. Sci. Tech. Inf. Process. **37**(5), 336–345 (2010)
8. Falkenhainer, B., Forbus, K., Gentner, D.: The structure-mapping engine: algorithm and examples. Artif. Intell. **41**, 1–63 (1989)
9. Eremeev, A., Varshavskiy, P., Alekhin, R.: Case-based reasoning module for intelligent decision support systems. In: Proceedings of the First International Scientific Conference Intelligent Information Technologies for Industry (IITI 2016), pp. 207–216 (2016). https://link.springer.com/chapter/10.1007%2F978-3-319-33609-1_18
10. User knowledge modeling data set. http://archive.ics.uci.edu/ml/datasets/User+Knowledge+Modeling

Efficient Feature Selection Algorithm Based on Population Random Search with Adaptive Memory Strategies

Ilya Hodashinsky(✉), Konstantin Sarin, and Artyom Slezkin

Tomsk State University of Control Systems and Radioelectronics, Tomsk, Russia
hodashn@gmail.com

Abstract. The effectiveness of classifier training methods depends significantly on the number of features that describe a dataset to be classified. This research proposes a new approach to feature selection that combines random and heuristic search strategies. A solution is represented as a binary vector whose size is determined by the number of features in a dataset. New solutions are generated randomly using normal and uniform distributions. The heuristic underlying the proposed approach is formulated as follows: the chance for a feature to be included into the next generation is proportional to the frequency of its occurrence in the previous best solutions. For feature selection, we have used the algorithm with a fuzzy classifier. The method is tested on several datasets from the KEEL repository. Comparison with analogs is presented. To compare feature selection algorithms, we found the values their efficiency criterion. This criterion reflects the accuracy of the classification and the speed of finding the appropriate features.

Keywords: Feature selection · Classification · Population random search Adaptive memory strategies

1 Introduction

Classification is one of the most extensively investigated problems of data mining and machine learning. Objects of classification can be scientific and industrial data, handwriting and multimedia content, biomedical and social media data. Such broad applicability is due to the fact that classification is aimed at relating a set of certain input variables (features) with a target output variable (class label). The most popular methods for data classification are decision trees, rule-based methods, probabilistic methods, support vector machines, and neural networks [1].

Fuzzy classifiers, which belong to rule-based methods, have significant advantages in terms of their functionality, as well as their subsequent analysis and design. A unique advantage of fuzzy classifiers is the interpretability of classification rules. In this case, classification accuracy is the main criterion to estimate the performance of a classifier, and it is often taken into account when comparing fuzzy classifiers with classifiers based on other principles [2].

The effectiveness of classifier training methods depends significantly on the number of features that describe a dataset to be classified. Feature selection makes it possible to

A. Abraham et al. (Eds.): IITI 2018, AISC 874, pp. 321–330, 2019.
https://doi.org/10.1007/978-3-030-01818-4_32

reduce the dimension of the input feature space by finding and eliminating redundant and irrelevant features. Feature selection is an NP-hard problem [3] where the optimal solution can be guaranteed only by exhaustive search. Metaheuristic methods allow one to obtain sub-optimal solutions to this problem without exploring the entire solution space.

2 Related Works

In this section, we discuss some works related to our research, which address feature selection, metaheuristics for feature selection, and adaptive memory-based search.

2.1 Feature Selection

Feature selection methods can be divided into two categories: filters and wrappers [3–5]. Filters are based on certain metrics, e.g., entropy, probability distribution, and mutual information, and do not use classification algorithms. In turn, wrappers employ classifiers to estimate feature subsets with the classifier itself being "wrapped," so to speak, in the feature selection loop. Filters and wrappers have their own pros and cons. The advantage of filters is their scalability and higher speed. Their common disadvantage is that the absence of communication with the classifier, as well as ignoring the inter-feature relationships, results in low classification accuracy, which, however, differs for different classifiers. The advantage of wrappers is that they work in cooperation with a particular classification algorithm and take into account the synergetic effect from the joint use of features selected. Their disadvantages are higher risk of overfitting and higher time cost, which is due to the necessity of estimating classification accuracy [6].

2.2 Metaheuristics for Feature Selection

Yusta [7] considered three metaheuristic strategies – GRASP, tabu search, and memetic algorithm – to solve the feature selection problem. These three strategies were compared with the genetic algorithm, which is the most popular metaheuristic for this problem [8], and some other feature selection methods, including sequential forward floating selection and sequential backward floating selection.

Aladeemy et al. [9] proposed a variation of the cohort intelligence algorithm for feature selection. The proposed algorithm was compared with the well-known metaheuristics: genetic algorithm, particle swarm optimization, differential evolution, and artificial bee colony.

Hodashinsky and Mekh [10] proposed a harmony search for feature selection. Based on discrete harmonic search, several feature subsets were generated using the wrapper scheme. The Akaike information criterion was employed to identify the best classifiers.

Gurav et al. [11] proposed a hybrid filter-wrapper algorithm, called the GSO-Infogain, for simultaneous feature selection with improved classification accuracy. The GSO-Infogain employs glowworm swarm optimization (GSO) with support vector machine as its internal learning mechanism and uses feature ranking based on information gain as a heuristic.

Marinaki et al. [12] proposed a method that uses honey bees mating optimization at the feature selection step and nearest neighbor-based classifiers at the classification step. The performance of the method was tested on a financial classification problem involving credit risk assessment.

2.3 Adaptive Memory-Based Search

A classification of metaheuristics proposed by Glover and Laguna [13] is based on three features: (1) the use of adaptive memory, (2) the kind of neighborhood exploration, and (3) the number of current solutions carried from one iteration to the next. Tabu search uses an adaptive memory, known as the tabu list, to keep track of the solutions (or solution attributes) that have been visited and should be avoided for a number of iterations; the tabu tenure determines how long a solution remains in the tabu list. The adaptive memory feature of tabu search enables the implementation of the procedures capable of searching the solution space efficiently and effectively. Adaptive memory also contrasts with the rigid memory designs typical for branch and bound strategies [13].

Adaptive memory programming [14] describes a class of metaheuristics as algorithms that use memory to store the information collected during the search process; this information is used to bias the solution construction and future search.

Hedar et al. [15] proposed an adaptive memory programming-based approach to optimize the input feature space of a solar radiation model. The proposed approach employs the tabu search attribute reduction method to select the best features and then applies a fuzzy classifier based on the features found.

3 Problem Statement

Suppose that we have a universum $U = (A, G)$, where $A = \{x_1, x_2, \ldots, x_n\}$ is a set of input features and $G = \{1, 2, \ldots, m\}$ is a set of class labels. Suppose also that $\mathbf{X} = x_1 \times x_2 \times \ldots \times x_n \times \Re^n$ is an n-dimensional feature space. An object in this universum is characterized by a feature vector. The classification problem consists in predicting the class of the object based on its feature vector.

The feature selection problem is formulated as follows: on the given set of features $\mathbf{X}$, find a feature subset that does not cause a significant decrease in classification accuracy, or even increases it, when the number of features decreases. The solution is represented as a vector $\mathbf{S} = (s_1, s_2, \ldots, s_n)^{\mathrm{T}}$, where $s_i = 0$ means that the ith feature is excluded from classification and $s_i = 1$ means that the ith feature is used by the classifier. Classification accuracy is estimated for each feature subset.

A fuzzy classifier can be represented as a function that assigns a class label to a point in the feature space with a certain evaluable confidence:

$$f : \Re^n \rightarrow [0, g]^m,$$

where $g \in \Re$ is determined by individual characteristics of a classifier.

The fuzzy classifier uses production rules of the form

R_i : IF $s_1 \wedge x_1 = A_{1i}$ AND $s_2 \wedge x_2 = A_{2i}$ AND … AND $s_n \wedge x_n = A_{ni}$ THEN $y = L_i$,

where $L_i \in \{1, 2, \ldots, m\}$ is the output of the *i*th rule ($i = 1, 2, \ldots, R$, where R is the number of rules); A_{ki} is a fuzzy term that characterizes the *k*th feature in the *i*th rule ($k = 1, 2, \ldots, n$); and $s_k \wedge x_k$ indicates the presence ($s_k = 1$) or absence ($s_k = 0$) of a feature in the classifier. The fuzzy classifier yields a vector $(\beta_1, \beta_2, \ldots, \beta_m)^{\mathrm{T}}$, where

$$\beta_j = \sum_{\substack{i = \overline{1,R} \\ L_i = j}} \prod_{k=1}^{n} \mu_{ki}(x_k),\; j = 1, \ldots, m;$$

here, $\mu_{kj}(x_k)$ is a membership function for the fuzzy term A_{ki} at the point x_k. Using this vector, the class is assigned based on the winner-takes-all principle:

$$\text{class} = \arg \max_{1 \le j \le m} \beta_j.$$

On an observations table $\{(\mathbf{x}_p, c_p), p = 1,2,\ldots, z\}$, the accuracy of the classifier can be expressed as follows:

$$E(\boldsymbol{\theta}, \mathbf{S}) = \frac{\sum_{p=1}^{z} \begin{cases} 1, & \text{if } c_p = \arg \max_{1 \le j \le m} f_j(\mathbf{x}_p; \boldsymbol{\theta}, \mathbf{S}) \\ 0, & \text{otherwise} \end{cases}}{z}, \tag{1}$$

where $f(x_p;\, \theta, \mathbf{S})$ is the output of the fuzzy classifier with the parameters $\boldsymbol{\theta}$ and features $\mathbf{S}$ at the point x_p. Thus, the problem of constructing fuzzy classifiers is reduced to finding the maximum of this function in the space of $\mathbf{S}$ and $\boldsymbol{\theta} = (\boldsymbol{\theta}^1, \theta^2, \ldots, \theta^D)$:

$$\begin{cases} E(\boldsymbol{\theta}, \mathbf{S}) \to \max \\ \theta^i_{\min} \le \theta^i \le \theta^i_{\max},\; i = \overline{1,D}, \\ s_j \in \{0, 1\}, j = \overline{1,n} \end{cases}$$

where $\theta^i_{\min}$ and $\theta^i_{\max}$ are the lower and upper boundaries of each parameter, respectively. In this paper, we solve this problem by using population random search with adaptive memory.

4 Algorithm Based on Population Random Search with Adaptive Memory Strategies

The proposed algorithm is based on the joint use of random and heuristic search strategies. The key element of the algorithm is the solution vector **S** that encodes features and acts as adaptive memory. At the first step of the algorithm, a population (set) of vectors **S** is generated (randomly or in some other way). The number of vectors in the population is a preset integer, which is also referred to as the population size. For each vector, classification accuracy E is estimated by Eq. (1). On each iteration, the vector with the maximum E (the best solution on the current iteration) is found. Another important element of the algorithm is the vector **B**, in which b_i stores the number of occurrences of the ith feature in the best solutions found on the previous iterations; the size of this vector coincides with that of the vector **S**. The vector **B** is used to implement the following heuristic: the chance for a feature to be included into the next population is proportional to the frequency of its occurrence in the previous best solutions. A new population is formed using this heuristic and random search. In the proposed algorithm, a normally distributed random variable $u \sim N(0,\sigma_g)$ governs the mechanism whereby features are added into and removed from the vector **S**, thus determining the number of added and deleted features. If u is greater than or equal to zero, then new features are added by increasing the number of ones in the vector **S**; otherwise, the number of features is reduced. The number of candidates for addition or removal is determined by the size n of the vector **S** and by the number of ones (let us denote it by r) in the vector **S**. If u is negative, then, among r elements of the vector **S** (which contain ones), l candidates for removal are randomly selected by the formula $l = \text{round}(r \cdot |\text{th}(u)|)$. The number of candidates for addition is calculated by the formula $l = \text{round}((n - r) \cdot |\text{th}(u)|)$. However, the very change in the value of the element s_i is determined by the value of $b_i \in \mathbf{B}$, i.e., the frequency of occurrence of the ith feature in the best solutions, which is found as b_i/t on the iteration $t \geq t_g$. Thus, $s_i = 1$ if rand $(0,1) \leq b_i/t$; otherwise, $s_i = 0$. The upper and lower boundaries for the frequency are p_g and $1 - p_g$, respectively.

The algorithm is executed iteratively; once a specified number of iterations is reached, the best vector is decoded into a solution, which is interpreted as the optimal one.

5 Experiments

5.1 Algorithm for Constructing Fuzzy Classifiers

This algorithm is based on fuzzy clusterization of training data. Each cluster is associated with a fuzzy rule. THEN-part of the rule specifies the class whose data instances form the cluster; membership functions of fuzzy terms are Gaussians, where the mean is given by the center of a cluster and the deviation reflects weighted average quadratic deviation of training data from the center. One class can be assigned to one or several clusters. For clustering, we employ the popular fuzzy *C*-means (FCM) algorithm [16].

5.2 Empirical Criterion to Estimate the Efficiency of the Algorithm

To compare feature selection algorithms, we need the corresponding efficiency criterion. When selecting features by the wrapper method, classification accuracy is regarded as an objective function that is evaluated by constructing a classifier on training data with a test set of features. The main computational load is associated with this evaluation process. An algorithm is considered efficient if it manages to find a feature set on which the classifier constructed shows the best classification accuracy. If accuracies coincide, then the algorithm with the minimum number of iteration steps is considered efficient. Since we compare population-based algorithms, we regard iteration as a computational unit. For all iterations to have the same computational complexity, we set the same population size for all algorithms.

Thus, we propose to estimate the efficiency of algorithms empirically. If algorithms are compared on one dataset, then, as an efficiency criterion, we can use the classification accuracy on the iteration t_{min}, on which the best solution is found by one of the algorithms. When using several datasets, we need to take into account the total efficiency of each algorithm on each dataset. The total estimate is based on the normalized efficiencies of an algorithm on each dataset. Let us introduce an empirical criterion ξ for estimating the efficiency of feature selection algorithms. This criterion is evaluated as follows.

Input: the number of feature selection algorithms (a) and the number of datasets (d).

Output: the value of the efficiency criterion for each algorithm, ξ_i (i = 1,…,a).

Step 1. Find, for each dataset, the iteration $t^j_{\min}$ with the best solution and the average accuracy of the ith algorithm for the jth dataset on this iteration:

$$E^i_{jt^j_{\min}}, i = \overline{1,a}, j = \overline{1,d}.$$

Step 2. Sum up the normalized accuracies:

$$\xi_i = \sum_{j=1}^{d} \frac{E^i_{jt^j_{\min}} - E^{minindex_j}_{jt^j_{\min}}}{E^{maxindex_j}_{jt^j_{\min}} - E^{minindex_j}_{jt^j_{\min}}}, \tag{2}$$

$$minindex_j = \arg \min_{i=\overline{1,a}} \left(E^i_{jt^j_{\min}} \right), \; maxindex_j = \arg \max_{i=\overline{1,a}} \left(E^i_{jt^j_{\min}} \right).$$

The larger the value of the criterion, the more efficient is the algorithm. This value varies in the interval [0, a].

5.3 Determining the Optimal Parameters of the Algorithm

To find the optimal values for the parameters σ_g, p_g, and t_g of the proposed algorithm, we evaluate the efficiency criterion for different values of the parameters and select those for which the algorithm yields the best results. For this purpose, we used 10 real datasets from the KEEL repository (http://keel.es/) (see Table 1). We tested the following parameter values: $\sigma_g \in \{1,2\}$, $p_g \in \{0.75, 0.85, 1\}$, and $t_g \in \{1, 50, 200\}$. To evaluate the efficiency criterion, classification accuracy on each iteration was measured as the average accuracy on 20 runs of the algorithm. The maximum number of iterations was T = 300 with the population size *popul* = 20. The best value of the efficiency criterion was obtained with the following parameters: p_g= 0.75, t_g= 1, and σ_g= 2. These values are optimal for comparison with other algorithms.

Table 1. Datasets used in experiments

Dataset	Number of variables	Number of classes	Sample size
Wine	13	3	178
Heart	13	2	270
Cleveland	13	5	297
Vowel	13	11	990
Australian	14	2	690
Vehicle	18	4	846
Hepatitis	19	2	80
Bands	19	2	365
Ionosphere	33	2	351
Dermatology	34	6	358

5.4 Comparison with Analogs

We compared the proposed binary random search with memory (BRSM) with the following algorithms: binary particle swarm optimization (BPSO) [17], binary cuckoo search (BCS) [18], and binary random search (BRS). For each algorithm, the maximum number of iterations was $T = 300$ with the population size $popul = 20$; classification accuracy on each iteration was measured as the average accuracy on 20 runs of the algorithm. The parameters of the BCS and BPSO were set as recommended in [18] and [17], respectively. Table 2 shows the classification accuracies on the iteration t_{min}; the parenthesized values represent t_{min}. In turn, Table 3 shows the values of the criterion ξ found by Eq. (2). It can be seen that, on each dataset, the BRSM yields the best or second-best results.

Classification accuracy versus the number of iterations is shown in Fig. 1. Classification accuracy was measured as the average accuracy on 20 runs of each algorithm. It can be seen that the BPSO and BRSM have the fastest convergence to the best solution.

Table 2. Classification accuracies on the iteration t_{min}

Dataset	BCS	BPSO	BRS	BRSM
Wine	0.879	0.881	**0.883 (279)**	0.882
Heart	0.757	**0.761 (94)**	0.751	0.759
Cleveland	0.589	0.59	0.586	**0.592 (54)**
Vowel	0.518	**0.576 (21)**	0.505	0.562
Australian	0.828	**0.837 (28)**	0.82	0.833
Vehicle	0.478	**0.532 (81)**	0.499	0.514
Hepatitis	0.937	0.936	0.925	**0.947 (294)**
Bands	0.571	0.628	0.63	**0.652 (216)**
Ionosphere	0.647	0.658	0.677	**0.921 (280)**
Dermatology	0.401	**0.53(296)**	0.402	0.524

Table 3. Values of the efficiency criterion ξ

Dataset	BCS	BPSO	BRS	BRSM
Wine	0	0.43	1	0.8
Heart	0.66	1	0	0.9
Cleveland	0.62	0.88	0	1
Vowel	0.18	1	0	0.81
Australian	0.5	1	0	0.78
Vehicle	0	1	0.38	0.67
Hepatitis	0.53	0.5	0	1
Bands	0	0.71	0.73	1
Ionosphere	0	0.04	0.11	1
Dermatology	0	1	0.01	0.95
ξ	2.49	7.57	2.23	**8.91**

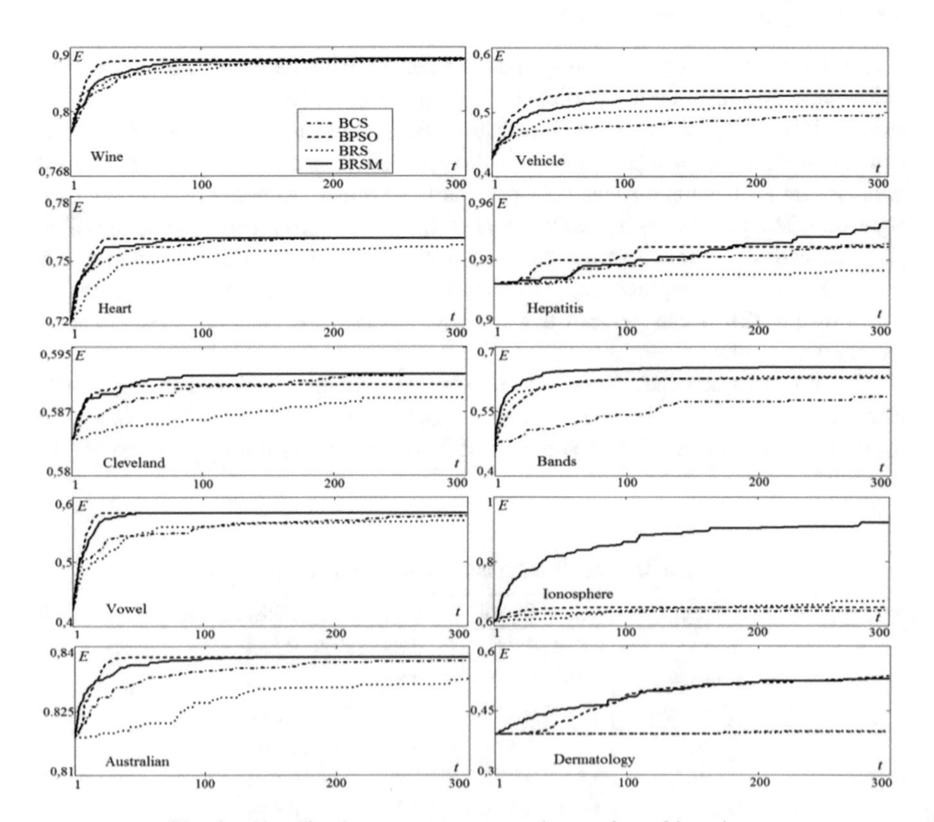

Fig. 1. Classification accuracy versus the number of iterations

6 Conclusions

In this paper, we have proposed an approach to feature selection based on population random search with adaptive memory. For feature selection, we have used the wrapper method with a fuzzy classifier. The feature selection algorithm is based on random search governed by the following heuristic: the chance for a feature to be included into the next population is proportional to the frequency of its occurrence in the previous best solutions. An important advantage of the proposed algorithm is the optimal tradeoff between diversification and intensification of search. Diversification is ensured by random search throughout the solution space, while intensification is provided by the proposed heuristic, which allows the best solutions to be preserved by taking the landscape of the search space into account. A method for selecting the parameters of the feature selection algorithm has been described. In addition, an algorithm for constructing fuzzy classifiers based on fuzzy clusterization of training data has been presented. The proposed feature selection algorithm has been compared with its well-known analogs; the comparison has confirmed the effectiveness of the proposed approach.

Further development of the proposed algorithm involves improving its diversification property by generating initial solutions based on quasi-random sequences and chaotic opposition. We intend to improve the adaptive capabilities of the algorithm by generating normally distributed numbers with variable deviations, as well as by introducing a variable parameter into the hyperbolic tangent function. Another possible modification of the algorithm is to introduce a list of previously found solutions, which will allow us both to reduce the runtime of the algorithm and to avoid its premature convergence.

Acknowledgements. This work was supported by the Russian Foundation for Basic Research, project no. 16-07-00034.

References

1. Aggarwal, C.C.: An introduction to data classification. In: Data Classification: Algorithms and Applications, pp. 2–36. CRC Press, New York (2015)
2. Hu, X., Pedrycz, W., Wang, X.: Fuzzy classifiers with information granules in feature space and logic-based computing. Pattern Recognit. **80**, 156–167 (2018)
3. Kohavi, R., John, G.H.: Wrappers for feature subset selection. Artif. Intell. **97**(1), 273–324 (1997)
4. Dash, M., Liu, H.: Feature selection for classification. Intell. Data Anal. **1**(1–4), 131–156 (1997)
5. Bolon-Canedo, V., Sanchez-Marono, N., Alonso-Betanzos, A.: Feature Selection for High-Dimensional Data. Springer, Heidelberg (2015)
6. Veerabhadrappa, R.L.: Multi-level dimensionality reduction methods using feature selection and feature extraction. Int. J. Artif. Intell. Appl. **1**(4), 54–68 (2010)
7. Yusta, S.C.: Different metaheuristic strategies to solve the feature selection problem. Pattern Recognit. Lett. **30**(5), 525–534 (2009)

8. Pedergnana, M., Marpu, P.R., Dalla Mura, M., Benediktsson, J.A., Bruzzone, L.: A novel technique for optimal feature selection in attribute profiles based on genetic algorithms. IEEE Trans. Geosci. Remote Sens. **51**(6), 3514–3528 (2013)
9. Aladeemy, M., Tutun, S., Khasawneh, M.T.: A new hybrid approach for feature selection and support vector machine model selection based on self-adaptive cohort intelligence. Expert Syst. Appl. **88**, 118–131 (2017)
10. Hodashinsky, I.A., Mekh, M.A.: Fuzzy classifier design using harmonic search methods. Program. Comput. Softw. **43**(1), 37–46 (2017)
11. Gurav, A., Nair, V., Gupta U., Valadi, J.: Glowworm swarm based informative attribute selection using support vector machines for simultaneous feature selection and classification. In: Panigrahi, B.K., et al. (eds.) SEMCCO 2014. LNCS, vol. 8947, pp. 27–37. Springer, Heidelberg (2015)
12. Marinaki, M., Marinakis, Y., Zopounidis, C.: Honey bees mating optimization algorithm for financial classification problems. Appl. Soft Comput. **10**, 806–812 (2010)
13. Glover, F., Laguna, M.: Tabu Search. Springer (1997)
14. Taillard, E.D., Gambardella, L.M., Gendreau, M., Potvin, J.-Y.: Adaptive memory programming: a unified view of metaheuristics. Eur. J. Oper. Res. **135**(1), 1–16 (2001)
15. Hedar, A., Abdel-Hakim, A.E., Almaraashi, M.: Granular-based dimension reduction for solar radiation prediction using adaptive memory programming. In: GECCO 2016 Companion Proceedings of the 2016 on Genetic and Evolutionary Computation Conference Companion, pp. 929–936. ACM, New York (2016)
16. Bezdek, J.C., Ehrlih, R., Full, W.: FCM: the fuzzy c-means clustering algorithm. Comput. Geosci. **10**(2–3), 191–203 (1984)
17. Kennedy, J., Eberhart, R.: A discrete binary version of the particle swarm algorithm. In: IEEE International Conference on System, Man, and Cybernetics, vol 5, pp. 4104–4108 (1997)
18. Pereria, L.A.M, Rodrigues, D., Almedia, T.N.S., Ramos, C.C.O., Souza, A.N., Yang, X.-S., Papa, J.P.: A binary Cuckoo search and its application for feature selection. In: Cuckoo Search and Firefly Algorithm. Studies in Computational Intelligence, vol. 516, pp. 141–154. Springer, London (2014)

Applying Fuzzy Computing Methods for On-line Monitoring of New Generation Network Elements

Igor Kotenko[1,2(✉)], Igor Saenko[1,2], and Sergey Ageev[1,2]

[1] St. Petersburg Institute for Informatics and Automation of the Russian Academy of Sciences (SPIIRAS), 14-th Liniya, 39, Saint-Petersburg 199178, Russia
{ivkote, ibsaen}@comsec.spb.ru, serg123_61@mail.ru

[2] St. Petersburg National Research University of Information Technologies, Mechanics and Optics (ITMO University), 49, Kronverkskiy prospekt, Saint-Petersburg, Russia

Abstract. New generation networks belong to the class of big sophisticated heterogeneous hierarchical geographically distributed systems. Their functional characteristics, defining reliability, are the main characteristics which provide the application of these networks for their intended purpose. The paper offers the method of on-line functional monitoring of technical states of the new generation network elements based on application of a hierarchical fuzzy logical inference. The method and the generalized algorithm of on-line functional monitoring of technical states are developed. For realization of the offered method, the technology of intelligent agents is used. The functional structure of the intelligent agent is offered. The order of its interaction with a network element is considered. Results of modeling have shown a high efficiency of the offered approach. The possibility of the hardware-software realization of the offered method and algorithm a near real time mode is shown.

Keywords: New generation network · Situation network · Monitoring
Fuzzy logical inference

1 Introduction

New generation networks (NGN) belong to the class of big sophisticated heterogeneous hierarchical geographically distributed systems. For similar systems the functional characteristics, defining their reliability, are ones of the main characteristics providing application of NGN on purpose [1].

Growth of the size of the network, its equipment complexity and extension of the list of its functionalities increase the responsibility of the network administrators for correctness and validity of the decisions made on network management. Network administrators, which experience determines the quality and reliability of NGN functioning, have rather small time resource for analysis of the current situation and decision making. Besides, they should make decisions in the conditions of incomplete information on technical states of network elements. This results in increasing

A. Abraham et al. (Eds.): IITI 2018, AISC 874, pp. 331–340, 2019.
https://doi.org/10.1007/978-3-030-01818-4_33

complexity of tasks, which need to be solved for maintenance of the network in workable conditions. In this situation there is a clear discrepancy between physical and functional possibilities of the network operators. In this regard development and deployment of elements of an intelligent operational decision-making system for on-line functional monitoring and diagnostics of technical states of NGN elements is an actual scientific and technical problem.

The existing variety of NGN key controlled parameters, their various physical nature and also various scales of their measurement determine high complexity for the problem of on-line monitoring of the states of network elements (NEs) in NGN by traditional methods, for example, statistical [2–4].

The paper offers a new approach to monitor the NE states in NGN based on use of the hierarchical fuzzy logical inference mechanism. The necessity of use in the offered approach of fuzzy data processing methods is caused by the following factors: (1) uncertainty of the reasons which can cause refusals of nodes and communication channels; (2) incompleteness of information on states of NEs and NGN in general; (3) lateness of transmitting data on NE states to processing nodes.

As it is well-known [5, 6], on-line support of decision-making in the conditions of uncertainty represents the solution of a set of semi-structured or unstructured tasks in the conditions of time restrictions. Characteristics of similar tasks are lack of methods of their solving on the basis of direct data transformation. Thus, decisions need to be made in the conditions of lack of full information. One of ways to solve the arising contradictions is the refusal of the traditional requirements imposed to the accuracy of input data. Similar requirements are the integral attribute of the strict mathematical analysis and the solution of clearly defined tasks. However, application of methods of fuzzy sets theory and fuzzy logical inference together with methods of the logical analysis in total allows us to realize adequate methods of on-line decision support in the conditions of uncertainty.

The theoretical contribution and novelty of the paper consists in the following: (1) the mechanism of a hierarchical fuzzy logical inference is offered; (2) for support of decision-making the approach based on use of intelligent agents is suggested; (3) the algorithm of training the fuzzy model which does not demand considerable computing resources is proposed. The further structure of the paper is as follows. In Sect. 2, the problem statement for on-line monitoring of states of NGN elements is given. In Sect. 3, the related works are discussed. Section 4 considers a method of assessment of NE states in NGN. In Sect. 5 the results of experimental assessment of the offered approach are analyzed. Section 6 contains the main conclusions and the directions of further research.

2 Problem Statement

The functioning of NEs in NGN can be represented as a sequence of time intervals of workable states and outages, including failures and recovery. The duration of these intervals is determined by various factors. Intervals can be considered mutually independent random variables having a certain distribution with average times.

Let n be a quantity of intervals, t_i be the duration of i-th time interval. Then the average time of no-failure operation T_0 (time between failures) is calculated as

$$T_0 = \sum_{i=1}^{n} t_i/n. \tag{1}$$

Let τ_i be the restoration time for i-th interval. Then the average recovery time is calculated as follows:

$$T_1 = \sum_{i=1}^{n} \tau_i/n. \tag{2}$$

NE reliability in the NGN is defined as probability of finding of NE in the workable state. It is equal to the mean time during which NE is in workable state. This definition is equivalent to a concept of the availability coefficient K_a. In this case, the following expression is true:

$$K_a = T_0/(T_0 + T_1). \tag{3}$$

or (for communication line)

$$K_a = \mu/(\mu + \lambda). \tag{4}$$

where $\lambda = 1/T_0$ is the rate of equipment failures; $\mu = 1/T_1$ is the rate of equipment recovery.

Analysis of expressions (3) and (4) shows that increasing of the value K_a corresponds to reduction of the recovery time of the controlled object T_1, which, in its turn, can be represented as follows:

$$T_1 = t_{det} + t_{ev} + t_{des} + t_{ex} \rightarrow \min, \tag{5}$$

where t_{det} is the time of detection of deviation from normative functioning mode; t_{ev} is the time of estimation of the new situation relatively to the state of the controlled NE; t_{des} is the time of decision making; t_{ex} is the time of decision realization.

Thus, the task of the decision making support (DMS) system is in production of such decision, in which the condition (5) is met. The time of decision realization is determined by the technical characteristics of the operations support subsystem. This time does not depend on DMS characteristics.

3 Related Work

Nowadays, the solution of the problem of NE state monitoring in NGN is implemented on the basis of the concept "the agent – the manager". This concept is considered in details in [2, 3, 7, 8]. According to this concept the agent at first accumulates information on the current state of NE, and then transfers it to the manager. The manager, in turn, transfers it to the network administrator. NGN management is implemented by the network administrator. At the heart of the known approaches the statistical methods are used to monitor the NE states [9].

Some works propose to use intelligent techniques to reduce a priori uncertainty and decrease the reaction time to change the NE state [5, 6, 10–12]. It is possible to realize the paradigm of "from the state detection to decision" on their basis. [12–14] propose to use neural networks and cognitive maps for network state monitoring. [15] suggests a dynamic evolutionary fuzzy logic system that implements adaptive training in a near real time. However, in the known works on monitoring of NE states the insufficient attention is paid to the DMS elements implementing methods for making optimal and rational decisions. At the same time, the experience of using the fuzzy inference to identify anomalous behavior and manage the security risks in the NGN [4, 16] allows to assert about appropriateness of its use for the NE state operational monitoring. Theoretical foundations of hierarchical fuzzy situational networks can be the basis for on-line functional control of technical states of the NGN elements [4–6, 11–13, 16, 17]. However, such categories as reference fuzzy situations are applied to decision-making in the known methods. With growth of the network size and, respectively, with growth of its dimension, application of this approach becomes extremely difficult and, often, impossible. For solving this problem it is suggested to unite hierarchical methods to assess a fuzzy situation of NE technical states with methods of fuzzy mathematical programming.

4 NE State Assessment in NGN

The paper offers to apply a fuzzy logical inference of Mamdani as the heart of creation of hierarchical fuzzy situational network [8, 9]. Let us consider this approach. After the block of fuzzification of the Mamdani fuzzy inference machine, the input variables, characterizing the NE state in NGN, take the form of linguistic input variables and are defined as follows:

$$Input = \langle x, T, U, G, M \rangle, \tag{6}$$

where x is variable's name; T is term-set, each element of which is determined by the fuzzy set on universal set U; G are syntax rules, generating the membership functions of terms' names; M are semantic rules, determining the membership rules on fuzzy terms, generated by the syntax rules from G.

Fuzzy logical inference for forming the situation estimations for NE state in NGN (based on the Mamdani fuzzy logical inference) has the following form [8, 9]:

$$(x_1 = a_{1j}\theta_j \ldots \theta_j x_n = a_{nj}) \times w_j \Rightarrow y_j = d_j, j = 1, \ldots, m, \tag{7}$$

where a_{ij} is a fuzzy term, by which the variable x_i in j-th rule of the knowledge base is estimated; d_j is the conclusion of j-th rule; m is the number of rules in the knowledge base; w_j are weight coefficients for each j-th rule of the knowledge base ($w_j \leq 1$); θ_j – is a logical operation, connecting premises in j-th rule of the knowledge base.

Generalizing the results received above, it is offered to implement the fuzzy logical inference having two-level hierarchical structure in DMS for assessment of the current situation.

The example of implementation of similar structure in the form of an intelligent agent (IA) is given in Fig. 1. In this figure the variant of interaction of an IA with a NE is presented.

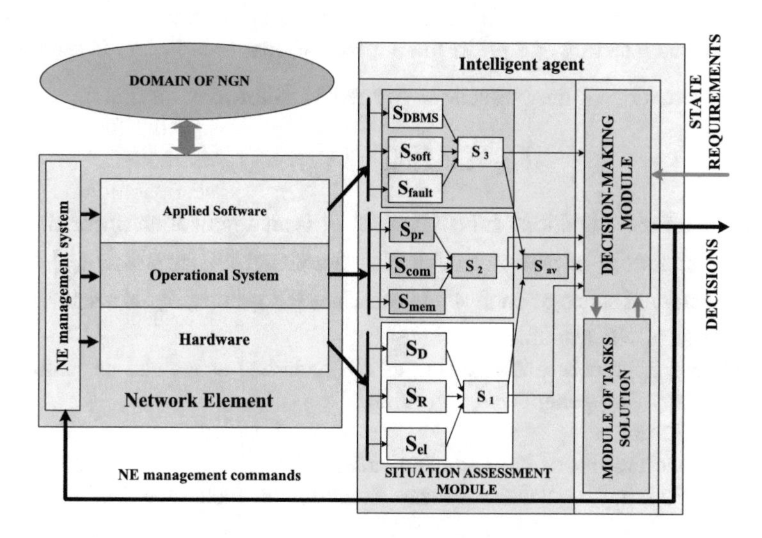

Fig. 1. Structure of an intelligent agent and its interaction with a network element.

In the presented structure the number of hierarchical levels has conditional character and can be changed in compliance with the solution of a specific objective. Each hierarchical level contains a fuzzy inference engine.

NE (for example, the router) is presented in this figure in the form of a set of hardware, an operating system, applied software, and a NE control system. The NE functions as a part of the NGN and interacts with the IA regarding procedures of on-line support of decision-making.

IA consists of three modules: assessment of the situation, decision-making and tasks solutions. The peculiarity of the IA structure is the lack of intermediate operations of defuzzification and fuzzification. These operations are carried out on the inputs and the output of DMS.

As input of the fuzzy logical inference engine of the first hierarchy level, the feature vectors $\{X_i\}$ are used for each controlled functional group of parameters defining the NE state. On the output of a hierarchical layer a set of estimates of fuzzy situations $\{S_i\}$ of NE states is formed concerning each functional group of parameters. At the following level of hierarchy these estimates are aggregated.

It should be noted that application of the variant of hierarchical structure on the basis of cluster analysis is possible. This structure can be applied at a large number of the input variables, characterizing technical NE states. It implements a hierarchy of methods of the fuzzy cluster analysis with the subsequent classification of the received results. Similar approach can be applied, for example, to control the productivity of solving a set of applied tasks by the NE processor module.

The structure of the classifier corresponds to the structure of the hierarchical fuzzy logical inference system considered above. The following clustering methods were chosen:

(1) Fuzzy k-means if the number of clusters a priori is known;
(2) Subtractive clustering if a priori the number of possible clusters is unknown.

A fuzzy situation of the NE state is formed as follows:

$$S^i_{\mathrm{NE}} = F_1(\{S^i_{\mathrm{fg}}\}, \{X^i_{\mathrm{fg}}\}, R^i_{\mathrm{fg}}), \tag{8}$$

where S^i_{NE} is a fuzzy situation of the NE state; F_1 is an aggregation operator; $\{S^i_{\mathrm{fg}}\}$ is a set of fuzzy situations of states of controlled functional NE groups; $\{X^i_{\mathrm{fg}}\}$ is a set of fuzzy parameters of states of controlled functional NE groups; R^i_{fg} is a set of functional and technological NE resources.

The offered approach to monitoring of NE technical states can be realized in the form of the following generalized algorithm.

STEP 1. Begin.

STEP 2. Monitoring of NE technical states.

STEP 3. Forming the values of fuzzy situations on each controlled functional NE group.

STEP 4. If $\mu(S^i_{\mathrm{fg}}) \geq \mu(S^i_{\mathrm{fg0}}) \forall i$, then the NE operation is standard.

STEP 5. If $\exists i$, $\mu(S^i_{\mathrm{fg\,adm}}) \leq \mu(S^i_{\mathrm{fg}}) < \mu(S^i_{\mathrm{fg0}})$, then the NE technical state is worsened, but is admissible. In this case, the following operations are performed: first, a decision variant for a case of further deterioration of the situation is prepared, second, an additional resource at the higher level of control is inquired.

STEP 6. If $\exists i$, $\mu(S^i_{\mathrm{fg\,adm}}) > \mu(S^i_{\mathrm{fg}})$ then NE technical states are very worse. Network functioning is not possible. The following operations are performed: the decision-making for NE refusal case; the inquiry of an additional resource at the higher level of control; redistribution of resources between other NEs. If the resources are received, then restoration of the NE, otherwise – extraction of the NE from the network structure. If after this the NE is restored, then continue monitoring and transit to Step 2. Otherwise – transit to Step 7.

STEP 7. End.

There are two decision making variants:

1. The IA makes the decision directly on the NE. Higher level of control is only notified on this decision. This variant is possible if higher level delegated the IA such powers.
2. The IA makes the decision which can be corrected by the higher level of control taking into account its preferences.

A computational complexity of the algorithm is equal $O(n^3 \cdot k)$, where n is the number of variables and k is the number of hierarchy labels.

5 Experimental Results

For modeling and experimental assessment of the proposed method, the NE "Router" has been chosen. The NE state was estimated in the experiment taking into account the following functional parameters: (1) electric parameters; (2) productivity; (3) software state.

As an example, Figs. 2 and 3 depict the characteristics of the fuzzy inference system for assessment of the fuzzy situation using the NE electric parameters.

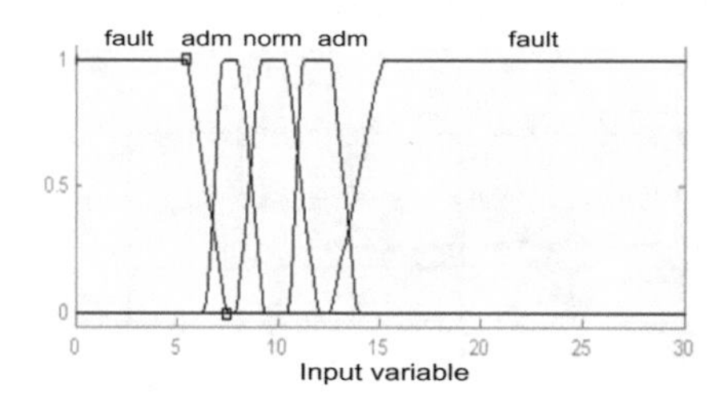

(a) member function for the "power supply" parameter

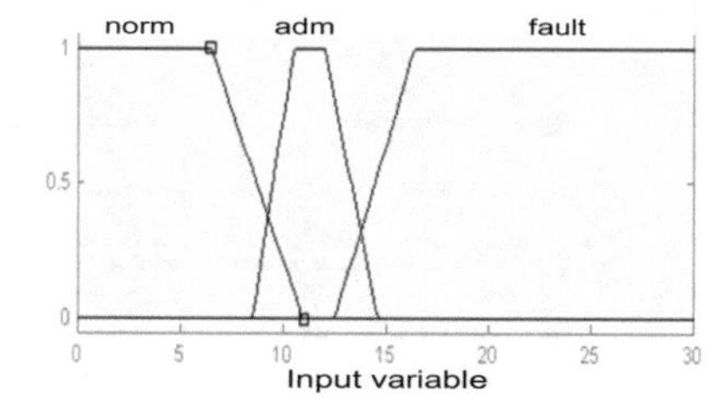

(b) member function for the "attenuation" parameter

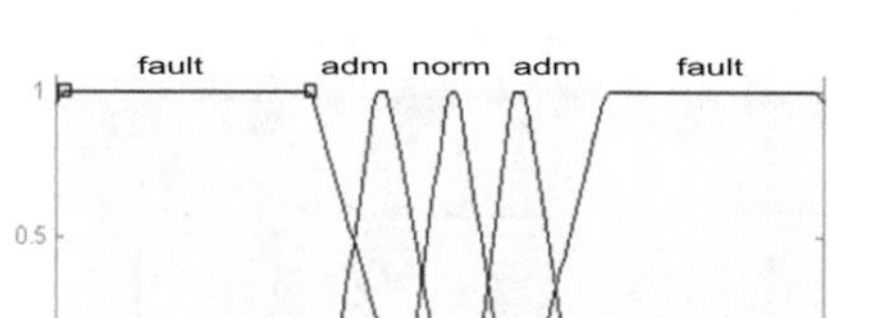

(c) member function for the "resistance of interfaces" parameter

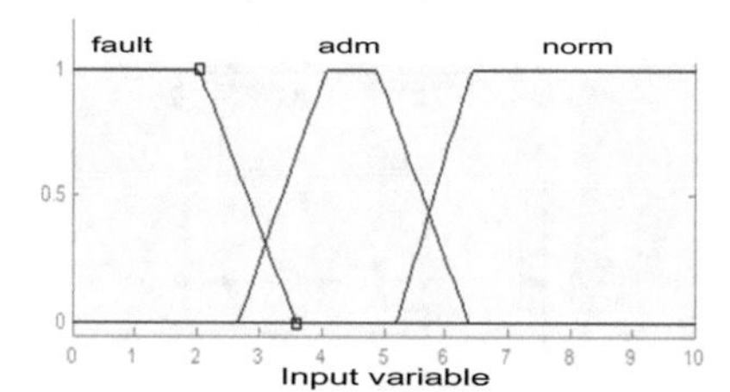

(d) member function for values of the fuzzy situation S_1

Fig. 2. Input and output member functions of the IA electric parameters.

Without loss of community, in this computing experiment all member functions are presented by trapezoidal functions. It is caused by simplicity of their implementation. The conducted various researches confirm their acceptable approximating properties [4, 16].

Functioning of the "Attenuation" and "Supply voltage" modules in a time domain is given as an example in Fig. 4. On the top and average graphics "1" designates the admissible level of the controlled parameter values, and "2" – their critical values. These levels are defined by α-sections of the corresponding member functions. In the lower graphic the value of the "Tag" parameter is depicted. If the situation is normal,

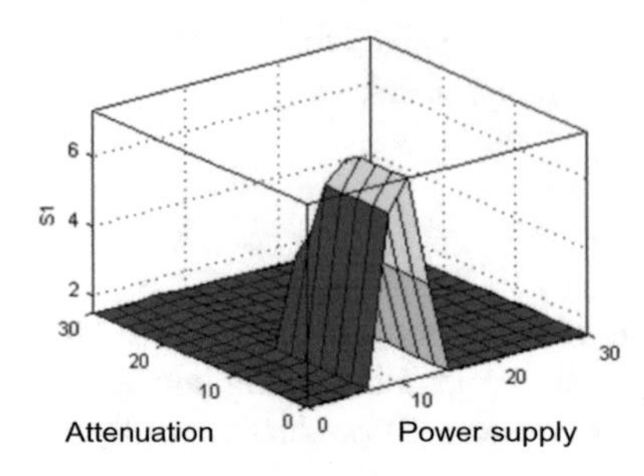

(a) member function for the "power supply – attenuation"

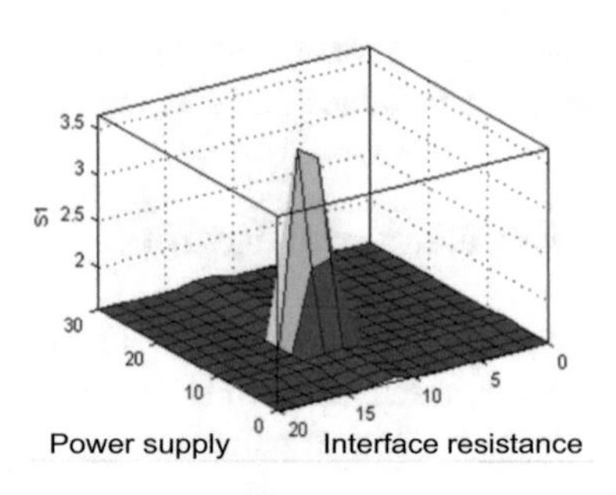

(b) member function for the "power supply – interface resistance"

Fig. 3. An example of two-dimensional member functions.

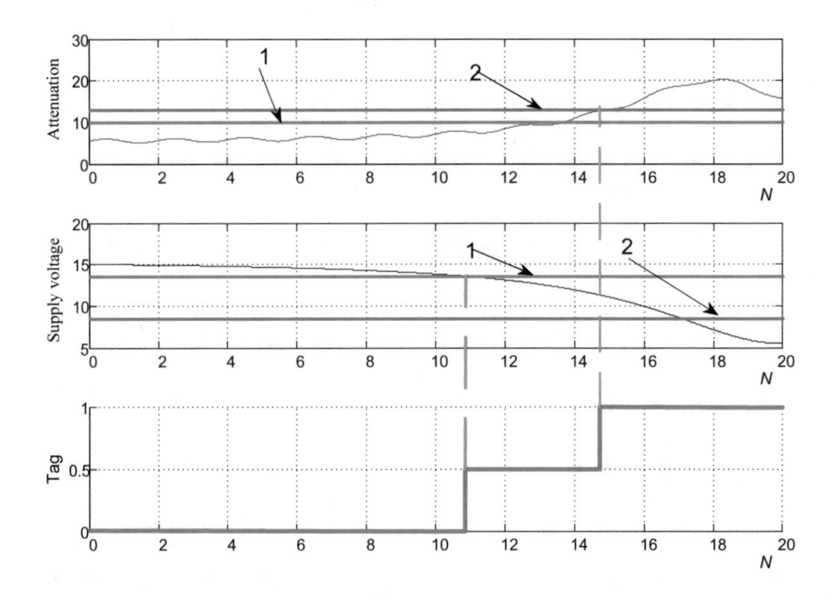

Fig. 4. Temporary charts for functioning of the hierarchical fuzzy logical inference.

the "Tag" value is equal 0. If the situation has worsened, but admissible, then the "Tag" is equal 0.5. If the situation is inadmissible, then the «Tag» is equal 1.

To evaluate the productivity in solving the applied tasks it is offered to use subtractive clustering. The results of evaluating the productivity of the processor module implementing this method are given in Fig. 5.

The analyzed clustering features for two cases are given in Fig. 5a. In the first case, 25 applied tasks were solved during T_1. In the second case, 35 tasks were solved during T_2. Location of the clusters' centers for various time intervals are assessed in Fig. 5b. The area A in both figures is an inadmissible area. It is characterized by the fact that time of solving the NE tasks taking into account the number of tasks and the processor

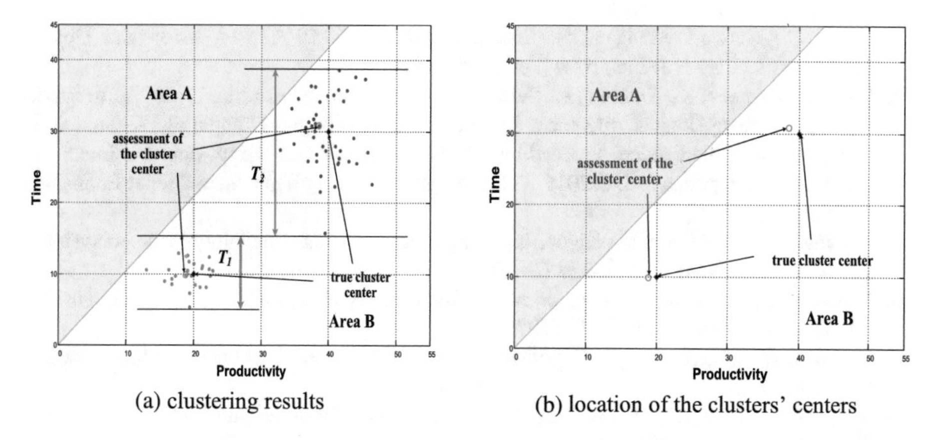

(a) clustering results

(b) location of the clusters' centers

Fig. 5. Results of modeling by subtractive clustering.

productivity is unacceptably high. After assessment of coordinates of the clusters' centers, by fuzzy logical inference, the NE productivity is estimated. In an experiment the value of the average relative assessment error for the clusters' centers is evaluated as $\delta \leq 7\%$.

The results of the experiment confirm high efficiency of the offered method to assess the NE states. The offered algorithms function in a near real time mode.

6 Conclusion

The paper has proposed a new method of online monitoring of NE states in NGN, applying fuzzy computing methods. On the basis of the carried-out analysis of methods for ensuring the NGN reliability the problem of on-line monitoring of NE states has been formulated. On the basis of the offered mechanism of fuzzy hierarchical inference the algorithm of on-line monitoring of NE states has been developed.

The analysis of experimental results for the developed algorithm has shown its high efficiency. Accuracy and reliability of the algorithm for assessment of NE states have been defined by characteristics of primary sources of the analyzed information.

The direction of further research is directed on applying the fuzzy inference for decision-making for NGN control.

Acknowledgements. This work was partially supported by grants of RFBR (projects No. 16-29-09482, 18-07-01369, 18-07-01488), by the budget (the project No. AAAA-A16-116033110102-5), and by Government of Russian Federation (Grant 08-08).

References

1. ITU-T: General principles and general reference model for next generation networks. Recommendation Y.2011, Geneva (2004)
2. RFC 1450: Management information base for version 2 of the simple network management protocol (SNMP v2). IETF (1993)

3. Black, U.: Network Management Standards: SNMP, CMIP, TMN, MIBs and Objects Libraries. McGraw-Hill Inc., New York City (1995)
4. Saenko, I., Ageev, S., Kotenko, I.: Detection of traffic anomalies in multi-service networks based on a fuzzy logical inference. In: Intelligent Distributed Computing X. Studies in Computational Intelligence. Proceedings of 10th International Symposium on Intelligent Distributed Computing - IDC'2016, vol. 678, pp. 79–88. Springer International Publishing (2016)
5. Mamdani, E., Efstathion, H.: Higher-order logics for handling uncertainty in expert systems. Int. Man-Mach. Stud. **3**, 243–259 (1985)
6. Mamdani, E., Assilian, S.: An experiment in linguistic syntheses with fuzzy logic controller. Int. Man-Mach. Stud. **7**(1), 1–13 (1975)
7. Stallings, W.: SNMP, SNMP v2, SNMP v3 and RMON 1 and 2, 3rd edn. Addison-Wesley, Reading (1998)
8. Harrington, D., Presuhn, R., Wijnen, B.: An architecture for describing SNMP management frameworks. IETF (1999)
9. Goel, A.L.: Software reliability models: assumptions, limitations, and applicability. IEEE Trans. Softw. Eng. **SE-11**(12), 1411–1423 (1985)
10. Zhang-Shen, R., McKeown, N.: Guaranteeing quality of service to peering traffic. In: Proceedings of the Joint Conference of the IEEE Computer and Communications Societies (INFOCOM 2008), pp. 1472–1480 (2008)
11. Zhang, J., Leung, Y.: Improved possibilistic c-means clustering algorithms. IEEE Trans. Fuzzy Syst. **12**, 209–217 (2004)
12. Yager, R., Filev, D.: Essentials of Fuzzy Modeling and Control. Wiley, Hoboken (1984)
13. Azruddin, A., Gobithasan, R., Rahmat, B., Azman, S., Sureswaran, R.: A hybrid rule based fuzzy-neural expert system for passive network monitoring. In: Proceedings of the Arab Conference on Information Technology (ACIT), pp. 746–752 (2002)
14. Souza, L., Barreto, G.: Nonlinear system identification using local ARX models based on the self-organizing map. Learn. Nonlinear Models - Rev. Soc. Bras. Redes Neurais (SBRN) **4** (2), 112–123 (2006)
15. Kasabov, N., Song, Q.: DENFIS: dynamic evolving neuro-fuzzy inference system and its application for time-series prediction. IEEE Trans. Fuzzy Syst. **10**(2), 144–154 (2002)
16. Kotenko, I., Saenko, I., Ageev, S.: Countermeasure security risks management in the internet of things based on fuzzy logic inference. In: Proceedings of the 14th IEEE International Conference on Trust, Security and Privacy in Computing and Communications (TrustCom-2015), 20–22 August 2015, Helsinki, Finland, p. 655–659 (2015)
17. Nikolaev, A.B., Sapego, Yu.S, Jakubovich, A.N., Bernerb, L.I., Stroganovc, V.Yu.: Fuzzy algorithm for the detection of incidents in the transport system. Int. J. Environ. Sci. Educ. **11** (16), 9039–9059 (2016)

An Approach to Similar Software Projects Searching and Architecture Analysis Based on Artificial Intelligence Methods

Yarushkina Nadezhda, Guskov Gleb(✉), Dudarin Pavel, and Stuchebnikov Vladimir

Ulyanovsk State Technical University, 432027 Ulyanovksk, Russian Federation
{jng,p.dudarin,ppnr}@ulstu.ru, guskovgleb@gmail.com

Abstract. Software engineers from all over the world solve independently a lot of similar problems. In this condition the problem of architecture reusing becomes an issue of the day. In this paper, two phase approach to determining software projects with necessary functionality and reusable architecture is proposed. This approach combines two methods of artificial intelligence: natural language clustering technique and a novel method for comparing software projects based on the ontological representation of their architecture automatically obtained from the projects source code. There are 3 metrics presented in this article that allow us to determine the measure of the relevance of the selected projects based on projects architecture indices.

Keywords: Ontology · Conceptual model
Natural language processing · Engineering design
Fuzzy hierarchical classifier · Clustering · Characteristic feature tree
Feature construction

1 Introduction

Well known that human resources in modern software development are the most valuable. Nevertheless, often happens that the same tasks are solved by independent engineers multiple times, this leads to an ineffective software development process organization. Quite often even within one enterprise after a change of project developers team the implemented software solutions are forgotten and not reused. This problem could not be solved by using version control system, because it has completely different purpose. Version control system provides storage of all the project versions with comments and ability to compare file versions, but not the modules' functionality and the way of implementation.

There are approaches for reusing source code at various stages of development. Basically, these approaches allow to reuse certain functions and classes only and do not allow reveal the project architecture. Knowledge of architecture,

A. Abraham et al. (Eds.): IITI 2018, AISC 874, pp. 341–352, 2019.
https://doi.org/10.1007/978-3-030-01818-4_34

gained from already implemented projects in the same subject area, allows to borrow large project parts and avoid conceptually incorrect solutions in the future.

An approach proposed in this paper consists of two phases. The first one is searching on open repositories. As long as information about software project collected from open sources and there is no common tag system or any commonly used classifier, the only way to group project is to perform clustering procedure. Firstly, projects filtered by keywords. This search could result in returning several thousand projects, which can not be handled by human. Project subject matter could be obtained by analyzing project description which is usually done in "readme.txt" file in versions control system repository and by scanning issue forums of the project. This task requires implementation of natural language processing (NLP) which is widely used in artificial intelligence and data science world and there are a lot of tools to perform NLP procedures for such a popular programming languages as Python, R, Java, C#. An interesting approach to forum analysis could be found in [1]. The state of the art technique in this area is word2vec models of natural language [2]. Nowadays there are many available pre-trained word2vec models [3,4]. This model transforms each word into vector. To get vector feature for sentence, vectors could be summarized and optionally normalized. This approach as a result of neuron network studying does not allow to take into account additional information about current task which could be provided by experts or by ontology for the specific field. Thus advanced technique should be used.

Previously authors have constructed hierarchical classifier from fuzzy graph obtained from KPI's dataset [5]. This hierarchical classifier could be corrected by experts and then treated as characteristic feature tree (CFT) for the further clustering procedure. This approach is close to the first phase of BIRCH algorithm [6], where CFT is constructed, but it was generalized for the case of non-vector features with similarity measure by using fuzzy graph clustering [7]. And then in [8] an approach to clustering procedure of short text fragments was proposed. In this paper this approach is adopted to the case of software projects intellectual search.

The intelligent search involves the selection of projects with the maximum level of architectural similarity. Selection is carried out among projects selected by experts from open repositories or projects from a repository of a large organization developing software in a specific subject area. To implement projects comparison based on their structure a tool for architectural concept extraction is needed. Commonly, the software project architecture is built at the design stage, which is prior to the development one. The UML language were developed to describe the project architecture with the required abstraction level. Based on the results of our previous research [9], could be concluded that developers use a lot of different types of structural elements. That is why ontology as a knowledge storage system could well act as a reference for the project analysis tool.

The minimal structural UML diagram elements, such as classes, interfaces, objects themselves, weakly convey the semantics and architectural solutions of the project. But combination of such an elements is much better describes the architecture. Stable combinations of structural elements are known as design patterns, this term exists in information technology for a long time but it is still relevant. Design patterns are actively used by the developer community, thus representing a reliable benchmark in the software project analysis. In addition, it makes sense to create local design patterns that solve specific task in a given subject area. A design pattern based on a specific subject area loses its main advantage - universality, but its greater semantic weight becomes more important characteristic for solving the problem of tool construction for searching and measure similarities between projects.

There are many works devoted to the integration of software engineering with ontologies at different levels: technical documents [10], maintenance and testing of the source code [11], UML diagrams [12,13]. There is a complete approach to development based on a domain known as development based on the subject area [14,15]. This article also proposes several metrics that allow the most efficient selection of relevant projects based on calculated indexes. A similar problem was solved by many other scientists. Different measures for evaluating of financial objects by means Hybrid Uncertainty were described in [16]. Technique providing a hierarchical aggregation of a large number of initial attributes into a smaller number of criteria, using various tools of verbal decision analysis was proposed in [17]. New correlation measures for measuring similarity and association of rating profiles obtained from bipolar rating scales were introduced in [18].

2 Software Projects Intellectual Search

2.1 Characteristic Feature Tree

As it were shown in paper [5] any set of words could be transformed into fuzzy graph [19]. Fuzzy graph by means of $\epsilon - clustering$ algorithms [20,21] could be transformed into hierarchical classifier. Any hierarchical classifier could be treated as characteristic feature tree (CFT) for the given data. The words are organized as a set of Characteristic Feature nodes (CF Nodes). The CF Nodes have a number of subclusters called Characteristic Feature subclusters (CF Subclusters) and these CF Subclusters located in the non-terminal CF Nodes can have CF Nodes as children. The CF Subclusters hold the necessary information for clustering which prevents the need to hold the entire input data in memory. This information includes:

1. Number of objects in a subcluster.
2. Minimal, average and maximal edge weight.
3. ϵ_i - related level of fuzzy graph clustering process.

In this paper a set of sentences from project descriptions, forum discussions of the project, set of comments from source code, etc. has been taken as an

Table 1. A set of sentences for software projects preliminary filtering

Project name	Sentence
VKCOM/ vk-java-sdk	Java library for VK API interaction, includes OAuth 2.0 authorization and API methods. Full VK API features documentation can be found here
dewarder/ HoldingButton	Button which is visible while user holds it. Main use case is controlling audio recording state (like in Telegram, Viber, VK)
korobitsyn/ VKOpenRobot VKOpenRobot	VK Open Bot is a library for bot creation for VK social network. Main features: mass friends collection, mass group searching and aggregation, user detailed information, user status detection
gleb-kosteiko/ vkb	Script allows you to automate the searching and participation in random reposts competitions in vk.com
PhoenixDev/ Phoenix-for-VK	First open-sourced VK client for Android inspired by Material Design
petersamokhin/ vk-bot-java-sdk	Comfortable and simple library for creating bots for VK
strelnikovkirill/ VKPhotoApp	Android OS + VK Api. VK application for a surfing in user news feed, but this news feed build only on posted photos
vladgolubev/ nowplayingVk	This app broadcasts currently playing song from last.fm account to vk.com status
shavkunov/ vk-analyzer	Application used to analyze wall of VK user or community and save results to internal database
asaskevich/ VK-Small-API	Small Java API used for work with VK. Example of using is in VK Example.java. Authorization Counters of new messages, friends, answers and groups
akveo/ cordova-vk	You can use this plugin to authenticate user via VK application rather than via webview. It makes use of official VkSDKs for iOS and Android
MLSDev/ DroidFM	This application shows how integrate the RxJava, Realm, VK API for information on popular artists, their songs and albums
Try4W/ VKontakteAPI	Simple, light weighted binding VK API for Java based on official Android SDK

input data. Projects with a set of keywords ("api", "java", "mobile", "sdk") from source code repositories like GitHub, GitLab has been selected. It total there were 490 projects in the input dataset. As long as there is not possibility to show all the input data, in Table 1 could be found the most demonstrative samples translated into English.

Each sentence was tokenized and lemmatized. Resulting terms were organized in fuzzy graph based on semantic similarity measure obtained from pre-trained word2vec model. And, finally, with a help of hierarchical fuzzy graph ϵ-clustering algorithm a fuzzy hierarchical classifier was obtained. See then resulting CFT in Fig. 1. On this figure only a two pieces of hierarchy are presented for the reason of space and clear visibility. These two sub-hierarchy shows two semantically related groups of words, the first one for the programming and API and the second one for the music. In the clustering result will be shown that these two groups lead to form clusters dedicated to VK APIs and VK Players respectively.

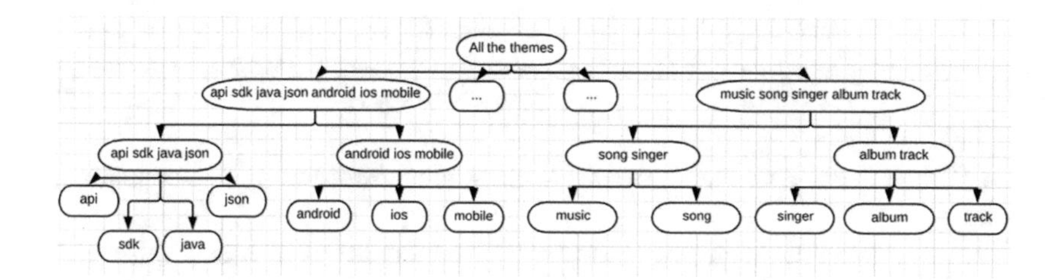

Fig. 1. Extract from hierarchical classifier for KPIs related to education

At this step CFT could be corrected by experts, so any CF Subcluster could be changed and this modification will influence the clustering result.

2.2 Feature Construction

The exact algorithm for feature construction was described in [8][1]. A function described bellow was used to transform sentences into vector form.

$$F : SL \to \mathbb{R}^n, \forall s \in SL F(s) = (s_1, s_2, ..., s_n),$$

$$s_i = R(s, v_i) * \frac{1}{\omega(v_i) * \sigma * \sqrt{2 * \pi}} * e^{-\frac{(\ln \omega(v_i) - \mu)^2}{2 * \sigma^2}},$$

where SL is a lemmatized set of sentences, R is a membership degree of sentence in current fuzzy classifier vertex.

$$\omega(v_i) = ||s \in SL \,|\, R(s, v_i) > 0||,\ n \in [1, ||HV||],\ v_i \in HV$$

with HV is an obtained classifier, parameter settings are: $\sigma = 1$, $\mu = 2.4$ in order to make function global maximum equal to 4 repetition of the word in the dataset. The next step is to find the most appropriate groups of similar software projects.

2.3 Projects Clustering

To perform clustering procedure an HDBScan [22] algorithm has been chosen. This clustering algorithm combined with features constructed from hierarchical

[1] Software implementation hosted on https://github.com/PavelDudarin/sentence-clustering.

classifier hierarchical classifier has been chosen for its ability to return quite accurate and pure clusters, mainly at the expense of precision. Metrics for precision, accuracy and purity are defined for each class as follows:

$$Precision_i = \frac{\max\limits_j CM_{i,j}}{\sum\limits_j CM_{i,j}} \quad Accuracy_i = \frac{\max\limits_j CM_{i,j}}{\sum\limits_k CM_{k,j_{\max}}} \quad Purity_i = \frac{\max\limits_j CM_{j,i}}{\sum\limits_j CM_{j,i}},$$

where CM $= [cm_{i,j} = (\omega_i \cap c_j), \omega_i$ - class with number i, c_j - cluster with number j].

General precision, accuracy and purity are calculated as mean values.

This method has shown quite a good performance results for a similar task which is discussed in paper [8].

During the clustering process 15 clusters we determined, main clusters with general description are presented on the Fig. 2. HDBScan algorithm was performed with parameter $min\_cluster\_size = 5$. Algorithm HDBScan as a true clustering algorithm always forms ‘noise cluster’ with number -1, where all the samples that could not be grouped are moved to.

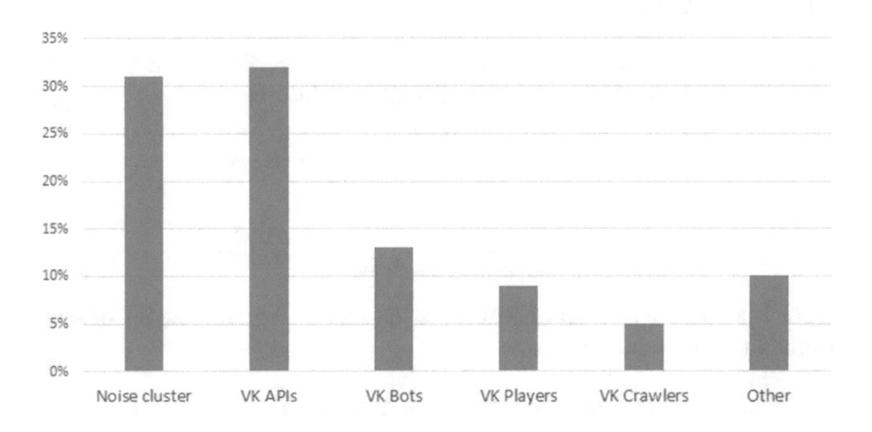

Fig. 2. Clustering results

In our example the ‘noise cluster’ is quite big, the main reason for this are too poor projects’ description and high dimensional clustering features. More detailed discussion for this could be found in [8]. Future studies will be dedicated to solving this problem.

Clustering labels for previously shown sentences could be found in Table 2. As it could be seen projects are grouped quite accurate. Calculated quality metrics based on expert evaluation are following: $precision = 0.7, accuracy = 0.98, purity = 0.98$. High level of accuracy and purity quite important in cases when expert evaluation follows the clustering process and homogeneous clusters are preferred to precise ones. In our case clustering has been made to facilitate the process of software projects filtering, but the final decision is on expert.

Table 2. Sample of clustering results

Cluster	−1	1	2	3	4
Project names	HoldingButton Phoenix-for-VK	vk-java-sdk VKPhotoApp VK-Small-API cordova-vk VKontakteAPI	VKOpenRobot vkb vk-bot-java-sdk	nowplayingVk DroidFM	vk-analyzer

A software projects selection strategy for the next phase could be different. The first strategy is to choose one example from each cluster in order to have a good variety of possible architectures. The opposite strategy is to chose a few entire clusters in case of getting some clusters that generally satisfy business purposes of a new software project. In the experiment part of this paper the second strategy has been chosen. For the further processing sentences from clusters with numbers 1, 2 and 3 have been chosen.

3 Software Design Ontology

The results presented in this paper are based on the work described in research [9]. The system described in that research made it possible to extract information from conceptual models and save it as an ontology of a certain format. UML-diagrams created at the design stage will be embodied in the source code of the project and then will update rarely. To track the project's active state, you need to analyze the source code of the project, which you can take from the version control system.

In order to be able to analyze and measure the projects structural similarity, it is necessary to transform information about projects from different sources to a single format. The most convenient way of presentation of the extracted information is a form of ontology using OWL format. OWL ontology format allows to preserve semantics of complex architectural solutions, to modify already existing data and to perform logical operations on statements.

If class diagram for software was built during design stage the structure analysis could be done. To complete this task ontology design approach was used and described below.

3.1 UML Meta-Model Based Ontology

As a target for storing knowlege from UML class diagrams has been chosen an OWL ontology format. OWL was chosen because this format is the most expressive in terms of representation of knowledge for complex subject areas. The class diagram elements should be translated into ontology as concepts with their semantics consideration. Semantics of the whole diagram is being formed from the semantics of the diagram elements and the semantics of their relations.

That is why the ontology was built on the basis of the UML meta-scheme, and not as a formal set of translated elements.

To solve the problem of intellectual analysis of project diagrams that ware included in the project documentation, it is necessary to have knowledge about formalized diagrams constructing.

Ontology contains concepts that describe the most basic elements of the class diagram, but it could be expanded if necessary. During translation of the UML meta-scheme the following notations were applied.

Formally, the ontology of project diagrams is represented as a set:

$$O^{prj} = \langle C^{prj}, R^{prj}, F^{prj} \rangle, \tag{1}$$

where: $C^{prj} = \{c_1^{prj}, ..., c_i^{prj}\}$ – is a set of concepts that define main UML diagram elements such as: "Class", "Object", "Interface", "Relationship" and others;
R^{prj} – the set of connections between ontology concepts. These relationships allow to describe correctly rules of UML notation.
F^{prj} – is the set of interpretation functions defined on the relationships R^{prj}.

3.2 Design Patterns as Structural Parts of Software Projects

Design patterns are inserted into ontology as a set of individuals based on the ontology concepts described above.

Design patterns can be formulated for a specific subject area. Thus, it will allow expressing the semantics of the project. Design patterns of the specific subject area are made by experts, taking into account the previously implemented projects. Semantic constraints and properties of design patterns are specified by the ObjectProperties and DatatypeProperties of OWL ontology.

One of the most commonly used design patterns is the Builder. In order to preserve this design pattern in the developed ontology, the following individuals belonging to relevant concepts are required.

- SimpleClass: Builder_Client, Builder_Director, Builder_ ConcreteBuilder, Builder_Product.
- AbstractClass: Builder_AbstractBuilder.
- Association: Builder_Client_AbstractBuilder, Builder_Client_Director, Builder_Client_IProduct, Builder_ConcreteBuilder_Product.
- Generalization: Builder_ConcreteBuilder_AbstractBuilder.
- Realization: Builder_Product_IProduct.

Ontological representation of the design pattern:

$$O_{tmp_i}^{prj} = \{inst(C_1^{prj}), ..., inst(r_1^{prj}), ..., r_{sameAs}\}, \tag{2}$$

In fact, the ontological representation of a single design pattern is a set of individuals of concepts and relations from the ontology of project diagrams.

To calculate the structural similarity of projects based on developed ontology, the following evaluation functions were proposed. The proposed metrics of project similarity allow us to take into account the architectural component of similarity since they are calculated on the basis of the elements of the UML language. Also, the proposed metrics contain a semantic component, since the design patterns for a specific subject area are consist of elements of the UML language. The first metric gives priority to the maximum single expressed design pattern in both diagrams:

$$\mu_{dc_\gamma,dc_\delta} = \bigvee_{tmp \in (dc_\gamma \cap dc_\delta)} \mu_{dc_\gamma \cap dc_\delta}(tmp), \tag{3}$$

where dc_γ and dc_δ is projects class diagrams presented as UML metamodel ontology Abox expressions,
$\mu_{dc_\gamma,dc_\delta(tmp)}$ - measure of expression the design pattern in project diagram.

The second metric considers the coincidence of all design patterns in equal proportions and does not considers design patterns with a measure of expression less than 0.3:

$$\mu_{dc_\gamma,dc_\delta} = (\sum_{tmp \in (dc_\gamma \cap dc_\delta) \geq 0.3} \mu_{dc_\gamma \cap dc_\delta})/N, \tag{4}$$

where N - count of design patterns with a measure of expression greater than 0.3 for both of projects.

The third metric works in the same way as the second one, but the contribution to the evaluation by design patterns depends on the number of elements in the design pattern (the design pattern with 20 elements means more than a design pattern with 5 elements):

$$\mu_{dc_\gamma,dc_\delta} = (\sum_{tmp \in (dc_\gamma \cap dc_\delta) \geq 0.3} \widetilde{\mu}_{dc_\gamma \cap dc_\delta})/N, \tag{5}$$

where $\widetilde{\mu}_{dc_\gamma \cap dc_\delta}$ - weighted measure of expression.

4 Structurally Similar Software Projects Searching Results

4.1 Searching Design Patterns in Projects

To determine the measure of similarity between two projects, it is necessary to calculate an expression degree for each design pattern in each project. The expression measure of the design pattern in the project can be calculated by mapping a project ontology Abox on a design pattern ontology Abox. The Table 3 contains expression degree for each design pattern in each project.

4.2 Results of Searching Structurally Similar Software Projects by Different Metrics

This estimations are normalized from 0 to 1. For the first metric estimations are always equal to 1. This could be easily explained because first metric chooses

Table 3. Expression of design patterns in projects

Project name/ Design pattern/ name	Delegator (3)	Adapter (8)	Builder (12)	Abstract/ superclass (3)	Interface (5)
Android-MVP	1.0	0.875	0.83	1.0	1.0
cordova-social-vk	1.0	0.875	0.83	1.0	0.8
cvk	1.0	0.875	0.83	1.0	0.8
DroidFM	1.0	0.875	0.92	1.0	1.0
VK-Small-API	1	0.625	0.42	0.33	0.6
VKontakteAPI	1.0	0.875	0.83	1.0	0.8
VK_TEST	1	0.75	0.58	0.66	0.6

Table 4. Similarity between projects by second and third metrics

Project 1/ Project 2	androidmvp	cordovavk	cvk	droidfm	vksmallapi	vkapi	vk_test
androidmvp	–	0.96 0.96	0.96 0.96	0.98 0.96	0.78 0.64	0.96 0.96	0.78 0.77
cordova-vk	0.96 0.96	–	1 0.99	0.94 0.93	0.85 0.67	1 0.99	0.83 0.80
cvk	0.96 0.97	1 0.99	–	0.94 0.93	0.85 0.67	1 0.99	0.83 0.80
droidfm	0.98 0.97	0.94 0.93	0.94 0.93	–	0.78 0.61	0.94 0.93	0.78 0.74
vksmallapi	0.78 0.64	0.85 0.67	0.85 0.68	0.78 0.61	–	0.85 0.67	0.96 0.87
vkapi	0.96 0.97	1 0.99	1 0.99	0.94 0.93	0.85 0.67	–	0.83 0.80
vk_test	0.79 0.77	0.83 0.80	0.83 0.80	0.78 0.74	0.95 0.87	0.83 0.80	–

the most expressed design pattern in both projects. Among the design patterns participating in the test there design patterns with a small number of elements, for example: Abstract superclass, interface and delegator. Such design patterns ware have measure of experssion equal 1 by first metric at both compared project. The estimations of similarity calculation between projects by second and third metrics are presented in the Table 4.

The results for the second and third metrics are also quite high. Design patterns with a expression degree less than 0.3 were excluded from consideration. All projects that participated in the comparison are downloaded from the open repository Github and realize interaction with the public API of the russian well-known social network vkontakte or with it's music or mobile API.

High estimations of similarity by second and third metrics are explained by applying the results of research first part an NLP procedures. Selection of projects was carried out by NLP procedures based on information obtained from: readme files, source code comments, version control system comments and issues excludes projects from other subject areas.

5 Conclusions

In this paper an approach to intellectual search and architecture analysis of software projects based on artificial intelligence methods is presented. NLP analysis and ontology construction allow to find and investigate projects with similar purposes and architecture. In the experimental part the proposed method was applied to different projects and proposed several similarity metrics to measure similarity between projects.

Moreover, the work presented in this paper have great potential for further research. Number of projects could be expanded. It is possible to include new design patterns in consideration. Ontologies obtained in the intermediate stages could be used separately in Protege editor. The results of this research correspond to the artificial intelligence and can be used to create intellectual systems.

Expanding the system by using ontologies of subject areas can significantly increase the relevance of the similar projects selection.

Acknowledgment. This study was supported Ministry of Education and Science of Russia in framework of project 2.1182.2017/4.6 and Russian Foundation of base Research in framework of project 16-47-732120 r_ofi_m.

References

1. Han, X., Ma, J., Wu, Y., Cui, C.: A novel machine learning approach to rank web forum posts. Soft Comput. **18**(5), 941–959 (2014)
2. Le, Q., Mikolov, T.: Distributed representations of sentences and documents. In: Proceedings of the 31st International Conference on Machine Learning, PMLR, vol. 32, no. 2, pp. 1188–1196 (2014)
3. Kutuzov, A., Kuzmenko, E.: WebVectors: a toolkit for building web interfaces for vector semantic models. In: Proceedings International Conference on Analysis of Images, Social Networks and Texts (AIST 2016), Moskow, Russian Federation, 27–29 July 2017, vol. 661, pp 155–161. Springer, Cham (2016)
4. Pelevina, M., Arefyev, N., Biemann, C., Panchenko, A.: Making sense of word embeddings. In: Proceedings of the 1st Workshop on Representation Learning for NLP Co-located with the ACL Conference, Berlin, Germany, 10 August 2017. arXiv:1708.03390
5. Dudarin, P.V., Yarushkina, N.G.: An approach to fuzzy hierarchical clustering of short text fragments based on fuzzy graph clustering. In: Proceedings of the Second International Scientific Conference Intelligent Information Technologies for Industry (IITI 2017). Advances in Intelligent Systems and Computing, vol 679. Springer, Cham (2018)
6. Zhang, T., Ramakrishnan, R., Livny, M.: BIRCH: an efficient data clustering method for very large databases. In: Proceedings of the 1996 ACM SIGMOD International Conference on Management of Data - SIGMOD 1996, pp. 103–114 (1996). https://doi.org/10.1145/233269.233324
7. Dudarin, P.V., Yarushkina, N.G.: Algorithm for constructing a hierarchical classifier of short text fragments based on the clustering of a fuzzy graph. Radio Eng. **2017**(6), 114–121 (2017)

8. Dudarin, P., Yarushkina, N.: Features construction from hierarchical classifier for short text fragments clustering. Fuzzy Syst. Soft Comput. **12**, 87–96 (2018). https://doi.org/10.26456/fssc26
9. Guskov, G., Namestnikov, A., Yarushkina, N.: Approach to the search for similar software projects based on the UML ontology. In: Proceedings of the Second International Scientific Conference Intelligent Information Technologies for Industry (IITI 2017), Varna, Bulgaria, 14–16 September 2017, vol. 680, pp. 3–10. Springer, Cham (2017)
10. Namestnikov, A., Guskov, G.: Ontological mapping for conceptual models of software system. In: Proceedings of Conference Open Semantic Technologies for Intelligent Systems, Minsk, Republic of Belarus, 16–18 February 2017, pp. 111–116 (2017)
11. Hossein, S., Sartipi, K.: Dynamic analysis of software systems using execution pattern mining. In: Proceedings 14th IEEE International Conference on Program Comprehension, ICPC, Athens, Greece, 14–16 June 2006, pp. 84–88. IEEE Computer Society (2006)
12. Zedlitz, J., Jorke, J., Luttenberger, N.: From UML to OWL 2. In: Proceedings of Third Knowledge Technology Week (KTW 2011), Kajang, Malaysia, 18–22 July 2011, vol. 295, pp. 154–163. Springer, Heidelberg (2011)
13. Bobillo, F., Straccia, U.: Representing fuzzy ontologies in OWL 2. In: Proceedings International Conference on Fuzzy Systems, Barcelona, Spain, 18–23 July 2010, pp. 2695–2700 (2010)
14. Wongthongtham, P., Pakdeetrakulwong, U., Marzooq, S.: Ontology annotation for software engineering project management in multisite distributed software development environments. In: Mahmood, Z. (ed.) Software Project Management for Distributed Computing, pp. 315–343. Springer, Cham (2017). ISBN 978-3-319-54325-3
15. Emdad, A.: Use of ontologies in software engineering. In: Proceedings of 17th International Conference on Software Engineering and Data Engineering, Los Angeles, California, USA, 30 June–2 July 2008, pp. 145–150 (2008)
16. Grishina, E., Wagenknecht, M., Yazenin, A.: On some models and methods of investment portfolio optimization by hybrid uncertainty. In: 19th Zittau Fuzzy Colloquium, Proceedings of East West Fuzzy Colloquium 2012 , Zittau, Germany, pp. 80–87 (2012)
17. Petrovsky, A., Royzenson, G.: Multi-stage technique PAKS for multiple criteria decision aiding. Int. J. Inf. Technol. Decis. Making **12**(5), 1055–1071 (2013)
18. Monroy-Tenorio, F., Batyrshin, I., Gelbukh, A., Rudas, I.: Correlation measures for bipolar rating profiles. Advances in Intelligent Systems and Computing (2018). https://doi.org/10.1007/978-3-319-67137-6-3
19. Rosenfeld, A.: Fuzzy graphs. In: Zadeh, L.A., Fu, K.S., Tanaka, K., Shimura, M. (eds.) Fuzzy Sets and Their Applications to Cognitive and Decision Processes, pp. 77–95. Academic Press, New York (1975)
20. Ruspini, E.H.: A new approach to clustering. Inf. Control **15**(1), 22–32 (1969)
21. Yeh, R.T., Bang, S.Y.: Fuzzy relation, fuzzy graphs and their applications to clustering analysis. In: Fuzzy Sets and their Applications to Cognitive and Decision Processes, pp. 125–149. Academic Press (1975). ISBN 9780127752600
22. Ester, M., Kriegel, H.P., Sander, J., Xu, X.: A density-based algorithm for discovering clusters in large spatial databases with noise. In: Proceedings of the 2nd International Conference on Knowledge Discovery and Data Mining, Portland, OR, pp. 226–231. AAAI Press (1996)

Analysis of the Dynamics of the Echo State Network Model Using Recurrence Plot

Emmanuel Sam[1], Sebastian Basterrech[2], and Pavel Kromer[3](✉)

[1] Nduom School of Business and Technology, Elmina, Central Region, Ghana
emsam@nsbt.edu.gh
[2] Department of Computer Science, Faculty of Electrical Engineering, Czech Technical University, Prague, Czech Republic
Sebastian.Basterrech@fel.cvut.cz
[3] Faculty of Electrical Engineering and Computer Science, VŠB-Technical University of Ostrava, Ostrava, Czech Republic
Pavel.Kromer@vsb.cz

Abstract. At the beginning of the 2000s, a specific type of Recurrent Neural Networks (RNNs) was developed with the name Echo State Network (ESN). The model has become popular during the last 15 years in the area of temporal learning. The model has a RNN (named reservoir) that projects an input sequence in a feature map. The reservoir has two main parameters that impact the accuracy of the model: the reservoir size (number of neurons in the RNN) and the spectral radius of the hidden-hidden recurrent weight matrix. In this article, we analyze the impact of these parameters using the Recurrence Plot technique, which is a useful tool for visualizing chaotic systems. Experiments carried out with three well-known dynamical systems show the relevance of the spectral radius in the reservoir projections.

Keywords: Recurrent Neural Network · Echo State Network Recurrence Plot · Chaotic systems · Time-series problems

1 Introduction

Recurrent Neural Networks (RNNs) are neural networks with cyclic path of connections among their neurons [1]. Due to this underlying property, they posses powerful computational and dynamical memory capabilities which make them suitable for modeling nonlinear relationships among sequential and temporal data. In the early 2000s, Echo State Network (ESN) [2] and a closely related approach known as Liquid State Machine (LSM) [3]), introduced a new computational framework for training RNN. This framework, which has lately become known as Reservoir computing (RC) [4], demonstrates that RNN can still perform significantly well even when only a subset of the network weights are trained. In this approach, a randomly initialized RNN, known as reservoir,

A. Abraham et al. (Eds.): IITI 2018, AISC 874, pp. 353–361, 2019.
https://doi.org/10.1007/978-3-030-01818-4_35

improves the linear separability of the input data. The reservoir (a matrix with the hidden-hidden weights) projects the input data in a feature space, then a supervised model is used to perform the outputs. Due to its simplicity, robustness, computational speed, and ease of implementation, RC has become popular in the Artificial Neural Network Community [5]. It has yielded successful results in many benchmark problems [5]. The ESN model and its variations have also been successfully applied on practical problems such as time series predictions [5,6] and pattern classification [2].

The computational power of ESN is based largely on the reservoir structure, and therefore the design of the reservoir and its characteristics have been the focus of many RC research over the years [7–9]. The standard ESN reservoir is influenced by a number of global parameters, which impact in the model accuracy. The most relevant ones are: sparsity and spectral radius of the reservoir matrix and dimension of the reservoir matrix [10]. The spectral radius of the reservoir matrix is related to a fundamental algebraic property, known as Echo State Property (ESN) [2], that ensures the state of the reservoir is suitable for good predictions. Guidelines on how the spectral radius can be tuned to guarantee good performance of ESN can be seen in [2]. In this study, we exploit the power of Recurrence Plot (RP) [11], a visual representation of the recurrences of dynamical systems, to investigate the effect of a given reservoir size and spectral radius combination on the dynamics of ESN reservoir. We generate several reservoir architectures with a given set of parameters and feed it with a benchmark signal. Then, we apply RP to analyze the reservoir projections. We experiment with three well-known benchmark datasets which include Henon, Lorenz, and Rossler dataset. A description of these benchmarks can be found in [7]. The possibility of exploring the dynamics of ESN and analyzing the stability of the recurrences has also been studied in [12]. The authors analyzed the effect of the input signal in the dynamics using RP and Recurrence Quantification Analysis (RQA) over two signals: sinusoidal waveform and Mackey-Glass time-series. In this article, we focus on analyzing the impact of the pair: reservoir size and spectral radius in the stability and accuracy of the ESN model using RP.

The rest of this paper is organized as follows. Section 2 describes the ESN model and its properties, and reviews relevant literature on ESN. Section 3 provides a description of the methodology for this study. Experimental results and their related explanations are presented in Sect. 4. We end with a discussion and recommendations for future work.

2 Description of Echo State Network

An ESN is made up of two main distinct structures: a random initialized and fixed hidden-hidden weights matrix called reservoir and a parametric mapping often a linear regression called readout. When its input neurons are driven by a signal, $\mathbf{s}(t)$ at any time t, the reservoir acts as a dynamical system that transforms the original input signal from an input space $\mathbb{R}^p$ into a larger space $\mathbb{R}^d$ with $p \ll d$, using a high dimensional feature map. Like kernel functions, this enhances the linear separability of the input data. Additionally, the recurrent matrix memorizes the sequence of input patterns, making the ESN suitable for solving temporal learning problems. The readout structure is a parametric mapping (often linear) from the feature map created by the reservoir and the output space. A characteristic of the model is that it does not train the hidden-hidden weights; only the readout parameters are trained. As a consequence, the model is fast and robust, and the problem of vanishing-exploding gradient [13] that is often presented in the training of RNNs is avoided. Several extensions and variations of the standard ESN model have been proposed in the literature. Examples include: intrinsic plasticity [14], BackPropagation-Decorrelation [15], Decoupled ESN [16], Leaky integrator [17], Evolino [18], and a recently introduced Echo State Network based on Queuing Theory [6,19].

In this study we consider an ESN with standard topology containing an input layer with p neurons connected to d hidden neurons, and a readout layer with o neurons. We assume that both input and output signals are real values, and we consider discrete time. The input weight matrix for the connections between the input neurons and reservoir neurons is denoted by $\mathbf{W}^{\text{in}}$ whilst the weight matrix for the internal connections inside the reservoir and the weight matrix for connections between the reservoir and readout layer are denoted by $\mathbf{W}^{\text{r}}$ and $\mathbf{W}^{\text{out}}$ respectively. The dimensions of these matrices are $d \times (1+p)$, $d \times d$, and $o \times (1+p+d)$, respectively. The first row of $\mathbf{W}^{\text{in}}$ and $\mathbf{W}^{\text{out}}$ contains a value corresponding to bias terms.

Given a training set composed of input signal $\mathbf{s}(t) \in \mathbb{R}^p$ the reservoir updates its activation state $\mathbf{x}(t) = (x_1(t), ..., x_N(t))$ using an activation function $g_h(\cdot)$ with parameters $\mathbf{W}^{\text{in}}$ and $\mathbf{W}^{\text{r}}$ as follows:

$$\mathbf{x}(t) = g_h(\mathbf{s}(t), \mathbf{x}(t-1), \mathbf{W}^{\text{in}}, \mathbf{W}^{\text{r}}). \tag{1}$$

Next, the parametric function shown below uses the actual reservoir states to execute the model output:

$$\hat{\mathbf{y}}(t) = g_o(\mathbf{x}(t), \mathbf{W}^{\text{out}}),$$

where $g_h(\cdot)$ is an activation function with parameters in $\mathbf{W}^{\text{out}}$. Although in the standard ESN model there are no connections between the input and readouts neurons [3,20], another readout form is the following:

$$\hat{\mathbf{y}}(t) = g_o(\mathbf{s}(t), \mathbf{x}(t), \mathbf{W}^{\text{out}}). \tag{2}$$

In this study, we used hyperbolic tangent $\tanh(\cdot)$ as the activation function $g_h(\cdot)$, and the dynamics was computed as:

$$\mathbf{x}(t) = \tanh(\mathbf{W}^{\text{in}}\mathbf{s}(t) + \mathbf{W}^{\text{r}}\mathbf{x}(t-1)). \tag{3}$$

The output of the model, $\hat{\mathbf{y}}(t)$ at a given time t is computed as follows:

$$\hat{\mathbf{y}}(t) = \mathbf{W}^{\text{out}}[\mathbf{s}(t); \mathbf{x}(t)], \tag{4}$$

where $[\cdot;\cdot]$ denotes a vector concatenation operation.

3 Methodology

We conducted experiments on three popular signals: Henon, Rossler, and Lorenz datasets. A detailed description of the datasets can be found in [7]. The samples in the dataset were initially normalized to lie between 0 and 1, and the resulting data was divided into two subsets: 80% was used to train the ESN and the remaining 20% was used to test it. As mentioned above, we consider a standard ESN where $\mathbf{W}^{\text{in}}$ and $\mathbf{W}^{\text{r}}$ are randomly initialized with uniformly distributed weights in the range $[-0.5, 0.5]$. For each dataset, the reservoir was configured with different combinations of reservoir size d and spectral radius ρ. We consider the pair (d, ρ) with values in the grid generated by $d = \{100, 250, 500\}$ and $\rho = \{0.1, 0.5, 0.99\}$. The reservoir states are updated with a leaking rate of 0.3 and the output weights are estimated by setting the regularization factor in the linear regression model to 1×10^{-3}. The trained ESN was ran in a generative mode (i.e. previous predictions were fed back into the reservoir as input for the next prediction), and the effect of a selected pair of parameters (d, ρ), on the accuracy of the ESN model was measured with Mean Square Error (MSE). The corresponding dynamical properties of the ESN reservoir was visualized using Recurrence Plot (RP) [11]. In the following we present a brief introduction of the RP technique.

We use a subset X_{sub} that collects n states during the training of the ESN model to generate the recurrence matrix R, to stand for the recurrences of the reservoir states. In the experimental results we set up the same value of $n = 500$ for all the signals. Given any multidimensional signal $\mathbf{x}(i)_{i=1}^{n}$, then the corresponding RP is based on the following matrix:

$$R_{i,j} = \begin{cases} 1 : \mathbf{x}(i) \approx \mathbf{x}(j), \\ 0 : \mathbf{x}(i) \not\approx \mathbf{x}(j), \end{cases} \quad i, j = 1, \ldots, n,$$

where n is the number of states considered. We say that $\mathbf{x}(i) \approx \mathbf{x}(j)$ if an arbitrary distance function is lower than an arbitrary value ϵ. In our case, we consider the sequence of reservoir states. Therefore we created the binary entries of R using the following rule [11]:

$$R_{i,j} = \begin{cases} 1, & \text{if } \|\mathbf{x}(i) - \mathbf{x}(j)\| < \varepsilon, \quad i, j = 1, \ldots n, \\ 0, & \text{otherwise}, \end{cases} \tag{5}$$

where $\|\cdot\|$ is the $L_2 - norm$(Euclidean norm), and ε is a threshold distance, computed using a percentage of the distance between the maximum $L_2 - norm$ and the minimum $L_2 - norm$ of X_{sub}. There are several variations of the RP technique, the main differences among them are the type of distance functions and epsilon values [11]. In the experimental results we visualize the reservoir matrix using three threshold distances: ε_1, ε_2, and ε_3, computed using the 10%, 50%, and 100% of the distance between the maximum and the minimum values in the $L_2 - norm$ of X_{sub}.

4 Experimental Results

In this section we present experimental results related to different configurations of the ESN reservoir. For each benchmark signal, we discuss the effect of each combination of reservoir size and spectral radius (d, ρ) on the accuracy of ESN, and interpret the dynamics of the reservoir using their associated RPs. Table 1 presents the accuracy for each of the analyzed dataset. The table shows the MSE according to the pairs (d, ρ). In the case of an ESN with a reservoir of 500 neurones and a spectral radius of 0.1 we obtained unstable results for the Rossler dataset, therefore it is not presented in the table.

Figures 1a and b present RPs representing the input signals from the Lorenz and Rossler data sets respectively. Figures 2a and b show the RPs and MSEs related to two different values of ρ with 100 neurons when driven by input signals from the Henon dataset. Though the recurrence matrices were obtained with similar ϵ (i.e. $\epsilon = 0.281$ in Fig. 2a and $\epsilon = 0.282$ in Fig. 2b), the MSEs are different and the RP in Fig. 2b is more sparse. Figures 3a and b present results based on the Lorenz dataset. The figures represent the RPs obtained with $\epsilon = 0.45$ (in Fig. 3a) and 0.55 (in Fig. 3b), for a reservoir with 100 neurons and spectral radius of 0.1 (in Fig. 3a) and 0.99 (in Fig. 3b). Note that, the RP shown in Fig. 3b was obtained with an ϵ value larger than the one shown in Fig. 3a. In spite of this, the RP created with $\rho = 0.1$ is much more dense than the RP visualization created with an ESN with spectral radius $\rho = 0.99$. Besides, the MSE for Fig. 3b is lower than that for Fig. 3a. Figures 4a and b represent the projections of the reservoir when the patterns are from the Rossler dataset. The reservoir has 100 neurons and the ϵ has values 0.47056 (in Fig. 4a) and 0.59753 (in Fig. 4b) and spectral radius of 0.1 (in Fig. 4a) and 0.99 (in Fig. 4b). Similar to the case of Lorenz dataset, the reservoir configuration which reaches lower MSE is the one that is more sparse when visualized using RP (i.e. Fig. 4b). Even though the ϵ of Fig. 4a is lower than ϵ that of Fig. 4b, the visualization presented in Fig. 4a is much more dense than the one presented in Fig. 4b. Figures 5a and b show visualizations of the reservoir projection for the Lorenz problem. Both figures have same spectral radius but different reservoir sizes. As a consequence, we can see how the reservoir size does not present a relevant impact in the recurrence dynamics.

Table 1. ESN accuracy using different network architecture.

Data	d	ρ	MSE	Data	d	ρ	MSE	Data	d	ρ	MSE
Henon	100	0.1	0.16013185	Lorenz	100	0.1	0.05212761	Rossler	100	0.1	0.06859436
	100	0.5	0.15717055		100	0.5	0.12274431		100	0.5	0.03451661
	100	0.99	0.14910897		100	0.99	0.09936286		100	0.99	0.02423478
	250	0.1	0.16173990		250	0.1	0.12022802		250	0.1	-
	250	0.5	0.16061450		250	0.5	0.08143574		250	0.5	0.06668869
	250	0.99	0.15219079		250	0.99	0.11635624		250	0.99	0.01436883
	500	0.1	0.16722330		500	0.1	0.13998824		500	0.1	-
	500	0,5	0.14815264		500	0.5	0.08130847		500	0.5	0.07337573
	500	0.99	0.15553267		500	0.99	0.07223923		500	0.99	0.00776130

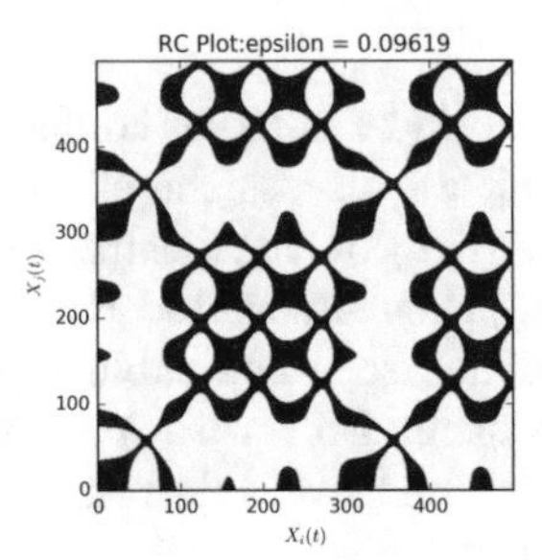

(a) Visualization using RP of a time-windows of the Lorenz dataset.

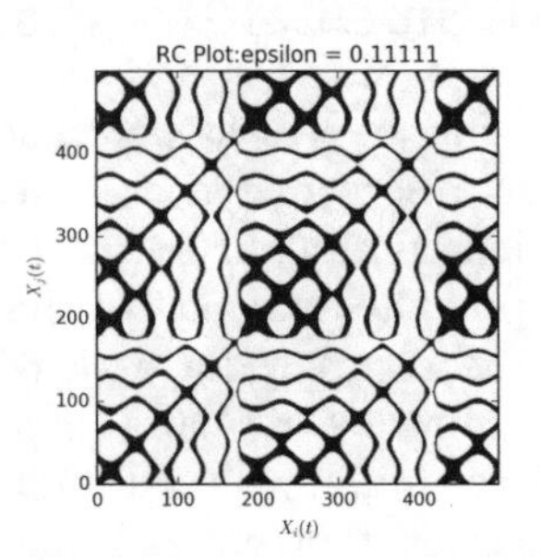

(b) Visualization using RP of a time-windows of the Rossler dataset.

Fig. 1. Visualization of the original sequential data.

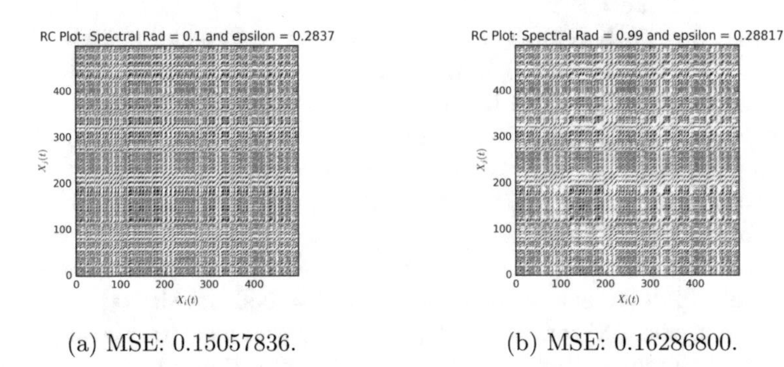

(a) MSE: 0.15057836.

(b) MSE: 0.16286800.

Fig. 2. Henon dataset: RPs created with 100 reservoir neurons and a spectral radius of 0.1 (Fig. 2a) and 0.99 (Fig. 2b), and epsilon of 0.281 (Fig. 2a) and 0.282 (Fig. 2b).

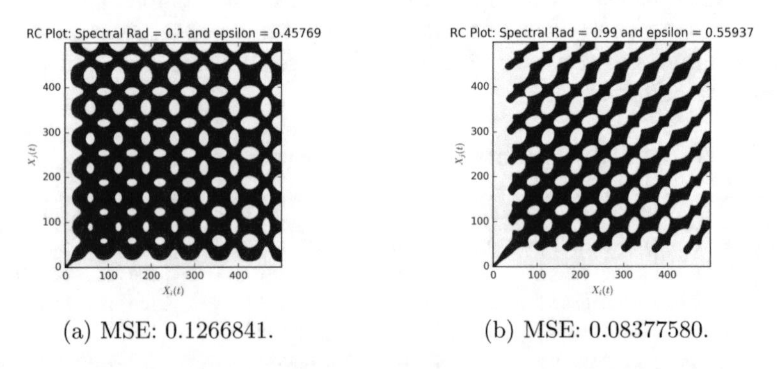

(a) MSE: 0.1266841.

(b) MSE: 0.08377580.

Fig. 3. Lorenz dataset: RPs created with 100 reservoir neurons and a spectral radius of 0.1 (left figure) and 0.99 (right figure), and epsilon of 0.45 (left figure) and 0.55 (right figure).

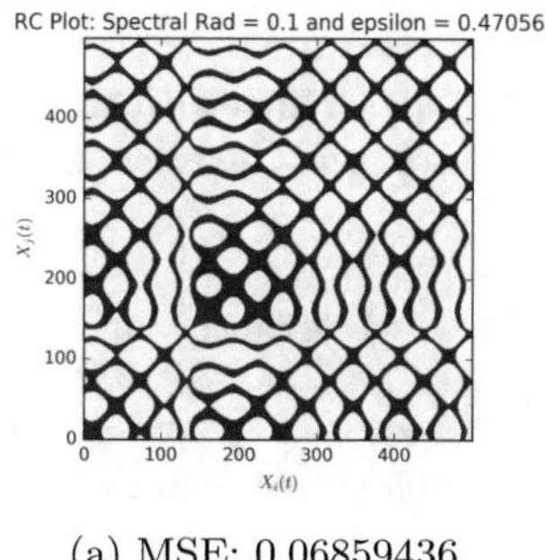

(a) MSE: 0.06859436.

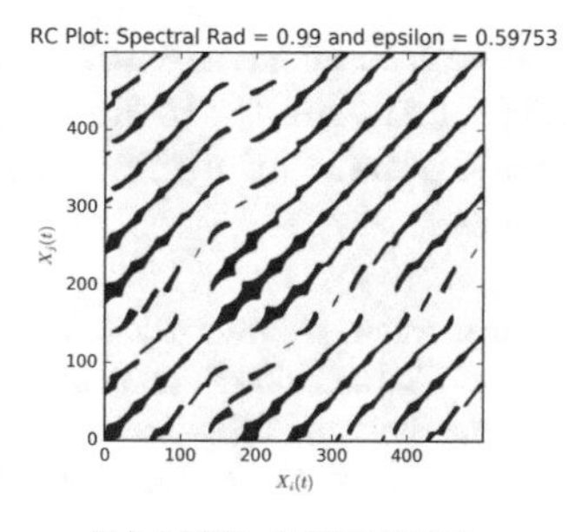

(b) MSE: 0.02423478.

Fig. 4. Rossler dataset: RPs created with 100 reservoir neurons and a spectral radius of 0.1 (Fig. 4a) and 0.99 (Fig. 4b), and epsilon of 0.47056 (Fig. 4a) and 0.59753 (Fig. 4b).

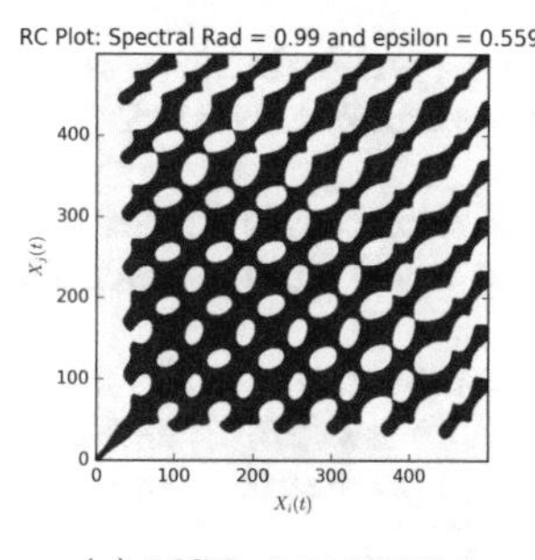

(a) MSE: 0.11635624.

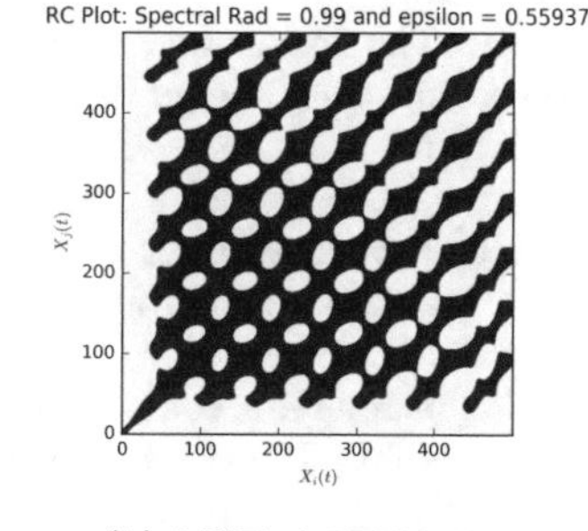

(b) MSE: 0.07223923.

Fig. 5. Lorenz dataset: RPs created with matrices with same spectral radius ($\rho = 0.99$), and different reservoir size. Figure 5a was made with a reservoir with 250 neurons and Fig. 5b was made with a reservoir of 500 neurons.

5 Conclusions and Future Work

In this paper, we have investigated the effect in the accuracy of the two main parameters of the Echo State Network (ESN) model. We used Recurrence Plots (RPs) for visualizing the recurrences generated by the phase space of the projections built by the hidden-hidden weight matrix (reservoir). We can infer from the experiments that, although both the reservoir size and spectral radius have significant effect on the accuracy of the model, the spectral radius of the reservoir matrix is much more relevant than the reservoir size as far as the projections are concerned. In other words, the sequence of reservoir states present similar characteristics, regardless of the reservoir size. However, the sequence of reservoir state is characterized by the spectral radius of the reservoir matrix. We noticed that, regardless of the epsilon used to obtain the RP, lower spectral radius (i.e. $\rho = 0.1$) lead to dense RP as well as higher MSE, and higher spectral radius ($\rho = 0.99$) lead to sparse RP as well lower MSE in most cases. Thus, another relevant result is that we found a relationship between the model accuracy and the RP matrix. In general, we obtain better results when the RP matrix is sparse and it has a form similar to the original input signal. In future work, we would like to analyze the relationship between the RP matrix and the memory capacity of the model.

Acknowledgment. This work was supported by the Czech Science Foundation under the grant no. GJ16-25694Y, and by the projects SP2018/126 and SP2018/130 of the Student Grant System, VSB-Technical University of Ostrava, and it has been supported by the Czech Science Foundation (GAČR) under research project No. 18-18858S.

References

1. Schmidhuber, J.: Deep learning in neural networks: an overview. Neural Netw. **61**, 85–117 (2015)
2. Jaeger, H.: The "echo state" approach to analysing and training recurrent neural networks. German National Research Center for Information Technology, Technical report, 148 (2001)
3. Maass, W., Natschläger, T., Markram, H.: Real-time computing without stable states: a new framework for a neural computation based on perturbations. Neural Comput. **14**, 2531–2560 (2002)
4. Verstraeten, D., Schrauwen, B., D'Haene, M., Stroobandt, D.: An experimental unification of reservoir computing methods. Neural Netw. **20**(3), 287–289 (2007)
5. Lukoševičius, M., Jaeger, H.: Reservoir computing approaches to recurrent neural network training. Comput. Sci. Revi. **3**, 127–149 (2009)
6. Basterrech, S., Rubino, G.: Echo state queueing networks: a combination of reservoir computing and random neural networks. Probab. Eng. Inf. Sci. **31**, 457–476 (2017). https://doi.org/10.1017/S0269964817000110
7. Basterrech, S.: Empirical analysis of the necessary and sufficient conditions of the echo state property. In: 2017 International Joint Conference on Neural Networks, IJCNN 2017, Anchorage, AK, USA, 14-19 May 2017, pp. 888–896 (2017)
8. Yildiza, I.B., Jaeger, H., Kiebela, S.J.: Re-visiting the echo state property. Neural Netw. **35**, 1–9 (2012)
9. Manjunath, G., Jaeger, H.: Echo state property linked to an input: exploring a fundamental characteristic of recurrent neural networks. Neural Comput. **25**(3), 671–696 (2013)
10. Lukoševičius, M.: A Practical Guide to Applying Echo State Networks. In: Montavon, G., Orr, G., Müller, K.-R. (eds.) Neural Networks: Tricks of the Trade. Lecture Notes in Computer Science, vol. 7700, pp. 659–686. Springer, Heidelberg (2012). https://doi.org/10.1007/978-3-642-35289-8_36
11. Marwan, N., Romano, M.C., Thiel, M., Kurths, J.: Recurrence plots for the analysis of complex systems. Phys. Reports **438**, 237–329 (2007)
12. Bianchi, F.M., Livi, L., Alippi, C.: Investigating echo-state networks dynamics by means of recurrence analysis. IEEE Trans. Neural Netw. Learn. Syst. **29**(2), 427–439 (2016)
13. Bengio, Y., Simard, P., Frasconi, P.: Learning long-term dependencies with gradient descent is difficult. IEEE Trans. Neural Netw. **5**(2), 157–166 (1994)
14. Schrauwen, B., Wardermann, M., Verstraeten, D., Steil, J.J., Stroobandt, D.: Improving reservoirs using intrinsic plasticity. Neurocomputing **71**, 1159–1171 (2007)

15. Steil, J.J.: Backpropagation-Decorrelation: online recurrent learning with O(N) complexity. In: Proceedings of IJCNN 04, vol. 1 (2004)
16. Xue, Y., Yang, L., Haykin, S.: Decoupled echo state networks with lateral inhibition. Neural Netw. **20**(3), 365–376 (2007)
17. Jaeger, H., Lukoševičius, M., Popovici, D., Siewert, U.: Optimization and applications of echo state networks with leaky-integrator neurons. Neural Netw. **20**(3), 335–352 (2007)
18. Schmidhuber, J., Wierstra, D., Gagliolo, M., Gomez, F.: Training recurrent networks by evolino. Neural Netw. **19**, 757–779 (2007)
19. Basterrech, S., Rubino, G.: Echo state queueing network: a new reservoir computing learning tool. In: 10th IEEE Consumer Communications and Networking Conference, CCNC 2013, Las Vegas, NV, USA, 11-14 January 2013, pp. 118–123 (2013). http://dx.doi.org/10.1109/CCNC.2013.6488435
20. Rodan, A., Tiňo, P.: Minimum complexity echo state network. IEEE Trans. Neural Netw. **22**, 131–144 (2011)

Query Answering over Some Extensions of Allen's Interval Logic

Gerald S. Plesniewicz(✉)

National Research University MPEI, Krasnokazarmennaya 14,
Moscow, Russian Federation
salve777@mail.ru

Abstract. We have considered a Boolean extension and a fuzzy Boolean extension of Allen's interval logic. We present, for extended logics, the complete systems of inference rules based on analytic tableaux. The methods of query answering over ontologies and fact bases written un these logics were developed.

Keywords: Temporal logics · Fuzzy logics · Allen's interval logic
Deduction · Query answering

1 Introduction

In 1983, J.A. Allen published the seminal paper "Maintaining knowledge about temporal intervals", where has proposed a simple temporal logic formalism [2]. He studied qualitative constraints with temporal intervals linked by elementary relations (like "before", "after", "during" and so on).

Allen's interval logic (denote it by **AL**) and its extensions were applied to various problems of designing intelligent systems (knowledge representation, common sense reasoning, natural language understanding, action planning, ontology modeling, etc.; see for example, [3–5]).

AL refers to the type of so-called interval logics. These logics formalize reasoning about relational structures with temporal intervals as primitive entities. Since modal logics can be considered as yet expressive languages for talking about relational structures, it is natural that logics with modalities, corresponding Allen's relations, were developed.

Table 1 shows the basic Allen's relations between temporal intervals A and B with their sense. There the names $\boldsymbol{b}$, $\boldsymbol{d}$, $\boldsymbol{e}$, $\boldsymbol{f}$, $\boldsymbol{m}$, $\boldsymbol{s}$, $\boldsymbol{o}$ denote the relations "before", "during", "equals", "finishes", "meets", "starts", "overlaps", and A^-, B^- denote the beginnings and A^+, B^+ denote the endings of the intervals A and B.

Let $\Omega = \{\boldsymbol{b}, \boldsymbol{d}, \boldsymbol{e}, \boldsymbol{f}, \boldsymbol{m}, \boldsymbol{s}, \boldsymbol{o}, \boldsymbol{b}^*, \boldsymbol{d}^*, \boldsymbol{f}^*, \boldsymbol{m}^*, \boldsymbol{s}^*, \boldsymbol{o}^*\}$ where θ^* ($\theta \in \Omega$) denotes the inversed relation (i.e., $X\,\theta^* Y \Leftrightarrow Y\,\theta\,X$). An arbitrary **AL** *sentence* is an expression of the form $A\ \omega\ B$ where $\omega \subseteq \Omega$. That sentence is equivalent to disjunction of atoms $A\ \theta\ B$ ($\theta \in \omega$). For example, the sentence $A\ \boldsymbol{sd}^*B$ is equivalent to the following formula (see second and sixth rows of Table 1): $(A^- < B^-) \wedge (A^+ = B^+) \vee (A^- < B^-) \wedge (B^+ < A^+)$.

A. Abraham et al. (Eds.): IITI 2018, AISC 874, pp. 362–372, 2019.
https://doi.org/10.1007/978-3-030-01818-4_36

Table 1. Basic Allen's relations

Relation	Illustration	Inequalities and equalities
$A\,\mathbf{b}\,B$	|====A====| |==B==|	$A^+ < B^-$
$A\,\mathbf{d}\,B$	|====A====| |==B==|	$A^- < B^-,\ B^+ < A^+$
$A\,\mathbf{e}\,B$	|====A====| |====B====|	$A^- = B^-,\ A^+ = B^+$
$A\,\mathbf{f}\,B$	|====A====| |==B==|	$A^- < B^-,\ A^+ = B^+$
$A\,\mathbf{m}\,B$	|====A====|==B==|	$A^+ = B^-$
$A\,\mathbf{s}\,B$	|====A====| |==B==|	$A^- < B^-,\ A^+ = B^+$
$A\,\mathbf{o}\,B$	|====A===| |==B==|	$A^- < B^-, B^- < A^+,\ A^+ < B^+$

A simple extension of Allen's logic is the Boolean Allen's logic (denote it by **BAL**). **BAL** sentences are Boolean combinations of atoms which are **AL** sentences or propositional variables. It is not difficult to construct a complete system of rules for inference in the logic **BAL** by means of analytic tableaux [1].

We call a **BAL** *ontology* a finite set $\boldsymbol{O}$ of **BAL** sentences. We can access ontologies with queries. By a *query* we mean an expression of the form $?x$: ψ where ψ is a **BAL** sentence in which one or several Allen's connectives are replaced with unknowns x. The *answer* to this query is the set of all θ such that $\boldsymbol{O} \vDash \psi(\theta)$ where $\theta \in \Omega$ and the sign '$\vDash$' denotes the relation of logical consequence. In Sect. 2 we show how to find the answers to such queries. We consider also other type of queries to **BAL** ontologies which have the form $?\psi$ where ψ is **BAL** sentence. The answer to the query $?\psi$ referred to the ontology $\boldsymbol{O}$ is YES or NO, and the answer is YES if and only if $\boldsymbol{O} \vDash \psi$.

Constraints (in particular, temporal constraints) in real applications sometimes are not exactly defined. Therefore, there is a need in fuzzifying Allen's interval logic. Several approaches to fuzzifying Allen's logic were proposed (see, for example [6–8]). Most approaches are based on different definitions of fuzzy temporal intervals having various membership functions. Our approach is more abstract: we do not use any membership functions, but introduce only estimates of truth degrees for **BAL** sentences.

Let $\boldsymbol{O} = \{\varphi_1, \varphi_2, \ldots, \varphi_n\}$ be any ontology in **BAL** and "_" be any fuzzy interpretation of the ontology, i.e., an assigning numbers "φ_i" $\in [0, 1] = \{x \mid 0 \leq x \leq 1, x$ is a real$\}$ to the sentences of the ontology $\boldsymbol{O}$. (The exact definition of a fuzzy interpretation will be given in Sect. 2. An *estimate* for a **BAL** sentence φ is an inequality one of the forms $\varphi \geq r$, $\varphi > r$, $\varphi \leq r$, or $\varphi < r$, where r is an rational number from [0, 1]. The interpretation "_" is extended to this estimates as follows: "φ" $\geq r$, "φ" $> r$, "φ" $\leq r$, "φ" $< r$. A *fact base* for an ontology $\boldsymbol{O}$ is a set $\boldsymbol{F}$ of lower estimates for its sentences, $\boldsymbol{F} = \{\varphi_1 \geq r_1, \varphi_2 \geq r_2, \ldots, \varphi_n \geq r_n\}$, where r_i are rational number from [0, 1]. The *answer* to a *query* $?x: \psi \geq x$, referred to a fact base $\boldsymbol{F}$, is the estimate $\psi \geq g$ such that $\boldsymbol{F} \models \psi \geq g$ and there is no $s > g$ with $\boldsymbol{F} \models \psi \geq s$. In other words, $\psi \geq g$ is the best low estimate that is followed from the fact base $\boldsymbol{F}$. For finding such answers, the method based on analytical tableaux can be applied.

Unfortunately, algorithms for finding answers through analytical tableaux have exponential computational complexity. However, if we fix the ontology $\boldsymbol{O}$ and the query $?x: \psi \geq x$, but we allow the fact base $\boldsymbol{F}$ to vary, then we can compute a general answer with parameters. From the general answer, we can obtain the specific answer to the query $?: \psi \geq x$ referred to any specific fact base by replacing the parameters with the specific values. Thus, from the general answer we can obtain very quickly the specific answers.

Let $a_1, a_2, \ldots$ be parameters that take values in [0, 1], and $\boldsymbol{O} = \{\varphi_1, \varphi_2, \ldots, \varphi_n\}$ be an **BAL** ontology. The *parametric fact base* for $\boldsymbol{O}$ is the set of parametric estimates $\boldsymbol{PF}(\boldsymbol{O}) = \{\varphi_1 \geq a_1, \varphi_2 \geq a_2, \ldots, \varphi_n \geq a_n\}$. The *general answer* to a query $?x: \psi \geq x$ refers to the parametric fact base $\boldsymbol{PF}(\boldsymbol{O})$. It is the estimate $\psi \geq g$ where $g = g(a_1, a_2, \ldots, a_n)$ is an expression with parameters a_i, such that the for any fact base $\boldsymbol{F} = \{\varphi_1 \geq r_1, \varphi_2 \geq r_2, \ldots, \varphi_n \geq r_n\}$ the estimate $\psi \geq g(r_1, r_2, \ldots, r_n)$ is the answer to the query $?x: \psi \geq x$ referred to the fact base $\boldsymbol{F}$. In Sect. 4 we show how to compute general answers to queries.

2 Interpretations of Ontologies

Let us introduce the following notation for any ontology $\boldsymbol{O}$ written in **BAL**: (1) I($\boldsymbol{O}$) and P($\boldsymbol{O}$) are the sets of all interval names and propositional names from $\boldsymbol{O}$; (2) E($\boldsymbol{O}$) = $\{A^-, A^+ \mid A \in \mathrm{I}(\boldsymbol{O})\}$; (3) A($\boldsymbol{O}$) is the set of all inequalities and equalities of the forms $X \leq Y$, $X < Y$ and $X = Y$ where X and Y are different names from E($\boldsymbol{O}$); (4) **AL**($\boldsymbol{O}$) and **BAL**($\boldsymbol{O}$) are the sets of all **AL** and **BAL** sentences with interval names from I($\boldsymbol{O}$).

An *interpretation* of an ontology $\boldsymbol{O}$ is a function "_" with the domain E($\boldsymbol{O}$) ∪ P($\boldsymbol{O}$) and the range $\mathbb{N} = \{0, 1, 2, \ldots\}$, such function that "$A^-$" < "$A^+$" for each $A \in \mathrm{I}(\boldsymbol{O})$, and "$p$" ∈ {0, 1} (here 0 and 1 are considered as Boolean values). The function "_" is extended naturally to A($\boldsymbol{O}$), **AL**($\boldsymbol{O}$) and **BAL**($\boldsymbol{O}$):

(i) The set {0, 1} of Boolean values is the range of the extended function "_";
(ii) "$X \leq Y$" = 1 ⇔ "X" ≤ "Y", "$X < Y$" = 1 ⇔ "X" < "Y", "$X = Y$" = 1 ⇔ "X" = "Y" for any $X, Y \in \mathbb{N}$;
(iii) "A ***b*** B" = "$A^+ < B^-$", "A ***d*** B" = ("$B^- < A^-$" ∧ "$A^+ < B^+$") and so on (see Table 1);
(iv) "A θω B" = "A θ B" ∨ "A ω B" (θ ∈ Ω, θ ⊆ Ω);
(v) "∼φ" = ∼"φ", "φ ∧ ψ" = "φ" ∧ "ψ", "φ ∨ ψ" = "φ" ∨ "ψ", "φ → ψ" = "φ" → ψ".

An interpretation "_" is a *model* of an ontology $\boldsymbol{O}$ if "φ" = 1 for all sentences $\varphi \in \boldsymbol{O}$. The relation '⊨' of logical consequence is defined using the notion of model: for any $\psi \in$ **BAL**($\boldsymbol{O}$) we have $\boldsymbol{O} \models \psi$ if and only if there is no model "_" of $\boldsymbol{O}$ such that "ψ" = 0. It is clear that the values "φ" of all sentences φ from **BAL**($\boldsymbol{O}$) is determined completely by the values of the function "_" on the set **A**(***O***). Of cause, not every function A(***O***) ↦ {0, 1} corresponds to any interpretation, since the interpretation should respect the semantics of equalities and inequalities in ℕ.

Let σ be a formula that is a Boolean combination of terms of the forms $U < V$, $U \leq V$ and $U = V$ where U and V denote integer variables. The set of all such formula is denoted by **F**. Every interpretation "_" defines the transformation of σ ∈ **F** into the formula "σ". The transformation is clear from following example. Let σ_0 be the formula $\sim(X = Y) \rightarrow (X < Y) \vee (Y < X)$. Then "$\sigma_0$" is the formula ~"$X = Y$" → "$X < Y$" ∨ "$Y < X$". In the formula σ_0 the variables X and Y take values in ℕ, but we assume that in the formula "σ_0" these variables take values in E(***O***).

Clearly, if a formula σ ∈ **F** contains the variables X_i ($i = 1, 2,\ldots, m$) and σ is true for all $X_i \in \mathbb{N}$, then, due (*i*), the formula "σ" is true (i.e., "σ" = 1) for all X_i ∈ E(***O***). Thus, if σ, τ ∈ **F** and σ → τ is true in ℕ, then "σ → τ" and "σ" → "τ" are true in E(***O***), i.e., "σ → τ" = 1 and "σ" ≤ "τ" for all values in E(***O***) of its variables. For example, we have ~"$X = Y$" ≤ "$X < Y$" ∨ "$Y < X$" for all X, Y ∈ E(***O***).

The above discussion motivate the following definition of fuzzy interpretations.

Definition. A *fuzzy interpretation* is any function "_": A(***O***) ∪ P(***O***) ∪ {0, 1} ↦ $[0, 1] = \{x \mid 0 \leq x \leq 1\}$ satisfying the conditions: (a) "φ" ≤ "ψ" for any formulas φ, ψ ∈ **F** such that the formula φ → ψ is true in ℕ; (b) "$A^- < A^+$" = 1 for any interval name A from ***O***; (c) "p" ∈ [0, 1], "0" = 0 and "1" = 1.

Fuzzy interpretations "_" are extended to **BAL** sentences by means of the above rules (*iii*), (*iv*) and (*v*) where the connectives ~, ∧, ∨ and → have the sense of Zadeh's logic: $\sim x = 1 - x$, $x \wedge y = \min\{x, y\}$, $x \vee y = \max\{x, y\}$ and $x \rightarrow y = \max\{1 - x, y\}$.

3 Query Answering over BAL Ontologies

We consider by example how to find the answers to queries referred to **BAL** ontologies.

Example 1. Let us assume that some agent can perform t/he actions *a*, *b* and *c*. Each action requires some time. Therefore, temporal intervals *A, B, C* are associated with the actions *a*, *b*, *c*. Also assume that there is a condition represented by the propositional variable *p*. Suppose it is known that:

(I) The action *b* is carried out only when the action *c* is carried out.
(II) If *p* is true then there is no time point at which both actions *a* and *b* are carried out;

Consider the question:

(III) What Allen's relations are impossible between the *A* and *C* if the conditions *p* holds?

It is clear that the (I) and (II) can be represented in **BAL** by the formulas *B* ***edfs*** *C* and *p* → *A* ***bb*****B*. The question (III) can be represented as the query ?*x*: *p* → ~*A x C* referred to the ontology ***O*** = {*p* → *A* ***bb*****B*, *B* ***edfs*** *C*}.

The answer to the query consists of relations *x* ∈ Ω such that ***O*** ⊨ *p* → ~*A x C* holds, i.e., when the set {*p* → *A* ***bb*****B*, *B* ***edfs*** *C*, ~(*p* → ~*A x C*)} is inconsistent.

Figure 1 shows the inference tree for proving inconsistency of this set. We built the tree starting with the initial branch {+*p* → *A*/ ***bb****B*, + *B* ***edfs*** *C*, − *p* → ∼*C x A*)} where the sign '+' means that a formula is true and the sign '−' that it is false.

There are labels for nodes of the tree. For example, the label '[1] T2(8)' is associated with the third node of the tree. This means that at step 1 the rule, located in Table 2 at the eight row, was applied to the formula −*p* → ∼*A x C*. As the result, two formulas +*p* and − ∼*A x C* with the label '1:' at the left were added one to another to the initial branch of the tree. The label '[3] T2(7)' is associated with the third node of the tree; this means that the rule, located in Table 2 at the seven row, was applied at step 3. In the result, the "fork" of two formulas −*p* and +*A* ***bb**** was added to the current branch of the tree.

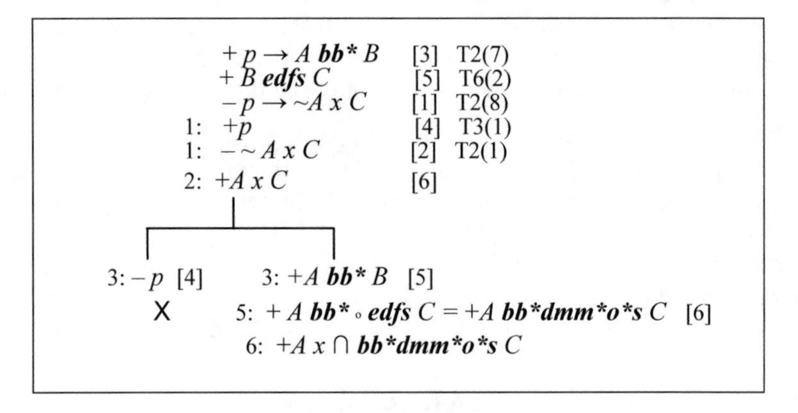

Fig. 1. Inference tree for Example 1

Table 2. Inference rules for Boolean connectives

Number	Antecedent	Consequents
1	$+\sim\varphi$	$-\varphi$
2	$-\sim\varphi$	$+\varphi$
3	$+\varphi \wedge \psi$	$+\varphi$ & $+\psi$
4	$-\varphi \wedge \psi$	$-\varphi \mid -\psi$
5	$+\varphi \vee \psi$	$+\varphi \mid +\psi$
6	$-\varphi \vee \psi$	$-\varphi$ & $-\psi$
7	$+\varphi \rightarrow \psi$	$-\varphi \mid +\psi$
8	$-\varphi \rightarrow \psi$	$+\varphi$ & $-\psi$

While constructing the inference tree, we applied the rules with making the preference for those of them contain no alternatives (i.e., the rules with numbers 1, 2, 3, 6, 8 in Table 2). At step 4 we applied the rule T3(1): $+\varphi, -\varphi \vdash$ X where X is the sign of contradiction. Thus, the first branch of the inference tree is inconsistent (closed). At step 5 we applied the rule T4(2) to the formulas +*B* ***edfs*** *C* and +*A* ***bb*B***. To obtain the result of applying this rule, we calculated the composition ***bb****$_\circ$ ***edfs*** of Allen's algebra: ***bb**** $_\circ$ ***edfs*** = ***b*** $_\circ$ ***e*** U ***b*** $_\circ$ ***d*** U ***b*** $_\circ$***f*** U ***b*** $_\circ$ ***s*** U ***b****$_\circ$ ***e*** U ***b****$_\circ$ ***d*** U ***b****$_\circ$***f*** U ***b****$_\circ$ ***s*** = ***b*** U ***bdmos*** U ***bdmos*** U ***b*** U ***b**** U ***b*dfm*o**** U ***b**** U ***b*dfm*o**** = ***bb*dfmm*oo*s***. Here we used the compositions from the Composition table presented by Allen in [2].

At step 5, the rule T4(3) was applied to +*A x C* and +*A* ***bb*dmm*o*s*** *C* with the result +*A x* ∩ ***bb*dmm*o*s*** *C*. This sentence is false in any interpretation if and only if *x* ∈ Ω***bb*dmm*o*s*** = ***d*ef*s***. Thus, the answer to query ?*x*: $p \wedge q \rightarrow \sim A\ x\ C$ consists of the relations ***d****, ***e***, ***f**** and ***s***. (*End of Example* 1.)

Table 3. Concluding rules

Number	Antecedents	Consequent
1	$+\varphi$, $-\varphi$	X
2	$\varphi \geq r$, $\varphi \leq s$, $r > s$	X
3	$\varphi \geq z$, $\varphi \leq r$	$z > r$
4	$\varphi \leq z$, $\varphi \geq r$	$z < r$

Plus the similar rules for other combinations of <, >, ≤ and ≥.

The result of applying the rules from the Table 4 to the **AL** ontology ***O*** is the transitive closure of ***O*** that contains logical consequences from ***O***. Unfortunately, the transitive closure not always gives all the logical consequences. H. Kautz gave the example of this situation [2]. Thus, the above sketched method for query answering is not complete.

In the next section we describe the complete method of finding the answers to queries referred to fact bases for ontologies written in fuzzy **BAL**. The method is based on using analytical tableaux. Note that the method is also suitable for query answering over **BAL** ontologies since a crisp ontology $\{\varphi_1, \varphi_2, \ldots, \varphi_n\}$ can be represent equivalently as the fact base $\{\varphi_1 \geq 1, \varphi_2 \geq 1, \ldots, \varphi_n \geq 1\}$.

Table 4. Allen's inference rules

Number	Antecedent	Consequents
1	$+A\ \omega^* B$	$+B\ \omega$ A
2	$+A\ \omega_1\ B$, $+B\ \omega_2\ C$	$+A\ \omega_1 \circ \omega_2\ B$
3	$+A\ \omega_1\ B$, $+A\ \omega_2\ B$	$+A\ \omega_1 \cap \omega_2\ B$
4	$-A\ \omega\ B$	$+A\ \Omega \backslash \omega\ B$

4 Query Answering over Fuzzy BAL Ontologies

We consider, by example, how to find the general answer to a query referred to the parametric fact base ***PF***(***O***) of an ontology ***O***. We applied the method of analytical tableaux with the inference rules located in Tables 2, 3, 4, 5, 6 and 7.

It is easy to prove that all these rules are sound. Let us take, for example, the rule T5(1): $\sim\varphi \geq t \vdash \varphi \geq 1 - t$. Suppose, the antecedent $\sim\varphi \geq t$ is true in some interpretation "–", i.e., "$\sim\varphi \geq t$" = 1. We have "$\sim\varphi \geq t$" = 1 $\Leftrightarrow$ "$\sim\varphi$" $\geq t \Leftrightarrow$ 1 − "φ" $\geq t$ $\Leftrightarrow$ "φ" $\geq 1 - t$ $\Leftrightarrow$ "$\varphi \geq 1 - t$" = 1. Hence, the consequent $\varphi \geq 1 - t$ is true. Therefore, the rule T5(1) is sound.

Table 5. Inference rules for inequalities with Boolean connectives

Number	Antecedent	Consequents
1	$\sim\varphi \geq t$	$\varphi \leq 1 - t$
2	$\sim\varphi \leq t$	$\varphi \geq 1 - t$
3	$\varphi \wedge \psi \geq t$	$\varphi \geq t \,\&\, \psi \geq t$
4	$\varphi \wedge \psi \leq t$	$\varphi \leq t \mid \psi \leq t$
5	$\varphi \vee \psi \geq t$	$\varphi \geq t \mid \psi \geq t$
6	$\varphi \vee \psi \leq t$	$\varphi \leq t \,\&\, \psi \leq t$
7	$\varphi \rightarrow \psi \geq t$	$\varphi \leq 1 - t \mid \psi \geq t$
8	$\varphi \rightarrow \psi \leq t$	$\varphi \geq 1 - t \,\&\, \psi \leq t$

Plus the same rules but with replacing $\geq$ and $\leq$ with $>$ and $<$, correspondingly.

Table 6. Inference rules for inequalities with Allen's connectives

Number	Antecedent	Consequents
1	$A\ \mathbf{b}\ B \geq t$	$(A^+ < B^-) \geq t$
2	$A\ \mathbf{b}\ B \leq t$	$(B^- \leq A^+) \geq 1 - t$
3	$A\ \mathbf{d}\ B \geq t$	$(B^- < A^-) \geq t \,\&\, (A^+ < B^+) \geq t$
4	$A\ \mathbf{d}\ B \leq t$	$(A^- - B^-) \geq 1 - t \mid (B^+ \leq A^+) \geq 1 - t$
5	$A\ \mathbf{e}\ B \geq t$	$(A^- = B^-) \geq t \,\&\, (A^+ = B^+) \geq t$
6	$A\ \mathbf{e}\ B \leq t$	$(A^- < B^-) \geq 1 - t \mid B^- < A^- \geq 1 - t \mid$ $(A^+ < B^+) \geq 1 - t \mid (A^+ < B^+) \geq 1 - t$
7	$A\ \boldsymbol{f}\ B \geq t$	$(B^- < A^-) \geq t \,\&\, (A^+ = B^+) \geq t$
8	$A\ \boldsymbol{f}\ B \leq t$	$(A^- \leq B^-) \geq 1 - t \mid (A^+ < B^+) \geq 1 - t \mid$ $(B^+ < A^+) \geq 1 - t$
9	$A\ \mathbf{m}\ B \geq t$	$(A^+ = B^-) \geq t$
10	$A\ \mathbf{m}\ B \leq t$	$(A^+ < B^-) \geq 1 - t \mid (B^- < A^+) \geq 1 - t$
11	$A\ \mathbf{s}\ B \geq t$	$(A^- = B^-) \geq t \,\&\, (A^+ < B^+) \geq t$
12	$A\ \mathbf{s}\ B \leq t$	$(A^- < B^-) \geq 1 - t \mid (B^- < A^-) \geq 1 - t \mid$ $(A^+ < B^+) \geq 1 - t$
13	$A\ \mathbf{o}\ B \geq t$	$(A^- < B^-) \geq t \,\&\, (B^- < A^+) \geq t \,\&\, (A^+ < B^+) \geq t$
14	$A\ \mathbf{o}\ B \leq t$	$(B^- \leq A^-) \geq 1 - t \mid (A^+ \leq B^-) \geq 1 - t \mid$ $(B^+ < A^+) \geq 1 - t$
15	$A\ \theta\omega\ B \geq t$	$(A\ \theta\ B) \geq t \mid (A\ \omega\ B) \geq t$

Let us take the rule T5(4): $\varphi \wedge \psi \leq t \vdash \varphi \leq t \mid \psi \leq t$. Suppose that the antecedent $\varphi \wedge \psi \leq t$ is true, i.e., "$\varphi \wedge \psi \leq t$" = 1. We have "$\varphi \wedge \psi \leq t$" = 1 $\Leftrightarrow$ "$\varphi \wedge \psi$" $\leq t \Leftrightarrow$ min{"φ","ψ"} $\leq t \Leftrightarrow$ "φ" $\leq t$ or "ψ" $\leq t \Leftrightarrow$ "$\varphi \leq t$" = 1 or "$\psi \leq t$" = 1.

Table 7. Transitivity rules for estimates with inequalities and equalities

Number	Antecedents	Consequent
1	$(X \leq Y) \geq s, (Y \leq Z) \geq t$	$(X \leq Z) \geq \min\{s, t\}$
2	$(X \leq Y) \geq s, (Y < Z) \geq t$	$(X < Z) > \min\{s, t\}$
3	$(X \leq Y) > s, (Y < Z) \geq t$	$(X < Z) > \min\{s, t\}$
4	$(X \leq Y) > s, (Y < Z) \geq t$	$(X < Z) > \min\{s, t\}$

Plus the same rules but with replacing $\geq s$ and $\geq t$ with $>s$ and $>t$.
Also plus the same rules but with replacing $X \leq Y$ with
$X = Y$ and/or $Y \leq Z$ with $Y = Z$

Let us take the rule T7(1): $(X \leq Y) \geq s, (Y \leq Z) \geq t \vdash (X \leq Z) \geq \min\{s, t\}$. Suppose, the antecedents $(X \leq Y) \geq s$ and $(Y \leq Z) \geq t$ are true in some interpretation "_", i.e., "$(X \leq Y) \geq s$" = 1 and "$(Y \leq Z) \geq t$" = 1. We have "$(X \leq Y)$ s" = 1, "$(Y \leq Z) \geq t$" = 1 $\Leftrightarrow$ "$X \leq Y$" $\geq s$, "$Y \leq Z$" $\geq t \geq$ min{"$X \leq Y$", "$Y \leq Z$"} $\geq \min\{s, t\}$. On the other hand, the formula $(X < Y) \wedge (Y < Z) \rightarrow X < Z$ is true in $\mathbb{N}$. Then, due the condition (b) in the definition of fuzzy interpretation (see Sect. 2), we have "$X < Y$" $\wedge$ "$Y < Z$" $\leq$ "$X < Z$" and, hence min{"$X \leq Y$", "$Y \leq Z$"} $\leq$ "$X < Z$". Thus, "$X < Z$" $\geq \min\{s, t\}$, i.e., "$(X < Z) \geq \min\{s, t\}$" = 1. Therefore, the rule T7(1) is sound.

Example 2. Let us take the ontology $\boldsymbol{O} = \{A\ \boldsymbol{d^*}D, D\ \boldsymbol{od}\ A \rightarrow B\ \boldsymbol{m}\ C, A\ \boldsymbol{bf}\ B\}$ and the query $?x$: $A\ \boldsymbol{b}\ C \vee D\ \boldsymbol{b}\ C \geq x$ referred to the parametric fact base $\boldsymbol{PF}(\boldsymbol{O}) = \{A\ \boldsymbol{d^*}D \geq a_1, D\ \boldsymbol{od}\ A \rightarrow B\ \boldsymbol{m}\ C \geq a_2, A\ \boldsymbol{bf}\ B \geq a_3\}$. Figure 2 shows the inference tree for the set of estimates $\boldsymbol{PF}(\boldsymbol{O}) \cup \{A\ \boldsymbol{b}\ C \vee D\ \boldsymbol{b}\ C < x\}$.

Let us write out from the second and the third branches all estimates for inequalities and equalities appending to them the always true estimates $(A^- < A^+) \geq 1$, $(B^- < B^+) \geq 1$, $(C^- < C^+) \geq 1$, $(D^- < D^+) \geq 1$:

$$S_1 = \{(A^- < D^-) \geq a_1, (D^+ < A^+) \geq a_1, (C^- \leq A^+) > 1-x, (C^- \leq D^+) > 1-x, (B^+ = C^-) \geq a_2, (A^+ < B^-) \geq a_3, (A^- < A^+) \geq 1, (B^- < B^+) \geq 1, (C^- < C^+) \geq 1, (D^- < D^+) \geq 1\},$$

$$S_2 = \{(A^- < D^-) \geq a_1, (D^+ < A^+) \geq a_1, (C^- \leq A^+) > 1-x, (C^- \leq D^+) > 1-x, (B^+ = C^-) \geq a_2, (B^- < A^-) \geq a_3, (A^+ = B^+) \geq a_3, (A^- < A^+) \geq 1, (B^- < B^+) \geq 1, (C^- < C^+) \geq 1, (D^- < D^+) \geq 1\}.$$

In the inference tree, the second and the three branches are closed if and only if the sets S_1 and S_2 are inconsistent.

There is an simple method for recognizing inconsistency of an arbitrary (finite) set S of estimates with equalities and inequalities. Let us define the directed labeled graph $\Gamma(S)$ as follows:

- The vertices of $\Gamma(S)$ are names A^- and A^+ such that A enters S;
- The arcs of $\Gamma(S)$ are triples (X, t, Y) such that estimates $(X\ \varepsilon\ Y) \geq t$ or $(X\ \varepsilon\ Y) > t$ belong to S, where $\varepsilon \in \{<, \leq, =, >, \geq\}$.

Denote by $P^{\#}$ the set of all estimates that correspond to arcs of the path.

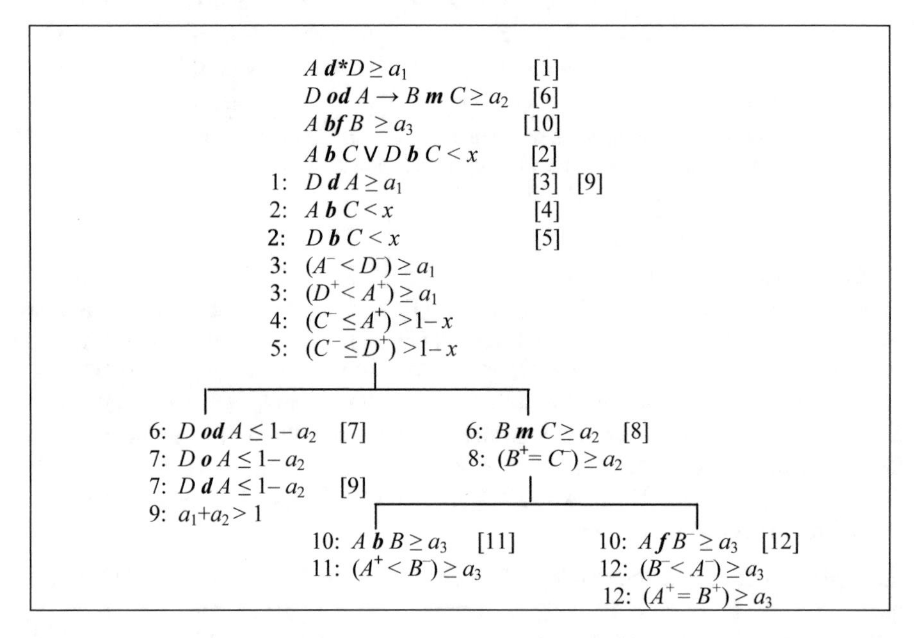

Fig. 2. Inference tree for Example 2

Figure 3 shows the graphs $\Gamma(S_1)$ and $\Gamma(S_2)$. (S_1 and S_2 are the above written sets of estimates.)

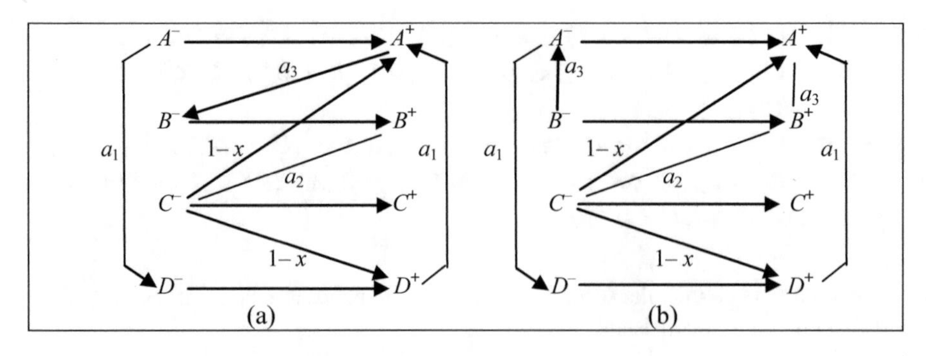

Fig. 3. Graphs $\Gamma(S_1)$ and $\Gamma(S_2)$ for Example 2

The graph $\Gamma(S_1)$ contains two simple cycles which form the base of the cycles in $\Gamma(S_1)$: $P_1 = \langle A^+, B^-, B^+, C^-, A^+\rangle$ and $P_2 = \langle A^+, B^-, B^+, C^-, D^+, A^+\rangle$. They have the corresponding sets of estimates: $P_1^\# = \{(A^+ < B^-) \geq a_3, (B^- < B^+) \geq 1, (B^+ = C^-)$ $a_2, (C^- \leq A^+) > 1 - x\}$, $P_2^\# = \{(A^+ < B^-) \geq a_3, (B^- < B^+) \geq 1, (B^+ = C^-) \geq a_2,$ $(C^- \leq D^-) > 1 - x, (D^- < A^+) \geq a_1\}$. Applying the transitivity rules to the first, the second and third estimates from $P_1^\#$, we get the estimate $(A^+ < C^-) \geq \min\{a_3, 1, a_2\} = \min\{a_2, a_3\}$, from it we get the estimate $(C^- \leq A^+) \geq 1 - \min\{a_2, a_3\}$. The pair $\{(C^- \leq A^+) \geq 1 - \min\{a_2, a_3\}. (C^- \leq A^+) > 1 - x\}$ is inconsistent if and only if $x \leq \min\{a_2, a_3\}$. Indeed, in any interpretation we have "$C^- \leq A^+$" $\geq 1 - \min\{a_2, a_3\}$ and "$C^- \leq A^+$" $> 1 - x$; therefore, there is contradiction if and only if $1 - \min\{a_2, a_3\} \geq 1 - x$. Similarly, from $P_2^\#$ we obtain the condition $x \leq \min\{a_1, a_2, a_3\}$. Thus, the second branch of the inference tree will be closed exactly for x that satisfy the condition $x \leq \min\{a_2, a_3\}$.

The graph $\Gamma(S_2)$ contains two cycles $P_3 = \langle A^+, B^+, C^-, D^+, A^+\rangle$ and $P_4 = \langle A^+, C^-, A^+\rangle$ From $P_3^\#$ and $P_4^\#$ we get the conditions $x \leq \min\{a_1, a_2\}$ and $x \leq \min\{a_2, a_3\}$ under which the third branch is closed. Thus, the third branch is closed under the condition $(x \leq \min\{a_1, a_2\}) \vee (x \leq \min\{a_2, a_3\})$ which is equivalent to the condition $x \leq \max\{\min\{a_1, a_2\}, \min\{a_2, a_3\}\}$. Therefore, the following condition closes the inference tree: $(a_1 + a_2 > 1) \wedge (x \leq \min\{a_2, a_3\}) \wedge (x \leq \max\{\min\{a_1, a_2\}, \min\{a_2, a_3\}\}$.). This condition is equivalent to the condition $(a_1 + a_2 > 1) \wedge (x \leq \min\{a_2, a_3\})$. Hence, the general answer to the query $?x$: A ***b*** $C \vee C$ ***m*** $D \geq x$ has the following bound: $g(a_1, a_2, a_3) =$ **if** $a_1 + a_2 > 1$ **then** $\min\{a_2, a_3\}$ **else** 0. (*End of Example* 2.)

5 Conclusion

In the paper we have showed how to get the answers to questions addressed to ontologies and fact bases written in the languages of Boolean and fuzzy Boolean extensions of the Allen's interval logic. The analytic tableaux method was used for such query answering. We considered facts only in the form of lower estimates for sentences of these logics. However, this method can be used for the case of upper estimates.

Acknowledgment. This work was supported by Russian Foundation for Basic Research (project 17-07-01332).

References

1. D'Aggostino, M., Gabbay, D., Hahnle, R., Possega, J. (eds.): Handbook of Tableaux Methods. Kluwer, Dordrecht (1999)
2. Allen, J.F.: Maintaining knowledge about temporal intervals. Commun. ACM **26**, 832–843 (1983)
3. Allen, J.F.: Towards a general theory of action and time. Artif. Intell. **23**(1), 123–154 (1984)

4. Allen, J.F., Ferguson, G.: Actions and events in interval temporal logic. J. Log. Comput. **4**, 531–579 (1994)
5. Allen, J.F., Kautz, H.A., Pelavin, R.N., Tenenberg, J.D.: Reasoning about Plans. Morgan Kaufmann, San Francisco, CA (1999)
6. Badaloni, S., Giacomin, M.: A fuzzy extension of Allen's interval algebra. LNAI, vol. 1792, pp. 155–165. Springer, Heidelberg (2000)
7. Ohlbach, H.J.: Relations between fuzzy time intervals. In Proceedings 11th International Symposium on Temporal Representation and Reasoning, pp. 44–51. Tatihou, France (2004)
8. Schockaert, S., De Cock, M., Kerre, E.E.: Fuzzifying Allen's temporal interval relations. IEEE Trans. Fuzzy Syst. **16**(2), 517–533 (2008)

Fuzzy Topological Approach to a Solid Control Task

Andrei Kostoglotov[1], Vladimir Taran[1], and Vladimir Trofimenko[1,2(✉)]

[1] RSTU, Rostov-on-Don, Russia
trofimvn@mail.ru
[2] DSTU, Rostov-on-Don, Russia

Abstract. Synthesis of the control algorithms based on varying structures allows increased operating effectiveness of the control systems. The paper presents an alternative combined control method for angular velocities stabilization of an axisymmetric solid based on optimal synthesis using the joint maximum principle and the predicting model method. The control structure transition procedure is implemented using fuzzy logic methods. The proposed approach effectiveness was verified using modeling results.

Keywords: Combined optimal control · Axisymmetric solid
The joint maximum principle · The predicting model · Modeling

1 Introduction

One of the requirements to modern technical systems is providing wide operating conditions range, which leads to various and even inconsistent operating modes. They are described by set of quality metrics, system dynamics and control limitations, disturbance nature and intensity, etc. For example, angular velocities control of a solid, based on the joint maximum principle [1] is realized in the limited controls condition, but results in sliding mode along the switching surface. Implementing the predicting model method in the task does not result in sliding mode and allows determining the set point of the phase space with high accuracy, though the control synthesis procedure does not consider control limitations [2].

Synthesis of the control algorithms based on varying structures allows increased operating effectiveness of the control systems. Defining a set of typical structures comes to solving a set of optimization problems, all of which are separate tasks. The control structure transition procedure may be implemented using fuzzy conclusion methods with modeling fuzzy topological space over set of algorithms with new typical structures [3, 4].

2 Problem Definition

The angular motion of a spacecraft considered as a solid with one axis of symmetry is described by following equations [5]:

A. Abraham et al. (Eds.): IITI 2018, AISC 874, pp. 373–381, 2019.
https://doi.org/10.1007/978-3-030-01818-4_37

$$\begin{aligned} \dot{\omega}_1 + A\omega_2\omega_3 &= u_1, & \omega_1(t)\Big|_{t=t_1} &= \omega_{s1}, \\ \dot{\omega}_2 - A\omega_1\omega_3 &= u_2, & \omega_2(t)\Big|_{t=t_1} &= \omega_{s2}, \\ \dot{\omega}_3 &= u_3, & \omega_3(t)\Big|_{t=t_1} &= \omega_{s3}, \end{aligned} \tag{1}$$

where A – reduced inertia moment; $\omega_1, \omega_2, \omega_3$ и u_1, u_2, u_3 – angular velocities and controls as time-dependent functions, respectively; the point over a variable designates derivation on time.

Let's make definitions for vectors of angular velocities and controls:

$$\omega = (\omega_1, \omega_2, \omega_3)^T, \quad u = (u_1, u_2, u_3)^T,$$

where ^{T} – a transposing symbol.

If necessary to specify some point of time we will put the time value in brackets behind the vector. For example, $\omega(t_1)$ – vector of angular velocities at the moment t_1.

For control quality estimation we will choose the operation speed and accuracy functionals.

We will consider a task for deriving control, which takes the system (1) from condition $\omega(t_1)$ to condition $\omega(t_2)$, while in the 'far' from the set point area of the phase space the control will be optimized using the operation speed criteria, and in the 'close' area using the accuracy criteria. As regular, the terms 'far' and 'close' have fuzzy definitions and require description using fuzzy conclusion methods. Figure 1 represents structure of the model (1).

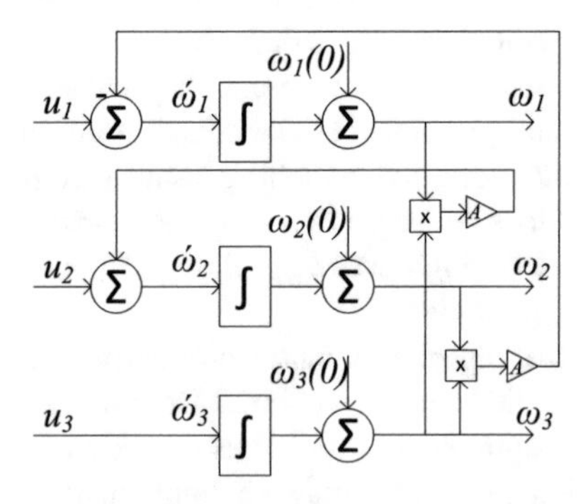

Fig. 1. Model of an axisymmetric solid angular movement

3 The Joint Maximum Principle Method

The joint maximum principle is based on the Lagrange method which states, that the movement Eq. (1) may be derived by usage of the extreme behavior of the functional [1, 6, 7]

$$J_2 = \int_{t_1}^{t_2} F(\omega)dt + \lambda \int_{t_1}^{t_2} (T + A)dt, \tag{2}$$

where T – system kinetic energy; A – control forces work; λ – Lagrange multiplier; $F(\omega)$ – positive definite function, defined by kinematic relation between angular velocities ω and generalized angular coordinates derivative. Let us find the functional (2) extremum from the permitted controls class. To do so, we will assume that $\tilde{u}$ – optimal control vector, so any deviation from that control will increase the functional increment ΔJ_2.

$$\Delta J_2 = \sum_{i=1}^{3} \left\{ \lambda [Q_i(u) - Q_i(\tilde{u})] + \left[\frac{\partial F(u)}{\partial \dot{q}_i} - \frac{\partial F(\tilde{u})}{\partial \dot{q}_i} \right] \right\} \dot{q}_i \Delta t$$
$$+ \int_{\tau}^{t_2} \left[-\dot{\lambda} \sum_{i=1}^{3} \frac{\partial T}{\partial \dot{q}_i} \delta q_i \right] dt + \int_{\tau}^{t_2} \sum_{i=1}^{3} \frac{\partial F(\tilde{u})}{\partial \dot{q}_i} \delta q_i dt > 0,$$

where Q_i – generalized forces of the corresponding generalized coordinates q_i, producing work

$$A = \sum_{i=1}^{3} \int_{q(t_1)}^{q(t_2)} Q_i(u) dq_i$$

Lagrange multiplier λ possesses sufficient freedom degree to assume $\dot{\lambda} = 0$. Then, considering optimality condition $\Delta J_2 \geq 0$, and continuity of kinematic function $F > 0$, we derive condition of the optimal control:

$$\lambda \sum_{i=1}^{3} [Q_i(u) - Q_i(\tilde{u})] \dot{q}_i + \sum_{i=1}^{3} \left[\frac{\partial F(u)}{\partial \dot{q}_i} - \frac{\partial F(\tilde{u})}{\partial \dot{q}_i} \right] \dot{q}_i \geq 0,$$

which results in

$$\sum_{i=1}^{3} \left[\lambda Q_i(\tilde{u}) + \frac{\partial F(\tilde{u})}{\partial \dot{q}_i} \right] \dot{q}_i = \max_{u \in \mathrm{U}} \sum_{i=1}^{3} \left[\lambda Q_i(u) + \frac{\partial F(u)}{\partial \dot{q}_i} \right] \dot{q}_i,$$

the optimal control law.

In going from generalized velocities $\dot{q}$ and coordinates q to angular velocities ω and controls, we derive the control law

$$\sum_{i=1}^{3} \left[\lambda \tilde{u} + \frac{\partial F}{\partial \omega} \right] \omega_i = \max_{u \in \mathrm{U}} \sum_{i=1}^{3} \left[\lambda u_i + \frac{\partial F(u)}{\partial \omega_i} \right] \omega_i.$$

For operation speed functional $F = 1$ and considering control limitation such as $u_i \leq |U_i|$, $i = \overline{1,3}$, the optimal control law transforms into simple switching [1]

$$\tilde{u}_i = U_i \cdot sign\left(\frac{\omega_i}{\lambda}\right).$$

Figure 2 represents the joint maximum principle based control device structure. Figure 3 shows the dynamics of angle velocities and control change on the ω_1 channel while using the joint maximum principle.

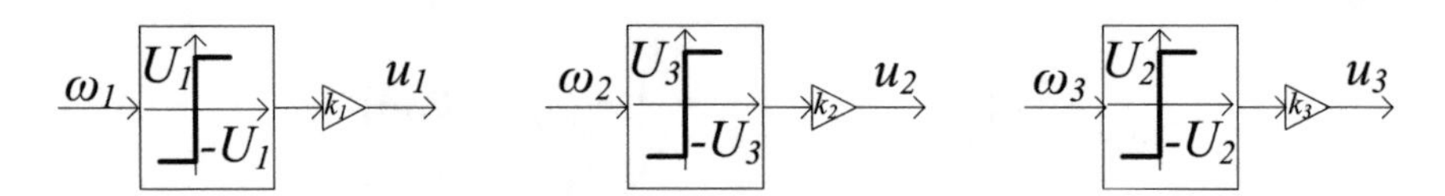

Fig. 2. The joint maximum principle based control device structure

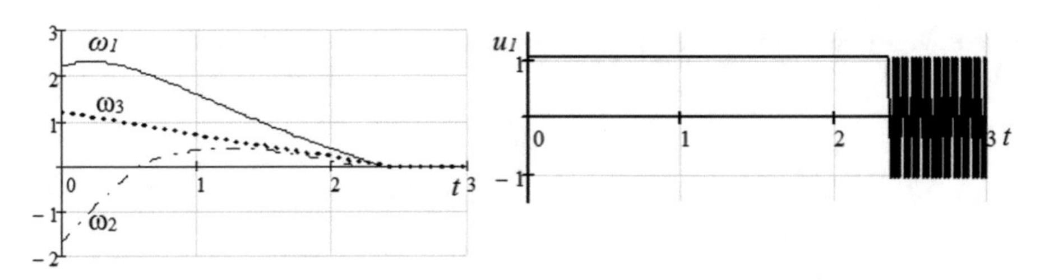

Fig. 3. The dynamics of angle velocities and control change while using the joint maximum principle

During the final control stage the system goes into the sliding mode, thus limiting precision of stabilization of solid angle velocities. The stabilization precision depends on minimal real width of the control pulse. Figure 4 represents a set of phase paths on the ω_1 channel and switch point areas in the phase space in sliding mode.

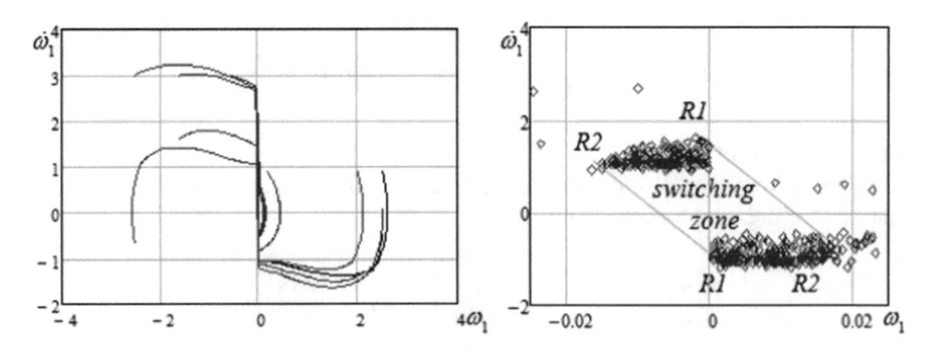

Fig. 4. A set of phase paths on ω_1 channel and switch point areas

4 The Predicting Model Method

As proven in paper [5], for the dynamic object described by the equation

$$\dot{x} = \varphi(x,t) + u, \quad u \in \mathrm{R}^m,\ x \in \mathrm{R}^n, \tag{3}$$

optimum control u_o is determined by minimizing the Krasovsky function (generalized work function)

$$J = V(x(t), x(t_2)) + \int_{t_1}^{t_2} Q(x,t)dt + \frac{1}{2}\int_{t_1}^{t_2} \left(u^{\mathrm{T}}k^{-1}u + u_o^{\mathrm{T}}k^{-1}u_o\right)dt \tag{4}$$

and is defined by the expression

$$u_o(t) = -k\left[\int_{t}^{t_2} \left(G^T(s,t)\dot{Q}_x\right)ds + G^T(t_2,t)\dot{V}_x\right]. \tag{5}$$

In expressions (3–5) x – object state vector, u – vector of the controls being optimized, u_o – required optimum control vector; $V(x(t),\, x(t_2))$, $Q(x,t)$ – the given positively definite differentiable functions, $k = \mathrm{diag}(k_1,\, k_2,\, \ldots k_m)$ – diagonal matrix of the given coefficients. The fundamental matrix $G(s,t)$ and vectors of partial derivatives of the corresponding functions $\dot{Q}_x$, $\dot{V}_x$ are defined on solutions of the unrestricted motion of the system

$$\dot{x} = \varphi(x,\, s),\ x(s)\Big|_{s\,=\,\mathrm{t}} = x(t),\ s \in [t,\, t_2], \tag{6}$$

and, a fundamental matrix of $G\ (s,\ t)$ is defined by the solution of the following equation:

$$\frac{\partial\, G(s,t)}{\partial\, s} = F_x \cdot G(s,t), \qquad G(s,t)\Big|_{s\,=\,t} = I, \tag{7}$$

where I – a unit matrix, F_x – a matrix of Jacobi for the object motion Eq. (3), and s – the parameter meaning the accelerated (predicted) time.

Therefore, the optimum control vector u_0 which transfers system (1) from a state $\omega(t_1)$ to a state $\omega(t_2)$ while minimizing function (4) is determined by the expression:

$$u_0(t) = -k\left[\int_{t}^{t_2} \left(G^{\mathrm{T}}(s,t)q\omega(s,t)\right)ds + G^{\mathrm{T}}(t_2,t)\cdot\frac{\partial V(\omega(t_2))}{\partial\omega(t_2)}\right], \tag{8}$$

where $\frac{\partial V}{\partial\omega}\Big|_{t=t_2} = \left[\frac{\partial V}{\partial\omega_{1\mathrm{k}}},\, \frac{\partial V}{\partial\omega_{2\mathrm{k}}},\, \frac{\partial V}{\partial\omega_{3\mathrm{k}}}\right]^{\mathrm{T}}$ – a terminant derivative on a terminal vector $\omega(t2)$.

The analytical control law for an axisymmetric solid was derived in [9]

$$u_1(t) = -k_1\Big\{\omega_1(q_1+q_2)(t_2-t) - \frac{q_1-q_2}{2\alpha}[\omega_1 \sin 2\gamma - \omega_2(\cos 2\gamma) - 1] + \omega_1 \times (m_1+m_2) + (m_1-m_2)(\omega_1 \cos 2\gamma + \omega_2 \sin 2\gamma) - 2m_1\omega_{1k}\cos\gamma + 2m_2\omega_{2k}\sin\gamma\Big\}, \tag{9}$$

$$u_2(t) = -k_2\Big\{\omega_2(q_1+q_2)(t_2-t) + \frac{q_1-q_2}{2\alpha}[\omega_2 \sin 2\gamma - \omega_1(\cos 2\gamma) - 1] + \omega_2 \times (m_1+m_2) + (m_1-m_2)(\omega_1 \sin 2\gamma - \omega_2 \cos 2\gamma) - 2m_1\omega_{1k}\sin\gamma - 2m_2\omega_{2k}\cos\gamma\Big\}, \tag{10}$$

$$u_3(t) = -k_3\Big\{\omega_3 q_3(t_2-t) + \frac{q_1-q_2}{4\alpha^2}\big[(\omega_2^2-\omega_1^2)(\sin 2\gamma - 2\gamma\cos 2\gamma) - 4\omega_1\omega_2(\cos 2\gamma + 2\gamma \sin 2\gamma - 1\big] + 2m_3(\omega_3-\omega_{3k}) + A(m_1-m_2)(t-t_2)\big[(\omega_2^2-\omega_1^2)\sin 2\gamma + 2\omega_1\omega_2 \times \cos 2\gamma\big] + 2A(t-t_2)[m_1\omega_{1k}(\omega_1\sin\gamma - \omega_2\cos\gamma) + m_2\omega_{2k}(\omega_1\cos\gamma + \omega_2\sin\gamma)]\Big\}. \tag{11}$$

In formulas (9–11) $\alpha = A\omega_3(t)$; $\gamma = A\omega_3(t)(t-t_2)$; matrices *k, q, m* have a diagonal appearance. *k* matrix terms in expressions (9–11) stand for the control system amplification coefficients. Figure 5 represents the predicting model based control device structure.

The feature of the predicting model based control is absence of the sliding mode and, therefore, ability to achieve the desired precision. Figure 6 represents a set of phase paths on the ω_1 channel, which initial conditions are switch point areas in the phase space in sliding mode (see Fig. 4). The graph shows that the whole set of phase paths in the origin of coordinates area is positioned in one line.

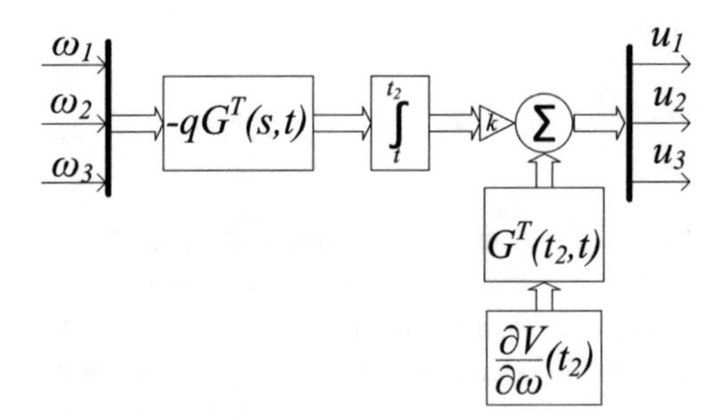

Fig. 5. The structure of the predicting model based control

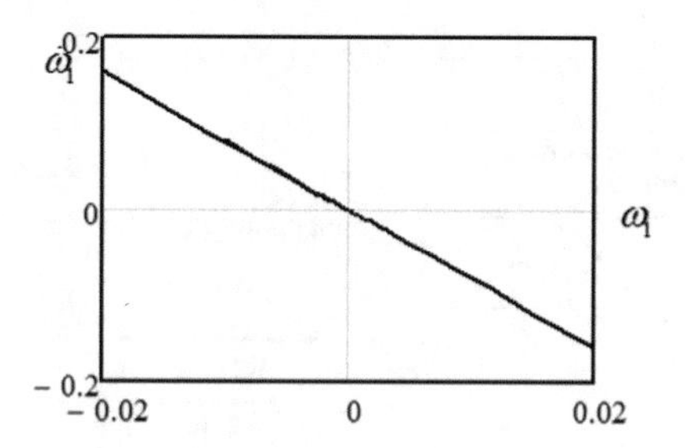

Fig. 6. A set of phase paths on ω_1 channel while using the predicting model based control (no sliding mode)

5 Fuzzy Topological Approach to a Solid Control Task

Computing experiment on researching of the joint maximum principle and the predicting model control laws results in completely different behavior in the phase space during transition between the two concepts. This results in appearance of topological manifolds having complex structure. The structures have a fuzzy ('nappy') character, and it is impossible to define the border using smooth curves, since such geometrical objects obviously have fractal character. Therefore, the transition between the joint maximum principle and the predicting model structures rightfully should be performed using the fuzzy control theory methods. It means that the fuzzy control module has to analyze the solid velocities using terms 'positive', 'negative', 'null', 'small negative', 'average negative', etc. This paper uses the double-granular control model [10]. R1: if (ω_i small), then use the predicting model. R2: if (ω_i great), then use the joint maximum principle (see Fig. 7).

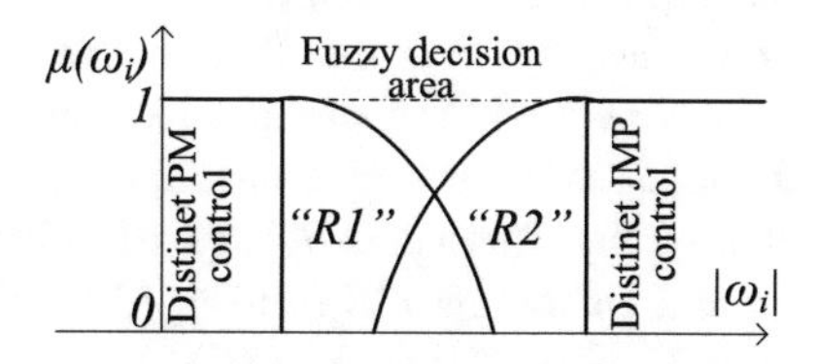

Fig. 7. Function showing belonging fuzzy decisions module over ω_i channel

Figure 8 represents structure of the combined control with fuzzy decision module based on the joint maximum principle (JMP) and the predicting model (PM).

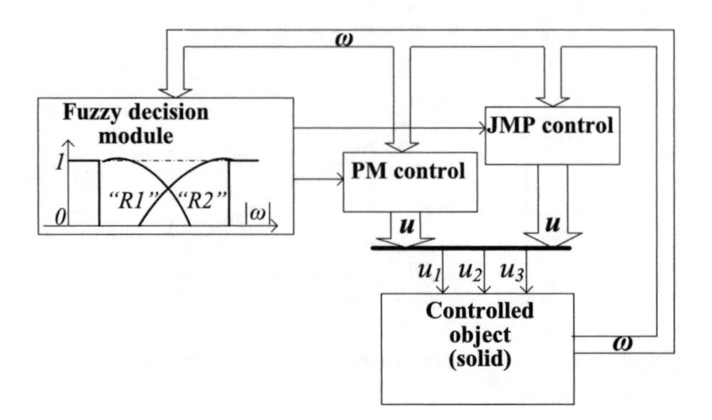

Fig. 8. Structure of the combined control based on the joint maximum principle (JMP) and the predicting model (PM)

6 Conclusion

As proved by the computing experiment, implementing varying structures control algorithms, including fuzzy control module based on the joint maximum principle and the predicting model algorithms allows to increase the effectiveness of a solid control.

For example, angular velocities average derivation in the sliding mode while using the JMP control in the tasks examined is $1.2 \cdot 10^{-2}$m/sec. Usage of the PM control decreased average derivation by more than an order. At the same time energy consumption for the varying structures control decreased for 19% on average.

The JPM control is being optimized using the operation speed criteria. However, as was shown by modeling, the sliding mode is typical for this control. First, the sliding mode does not allow achieving the desired control precision. Second, it results in unreasonable energy consumption of the controller. Using fuzzy switching in the switching point areas of the phase space at the final control stage to switch from the JMP to the PM control allows avoiding the sliding mode and, respectively, increases the control precision and decreases the energy consumption.

This work was supported by Russian Ministry of Education and Science in accordance to the Government Decree № 218 from April 9, 2010 (project number № 074-11-2018-013 from May 31, 2018 (03.G25.31.0284)).

References

1. Kostoglotov, A.A., Taran, V.N., Trofimenko, V.N.: Control algorithms adaptation based on the predictive model and the united maximum principle methods. Vestn. RGUPS **4**, 124–132 (2017)
2. Zadeh, L.A.: Calculus of fuzzy restriction. In: Fuzzy Sets and its Application to Cognitive and Division Processes, vol. 4, pp. 1–39 (1975)
3. MathWorks. https://www.mathworks.com/help/fuzzy/examples/using-lookup-table-in-simulink-to-implement-fuzzy-pid-controller.html. Accessed 10 Apr 2018

4. Tikhonravov, M.K., Yatsunsky, I.M., Maximov, G.Yu., Bazhinov, I.K., Gurko, O.V.: Bases of the theory of flight and elements of design of artificial Earth satellites. Mashinostroeniye, Moscow (1967)
5. Kostoglotov, A.A.: Method for synthesis of optimal attitude stabilization algorithm based on joint maximum principle. Autom. Control Comput. Sci. **5**(36), 21–28 (2002)
6. Kostoglotov, A.A., Lazarenko, S.V., Kuznetsov, A.A., Deryabkin, I.V., Losev, V.A.: Structural synthesis of discrete adaptive tracking systems based on the combined maximum principle. Vestn. DSTU **17**(1), 105–112 (2017)
7. Kostoglotov, A.A., Lyashchenko, Z.V., Lazarenko, S.V., Deryabkin, I.V., Manaenkova, O. N.: Synthesis of adaptive multi-mode control on basis of combined control joint maximum principle. Vestn. RGUPS **3**, 124–132 (2016)
8. Taran, V.N.: Maksimalno's a plausible assessment of a condition of optimum operated system. Autom. Equip. Telemech. **8**, 101–108 (1991)
9. Taran, V.N., Trofimenko, V.N.: Transport systems intellectualization based on analytical control synthesis of angular velocities for the axisymmetric spacecraft. In: Proceedings of the Second International Scientific Conference on "Intelligent Information Technologies for Industry" (IITI 2017), vol. 2, pp. 154–160 (2017)
10. Piegat, A.: Fuzzy Modeling and Control. Physica-Verl, Heidelberg (2001)

Time Series Grouping Based on Fuzzy Sets and Fuzzy Sets Type 2

Anton Romanov(✉) and Irina Perfilieva

Institute for Research and Applications of Fuzzy Modeling, University of Ostrava, Ostrava, Czech Republic
romanov73@gmail.com, irina.perfilieva@osu.cz

Abstract. The contribution is focused on a new method of grouping time series according to their local tendency indicator that is expressed by a linear coefficient of the F^1-transform. The useful consequence of grouping is an effective procedure of forecasting such that only one time series from a group is forecasted. Our approach for the analysis and forecasting of the time series of software development is used.

1 Introduction

Previously applied approach based on the method of F-transform to decomposition time series [8] allows to build adequate time series model. In practice there are indicators characterized by similar changes. An example of this situation is the dependence of the group of indicators in the economy of the common factor. Performing analysis to identify groups of similar time series of indicators, we can identified information for each indicator relative to this factor. The advantages of this analysis are:

1. Reducing the forecasting time in general by simplifying computing operations.
2. Identification of groups of similar processes purpose of obtaining information about the factors that grouping these processes.

This paper considers two array of time series. The first array - time series of statistics collected between 1970-ies and 2000-th in the RSFSR and Russia in the areas of economics, production, social, health, culture. The second array are time series of competition NN3.

We define the concepts that will appear in the text. *Similar time series* - the series with the linear correlation coefficient within a given threshold. *Time series group* - set of time series, which have a joint linear correlation coefficient in the specified limits.

2 Software Time Series

Another application of proposed technique is in software development. One of the factors of cost-effective activity of design organizations is the constant analysis

A. Abraham et al. (Eds.): IITI 2018, AISC 874, pp. 382–389, 2019.
https://doi.org/10.1007/978-3-030-01818-4_38

of many projects of a large project organization throughout the life cycle. The task of measuring the characteristics of the project activity should be considered as dependent on the creation of a means for project management [1]. This tool automates the clustering processes by the similarity of all available enterprise project events for the subsequent forecast of values. Project metrics are exported from the version control system.

The model of analysis and management of a set of projects in the process of project activity is developed.

$$\{C_t, R_t, B_t, I_t, F_t, R^{BI}, R^{IF}\},$$

where C_t – commits time series, R_t – release time series, B_t – bugs time series, I_t – improvement time series, F_t – new feature time series, R^{BI} – dependence of the number of bugs on improvements, R^{IF} – the dependence of new functional properties on the number of improvements (new features from improvements).

We define these types of time series for detailed structurisation of software production processes and their management. We propose use our techniques for extracting dependencies between represented types of time series.

3 Fuzzy Sets of Type 2

In most cases for fuzzy systems implementation, fuzzy sets of the first order are used. In 1975, Lotfi Zadeh presented fuzzy sets of the second order (type 2) and fuzzy sets of higher orders, to eliminate the disadvantages of fuzzy sets of type 1. These disadvantages can be attributed to the problem that membership functions are mapped to exact real numbers. This is not a serious problem for many applications, but in cases where it is known that these systems are uncertain.

The solution to the above problem can be the use of fuzzy sets of type 2, in which the boundaries of the membership areas themselves are fuzzy, which in themselves are fuzzy [16].

It can be concluded that this function represents a fuzzy set of type 2, which is three-dimensional, and the third dimension itself adds a new degree of freedom to handle uncertainties. In [16] Mendel defines and differentiates two types of uncertainties, random and linguistic. The first type is characteristic, for example, for the processing of statistical signals, and the characteristic of linguistic uncertainties is contained in systems with inaccuracies based on data determined, for example, through expert statements.

To illustrate, we note the main differences between fuzzy sets of type 1 and fuzzy sets of type 2. Let us turn to Fig. 1, which illustrates a simple triangular membership function.

Figure 1(a) shows a clear assignment of the degree of membership. In this case, to any value of x there corresponds only one point value of the membership function. If you use a fuzzy membership function of the second type, you can graphically generate its designation as an area called the footprints of uncertainty (FOU). In contrast to the use of the membership function with clear boundaries, the values of the membership function of type 2 are themselves fuzzy functions.

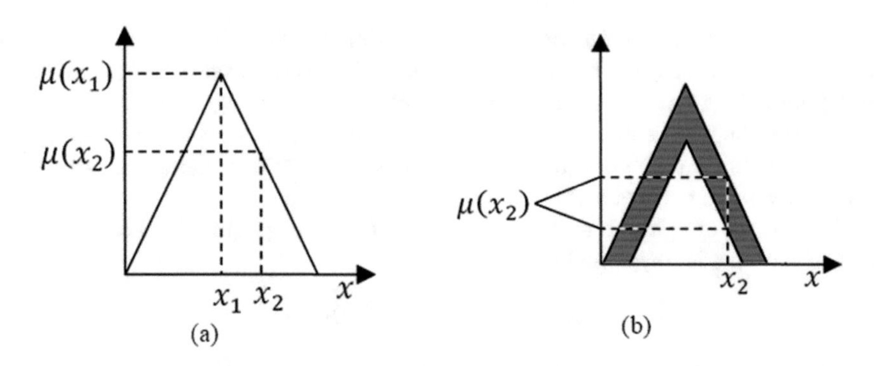

Fig. 1. The type of fuzzy sets of the 1st (a) and the 2nd (b) types.

This approach gave the advantage of approximating a fuzzy model to a verbal one. People can have different estimates of the same uncertainty. Especially it concerns estimated expressions. Therefore, it became necessary to exclude a unique comparison of the obtained value of the degree of the membership function. Thus, when an expert assigns membership degrees, the risk of error accumulation is reduced because of the non-inclusion of points located near the boundaries of the function and under doubt.

Imagine the object and show the difference in approaches, when the characteristics are not only quantitative, but also qualitative. For example, let's represent the statement typical for modeling by sets of type 1: **If the function in the program has many lines, then its decomposition is required**.

Here it is possible to construct a function of belonging to the linguistic variable **"the length of a function in rows"**. It is characterized by the values [**small, acceptable, many**]. However, in practice, the acceptable number of rows may be object clear - in the requirements for the construction of the limit specified explicitly code an acceptable value or is controlled by the rules (e.g., body function must fit on the screen). However, this option requires the involvement of additional information in the form of parameters or rules, which does not completely reflect the nuances in the model. Solving the problem through fuzzy sets of type 1 can partially reduce the gap between the model and reality. But in this case we get a share of subjectivism in the processes of fuzzification/defuzzification and deprives the possibility of a more flexible approach: the ability to take into account the greater number of uncertainties. Let's try to formulate such expressions: **If the function in the code has many lines, then its decomposition is probably required**.

If the function in the code has a lot of lines, then its decomposition is required. Here stand the fuzzy variable **"characteristic line spacing"**, which characterizes the previous fuzzy variable **"length function in the lines"** with a quality side and allows you to create new rules for the handling of the situation. This moment will be taken into account in the model in the form of a fuzzy set of type 2, where the primary variable is the **"length of the function in strings"**, and its value is a fuzzy function **"characteristic of the line interval"**.

4 F-Transform of a Higher Degree

For each time series are given fuzzy partition functions, apply F-transform zero degree (F^0-transform). To compute the first degree F-transform [7] (F^1-transform) by $F_k^1 = c_{k,0}^1 + c_{k,1}^1(x - x_k)$ is need to define $c_{k,0}^1$ and $c_{k,1}^1$. Coefficients $c_{k,0}^1$ are F^0-transform components. Coefficients $c_{k,1}^1$ are defined as:

$$c_{k,1}^1 = \frac{\sum_{t=x_k}^{x_{k+1}} f(t)(t - x_k)A_k(t)}{\sum_{t=x_k}^{x_{k+1}} (t - x_k)A_k(t)} \tag{1}$$

where $f(t)$ - time series value at time t, $A_k(t)$ - fuzzy partition functions F^1-transform at time t_k. Coefficients $c_{k,1}^1$ are equal weighted mean tangent of the piecewise linear trend of the time series. This coefficients will be the basis of time series groups analysis. F^1-transform components are vectors and describe the local trends of the time series. As an example Fig. 2 shows the time series #104 NN3 and component for the last segment of the partition.

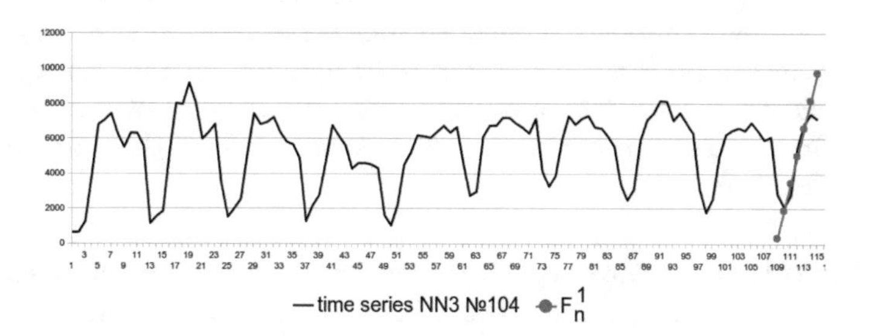

Fig. 2. F_n^1 component for time series example

5 Creating Groups of Time Series

Grouping time series will be create by analyzing coefficients $c_{k,1}^1$. For each pair of time series calculated the Pearson correlation coefficient (linear correlation coefficient):

$$r_{f_1(x),f_2(x)} = \frac{\sum_{i=1}^{m}(f_1(x_i) - \overline{f_1(x)})(f_2(x_i) - \overline{f_2(x)})}{\sqrt{\sum_{i=1}^{m}(f_1(x_i) - \overline{f_1(x)})^2 \sum_{i=1}^{m}(f_2(x_i) - \overline{f_2(x)})^2}} \tag{2}$$

where $f_1(x_i), f_2(x_i)$ are the values of the coefficients $c^1_{k,1}$ for the pair time series.

An important requirement for the analysis is equal length of the time series. Before computing components F-transform normalize time series values to the interval $[0, 1]$ in absolute value. Normalization is necessary for the similarity the range of values of the coefficients.Grouping of time series has the following problems:

1. Need to select the number of points that covers the basic function. This value directly affects the correlation of time series by coefficient $c^1_{k,1}$. Greater number of points covered by basic function, the smoother will be the time series, and correlation will be calculated on the general tendencies of the time series.
2. Setting the threshold affects the number similar to the correlation coefficient time series. It is important to determine what more necessary: include in the group more time series or select only the closest time series.

6 Application of the Approach

Consider two experiments. The first experiment performed on a set of time series statistics described above. Have been selected time series with the same length (14 time series, their lengths are 37). Known in advance that the series in this set have a similar type of behavior. The second experiment was performed on time series competition NN3 (number of time series is equal to 8, lengths - 126 points). Number of points, which covers basic function, was chosen equal 7 that F-transformation components smoothed acute fluctuations of time series. The groups of time series with correlation coefficient > 0.9 for the first experiment shown on the Table 1.

Table 1. Count similar time series with correlation coefficient > 0.9

Time series	1	2	3	4	5	6	7	8	9	10	11	12	13	14
Count similar	10	10	7	2	4	1	10	8	9	8	9	8	10	8

One of the options for forming groups according to the results - select time series that have the maximum number of similar and include in list all participating time series. For example, can select the next group of rows: $1, 2, \ldots, 14$. Samples of selected time series shown on Fig. 3.

Not included in a group of similar time series shown on the Fig. 4.

The results of the same experiment for time series NN3 in Table 2, experiment #2.

In group of similar time series are included: 101, 105, 106, 107, Fig. 5.

Not similar time series: 102, 103, 110, 111, Fig. 6.

For one of the members of the group make forecast coefficient $c^1_{k,1}$. Example consider at the group of statistical time series. Make forecast coefficient $c^1_{k,1}$ for

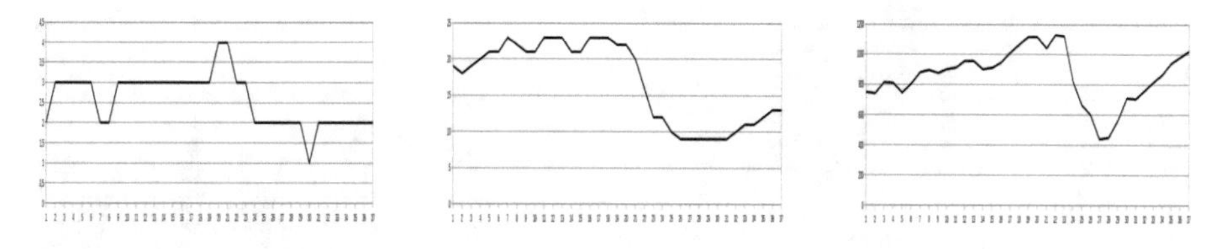

Fig. 3. Samples of grouped time series. Experiment #1.

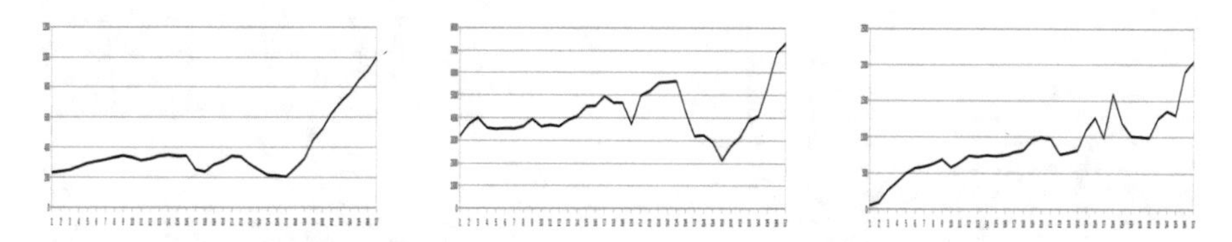

Fig. 4. Samples of not grouped time series. Experiment #1.

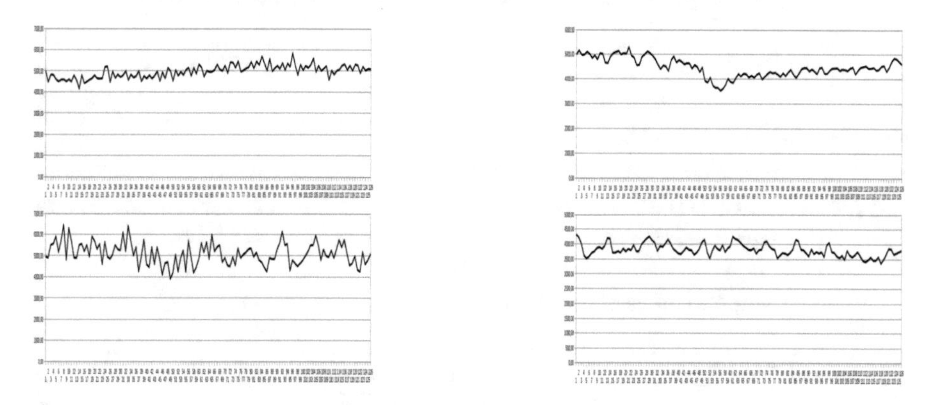

Fig. 5. Samples of grouped time series NN3. Experiment #2.

Table 2. Count similar time series with correlation coefficient > 0.9

Time series	101	105	106	107
Count similar	3	3	3	3

the time series #1 of the selected group. Now we can make a forecast for other time series group. As an example, we make forecast for one component of the second and third time series (see Fig. 7).

In this example was used existing coefficients $c^1_{k,0}$ and predictive value of $c^1_{k,1}$ of the first time series of the group. But even if the predict coefficient $c^1_{k,0}$, the local trend retains its direction. The Fig. 8 shows this case for the 2 and 3 time series of the group.

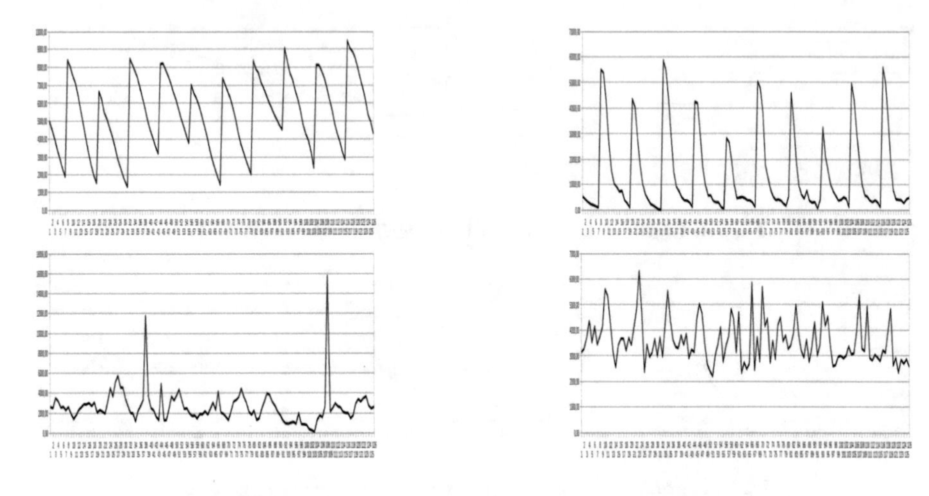

Fig. 6. Samples of not grouped time series NN3. Experiment #2.

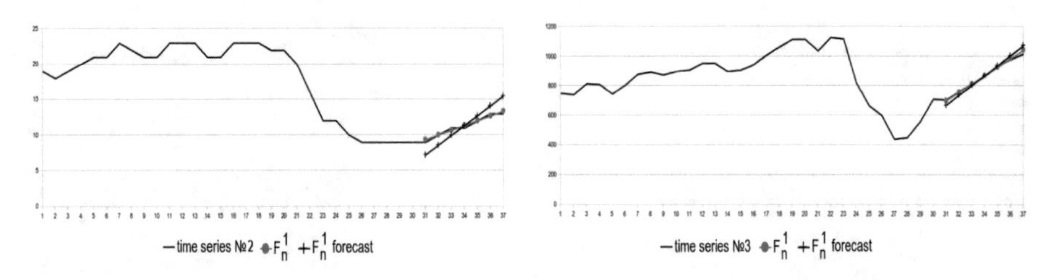

Fig. 7. Forecasting local tendencies of grouped time series

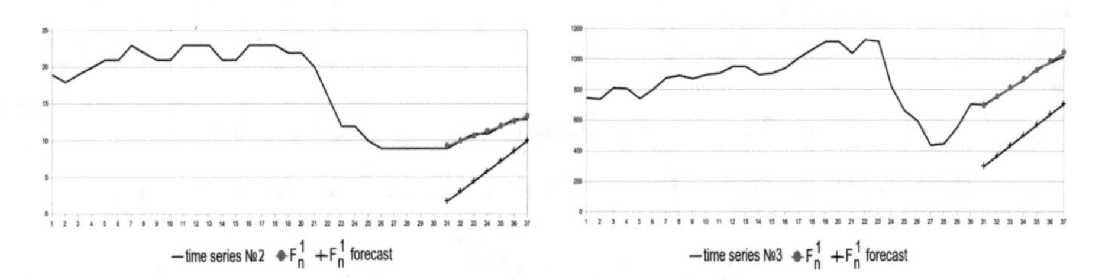

Fig. 8. Forecasting local tendencies of grouped time series

7 Conclusion

Was considered the Approach using of F-component transformation of higher degrees, namely zero and first to solve the problem of grouping and time series prediction. Experiments shown that using coefficients F^1-transform allows to combine time series having similar local tendencies. At the same time formed groups of time series allow to simplify forecasting by reducing the computational operations and make forecast for others time series in group using local trends predicted for only one time series.

Acknowledgements. The authors acknowledge that the work was supported by the framework of the state task of the Ministry of Education and Science of the Russian Federation No. 2.1182.2017/4.6 "Development of methods and means for automating the production and technological preparation of aggregate-assembly aircraft production in the conditions of a multi-product production program" and RFFI-16-47-732070.

References

1. Moshkin, V.S.: Intelligent data analysis and ontological approach in project management. In: Moshkin, V.S., Pirogov, A.N., Timina, I.A., Shishkin, V.V., Yarushkina, N.G. (eds.) Automation of Management Processes, vol. 4, no. 46, pp. 84–92 (2016). (in Russian)
2. Box, G., Jenkins, G.: Time Series Analysis: Forecasting and Control. Holden-Day, San Francisco (1970)
3. Hwang, J.R., Chen, S.M., Lee, C.H.: Handling forecasting problems using fuzzy time series. Fuzzy Sets Syst. **100**, 217–228 (1998)
4. Novák, V., Štěpnička, M., Dvořák, A., Perfilieva, I., Pavliska, V.: Analysis of seasonal time series using fuzzy approach. Int. J. Gen. Syst. **39**, 305–328 (2010)
5. Perfilieva, I.: Fuzzy transforms: theory and applications. Fuzzy Sets Syst. **157**, 993–1023 (2006)
6. Perfilieva, I.: Fuzzy transforms: a challenge to conventional transforms. In: Hawkes, P.W. (ed.) Advances in Images and Electron Physics, vol. 147. Elsevier Academic Press, San Diego (2007)
7. Perfilieva, I., Danková, M., Bede, B.: Towards a higher degree F-transform. Fuzzy Sets Syst. **180**, 3–19 (2011)
8. Perfilieva, I., Yarushkina, N., Afanasieva, T., Romanov, A.: Time series analysis using soft computing methods. Int. J. Gen. Syst. **42**(6), 687–705 (2013)
9. Sarkar, M.: Ruggedness measures of medical time series using fuzzy-rough sets and fractals. Pattern Recognit. Lett. Arch. **27**, 447–454 (2006)
10. Song, Q., Chissom, B.: Fuzzy time series and its models. Fuzzy Sets Syst. **54**, 269–277 (1993)
11. Song, Q., Chissom, B.: Forecasting enrollments with fuzzy time series. Part I. Fuzzy Sets Syst. **54**, 1–9 (1993)
12. Wold, H.: A Study in the Analysis of Stationary Time Series. Almqvist and Wiksel, Stockholm (1938)
13. Yarushkina, N.G.: Principles of the Theory of Fuzzy and Hybrid Systems. Finances and Statistics, Moscow (2004)
14. Romanov A.A., Yarushkina N.G., Perfilieva, I.: Time series grouping on the basis of F1-transform. In: IEEE International Conference on Fuzzy Systems, pp. 517–521 (2014)
15. Yarushkina, N., Afanasieva, T., Igonin, A., Romanov, A., Shishkina, V., Perfilieva, I.: Time series processing and forecasting using soft computing tools. Lecture Notes in Computer Science, vol. 6743, pp. 155–162 (2011)
16. Mendel, J.M., John, R.I.B.: Type-2 fuzzy sets made simple. IEEE Trans. Fuzzy Syst. **10**(2), 117–127 (2002)

Using the Concept of Soft Computing to Solve the Problem of Electromagnetic Compatibility Control

Vladimir Taran, Aleksey Shandybin(✉), and Elena Boyko

Rostov State Transport University, Rostov-on-Don, Russia
shav850@mail.ru

Abstract. This article proposes an intelligent electromagnetic compatibility management system based on the soft computing concept. The system is designed to improve the efficiency of the rolling stock usage. The article presents the algorithms for countercurrent generation, which are found from the condition of minimum of the mean-square error of the difference between the induced and compensating currents. The countercurrents are formed using the kernels of linear integral transformations of the first and second orders. These kernels are the result of processing large databases using soft computing.

Keywords: Soft computing · Intellectual system
Compensation of electromagnetic fields
Kernels of linear integral transformations

1 Introduction

The efficiency of the rolling stock usage is one of the component (key) metrics of the entire rail transport complex functioning. It is formed by a number of explicit and implicit factors. The fullest use of multifactor data is possible only on the basis of modern intellectual information and analytical algorithms [1, 2].

The basis for increasing the efficiency of the transport usage is the general trend related to its intellectualization. The implemented systems are designed not only to increase productivity, but to create safe working conditions and to ensure the safety of freight [3, 4]. No less important problem is to improve the electromagnetic compatibility of various radio electronic devices, as well as of the automation, telemechanics and communication systems.

The electromagnetic compatibility (EMC) is the basic concept when considering issues of the mutual influence through electromagnetic fields (EMF). Only with the use of intelligent systems it is possible to provide a level of electromagnetic interaction, when the information security will fully provide all cognitive functions.

When dealing with EMC the particular attention is paid to work with high-current or highly sensitive systems. The main examples are power lines, traction power systems for railways, medical and measuring tools.

Prospects for further development of railway transport at the moment do not imply a significant reduction of the energy consumption. Consequently, the EMC related

A. Abraham et al. (Eds.): IITI 2018, AISC 874, pp. 390–400, 2019.
https://doi.org/10.1007/978-3-030-01818-4_39

problems will only become sharper and more significant, for example in the designed Hyperloop system.

Hyperloop is the transport system of the future, designed to transport people and freight through pipelines where the pressure of about 100 Pa is maintained. A separate capsules are used as its transport units. The movement speed can reach more than 1000 km/h. This speed is due to the linear electric motors, carrying cushion and almost total absence of friction during the movement. Acceleration and braking is carried out using an aluminum rail of the length of 15 m, located on the bottom of the pipe [5].

Currently several options for using the cushion are being considered. The first is the use of a magnetic cushion. The second is based on the air cushion, which is formed due to air residues in the pipe. Their integrated use is also possible.

In all cases considered the strong electromagnetic fields will inevitably be formed. They will affect the capsule as a whole and also will penetrate into the internal space of the module. The presence of these fields will cause unwanted currents in all conductive materials. The induced effects can cause damage to the freight being carried, especially to the sensitive electronic devices. To prevent this, it is proposed to use a double capsule shell, the inner part of which should be equipped with an intelligent protection system that will compensate the induced fields. The hypothetical scheme of integration of such a device is considered in [6].

This system should detect the electromagnetic influence $I_{out}(t)$, induced in the outer shell, then process it, determine the phase and levels, and feed the signal to the inner shell in the opposite phase of $I_{comp}(t)$. Thus, the external influencing field will be compensated down to safe levels.

The article continues the authors' research [6–9]. The experimentally measured mains current spectrum of the electric locomotive 2ES6-001 (Fig. 1) is presented in the paper [9]. As a result of the analysis, the linear active compensator scheme is proposed (Fig. 2). The experimental impulse function calculation for the compensator kernel is provided (Fig. 3).

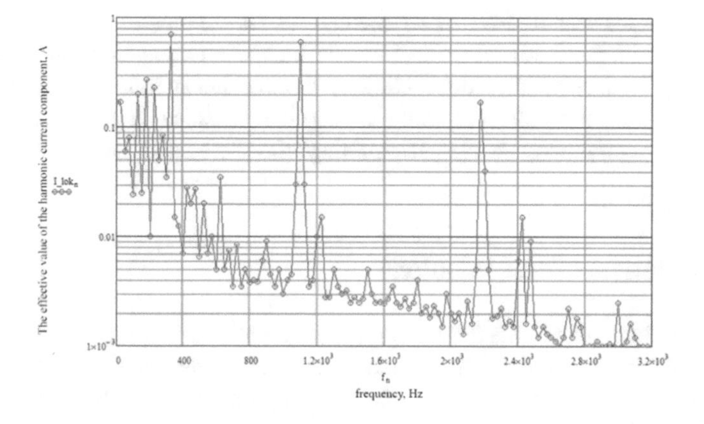

Fig. 1. Spectral density of the induced currents.

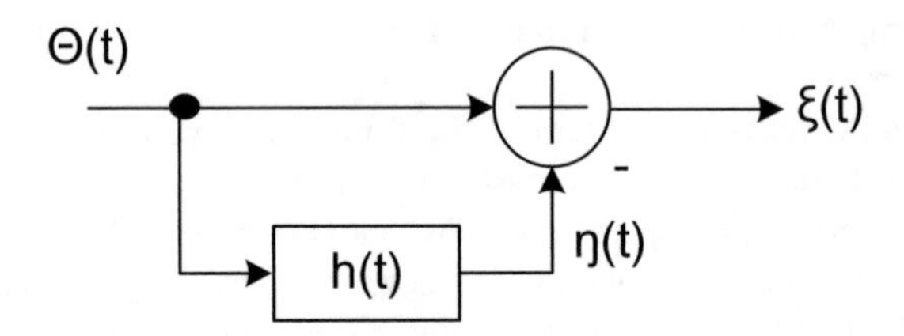

Fig. 2. Schematic diagram of the compensator

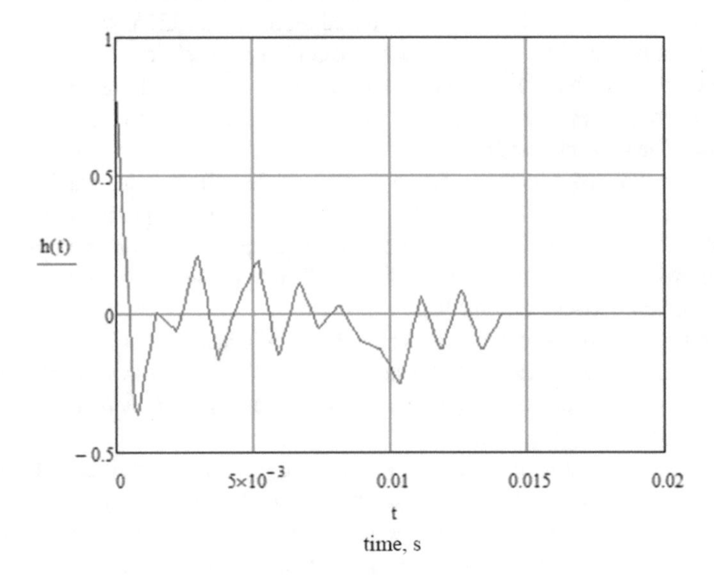

Fig. 3. The experimental impulse function

Developing the works [6–9], new structural results are proposed. Comparing to the paper [9], considerable advancement towards increasing compensation effectiveness due to taking in account the contribution of the non-linear influence components is achieved. Besides, the soft computing conception is added in order to reduce the computational complexity while determining the first- and second-order weight functions.

2 Formulation of the Problem

We represent the model of the induced currents by a stochastic process. This process is characterized by complex random phenomena and is described by corresponding correlation functions. The most complete description is given by the distribution functional [10, 11].

The problem is to compensate the induced influences. In other words, it is required to develop an intelligent device that analyzes the statistical properties of the interacting fields and synthesizes the compensating currents.

To be more specific, we put the mathematical expectation of the induced currents equal to zero, and the correlation functions existing up to the n-th order inclusive.

Let us denote the mean-square error of compensation as

$$M\left[\xi(t)^2\right] = M\left[(\Theta(t) - \eta_{\text{sum}}(t))^2\right], \tag{1}$$

where M is the mathematical expectation operator; $\Theta(t)$ is the signal of the induced influence; $\eta_{\text{sum}}(t)$ is the total compensating current (countercurrent); $\xi(t)$ is the compensation error.

Compensating devices form the currents of the first $\eta_1(t)$ and second $\eta_2(t)$ orders, which are described by following integral transformations:

$$\eta_1(t) = \int_{-\infty}^{\infty} h_1(\tau)\Theta(t-\tau)d\tau; \tag{2}$$

$$\eta_2(t) = \int_{-\infty}^{\infty}\int_{-\infty}^{\infty} h_2(\tau_1,\tau_2)\Theta(t-\tau_1)\Theta(t-\tau_2)d\tau_1 d\tau_2, \tag{3}$$

where $h_1(\tau)$ is impulse function of the first order; $h_2(\tau_1,\tau_2)$ is impulse function of the second order; t, τ, τ_1, τ_2 are the time variables, where $t \geq \tau_i$ due to the cause-effect relations.

The countercurrents of higher order are defined similarly.

Then the total countercurrent is defined by the following relation:

$$\eta_{\text{sum}}(t) = \sum\nolimits_i \eta_i(t). \tag{4}$$

The problem is to minimize the mean-square error by determining the kernels of linear, quadratic and higher order compensating devices [9].

Depending on the weight $\eta_i(t)$ the series can be limited to a few first members.

3 The Solution Method

To determine (2) and (3) we expand the functional (1) taking into account (4). Then we obtain:

$$M\left[\xi(t)^2\right] = M\left[(\Theta(t) - \eta_1(t) - \eta_2(t) - \cdots - \eta_n(t))^2\right]. \tag{5}$$

Usually the contribution of the first two terms is predominant and amounts to about 90% [12]. So we can restrict ourselves to the first two components $\eta_1(t)$ and $\eta_2(t)$. The squaring leads to the appearance of the following terms:

$$M\left[\xi(t)^2\right] = M[\Theta^2(t) - 2\Theta(t)\eta_1(t) - 2\Theta(t)\eta_2(t) + \rightarrow \\ \rightarrow + 2\eta_1(t)\eta_2(t) + \eta_1^2(t) + \eta_2^2(t)]. \tag{6}$$

We perform further calculations by each term separately. Substituting expressions (2) and (3) into (6) we obtain a system of functionals:

$$\begin{cases} \Phi_1 = K_\Theta(t,t); \\ \Phi_2[h_1] = -2\int_{-\infty}^{\infty} h_1(\tau)K_\Theta(t,t-\tau)d\tau; \\ \Phi_3[h_2] = -2\int_{-\infty}^{\infty}\int_{-\infty}^{\infty} h_2(\tau_1,\tau_2)K_\Theta(t,t-\tau_1,t-\tau_2)d\tau_1 d\tau_2; \\ \Phi_4[h_1,h_2] = 2\int_{-\infty}^{\infty}\int_{-\infty}^{\infty}\int_{-\infty}^{\infty} K_\Theta(t-\tau,t-\tau_1,t-\tau_2)h_1(\tau)h_2(\tau_1,\tau_2)d\tau d\tau_1 d\tau_2; \\ \Phi_5[h_1] = \int_{-\infty}^{\infty}\int_{-\infty}^{\infty} K_\Theta(t-\tau_3,t-\tau_4)h_1(\tau_3)h_1(\tau_4)d\tau_3 d\tau_4; \\ \Phi_6[h_2] = \int_{-\infty}^{\infty}\int_{-\infty}^{\infty}\int_{-\infty}^{\infty}\int_{-\infty}^{\infty} \begin{bmatrix} K_\Theta(t-\tau_5,t-\tau_6,t-\tau_7,t-\tau_8) * \rightarrow \\ \rightarrow * h_2(\tau_5,\tau_6)h_2(\tau_7,\tau_8) \end{bmatrix} d\tau_5 d\tau_6 d\tau_7 d\tau_8. \end{cases} \tag{7}$$

Here the correlation functions K_Θ are defined by the following relation [11]:

$$K_\Theta(t_1,t_2,\ldots,t_n) = M[\Theta(t_1),\Theta(t_2),\ldots,\Theta(t_n)].$$

To find the minimum value of the mean-square error it is necessary to find such weight functions h_1, h_2, that deliver an extremum to the functional (1). It is well known that such value can be obtained by determining the functional derivatives with respect to the corresponding arguments (functions) and equating them to zero.

Here the functional derivative is understood as the limit of the form [13–15]

$$\lim_{\substack{\Delta t \to 0 \\ \max\limits_{\tau} \delta h_1(\tau) \to 0}} \frac{\Phi[h_1 + \delta h_1] - \Phi[h_1]}{\int_t^{t+\Delta t} \delta h_1(\tau)d\tau} = \frac{\delta\Phi[h_1]}{\delta h_1}. \tag{8}$$

Specifically to the considered example the derivatives of (7) with respect to h_1 take the form:

$$\begin{cases} \frac{\delta\Phi_1}{\delta h_1} = 0; \\ \frac{\delta\Phi_2[h_1]}{\delta h_1} = -2K_\Theta(t,t-\tau); \\ \frac{\delta\Phi_3[h_2]}{\delta h_1} = 0; \\ \frac{\delta\Phi_4[h_1,h_2]}{\delta h_1} = 2\int_{-\infty}^{\infty}\int_{-\infty}^{\infty} K_\Theta(t-\tau,t-\tau_1,t-\tau_2)h_2(\tau_1,\tau_2)d\tau_1 d\tau_2; \\ \frac{\delta\Phi_5[h_1]}{\delta h_1} = 2\int_{-\infty}^{\infty} K_\Theta(t-\tau_3,t-\tau_4)h_1(\tau_3)d\tau_3; \\ \frac{\delta\Phi_6[h_2]}{\delta h_1} = 0. \end{cases} \tag{9}$$

Combining the results we obtain the derivative of the functional (6) with respect to the first-order kernel. In accordance with the stated above the expression obtained must be equated to zero. Finally we get the equation:

$$\begin{gathered}\int_{-\infty}^{\infty} K_{\Theta}(t-\tau_3, t-\tau_4) h_1(\tau_3) d\tau_3 + \rightarrow \\ \rightarrow + \int_{-\infty}^{\infty} \int_{-\infty}^{\infty} K_{\Theta}(t-\tau, t-\tau_1, t-\tau_2) h_2(\tau_1, \tau_2) d\tau_1 d\tau_2 = K_{\Theta}(t, t-\tau).\end{gathered} \tag{10}$$

The functional derivatives for a second-order kernel are determined similar to (8) by prolongation onto the two-dimensional Hilbert space:

$$\lim_{\substack{\Delta T \to 0 \\ \Delta t \to 0 \\ \max\limits_{\tau_1,\tau_2} \delta h_2(\tau_1,\tau_2) \to 0}} \frac{\Phi[h_2 + \delta h_2] - \Phi[h_2]}{\int_T^{T+\Delta T} \int_t^{t+\Delta t} \delta h_2(\tau_1, \tau_2) d\tau_1 d\tau_2} = \frac{\delta \Phi[h_2]}{\delta h_2}. \tag{11}$$

Applying the algorithm (11) to the functionals (7) we obtain the system:

$$\begin{cases} \frac{\delta \Phi_1}{\delta h_2} = 0; \\ \frac{\delta \Phi_2[h_1]}{\delta h_2} = 0; \\ \frac{\delta \Phi_3[h_2]}{\delta h_2} = -2K_{\Theta}(t, t-\tau_1, t-\tau_2); \\ \frac{\delta \Phi_4[h_1,h_2]}{\delta h_2} = 2\int_{-\infty}^{\infty} K_{\Theta}(t-\tau, t-\tau_1, t-\tau_2) h_1(\tau) d\tau; \\ \frac{\delta \Phi_5[h_1]}{\delta h_2} = 0; \\ \frac{\delta \Phi_6[h_2]}{\delta h_2} = 2\int_{-\infty}^{\infty} \int_{-\infty}^{\infty} K_{\Theta}(t-\tau_5, t-\tau_6, t-\tau_7, t-\tau_8) h_2(\tau_5, \tau_6) d\tau_5 d\tau_6. \end{cases} \tag{12}$$

For the second-order kernel we obtain equation of the following form:

$$\begin{gathered}\int_{-\infty}^{\infty} K_{\Theta}(t-\tau, t-\tau_1, t-\tau_2) h_1(\tau) d\tau + \rightarrow \\ \rightarrow + \int_{-\infty}^{\infty} \int_{-\infty}^{\infty} K_{\Theta}(t-\tau_5, t-\tau_6, t-\tau_7, t-\tau_8) h_2(\tau_5, \tau_6) d\tau_5 d\tau_6 = \rightarrow \\ \rightarrow = K_{\Theta}(t, t-\tau_1, t-\tau_2).\end{gathered} \tag{13}$$

Applying the Kotelnikov theorem and omitting the intermediate transformations, we obtain the discrete analogs of the corresponding Eqs. (10) and (13):

$$\sum_{i=0}^{t} A_{t,i,j} \cdot x_i + \sum_{l=0}^{t} \sum_{m=0}^{t} B_{t,k,l,m} \cdot y_{l,m} = C_{t,k}; \tag{14}$$

$$\sum_{k=0}^{t} B_{t,k,l,m} \cdot x_k + \sum_{n=0}^{t} \sum_{p=0}^{t} D_{t,n,p,q,r} \cdot y_{n,p} = E_{t,l,m}, \tag{15}$$

where x_i, x_k are the discrete analogs of the first order kernel, in particular $x_i = h_1(\Delta \cdot i)$, the other quantities similarly; $y_{l,m}$, $y_{n,p}$ are respectively the discrete analogs of the second order kernel; $A_{t,i,j}$, $B_{t,k,l,m}$, $C_{t,k}$, $B_{t,k,l,m}$, $D_{t,n,p,q,r}$, $E_{t,l,m}$ are the respective analogs of the correlation functions.

Analysis of algebraic Eqs. (14) and (15) shows that the given system is overdetermined, that is, the number of unknowns is substantially less than the number of equations. Therefore, this problem is ill-posed.

To regularize this system, we apply the method of minimum deviation of the solution from the averaged right-hand side. To do this, it is necessary to determine the square of the difference between the left and right sides of Eqs. (14) and (15), followed by summation over free indices *t, j, k, l, m, q, r*.

Thus, the problem can be reduced to solving a linear system of algebraic equations of larger dimension.

4 The Concept of Soft Computing

To make the concept of soft computations more concrete [16], we consider systems of Eqs. (14) and (15) reduced to the standard form.

This system of linear equations is characterized by a large number of variables, in particular, for Eqs. (14) and (15) we have $N + N^2$ unknowns

$$\begin{aligned} a_{11}x_1 + a_{12}x_2 + \cdots a_{1n}x_n &= b_1; \\ a_{21}x_1 + a_{22}x_2 + \cdots a_{2n}x_n &= b_2; \\ \cdots & \\ a_{n1}x_1 + a_{n2}x_2 + \cdots a_{nn}x_n &= b_n, \end{aligned} \tag{16}$$

where $x_1 \cdots x_N$ are the first order kernels (analogues of x_i and x_k from Eqs. (14) and (15)); $x_{N+1} \cdots x_n$ are the second order kernels (analogues of $y_{l,m}$ and $y_{n,p}$ from Eqs. (14) and (15)); $a_{11} \cdots a_{nn}$, $b_1 \cdots b_n$ – are the respective analogs of the correlation functions from Eqs. (14) and (15).

It is required to find a solution using the methodology of purposeful systems [17, 18].

Let us introduce two n-dimensional spaces. We call the space **X** the solution space with the usual Euclidean metric ρ.

Similarly, we introduce a space **Y** with the same metric and call it a space of congruences. This space is formed by the vectors of the form

$$\begin{aligned} y_1 &= a_{11}x_1 + a_{12}x_2 + \cdots a_{1n}x_n; \\ y_2 &= a_{21}x_1 + a_{22}x_2 + \cdots a_{2n}x_n; \\ \cdots & \\ y_n &= a_{n1}x_1 + a_{n2}x_2 + \cdots a_{nn}x_n. \end{aligned} \tag{17}$$

In the solution space, we randomly define the first set of vectors $X_1^{(1)} X_2^{(1)}, \ldots, X_M^{(1)}$. $X^{(1)} \in \mathbf{X}$.

Similarly, in accordance with (16) and (17), we define the first set of points in the space of congruences $Y_1^{(1)} Y_2^{(1)}, \ldots, Y_M^{(1)}$. $Y^{(1)} \in \mathbf{Y}$.

The space of congruences should undoubtedly be associated with the goal. Namely, comparing the current state with the goal, we can develop mechanisms developed to achieve the goal.

Estimates of the mathematical expectation and roof-mean-square deviation will have the form

$$m_x^* = \frac{1}{M}\sum\nolimits_{m=1}^{M} X_m^{(1)}; \tag{18}$$

$$\left(\delta_x^*\right)^2 = \frac{1}{M-1}\sum\nolimits_{m=1}^{M} (X_m^{(1)} - m_x^*)^2. \tag{19}$$

Similarly, we define the corresponding parameters in the space of congruences

$$m_y^* = \frac{1}{M}\sum\nolimits_{m=1}^{M} Y_m^{(1)}; \tag{20}$$

$$\left(\delta_y^*\right)^2 = \frac{1}{M-1}\sum\nolimits_{m=1}^{M} (Y_m^{(1)} - m_y^*)^2. \tag{21}$$

We call the vector $B^T = \{b_1, b_{2,} \ldots, b_n\}$ "goal" and determine the distance between the goal and the mean value m_y^*, $\rho\left(m_y^*, B\right)$. In addition, we will define a point Y_K ranged from the goal for a minimum distance

$$Y_K = arg(\min_m \rho(Y_m, B)), \quad m \in [1, M], \tag{22}$$

and it is evident that

$$\rho\left(m_y^*, B\right) \geq \rho(Y_m, B) \tag{23}$$

It is quite clear that single-valued correspondences (isomorphisms) are established between points $X_1^{(1)} X_2^{(1)}, \ldots, X_M^{(1)}$ and points $Y_1^{(1)} Y_2^{(1)}, \ldots, Y_M^{(1)}$.

For any index from the first series of vectors $X_K^{(1)}$ it is possible to find the vector $Y_K^{(1)}$ and vice versa, based on vector $Y_j^{(1)}$ one can specify the vector $X_j^{(1)}$.

This is due to the fact that when we generate the random points, they are bound to the indices as they appear.

Between the dispersion in the solution space $\left(\delta_x^*\right)^2$ and the corresponding quantity in the space of congruences $\left(\delta_y^*\right)^2$ we can determine the proportionality factor

$$K^2 = \frac{\left(\delta_x^*\right)^2}{\left(\delta_y^*\right)^2}. \tag{24}$$

Now we use the Chebyshev inequality

$$P\left\{Y_j \not\subset \bigcup\nolimits_{\rho\left(m_y, Y_i\right) < \alpha} Y_i\right\} \leq \frac{D}{\alpha^2}, \tag{25}$$

where $\alpha = \rho\left(m_y^*, Y_K\right)$ is the hypersphere with the center m_y^*; $\bigcup_{\rho\left(m_y,Y_i\right)<\alpha} Y_i$ is the set of points inside the hypersphere with the center at the point m_y^*; $Y^{(1)} \setminus \bigcup_{\rho\left(m_y,Y_i\right)<\alpha} Y_i$ is the set of points outside the hypersphere; D is the corresponding sample variance.

From the analysis of this inequality, it follows that there is a non-zero probability of the appearance of a random point X_ρ, and because of isoformism the point Y_ρ such that the distance from the goal B to the point Y_ρ is less than $\rho(Y_K, \mathrm{B})$.

In other words, if we move the distribution parameter m_x^* to the point X_K i.e. $m_x^* = X_K$ then $m_Y^* = Y_K$ is changes as well. Then we can perform again statistical tests of the vectors $X_1^{(2)} X_2^{(2)}, \ldots, X_M^{(2)}$.

Unlike the previous series, the selective mathematical expectation will already be located closer to the solution than the previous ones. This conclusion is true for the following reasons.

First, the mathematical expectation in the space of congruences m_Y^* is closer because of the choice made above.

Secondly, due to the Chebyshev inequality the probability of appearance of the point closer to the goal will only increase, since the mathematical expectation is closer to the goal.

Let the nearest point in the series $Y_1^{(2)} Y_2^{(2)}, \ldots, Y_M^{(2)}$ is Y_P.

Thus, step by step, moving in the solution space, we come to solution fulfilling at the step Q the inequality $\rho\left(X_q^{(Q)}, X^*\right) \leq \varepsilon$ which follows from the analogous inequality in the space of congruences $\rho\left(Y_q^{(Q)}, B\right) \leq \delta$.

Computational experiments with sufficiently small dimensions of the problem showed satisfactory results, which allowed to obtain a certificate of state registration of the computer program [19].

5 The Discussion of the Results

Equations (10) and (13) show that the first and second order kernels should be determined simultaneously.

In the process of functioning of the intelligent compensation system it is necessary to process statistical information on the currents for determining the multipoint correlation functions.

The structural diagram of the protection system is presented in Ref. [6].

The structure includes: the detector of electromagnetic fields induced on the outer shell; the intelligent compensating signal generators working according to the algorithms: (2) - the first-order compensation device CC 1 and (3) - the second-order compensation device CC 2; the adder; the generator-converter of the compensation currents.

The device will perform statistical processing of the electromagnetic environment in order to intelligently adjust the values of $h_i(t)$. For this purpose the algorithm given in [7, 8] can be used. The dimensionality of the system (17) is extremely large and therefore it is not possible to perform full scale computational experiment because of the limited computing resources. The possibility to implement the intelligent system of active compensation of the inducings was substantiated using the example of a system with a first-order kernel [9]. The realizability of systems with higher-order kernels can be confirmed using the Big Data calculation techniques.

6 Conclusion

To successfully design and operate the proposed system it is necessary to fully use the presentation models and methods of obtaining knowledge. They will allow to extract information about the required correlation functions as a result of the intelligent analysis of the measured data using the concept of soft computing.

The software of the proposed intelligent system can be built using modern hybrid, fuzzy and granular calculations, which in some cases will significantly reduce the amount of calculations and increase the speed of reflection of the entire system.

References

1. Pearl, J.: Probabilistic Reasoning in Intelligent Systems: Networks of Plausible Inference. Morgan Kaufmann, San Francisco (1988)
2. Chinakal, V.O.: Intellectual systems and technologies. Workbook, RUDN, Moscow (2008)
3. Gavrilov, A.V.: Hybrid Intelligent Systems. NGTU, Novosibirsk (2002)
4. Makarov, I.M., Lokhin, V.M., Man'ko, S.V., Romanov, M.P.: Artificial Intelligence and Intelligent Control Systems. Nauka, Moscow (2006)
5. «Hyperloop». Wikipedia, The Free Encyclopedia. https://en.wikipedia.org/wiki/Hyperloop. Accessed 16 Feb 2018
6. Taran, V.N., Kulbikayan, Kh.Sh., Shandybin, A.V.: Method of suppression of electromagnetic fields in pipeline transport systems. In: Proceedings of the All-Russian National Scientific and Practical Conference "Modern Development of Science and Technology" (Science 2017), pp. 116–118. RGUPS, Rostov-on-Don (2017)
7. Kulbikayan, Kh.Sh., Taran, V.N., Shandybin, A.V., Kulbikayan, B.Kh.: Estimation of the probability of the density of induced currents in the shells of cable communication networks. RGUPS Bull. **4**(60), 41–48 (2015)
8. Shandybin, A.V., Taran, V.N., Kulbikayan, Kh.Sh., Kulbikayan, B.Kh.: Certificate of state registration of the computer program #2015660049. Program for the implementation of the algorithm for estimating the probability density and the distribution function by the criterion of a minimum of the quality functional, 21 September 2015
9. Shandybin, A.V.: Method of active compensation of induced currents. Izvestia of SFedU. Techn. Sci. **11**(172), 109–119 (2015)
10. Levin, B.R.: Theoretical Foundations of Statistical Radio Engineering, 3rd edn. Radio and Communications, Moscow (1989)
11. Mandel, L., Wolff, E.: Optical Coherence and Quantum Optics. Nauka, Moskow (2000). Transl. from English, under sup. of V.V. Samartsev
12. Pankratov, E.L., Bulaeva, E.A.: Rows. Educational-Methodical Guide, Nizhny Novgorod State University, Nizhny Novgorod (2015)
13. Dubrovin, B.A., Novikov, S.P., Fomenko, A.T.: Modern Geometry: Methods and Applications. Nauka, Moskow (1979)

14. Taran, V.N.: Functional equation of a long line. Radio Eng. Electron. **36**, 1497 (1991)
15. Tabor, M.: Chaos and Integrability in Nonlinear Dynamics. Editorial URSS, Moscow (2001)
16. Kumar, P., Singh, Asheesh K.: Soft computing techniques for optimal capacitor placement. In: Complex System Modelling and Control Through Intelligent Soft Computations. STUDFUZZ, vol. 319, pp. 597–625 (2015)
17. Ackoff, R., Emery, F.O.: On Purposeful Systems. Sov. Radio, Moskow (1974). Transl. from English, under sup. of I.A. Ushakov
18. Fletcher, R.: A new approach to variable metric algorithms. Comput. J. **13**(3), 317–322 (1970). https://doi.org/10.1093/comjnl/13.3.317
19. Taran, V.N., Boyko, E.Yu., Dolzhenko, A.M.: Certificate of state registration of the computer program #2017611577. Program for the implementation of the metric algorithm for solving systems of linear algebraic equations, 06 February 2017

Generation of Efficient Cargo Operation Schedule at Seaport with the Use of Multiagent Technologies and Genetic Algorithms

Olga Vasileva[1(✉)] and Vladimir Kiyaev[2]

[1] Saint-Petersburg State University, Saint-Petersburg, Russia
vasiljevaa@mail.ru
[2] Saint-Petersburg State University of Economics, Saint-Petersburg, Russia

Abstract. Seaport is a complex economic techno - technological facility. In the modern meaning of this concept it is a specially built and equipped enterprise on the coast, designed for sheltering, loading/unloading and servicing of ships. Information support of such object's operations is a rather difficult information-computational task for many reasons - the main one being the generation of an efficient schedule for loading and unloading operations. For these reasons, there is a need to develop an automated system capable of generating an optimal schedule of loading/unloading in a seaport, taking into account the dynamic effect of external factors on its efficient operation. The approach to scheduling is based on multi-agent technologies and a genetic algorithm for generating a schedule for port operation improving the quality of the obtained schedule. It is supposed to use the data received from different agents and corrected during the interaction of such agents. The scheduling is carried out with the genetic algorithms.

Keywords: Seaport work schedule · Multiagent technologies
Genetic algorithms

1 Introduction

The problem of optimizing the seaport's work schedule has a long history, and acquired special popularity in the 1980's. A huge number of scientific researches were devoted to the optimal arrangement of the shipping fleet in the continuous planning framework [8, 9]. These researches also included some guidelines for development of an interaction schedule between different transport types in the seaport [10]. However, with the development of technological progress and the growth of factors' number, that are affecting the port's work, the task of scheduling becomes more complicated and time consuming. For these reasons, the need to consider new methods to optimize the performance of the port has been occurred.

Seaport is a complex economic techno- technological facility. In the modern meaning of this concept it is a specially built and equipped enterprise on the coast, designed for sheltering, loading/unloading and servicing of vessels. Information support of such object's operations is a rather difficult information-computational task for many reasons - the main one being the generation of an efficient schedule for loading

A. Abraham et al. (Eds.): IITI 2018, AISC 874, pp. 401–409, 2019.
https://doi.org/10.1007/978-3-030-01818-4_40

and unloading operations. Due to optimal work schedule, correct allocation of resources and continuous monitoring of operations execution, the probability of successfully achieving set goals increases [1, 5].

In the present paper the schedule of the seaport activities means the dynamic schedule of loading/unloading and logistics operations needed for cargo handling in the port. Non-availability of idle time in the port's work imposes significant limitations on the schedule generation. Considering that the information on incoming ships is provided on an ongoing basis, there is a need for continuous work on schedule generation with an increasing planning horizon. Therefore, the manual compilation and current adjustment of the schedule is an extremely time-consuming task, meaning that the schedule should ensure the most efficient port operation by the principle of Just-in-Time and eliminate delays in loading/unloading, discrepancy between the weight and volume of delivered containers and the capacity of transport units (trucks and railway platforms), suboptimal routes in the port territory, address errors in the delivery of goods, etc. [2–4].

For these reasons, there is a need to develop an automated system capable of generating an optimal schedule of loading/unloading in a seaport, taking into account the dynamic effect of external factors on its efficient operation. The approach to scheduling is based on multi-agent technologies and a genetic algorithm for improving the quality of the obtained schedule. It is supposed to use the data received from different agents and corrected during the interaction of such agents, to generate a schedule for port operation. The scheduling is carried out with the genetic algorithms [11].

2 Agent Models

The following meta-agents are supposed to be used in the port scheduling:

- Vessel. This agent is a system of agents tracking various indicators, such as: fuel stock, drinking water reserve, ship draft, on-board situation, ship's course, weather conditions, availability of space and the possibility of placing additional cargo.
- Port. This is a set of agents including the agents monitoring the condition of port cranes, refuelers of fuel and water, free areas in warehouses.
- Transport. It is a group of agents with the task of monitoring the performance of road and rail transport involved in loading/unloading vessels.

The description of some agents used in the algorithm are presented as an example.

The ship's agent. This agent is an aggregator of information coming from agents performing constant monitoring of the ship's condition. Based on this information, the agent warns the captain about the occurrence of potential risks in the vessel operation and performs the functions that are taken for consideration in this task - forwards information to the port about the required operations performed with the vessel. The algorithm of the agent:

```
procedure ShipInfo
tm = current time;
ld = false; // free space on the ship
maxld = maximum cargo sizerequired for unloading at the
port;
maxt = maximum time available for unloading in port;
track = ship location;
ft = condition of fuel tanks;
wt = condition of drinking water tanks;
shdep = aggregate risk indicator for the vessel, related
to weather conditions, draft and airborne instrument
readings;

function ReportIn
track = current GPS sensors indication;
ld = current loading sensors indication;
ft = current fuel tanks indication;
wt = current drinking water tanks indication;
maxld = calculation of the maximum size of the cargo to
be unloaded at the nearest port;
shdep = calculation of risks based on ship's sensors
if shdep > maximum level of risk
Send a SOS message
end if
end function

function ReportOut
if ft < the required amount of fuel that is needed to get
to the port, which should enter the vessel after visiting
the nearest point of the route +10%
then
Message about the required refueling
end if
if wt < the required amount of drinking water that is
needed to get to the port, which should enter the vessel
after visiting the nearest point of the route +10%
then
Message about the required refueling
end if
Message with the information about maxld, maxt
end function

repeat
every 30 minutes ReportIn
if Distance to port < 2 days
then every 5 hours ReportOut
end if
until
Vessel entered port
end procedure
```

The exchange of information between these agents results in the data adjustment [6]. Then the data is transferred to the schedule calculation module generating the schedule based on the genetic algorithm.

3 General Elements of the Port Scheduling Algorithm

During the analysis of the port's schedule requirements, the following basic principles of algorithm construction are identified [7]:

- the possibility of expanding the list of requirements for the port schedule;
- the ability of regulating the priorities for meeting specific requirements when generating schedules;
- the possibility of constantly adding the information about the need for extraordinary servicing of certain vessels.

The basic principle of the algorithm is to assess the freedom of ship's approach to some terminals of the port. Vessels, for unloading/loading which require, for example, heavy-duty port cranes, can be serviced only in a limited number of terminals. Vessels, for unloading/loading of which railway transport is required, require servicing in terminals to which railroads are conducted - otherwise substantial resource costs will arise for the movement of goods inside the port. Thus, these parameters affect the freedom of vessels' approach to some terminals of the port.

It is advisable to start forming the port's work schedule with the addition of servicing of ships, the requirements for unloading which have the greatest number of restrictions.

The steps of this algorithm.

1. Forming a system of importance scales of certain parameters for assessing the performance of port services.
2. Information on ships planning to unload in the port. This information includes the data on the port arrival time, the waiting period in the roadstead, the volume and number of cargo units (containers), the nature of the cargo, the type of land transport that this cargo should be taken out of the port, the amount of cargo that must be temporarily transferred to the port warehouses, requirements for maintenance of the vessel (the amount of water, fuel, food required for refueling).
3. Based on the processing of the information received, an estimate is made of the time necessary for the optimum servicing of the vessel in the port.
4. To form a time estimate, let us consider the set of parameters that will be used.

$$A = \{Z\{N, V_i, N_i(t)\}, T(R, R(t), L, L(t)), S(F, F(t), W, W(t), FS, FS(t))\}$$

where A – terminal's parameters, $Z\{N, V_i, N_i(t)\}$ – set of port cranes with a separation by load capacity, N – number of port cranes at the terminal, V_i – maximum permitted load capacity of i crane, $N_i(t)$ – throughput of i crane, $T(R, R(t), L, L(t))$ – loading/unloading possibility by various types of land transport, where $R, R(t)$ – availability and capacity of railway transport at the terminal, $L, L(t)$ – availability and capacity of freight transport at the terminal, $S(F, F(t), W, W(t), FS, FS(t))$ – possibility of vessel servicing at the terminal, where $F, F(t)$ – availability and speed of refueling the vessel with fuel at the terminal, $W, W(t)$ – availability and speed of filling the vessel with water at the terminal, FS, $FS(t)$ – availability and speed of fueling the vessel with food at the terminal.

$$B = \{D, D(t), C\},$$

where B – parameters of cargo movement in port, D – free space in the warehouse, $D(t)$ – time of cargo movement from the warehouse when loading/unloading the vessel, C – a set of customer-recipients of the cargo, ready to pick it up without using a port warehouse.

$$G = \{G_d, G_u, S_s\{F_s, W_s, FS_s\}, P\},$$

where G – service parameters of the vessel entering the port, G_d – the amount of cargo to be unloaded, G_u – the amount of cargo to be loaded, $S_s\{F_s, W_s, FS_s\}$ – vessel's service parameters, F_s – amount of fuel, that should be loaded, W_s – amount of drinking water, that should be loaded, FS_s – the amount of food, that should be loaded, P – priorities of the ship's cargo.
The considered parameters are basic for generating temporary parameters of the vessel's service. For generating these parameters let's consider the general parameter T that estimates the time of the vessel's stay in the port.

$$T = \begin{cases} T_1 + T_2, \ k = 0 \\ \max(T_1, T_2), \ k = 1 \end{cases}$$

where T_1 – estimation of time expenditure for loading and unloading a vessel, T_2 – estimate of the time expenditure for servicing the vessel in port, k – a services compatibility parameter that takes on a value from the set of $\{0, 1\}$, with $k = 0$ it is impossible to simultaneously combine loading/unloading work on the vessel with its servicing, whereas with $k = 1$, these procedures may be combined.

$$T_1 = \frac{(K_{lr} + K_{sr})R(t)}{R} + \frac{(K_{ll} + K_{sl})L(t)}{L} + \frac{(K_{ls} + K_{ss})D(t) + K\sum_{i=1}^{M} N_i(t)}{M},$$

where K_{lr}, K_{sr} – the quantity of goods required for loading and unloading, which will be handled by means of railway transport, K_{ll}, K_{sl} – the quantity of goods required for loading and unloading, which will be handled by means of freight transport, K_{ls}, K_{ss} – the quantity of goods required for loading and unloading, which logistics are related to the warehouse capacities of the port, K – total amount of cargo to be loaded and unloaded from the vessel, i.e.

$K = K_{lr} + K_{sr} + K_{ll} + K_{sl} + K_{ls} + K_{ss} = G_d + G_u$, M – the number of port cranes suitable for unloading a ship that has come to be unloaded in a terminal.

$$T_2 = \begin{cases} F_s F(t) + W_s W(t) + FS_s FS(t), \ j = 0 \\ \max(F_s F(t), W_s W(t), FS_s FS(t)), \ j = 1 \end{cases},$$

where j – the compatibility parameter of the ship's service processes, which takes a value from the set $\{0, 1\}$, with $j = 0$ it is impossible to simultaneously combine loading and unloading work on a ship and servicing of the vessel, and with $j = 1$ these procedures may be combined.
It should be taken into account that when generating the port's working schedule, a set $\{T\}$ is generated for each vessel – estimates of the ship's stay in the port time, while servicing the vessel at various terminals, if their characteristics are significantly different.

5. After the assessment of the servicing time of the vessel in the port, an estimate of the location latitude for the vessel's service in the schedule is generated. At the same time, it is necessary to take into account the fact of the continuity of the port operation, the constantly incoming information about the ships and the incoming information about the consignees. Thus, it is advisable to consider two sets of ships - the set $S_1 = \{s_i | s_i \in G\}$ – a set of ships that are already on the port's schedule, and $S_2 = \{s_i | s_i \in G\}$ – a set of ships, information about which has already been received, but they are not yet included in the schedule.
The port's schedule is generated for 14 days. It is made on the basis of set S_2 when building an assessment of the location latitude for vessel's service in the schedule. Such an estimate can be represented in the form

$$Y_i = \frac{T_i}{\max\limits_{S_2}(T_i)},$$

where Y_i – assessment of the location latitude of ship's service in the schedule, T_i – estimate the time of stay of the i ship in port.

6. After assessing the latitude of i ship's service location in the schedule, the list of vessels S_2 is sorted from the set in ascending order of their estimates of location latitude $Y_i \leq Y_{i+1}, i = 1..|S_2|$. Then, the resulting set of estimates is divided into subsets according to the priority ranking of the goods P.
7. After sorting, the ships with the lowest latitude estimate and the highest cargo priority are added to the schedule in the first place, taking into account the following factors:

- the terminal has cranes of the required load-carrying capacity;
- the terminal has the necessary infrastructure for transportation and maintenance of the vessel;
- the terminals do not overlap.

If the mandatory conditions are met, an assessment of the quality of the vessel's service location in the schedule is taken according to the following criteria:

- occurrence of unloading time (windows) in the operation of terminals;
- redundancy of loading and unloading capacities of the terminal in relation to the needs of the vessel;
- carrying out loading and unloading works at night;
- the possibility of combining work on vessel maintenance and loading and unloading operations;
- speed of loading and unloading operations;
- the disappearance of windows in the terminals work;
- the absence of idle terminals in the presence of vessels requiring loading and unloading operations.

Each of the criteria for assessing the quality of the vessel's service location in the schedule is implemented as a separate specialized function, which makes it easy to add new schedule criteria.

8. To obtain an estimate of the quality of the vessel's service location, we use:

$$Q_{il} = \sum_{j=1}^{h} w_j k_{jl},$$

where Q_{il} – service quality score of i ship on the position l in schedule, w_j – weight coefficient of j assessing quality criteria, k_{jl} the value obtained by j assessing quality criteria of vessel's service location in the l position in the schedule, h – number of quality assessment criteria.

After assessing the quality of all possible options for the location of vessel maintenance in the schedule, the option is chosen in which the maximum value of the location quality assessment is reached:

$$Q_i = \max_{l=1..m}(Q_{il}),$$

where Q_i – service quality indicator of the i ship in the schedule, m – number of possible options for the location of the vessel's service in the schedule.

After generating the schedule, the quality of the schedule is assessed. To assess the quality of the schedule, the sum of the quality ratings of all vessels in the schedule is used.

To evaluate the quality of the schedule, we use:

$$Q = \sum_{i=1}^{|S_2|} Q_i,$$

where Q – indicator of the schedule quality.

9. The schedule is generated with other elements from the sets $\{T\}$, a schedule is generated with the different estimates of the ship's stay time in the port, while servicing the vessel at various terminals, if their characteristics are significantly different
10. The obtained results of the algorithm work are provided to the administrator.

11. A set of quality indicators of the schedule Q for each set of elements from sets $\{T\}$ as well as a general estimate of the time spent by the port for each set of such items are presented to the administrator. Then, the administrator decides whether to re-generate the schedule with new adjustment factors w_j or modify any of the received schedules manually for the further use.

4 Improving the Quality of the Schedule Based on Genetic Algorithms

To improve the quality of the obtained schedule, the use of a genetic algorithm is proposed, which will be used to select the weight coefficients of the criteria for assessing the quality of the schedule [9]. Thus, estimates of the vessels anchorage time in the port are considered as the genes, and the chromosome of the genetic algorithm will have the form of a set of real numbers:

$$\{w_1, \ldots, w_h\}$$

where $w_1, \ldots, w_h$ – chromosome genes, tuning factors of the scheduling algorithm; h – the number of weighting factors of the quality assessment criterion.

Creation of the Initial Population

To create the initial population, it is necessary to simulate a random set of coefficients corresponding to the number of these coefficients in the problem under consideration. Then the evolutionary process can be implemented.

The life cycle of a population is a few random crosses (through a crossover) and mutations, resulting in the addition of new individuals to the population. The work of the genetic algorithm consists of continuing actions of selection, crossing, mutation and verification for the criterion of stopping the operation of the algorithm.

Selection. We select the best half of the individuals from the population sorted by the quality indicator of the compiled schedule Q.

Crossing. Individuals are crossing in pairs (1st with 2nd, 3rd with 4th, etc.). The result is also usually two individuals with components taken from their parents. A single-point crossing is used, by arbitrarily choosing the crossing point. After crossing, we make a substitution of parents for descendants.

Mutation. The mutation operator simply changes an arbitrary gene on the chromosome to another arbitrary number.

Stopping Criteria. The work stops when there is only one individual in the population.

As a result of crossing and mutations with a given probability and population size, new sets of tuning coefficients will be obtained and as a consequence, the results of the algorithm will be different. In this case, selection in the population is carried out taking into account the values w_j set by the administrator. The algorithm will always finish its work, giving the best result because the set of distribution possibilities is finite (the number of ships and the number of terminals are finite). At the same time, local minimum will not been exceeded as mutation is used in the algorithm. Thus, we consider the possibility of generating the most satisfactory schedule for the quality estimations as viewed by the administrator with using slightly changed quality estimates in the course of its generation.

5 Conclusion

The generation of an efficient cargo-handling schedule in a seaport by using multi-agent technologies and the genetic algorithm described in this article allows us to establish the work of the port in the most efficient way, taking into account only the necessary criteria. In addition, the construction of the genetic algorithm makes it possible to expand the number of criteria affecting the schedule quality.

References

1. Baniamerian, A., Bashiri, M., Zabihi, F.: Two phase genetic algorithm for vehicle routing and scheduling problem with cross-docking and time windows considering customer satisfaction. J. Ind. Eng. Int. **14**(1), 15–30 (2018)
2. Borumand, A., Beheshtinia, M.A.: A developed genetic algorithm for solving the multi-objective supply chain-scheduling problem. Kybernetes (2018)
3. Changan, R., Zhao, J., Chen, L.: A fast information scheduling algorithm for large scale logistics supply chain. J. Discret. Math. Sci. Cryptogr. **20**(6–7), 1459–1463 (2017)
4. He, Z., Guo, Z., Wang, J.: Integrated scheduling of production and distribution operations in a global MTO supply chain. Enterp. Inf. Syst., 1–25 (2018)
5. Hollan, J.H.: Adaptation in Natural and Artificial Systems. MIT Press, Cambridge (1975)
6. Ivaschenko, A., Minaev A.: Multi-agent solution for adaptive data analysis in sensor networks at the intelligent hospital ward. In: International Conference on Active Media Technology, pp. 453–463. Springer (2017)
7. Liu, J., Luo, Z., Duan, D., Lai, Z., Huang, J.: A GA approach to vehicle routing problem with time windows considering loading constraints. High Technol. Lett. **23**(1), 54–62 (2017)
8. Qing, C.: Vehicle scheduling model of emergency logistics distribution based on internet of things. Int. J. Appl. Decis. Sci. **11**(1), 36–54 (2018)
9. Shibaev, A.G.: Improvement of methods of chart optimization the sea cargo ships' work. Moscow (1984)
10. Sologub, N.K., Sharov, V.A., Abramov, A.A.: Plan development for the interaction of different transport's types in a node. A manual on the course "ETS and the basis for the interaction of various modes of transport" for training specialists in the field of transport communications, Moscow (1982)
11. Stone, P., Veloso, M.: Multiagent systems: a survey from a machine learning perspective. Auton. Robot. **8**, 345–383 (2008)

Resource Managing Method for Parallel Computing Systems Using Fuzzy Data Preprocessing for Input Tasks Parameters

Anastasia Voitsitskaya[1], Alexader Fedulov[2], and Yaroslav Fedulov[2(✉)]

[1] National Research University "Moscow Power Engineering Institute", Moscow 111250, Russia
[2] Smolensk Branch of National Research University "Moscow Power Engineering Institute", Smolensk 214013, Russia
fedulov_yar@mail.ru

Abstract. In this paper, the method of dispatching and optimal distribution of resources of various types in parallel computing systems is considered, based on preliminary processing of the individual problems parameters, construction of fuzzy evaluation systems and hybrid neural-fuzzy production systems. The application of this method provides advantages in conditions of inaccurate, incomplete and difficult to formalize information about the characteristics of performed tasks, taking into account initially established preferences and achieving the desired performance indicators for the tasks and selected planning strategy.

Keywords: Fuzzy evaluation models · Resource management
Neuro-fuzzy production models · Data preprocessing
Parallel computing systems

1 Introduction

At present, the implementation process of parallel high-performance computing systems (PCS) of various types (computational clusters, multicomputers, grid systems) is actively developing in various fields of production and technology, as well as in the field of education with the aim of mastering and using methods and technologies for parallelizing program execution, suitable equipment is being purchased and installed, corresponding courses and laboratory classes are created.

For National Research Universities, in addition to education, high-performance PCS are also involved in solution of scientific objectives, usually requiring a considerable amount of resources and time for implementation.

To solve the problem of efficient and flexible incoming tasks assignment to available computing resources, PCS uses planning systems (resource managers).

A. Abraham et al. (Eds.): IITI 2018, AISC 874, pp. 410–419, 2019.
https://doi.org/10.1007/978-3-030-01818-4_41

When deploying and maintaining high-performance PCS, the main problem is the complexity of setting up scheduling programs that perform tasks assignment for computing resources, which, as a rule, are characterized by the following features:

- Hardware computing nodes heterogeneity and dynamically changing load require special accounting.
- Different types of characteristics assignments, characteristics mutual influences, characteristics significance and resource requirements.
- The requirement to take into account the user's preferences and achieve the desired performance parameters of tasks execution.
- The presence of inaccurate, incomplete and difficult to formalize information about the performed tasks parameters.

In the conditions of an increase in the number of applied problems and a significant growth of loads to the parallel computing systems nodes, the considered features significantly complicate intelligent decision support for planning the effective computing resources allocation with traditional methods used in the most common nowadays scheduling programs (resource managers Slurm, Torque; job planners Moab, Maui; combined systems HTCondor and DIET).

The requirements of reducing the time costs for solving applied problems, simplifying the PCS maintenance procedures, taking into account the considered features, justify the urgency of developing new planning methods using fuzzy logic and fuzzy sets.

In this paper, a method for planning the high-performance PCS resources based on the incoming information preprocessing by using fuzzy evaluation models and the subsequent hybrid neural-fuzzy production output system for each of the incoming tasks is proposed.

2 General Method of Resource Management

The proposed method consists in monitoring the resource loading current state and the iterative process of assigning the optimal amount of free resources to the received task, based on the data obtained from the hybrid neural-fuzzy production model, which receives individual task characteristics fuzzy estimates from developed fuzzy evaluation models.

The incoming to the planning system tasks consist of a subtasks set that allow one or another level of parallelization. Subtasks are an atomic unit of planning within a single task and can be independent or organized into a dependency tree. Dependencies determine the order of job execution and data flows between tasks.

The scheme of the general method of resource management is shown in the Fig. 1.

The tasks flow enters the input queue. The planning system considers each task from the queue, evaluates all its integral characteristics with the help of the corresponding fuzzy evaluation models, then the resulting fuzzy characteristics estimates arrive at the neural-fuzzy production model input, whose outputs determine the optimal indicator of each type of available PCS resource required for the current task.

The obtained values of the resources required for solving each problem are used in the procedure for finding the optimal combination, which selects the best order of the tasks execution from the point of view of the user-defined strategy (taking into account the possibility of parallelization at the subtask level). The execution strategy can be defined as the requirements for minimizing the time or resource costs for implementation, as well as taking into account the entered priorities for individual tasks.

Based on the results of the optimal combination procedure, a task schedule is created, and the resource management system begins to monitor their implementation while simultaneously controlling and distributing available PCS resources, if necessary.

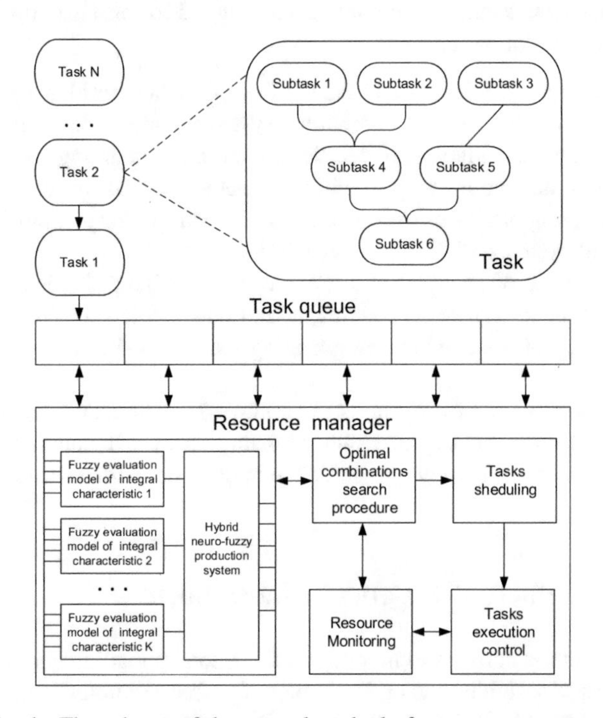

Fig. 1. The scheme of the general method of resource management

The described method steps are performed iteratively until the flow of input tasks runs out; thereby both a statistical and dynamic approach to planning is implemented.

From the presented generalized method it is reflected that the role of preliminary data processing on incoming tasks is extremely important for subsequent intellectual execution planning. Further, the method of direct evaluation with fuzzy evaluation models and hybrid neural-fuzzy production models for preprocessing incoming data is considered directly in this paper.

3 Fuzzy Evaluation

3.1 Fuzzy Evaluation Models

In general, the problem of constructing the proposed fuzzy evaluation models for the estimation of input tasks individual indicators is formulated as follows. Let there be a set of characteristics with the values that represent the results of measurement/evaluation of the corresponding tasks properties from the queue. It is required to construct the fuzzy evaluation model based on multilevel evaluation structure, various significance of characteristics and fuzzy coherence relationships between characteristics at each level of the model hierarchy.

The whole set of indicators is divided by levels of hierarchy. At each level of the hierarchy characteristics form a subsets, each of which correspond to the characteristic adjacent to it at higher hierarchy level. Beginning with the second level of the hierarchy, characteristics can exist without forming subsets at the lower level ("leaves"). On the first level of the hierarchy is a subset of a single (generalized) characteristic. Each characteristic is assigned to the weight. Characteristics belonging to the same subset, form fuzzy coherence relation.

Suggested fuzzy evaluation models are characterized by the following features: flexible hierarchical structure of characteristics, allowing to reduce the problem of multicriterial evaluation to one criterion; allow fuzzy representation of characteristics and coherence relations between them; consider various significance of characteristics; contain the required set of formalization for a software implementation.

3.2 Fuzzy Direct Evaluation Method

The proposed direct fuzzy evaluation method consists of the following stages.

Stage 1. Building the fuzzy evaluation model, includes formation of a hierarchical structure for characteristics evaluation, definition of weights and fuzzy coherence relationship between characteristics at each level of the model hierarchy.

Depending on the specific estimation tasks, *coherence* can be interpreted as correlation, interference of particular characteristics, and simultaneous attainability of values for compared particular characteristics.

Stage 2. Determination of the coherence levels $c^{(j)}_{q,\,kl} \in [0\,,1]$ *for aggregated characteristics* $p^{(j)}_{q,\,k}$ and $p^{(j)}_{q,\,l}$ $(k, l = 1, \ldots, n$, where n – number of characteristics) in fuzzy coherence relation $\tilde{R}^{(j)}_q = \{((p^{(j)}_{q,\,k},\, p^{(j)}_{q,\,l})/c^{(j)}_{q,\,kl})\}$, that can be set directly by the experts themselves or obtained through experiments.

The values $c^{(j)}_{q,\,kl}$ can be compared with the criterial coherence degrees, sorted in ascending order, for example, C = {NC – «No coherence», LC – «Low coherence», MC – «Medium coherence», HC – «High coherence», FC – «Full coherence»}:

$$c_{kl} \leftrightarrow c_u \in C = \{NC,\ LC,\ MC,\ HC,\ FC\}, \tag{1}$$

where u – element index of set C, $k, l = 1, \ldots, n$.

Fuzzy coherence relation between characteristics of the subsets can be represented in the form of fuzzy oriented graphs and moving on to more clearly their designation: $\tilde{G} = (\tilde{P}, \tilde{R})$, where $\tilde{P} = \{(p_i/\mu_P(p_i)\}$ – fuzzy set of characteristics (vertices), $p_i \in P$ $i \in \{1,\ldots,n\}$; $\mu_P(p_i) \in [0, 1]$ – membership degree to the base set for characteristics p_i; $\tilde{R} = \{(p_k, p_l)/c_{kl})\}$, $k, l = 1,\ldots,n$, fuzzy set of oriented arcs, where each arc (p_k, p_l) mapped to the corresponding coherence level $c_{kl} \in [0, 1]$ for characteristics p_k and p_l.

Fuzzy characteristics representation allows using the theory of fuzzy sets and numbers for evaluation. Representation of characteristics compatibility degrees on the basis of fuzzy coherence relations enables to apply the theory of fuzzy relations methods.

Figure 2 shows an example of the constructed fuzzy quality evaluation model considering significance and coherence degrees of particular characteristics {p1 – cyclic structures fraction, p2 – redundancy, p3 – subtasks branching}.

Stage 3 and 4. Justification of convolution operations and their comparison with coherence levels for aggregated characteristics.

Based on the analysis results, as the characteristics convolution operations, compared with extreme evaluation strategies (the achievement of the *lowest values for all* of the characteristics or *maximum values for at least one* of the characteristics) operations min and max, are chosen.

The whole set of compromise evaluation strategies is provided with the parameterized family of convolution operations: $\text{med}(p_k, p_l; \alpha), k, l \in \{1,\ldots,n\}, \alpha \in [0, 1]$ where α – parameter which characterizes the coherence level of characteristics.

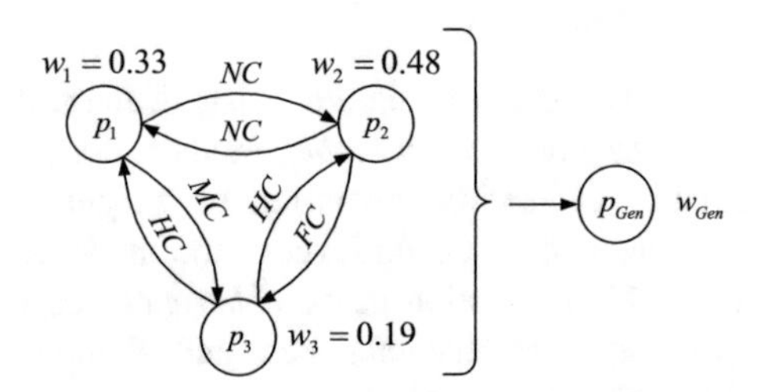

Fig. 2. Fuzzy quality evaluation model of the product

Table 1 provides an example of comparison between the convolution operations and criterial coherence levels (1) of characteristics.

Table 1. Comparison of aggregation operations with criterial levels of characteristics coherence

№	Operation characteristics aggregation of p_k and p_l	Font size and style	Description of criterion level of coherence
1	$\min(p_k, p_l)$	*NC*	No coherence
2	$\text{med}(p_k, p_l;\ 0.25)$	*LC*	Low coherence
3	$\text{med}(p_k, p_l;\ 0.5)$	*MC*	Medium coherence
4	$\text{med}(p_k, p_l;\ 0.75)$	*HC*	High coherence
5	$\max(p_k, p_l)$	*FC*	Full coherence

Stage 5. Specification of the evaluation strategy depends on the decision maker's preferences and features of an estimated object. It may be divided into two stages:

- *firstly*, assignment of coherence levels viewing order, which determines characteristic aggregation sequence in model;
- *secondly*, setting of the recalculation procedure of characteristics coherence levels during their serial convolution.

There can be set two main fuzzy evaluation strategies: from the least to the most coherent characteristics; from the most coherent to least coherent characteristics. Moreover, the evaluation strategy can be set for the entire model, as well as separately for each of the characteristics subsets.

Stage 6. Splitting of fuzzy compatibility relation into coherence classes and selection of the corresponding convolution operations. A fuzzy compatibility relation of characteristics can be divided into so-called coherence classes regarding the criterion coherence levels.

The case of evaluation strategy from the least to the most coherent characteristics is considered. Figure 3 shows that in a given example fuzzy compatibility relation with respect to criterion level *NC* is divided into two coherence classes.

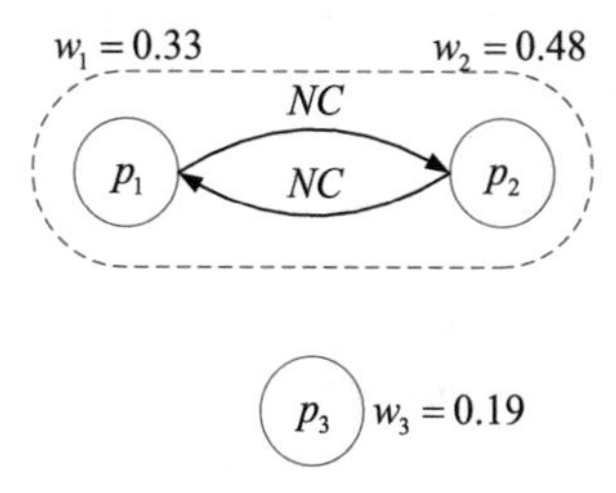

Fig. 3. Coherent classes for fuzzy compatibility relation

For the aggregation of a single coherence class the same operation corresponding to a predetermined criterial level is used. The order of the characteristics convolution within a single class is irrelevant.

Step 7. Modification offuzzy compatibility relation is performed after characteristics convolution with the change of characteristics coherence degree (level) to reflect the new characteristic, the weight of which is equal to the sum of the weights of aggregates (see Fig. 4).

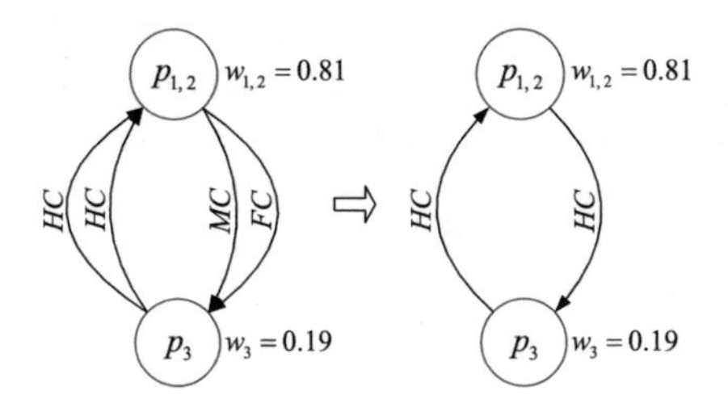

Fig. 4. Modification of fuzzy compatibility relation

Step 8. Structure formation of characteristics convolution for fuzzy evaluation model. Steps 6 and 7 are repeated on all hierarchy levels of fuzzy evaluation model, starting from the bottom, and at each level of the hierarchy – for all subsets of characteristics. This results in a structure of characteristics convolution: $h^*(p_1,\ldots,p_n) = h_u(h_y(\ldots(h_t(p_1,p_2),\ldots),p_{n-1}),p_n)$ where t, y, u – convolution operations indexes.

For given example of a product quality assessment, characteristics convolution takes the form: $p_{Gen} = \mathrm{med}((\min(p_1,\ p_2)),\ p_3;\ 0.75)$.

Step 9. Assignment of the characteristics weighted values and fuzzy evaluation of task parameters. Generally, fuzzy values of characteristics may be represented as the fuzzy sets (numbers), and in particular – distinct values. Immediately before the direct fuzzy evaluation it is required to take into account the significance of various characteristics.

4 Hybrid Neural-Fuzzy Production Model

4.1 Constructing of the Neural-Fuzzy Production Model

The mechanism of the hybrid neural-fuzzy product model by analogy with fuzzy logical inference is based on the knowledge base formed by the subject area specialists in the form of a fuzzy predicate rules system, such as:

$$\Pi_i : IF\ \tilde{X}_1^k\ is\ A_{ij}\ and\ \ldots\ \tilde{X}_n^k\ is\ A_{jn}, THEN\ \tilde{Y} = B_i \tag{2}$$

where $\widetilde{X}_1,\ldots,\widetilde{X}_n$ – fuzzy input variables, $\widetilde{Y}$ – fuzzy output variable, A_i, B_i – the values of linguistic terms that characterize the corresponding membership functions.

The activation level of the i-th rule with respect to the considered k-th problem is interpreted as the membership level characterizing the values of the corresponding resource required to solve it.

Usually, the activation level is designated as α_i, and is calculated as follows:

$$\alpha_i = \mathrm{T}(\mu_{A_{ij}}(\widetilde{X_1^k}),\ \ldots,\ \mu_{A_{jn}}(\widetilde{X_n^k})), \tag{3}$$

where T – the t-norm operator modeling the logical "AND" operator.

As an example, the procedure for constructing a neural-fuzzy product model to determine the optimal value of each of the $P = 1,\ 2,\ \ldots,p$ available PCS resources required to perform each of the $N = 1,\ 2,\ \ldots,n$ input tasks is considered. Let the fuzzy partition for each input characteristic assume the presence of three linguistic terms "low", "medium" and "high", which are represented by Gaussian membership functions. For the initial input characteristics space fuzzy partitioning, rules that represent the basis of the a priori data can be generated.

One of the possible structures of the hybrid neural-fuzzy production model for the example in question is presented in the Fig. 5.

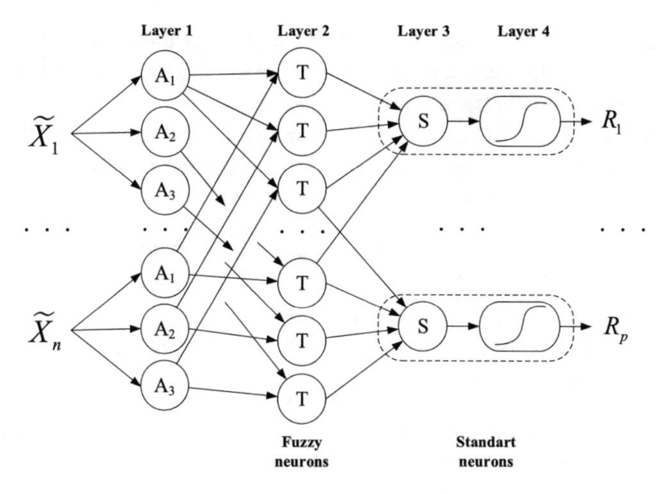

Fig. 5. The structure of the hybrid neural-fuzzy production model of the example

Layer 1. At output elements of this layer, the membership levels of the input variables to fuzzy sets are formed.

$$\mu_{A_l}(\widetilde{x_p}) = \exp\left(-\left(\widetilde{x_p} - a_{ij}\right)^2 / 2 \cdot b_{ij}^2\right). \tag{4}$$

Layer 2. The signals and weights of the fuzzy neurons "T" of this layer are combined using the S-norm operation, and the output value is aggregated by the T - norm operation using formula (3).

As operations of the T (S) - norm one of the standard or extended operations of intersection (union) over fuzzy sets can be chosen. The outputs of this layer are designated as $\beta(A_i, B_i) = \mathrm{T}(\mu_{1i}, \mu_{2i})$.

Layers 3–4. The number of neurons in these layers is equal to the number of available resources. Elements of these layers are ordinary (standard) neurons. Neurons of layer 3 are intended for weighted summation of the elements outputs values of the previous layer $S_p = \sum \omega_{ijp} \cdot \beta\left(A_i,\ B_j\right)$. Weights ω_{ijp} are configurable parameters.

The outputs elements values of layer 4 are formed using activation functions of the sigmoidal type $\mu_{K_p}(y) = \left(1 + e^{(-d_p \cdot (y - c_p))}\right)^{-1}$, where d_p and c_p – activation functions parameters of the sigmoidal type for determining the membership levels for the presented task to the corresponding resource.

4.2 Training of the Neural-Fuzzy Production Model

Due to the fact that the hybrid neural-fuzzy production model is presented as a multilayer structure with direct signal propagation, and the value of the outputs can be changed by adjusting the parameters of the layer elements, then to train the model of this type the back propagation algorithm can be used. This learning algorithm belongs to the class of gradient algorithms, the idea of which is to reduce the previous value of the tuned parameter by the magnitude of the derivative of the error measure multiplied by some coefficient. The process must last until the error at the output of the system has reached an a priori established minimum value.

Parametric layers of the neuro-fuzzy classifier, i.e. layers, the parameters of the elements in which will be adjusted during training, are the first, third and fourth, and the parameters configured in the learning process are:

- in the first layer – nonlinear parameters a_{ij}, b_{ij} of membership function fuzzy sets $\mu_{A_l}(\tilde{x}_p)$ of rules preconditions;
- in the third layer – weighting coefficients ω_{ijp};
- in the fourth layer – nonlinear parameters d_p and c_p of membership function fuzzy sets $\mu_{K_p}(y)$ of rules conclusions.

In this form, the learning algorithm of the hybrid neural-fuzzy production model is similar to the learning algorithms for neural-fuzzy ANFIS (Adaptive Network-based Fuzzy Inference System) networks.

5 Conclusion

The method of dispatching and optimal distribution of resources of various types in parallel computing systems, based on preliminary processing of the individual problems parameters, construction of fuzzy evaluation systems and hybrid neural-fuzzy production systems provides advantages in conditions of inaccurate, incomplete and difficult to formalize information about the characteristics of performed tasks comparing to traditional methods.

Proposed method allows reducing the problem of multicriterial evaluation to one criterion for each input task, procures fuzzy representation of task characteristics, various significance and coherence relations between them, takes into account initially established preferences and helps to achieve the desired performance indicators.

Described mathematical apparatus contains the required set of formalization for the efficient software implementation.

In particular, series of experiments conducted on test equipment (a computational cluster with three nodes with the CentOS operating system) confirmed the average increase in the performance level 5.7% compared to the Slurm scheduling system for the same set of input tasks and initial resource utilization.

References

1. Blaiewicz, J., Drozdowski, M., Markiewicz, M.: Divisible task scheduling – concept and verification. Parallel Comput. **25**, 87–98 (1999)
2. Borisov, V.V., Fedulov, A.S., Fedulov, Y.A.: "Compatible" fuzzy cognitive maps for direct and inverse inference. In: Proceedings of the 18th International Conference on Computer Systems and Technologies, CompSysTech 2017, Ruse, Bulgaria, 23–24 June. ACM International Conference Proceeding Series, vol. 1369 (2017)

3. Fan, G., et al.: A hybrid fuzzy evaluation method for curtain grouting efficiency assessment based on an AHP method extended by D numbers. Expert Syst. Appl. **44**, 289–303 (2016)
4. Golubev, I.A., Smirnov, A.N.: Clustering and classification tasks adaptation to cloud environment. In: IEEE RNW Section Proceedings, vol. 2. IEEE (2011)
5. HTCondor Version 8.0.0 Manual. University of Wisconsin–Madison: Center for High Throughput Computing (2013)
6. Neuman, B., Rao, S.: Resource management for distributed parallel systems. In: Proceedings of 2nd International Symposium on High Performance Distributed Computing (1993)
7. Rauber, T., Runger, G.: Parallel Programming: For Multicore and Cluster Systems. Springer, Heidelberg (2013)
8. Tang, W., Feng, W.: Parallel map projection of vector-based big spatial data: coupling cloud computing with graphics processing units. Comput. Environ. Urban Syst. **61**, 187–197 (2017)
9. Torque v.4.2.4 Administrator Guide. Adaptive Computing Enterprises (2013)
10. Yuan, Z.-W., Wang, Y.-H.: Research on K nearest neighbor non-parametric regression algorithm based on KD-tree and clustering analysis. In: Proceedings of the 2012 Fourth International Conference on Computational and Information Sciences, ICCIS 2012. IEEE Computer Society, Washington, DC (2012)

Cognitive Technologies on the Basis of Sensor and Neural Networks

Post-processing of Numerical Forecasts Using Polynomial Networks with the Operational Calculus PDE Substitution

Ladislav Zjavka(✉) and Stanislav Mišák

ENET Centre, VŠB-Technical University of Ostrava, Ostrava, Czech Republic
ladislav.zjavka@vsb.cz

Abstract. Large-scale weather forecast models are based on the numerical integration of systems of differential equation which can describe atmospheric processes in light of physical patterns. Meso-scale weather forecast systems need to define the initial and lateral boundary conditions which can be supplied by global numerical models. Their overall solutions, using a large number of data variables in several atmospheric layers, represent the weather dynamics on the earth scale. Post-processing methods using local measurements were developed in order to adapt numerical weather prediction model outputs for local conditions with surface details. The proposed forecasts correction procedure is based on the 2-stage approach of the Perfect Prog method using data observations to derive a model which is applied to the forecasts of input variables to predict 24-h series of the target output. The post-processing model formation requires an additional initial estimation of the optimal number of training days in consideration of the latest test data. Differential polynomial network is a recent machine learning technique using a polynomial PDE substitution of Operational calculus to form the test and prediction models. It decomposes the general PDE into the 2nd order sub-PDEs in its nodes, being able to describe the local weather dynamics in the surface level. The PDE sum models represent the current local data relations in a sort of settled weather which allow improvements in local forecasts corrected with NWP utilities in the majority of days.

Keywords: Polynomial neural network · General partial differential equation Polynomial substitution · Operational calculus · Post-processing model

1 Introduction

Global Numerical Weather Prediction (NWP) systems and their sub-forms meso-scale models do not reliably determine weather conditions near the ground level. The NWP models are based on atmospheric physics which allow forecasting large-scale weather patterns and succeed in forecasting upper air patterns. They are too crude to account for local variations in surface weather and possess the advantage of allowing for partial detail processes to occur within grid cells. Statistical models, based on another approach, relate surface inputs → output observations to represent specific local characteristics and climatological conditions. They are excellent at forecasting idiosyncrasies in local

A. Abraham et al. (Eds.): IITI 2018, AISC 874, pp. 423–433, 2019.
https://doi.org/10.1007/978-3-030-01818-4_42

weather but their predictions are usually worthless beyond about several hours. Post-processing methods, called Model Output Statistics (MOS), apply measurements and 3-dimensional field outputs from NWP models to describe statistical relations between model errors and the parameters. MOS typically derive a set of linear equations to relate forecasts to actual local observations at a certain time to detail specific surface features effects and eliminate systematic forecast errors [2]. The biases of NWP models are induced due to the parameterization of weather events and the inability to account for physical processes at a scale smaller than the grid used in numerical equations [1]. Additional NWP model errors may arise from computational limitations [7]. The non-systematic component of numerical forecast errors can be estimated assuming it is linearly dependent on a certain combination of the initial fields, tendency and end-time forecasts [6]. The linearity of traditional MOS equations may be a limitation, thus nonlinear generalization of multiple regression is used to model the interactions of the underlying processes. Inverse solutions of the Lagrange interpolation polynomials, whose coefficients are determined by past model performance, can express the NWP errors in respect to the length of previous multi-time data observations [8]. Another statistical post-processing method, commonly used in the atmospheric sciences, is referred to as Perfect-Prog (PP). PP applies forecasted variables to its analog or regression model, developed with the corresponding observation quantities [4]. Hybrid models can combine additional unbiased data of an independent observing system with NWP model forecasts. Higher-resolution models can apply 2 down-scaling methods:

- Physical - using boundary conditions provided by global NWP models.
- Statistical - develops statistical relationships between large-scale predictors and local output variables to apply them to global NWP model forecasts.

Artificial neural networks (ANN) are usually simple 1-layer structures which do not allow modelling complex weather systems described by a mass of data. ANN require data pre-processing to reduce the number of input variables, which can lead to the models over-simplification. For example, only several input variables from hundreds of quantities remain for the ANN classification of ceiling and visibility [3]. The number of parameters in polynomial regression grows exponentially with the number of variables. Polynomial Neural Networks (PNN) decompose the general connections between inputs and output variables expressed by the Kolmogorov-Gabor polynomial (1).

$$Y = a_0 + \sum_{i=1}^{n} a_i x_i + \sum_{i=1}^{n}\sum_{j=1}^{n} a_{ij} x_i x_j + \sum_{i=1}^{n}\sum_{j=1}^{n}\sum_{k=1}^{n} a_{ijk} x_i x_j x_k + \ldots \tag{1}$$

n - number of input variables x_i a_i, a_{ij}, a_{ijk}, ... - *polynomial parameters*

The Group Method of Data Handling (GMDH) evolves a multi-layer PNN in successive steps, adding one last layer at a time, whose nodes are selected and parameters calculated. PNN nodes decompose the complexity of a system into a number of simpler relationships, each described by low order polynomial transfer functions (2) for every pair of input variables x_i, x_j [5].

$$y = a_0 + a_1x_i + a_2x_j + a_3x_ix_j + a_4x_i^2 + a_5x_j^2 \quad (2)$$
x_i, x_j - *input variables of polynomial neuron nodes*

Differential polynomial neural network (D-PNN) is a new type of neural network which extends the complete PNN structure to form the 2nd order sub-PDEs in its nodes that decompose the complete general PDE (Partial Differential Equation). It uses the polynomial PDE substitution of Operational Calculus leading to rational fraction functions which can represent the Laplace transforms of unknown partial node functions. The inverse Laplace transformation is applied to the function images to solve the node sub-PDEs whose sum models the separable searched function. D-PNN expands the general derivative formula into a number of node specific sub-PDEs analogous to the GMDH decomposition of the general connection polynomial (1). Each neuron, i.e. a node PDE solution in this context, can be directly included in the total network output sum. The D-PNN multi-layer structure complexity is proportional to the number of input variables (in contrast to the ANN) as nodes in additional layers can form all the possible derivatives of applicable node PDEs in respect of variables of the previous layers using the composite polynomials (2).

The proposed NWP data correction procedure uses spatial inputs → output observations from the last few days to elicit models which can post-process daily NWP model forecasts of the input variables to calculate the target output prediction series, analogous to PP [4]. The D-PNN prediction models additionally require different initialization, i.e. specific time-intervals whose data samples are used to adapt the parameters. The training is stopped at the optimal daily training errors to compensate for the weather dynamics and inaccuracies of processed forecasts [10]. The D-PNN models represent current weather conditions which are supposed to be actual in the prediction time-horizon to allow improvements in 24-h local forecasts of regional NWP models [9]. The post-processing model is not valid in the case of an overnight weather change although these days are less frequent as more or less settled conditions prevail. These flawed models can be detected in advance and not applied to NWP data.

2 Post-processing of NWP Data Based on Perfect Prog

The proposed method is a 2-stage process analogous to the PP approach. In the 1st stage the training is made on real observations to derive the PDE model which is then applied to the forecasts in the 2nd stage. D-PNN is trained with inputs → output data observations over the last few days (Fig. 1 left) to form daily PDE models which post-process forecasts of the input variables (Fig. 1 right) to calculate 24-h output prediction series at the corresponding time. The 1st PP stage is extended into 2 steps:

1. The optimal number of training days and training error, used to elicit the prediction model, are estimated by an assistant test model.
2. Parameters of the prediction model are adapted for inputs → output data samples to process forecasts of the training input variables.

The assistant model is an additional D-PNN test model formed initially in the 1st step with observations from increasing number of the previous days *2, 3, …, x*. The assistant model processes NWP data from the previous day to compare its output with the latest 6-h testing observations. The number of training days and training error, giving the minimal test error, are applied in the daily prediction model formation.

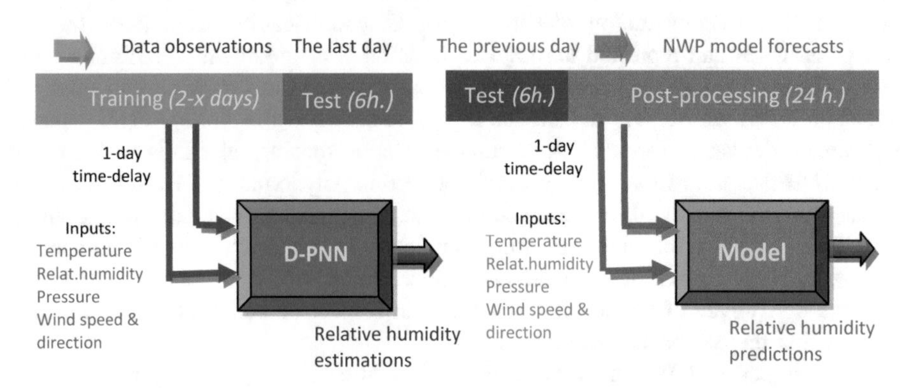

Fig. 1. D-PNN is trained with data observations (blue-left) over the last few days to form daily prediction PDE models that process forecast series of the input variables (red-right)

The D-PNN models can post-process and improve forecasts in more or less settled weather which tend to prevail and whose patterns do not change fundamentally within short-time intervals. If a significant change in the atmosphere state occurs, e.g. passing an atmospheric front, the model derived from the previous last days is not actual because there are different data relationships and its NWP revision fails.

3 The PDE Polynomial Substitution Using Operational Calculus

D-PNN defines and substitutes for the general linear PDE (3) whose exact form is not known and which can describe unknown complex dynamic systems. It decomposes the general PDE into 2nd order specific sub-PDEs in the PNN nodes to model unknown node functions u_k whose sum gives the complete *n*-variable model *u* (3).

$$a + bu + \sum_{i=1}^{n} c_i \frac{\partial u}{\partial x_i} + \sum_{i=1}^{n}\sum_{j=1}^{n} d_{ij} \frac{\partial^2 u}{\partial x_i \partial x_j} + \ldots = 0 \qquad u = \sum_{k=1}^{\infty} u_k \tag{3}$$

$u(x_1, x_2, \ldots, x_n)$ - unknown separable function of n-input variables
$a, b, c_i, d_{ij}, \ldots$ - weights of terms *u_i - partial functions*

Considering 2-input variables of the PNN nodes, the derivatives of the 2nd order PDE (4) correspond to variables of the GMDH polynomial (2). This type of the 2nd order PDE is most often used to model physical or natural systems non-linearities.

$$\left(\sum\frac{\partial u_k}{\partial x_1},\sum\frac{\partial u_k}{\partial x_2},\sum\frac{\partial^2 u_k}{\partial x_1^2},\sum\frac{\partial^2 u_k}{\partial x_1 \partial x_2},\sum\frac{\partial^2 u_k}{\partial x_2^2}\right) \tag{4}$$

u_k - node partial sum functions of an unknown separable function u

The polynomial conversion of the 2nd order PDE (4) using procedures of Operational Calculus is based on the proposition of the Laplace transform (L-transform) of the function n^{th} derivatives in consideration of initial PDE conditions (5).

$$L\{f^{(n)}(t)\} = p^n F(p) - \sum_{k=1}^{n} p^{n-i} f_{0+}^{(i-1)} \qquad L\{f(t)\} = F(p) \tag{5}$$

$f(t), f'(t), \ldots, f^{(n)}(t)$ – originals continuous in <0+, ∞> *p, t - complex and real variables*

This polynomial substitution for the *f(t)* function n^{th} derivatives in the PDE leads to algebraic equations from which the L-transform image *F(p)* of an unknown function *f(t)* is separated in the form of a pure rational function (6). These fractions represent the L-transforms *F(p)*, expressed with the complex number *p*, so that the inverse L-transformation is applied to them to obtain the original functions *f(t)* of a real variable *t* (6) described by the specific 2nd order PDE (7).

$$F(p) = \frac{P(p)}{Q(p)} = \sum_{k=1}^{n} \frac{P(\alpha_k)}{Q_k(\alpha_k)} \frac{1}{p - \alpha_k} \qquad f(t) = \sum_{k=1}^{n} \frac{P(\alpha_k)}{Q_k(\alpha_k)} e^{\alpha_k \cdot t} \tag{6}$$

α_k - simple real roots of the multinomial Q(p) *F(p) - L- transform image*

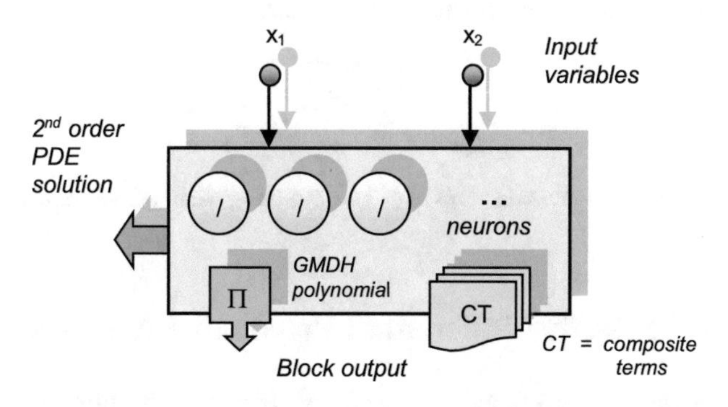

Fig. 2. A block of derivative neurons - 2nd order PDE solutions in the PNN nodes

The inverse L-transformation is applied to the selected neurons, i.e. polynomial terms (6) produced in each D-PNN node block (Fig. 2), to substitute for the specific sub-PDEs (7) and obtain the node originals u_k whose sum gives the modelled unknown separable output function *u* (3). Each block contains a single output polynomial (2) which is used to form neurons that can be selected and directly included in the total network output sum of a general PDE solution.

$$F\left(x_1, x_2, u, \frac{\partial u}{\partial x_1}, \frac{\partial u}{\partial x_2}, \frac{\partial^2 u}{\partial x_1^2}, \frac{\partial^2 u}{\partial x_1 \partial x_2}, \frac{\partial^2 u}{\partial x_2^2}\right) = 0 \tag{7}$$

where $F(x_1, x_2, u, p, q, r, s, t)$ is a function of 8 variables

While using 2 input variables in the PNN nodes the 2nd order PDE can be expressed in the equality of 8 variables (7), including derivative terms formed in respect of all variables of the GMDH polynomial (2).

$$y_1 = w_1 \frac{a_0 + a_1x_1 + a_2x_2 + a_3x_1x_2 + a_4sig(x_1^2) + a_5sig(x_2^2)}{b_0 + b_1x_1} \cdot e^{\varphi} \tag{8}$$

$$y_3 = w_3 \frac{a_0 + a_1x_1 + a_2x_2 + a_3x_1x_2 + a_4sig(x_1^2) + a_5sig(x_2^2)}{b_0 + b_1x_2 + b_2sig(x_2^2)} \cdot e^{\varphi} \tag{9}$$

$$y_5 = w_5 \frac{a_0 + a_1x_1 + a_2x_2 + a_3x_1x_2 + a_4sig(x_1^2) + a_5sig(x_2^2)}{b_0 + b_1x_1 + b_2x_{12} + b_3x_1x_2} \cdot e^{\varphi} \tag{10}$$

$\varphi = arctg(x_1/x_2)$ - phase representation of 2 input variables x_1, x_2
a_i, b_i - polynomial parameters w_i - weights   sig - sigmoidal transformation

Each D-PNN block can form *5* simple derivative neurons in respect of single x_1, x_2 (8) squared x_1^2, x_2^2 (9) and combination x_1x_2 (10) derivative variables, which can solve the specific 2nd order sub-PDEs in the PNN nodes (7). The Root Mean Squared Error (RMSE) is calculated in the simultaneous polynomial parameter optimization, node block 2-input combination and neuron + CT selection (11).

$$RMSE = \sqrt{\frac{\sum_{i=1}^{M}\left(Y_i^d - Y_i\right)^2}{M}} \rightarrow \min \tag{11}$$

Y_i - produced and Y_i^d - desired D-PNN output for i^{th} training vector of M-data samples

4 Backward Selective Differential Polynomial Network

Multi-layer networks form composite functions (Fig. 2.). The blocks of the 2nd and next hidden layers can produce additional Composite Terms (CT) equivalent to the neurons. The CTs substitute for the derivatives of node sub-PDEs with respect to variables of back-connected blocks of the previous layers according to the composite function (12) partial derivation rules (13).

$$F(x_1, x_2, \ldots, x_n) = f(z_1, z_2, \ldots, z_m) = f(\phi_1(X), \phi_2(X), \ldots, \phi_m(X)) \tag{12}$$

$$\frac{\partial F}{\partial x_k} = \sum_{i=1}^{m} \frac{\partial f(z_1, z_2, \ldots, z_m)}{\partial z_i} \cdot \frac{\partial \varphi_i(X)}{\partial x_k} \quad k = 1, \ldots, n \tag{13}$$

The 2nd layer blocks can form and select one of their neurons or additional 10 CTs using applicable neurons of the 1st layer 2 blocks in the products (13) for composite sub-PDE solutions (14). The 3rd layer blocks can select from additional 10 + 20 CTs using neurons from 2 and 4 blocks in the previous 2nd and 1st layers (20), etc. The number of possible block CTs doubles along with each joined preceding layer (Fig. 3). The L-transform image *F(p)* is expressed in the complex form, so the phase of the complex representation of 2-variables is applied in the inverse L-transformation.

$$y_{2p} = w_{2p} \cdot \frac{a_0 + a_1x_{11} + a_2x_{13} + a_3x_{11}x_{13} + a_4x_{11}^2 + a_5x_{13}^2}{x_{11}} \cdot e^{-\varphi_{21}} \cdot \frac{b_0 + b_1x_1 + b_2x_1^2}{x_{11}} \cdot e^{\varphi_{11}} \quad (14)$$

$$y_{3p} = w_{3p} \cdot \frac{a_0 + a_1x_{21} + a_2x_{22} + a_3x_{21}x_{22} + a_4x_{21}^2 + a_5x_{22}^2}{x_{21}} \cdot e^{-\varphi_{31}} \cdot c_{21} \cdot \frac{b_0 + b_1x_2}{x_{11}} \cdot e^{\varphi_{11}} \quad (15)$$

y_{kp} - p^{th} Composite Term (CT) output *$\varphi_{21} = arctg(x_{11}/x_{13})$* *$\varphi_{31} = arctg(x_{21}/x_{22})$*
c_{kl} - complex representation of the lth block inputs x_i, x_j in the kth layer

The CTs include the L-transformed fraction of the external function sub-PDE (left) in the starting block and the selected neuron (right) from a block in one of the preceding layer (14). The complex representation of 2-inputs (16) of blocks in the CTs inter-connected layers can substitute for the node internal function PDEs (15). The pure rational fractions (6) correspond to the amplitude *r* (radius) in Eulers's notation of a complex number c (16) in polar coordinates.

$$c = \underbrace{x_1}_{\text{Re}} + i \cdot \underbrace{x_2}_{\text{Im}} = \sqrt{x_1^2 + x_2^2} \cdot e^{i \cdot \arctan\left(\frac{x_2}{x_1}\right)} = r \cdot e^{i \cdot \varphi} = r \cdot (\cos\varphi + i \cdot \sin\varphi) \quad (16)$$

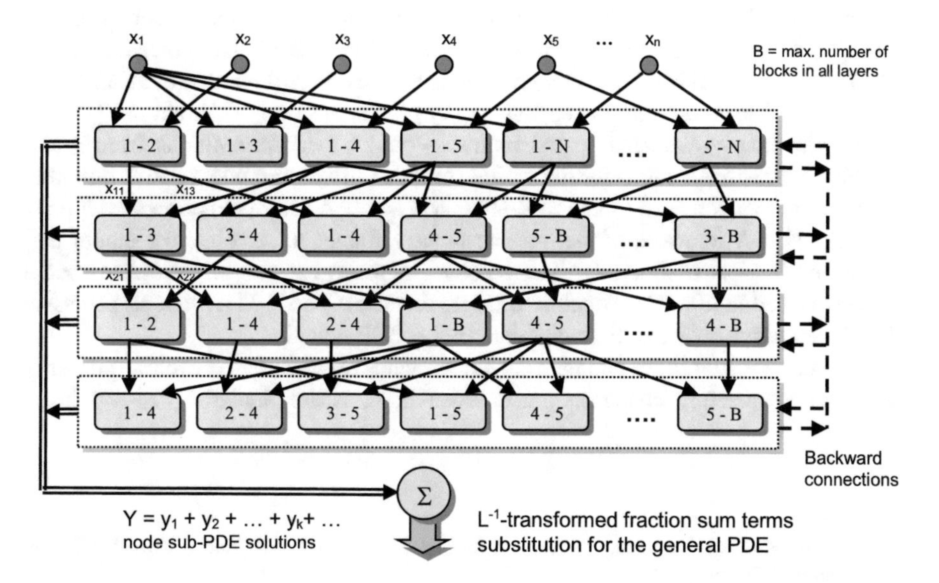

Fig. 3. D-PNN selects from possible 2-variable combination blocks in each hidden layer

N-variable D-PNN must select from the possible 2-combination node blocks in each layer (analogous to GMDH) as the number of the input combination couples grows exponentially in each next layer. The D-PNN complete PNN structure is optimized as a whole, i.e. initialized with an estimated number of layers and node blocks, which are not added one by one. A convergent combination of selected neurons and CTs, which can form a general PDE solution, is not able to accept a disturbing effect of the rest of the node sub-solutions (able to form other local error solutions) in the parameter optimization. The D-PNN total output Y is the arithmetic mean of active neurons + CTs output values so as to prevent their changeable number in a combination from influencing the total network output (17).

$$Y = \frac{1}{k}\sum_{i=1}^{k} y_i \qquad k = \text{the number of active neurons + CTs (node sub-PDEs)} \tag{17}$$

2 simultaneous processes of the optimal block 2-inputs and neurons + CTs combination selection are finished gradually in the initial D-PNN structure formation and primary PDE definition. They are randomly selected and performed along with the continuous polynomial parameters and term weights optimization using the Gradient Steepest Descent (GSD) method in each iteration step to decrease maximally the training RMSE. The binary Particle Swarm Optimization (PSO), being able to solve large combinatorial problems, can optimize the neurons + CTs selection [10].

5 NWP Post-processing Using Daily PDE Correction Models

National Oceanic and Atmospheric Administration (NOAA) provides, among other services, free tabular 24-h forecasts of temperature, relative humidity, wind speed and direction at selected localities[1]. The atmospheric pressure tendency prognosis is missing but the WU hourly forecasts[2] can supply it. The Global Forecast System (GFS) provides initial and boundary conditions for the regional North American Mesoscale (NAM) forecast system to generate hourly predictions every 6th hour. D-PNN was trained with hourly averaged spatial inputs → output data observations from 3 surrounding + 1 central stations for the periods of 2 to 11 days to model relative humidity at the central location at the same time (Fig. 4). Next, the daily prediction model post-processes NOAA forecasts of the training input variables to calculate the output predictions each 1–24 h (Sect. 2). The results are presented for initial NAM model forecasts issued at 00 of the local time (Figs. 5 and 6). The D-PNN models represent relative derivative changes of spatial data in the PDE equality.

The relative humidity temporal behavior tends toward daily cycles, primarily inverse to temperature changes. The doubled input vector, including 24-h delayed

---

[1] NAM forecasts http://forecast.weather.gov/MapClick.php?lat=46.60683&lon=-111.9828333&lg=english&&FcstType=digital.

[2] WU forecasts www.wunderground.com/hourly/us/mt/great-falls?cm_ven=localwx_hour.

Fig. 4. The forecasted central locality and 3 surrounding stations - observation and forecast data

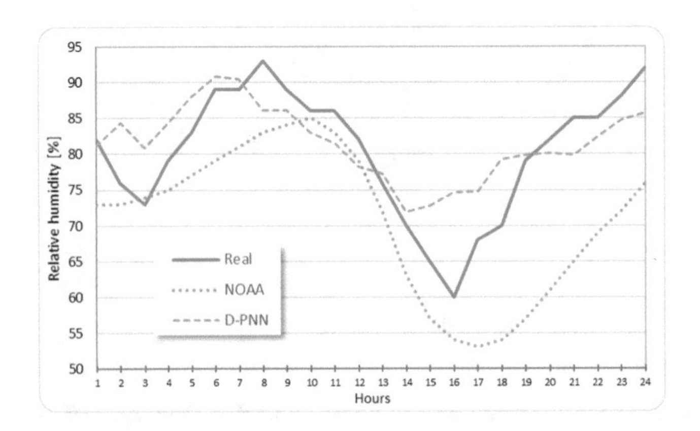

Fig. 5. 27.10.2015, Great Falls - RMSE: NOAA = 11.53, D-PNN = 5.74

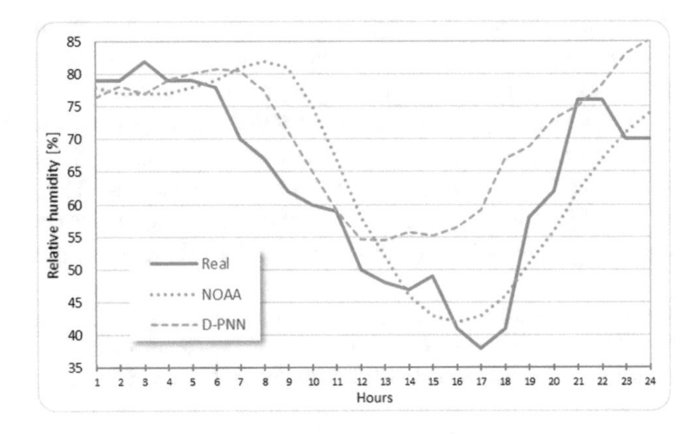

Fig. 6. 29.10.2015, Great Falls - RMSE: NOAA = 8.11, D-PNN = 10.30

observation and forecast data in the training and post-processing, improves the prediction accuracy in modelling the daily periodical quantities, e.g. power load.

The optimal numbers of training days were estimated initially by additional D-PNN assistant models trained analogously to the prediction models and tested with the previous day forecasts to compare the outputs with the latest 6-h observations. NOAA provides free historical weather data archives[3], shared also by WU[4], and current 2-day hourly tabular observations[5] for many land-based stations. Figure 7 shows the relative humidity daily average prediction RMSEs of the original NOAA, NAM and D-PNN models in a week period (the x-axis represents the real data).

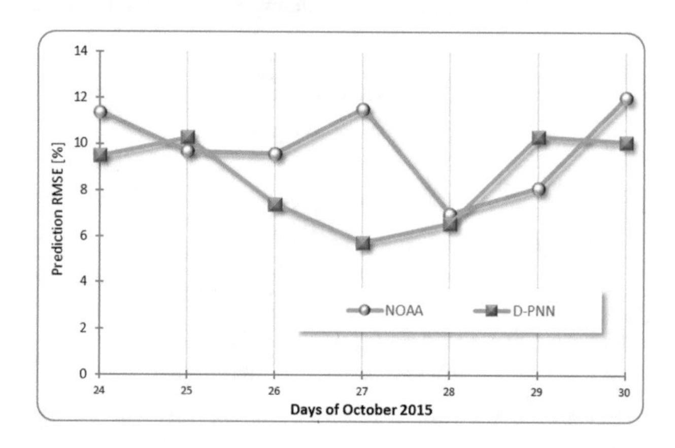

Fig. 7. 1-week relative humidity average prediction RMSE: NOAA = 9.90, D-PNN = 8.56

6 Conclusions

Weather conditions mostly do not change fundamentally within short time periods of several days which are followed by overnight changes. The D-PNN models are not actual in the prediction day in this case as they represent different spatial data relations in the previous days and their post-processing of NWP data fails (Fig. 6). The assistant model test errors or an appropriate NWP data comparative analysis can indicate the sporadic days of frontal breaks to discard the prediction models. The D-PNN daily corrections of relative humidity forecasts obtain the best results in comparison to other weather quantities, e.g. wind speed or the most problematic temperature.

Acknowledgement. This paper was supported by the following projects: LO1404: Sustainable Development of ENET Centre; CZ.1.05/2.1.00/19.0389 Development of the ENET Centre Research Infrastructure; SP2018/58 and SP2018/78 Student Grant Competition and TACR TS777701, Czech Republic.

[3] NOAA National Climatic Data Center archives www.ncdc.noaa.gov/orders/qclcd/.

[4] WU data www.wunderground.com/history/airport/KGTF/2015/10/3/DailyHistory.html.

[5] NOAA data observations www.wrh.noaa.gov/mesowest/getobext.php?wfo=tfx&sid=KGTF.

References

1. Durai, V.R., Bhradwaj, R.: Evaluation of statistical bias correction methods for numerical weather prediction model forecasts of maximum and minimum temperatures. Nat. Hazards **73**, 1229–1254 (2014)
2. Klein, W., Glahn, H.: Forecasting local weather by means of model output statistics. Bull. Am. Meteorol. Soc. **55**, 1217–1227 (1974)
3. Marzban, C., Leyton, S., Colman, B.: Ceiling and visibility forecasts via neural networks. Weather Forecast. **22**, 466–479 (2007)
4. Marzban, C., Sandgathe, S., Kalnay, E.: MOS, perfect prog, and reanalysis. Mon. Weather Rev. **134**, 657–663 (2006)
5. Nikolaev, N.Y., Iba, H.: Adaptive Learning of Polynomial Networks. Genetic and Evolutionary Computation. Springer, New York (2006)
6. Shao, A.M., Xi, S., Qiu, C.J.: A variational method for correcting non-systematic errors in numerical weather prediction. Earth Sci. **52**, 1650–1660 (2009)
7. Vannitsem, S.: Dynamical properties of MOS forecasts: analysis of the ECMWF operational forecasting system. Weather Forecast. **23**, 1032–1043 (2008)
8. Xue, H.-L., Shen, X.-S., Chou, J.-F.: A forecast error correction method in numerical weather prediction by using recent multiple-time evolution data. Adv. Atmos. Sci. **30**, 1249–1259 (2013)
9. Zjavka, L.: Wind speed forecast correction models using polynomial neural networks. Renew. Energy **83**, 998–1006 (2015)
10. Zjavka, L.: Numerical weather prediction revisions using the locally trained differential polynomial network. Expert Syst. Appl. **44**, 265–274 (2016)

Visibility Loss Detection for Video Camera Using Deep Convolutional Neural Networks

Alexey Ivanov and Dmitry Yudin(✉)

Belgorod State Technological University named after V.G. Shukhov,
Kostukova Str. 46, Belgorod 308012, Russia
ydin.da@bstu.ru

Abstract. The article describes the application of various machine learning methods for the analysis of images obtained from a video camera with the purpose of detection its partial or total visibility loss. Computational experiments were performed on a data set containing more than 6800 images. Support vector machine, categorical boosting and simplified modifications of VGG, ResNet, InceptionV3 architectures of neural networks are used for image classification. A comparison of the methods quality is presented. The best results in terms of classification accuracy are obtained using ResNetm and InceptionV3m architectures. The recognition accuracy is on the average more than 96%. The processing time per reduced input frame is 8–12 ms. The obtained results confirm the applicability of the proposed approach to the detection of camera visibility loss for real tasks arising in on-board machine vision systems and video surveillance systems.

Keywords: Image recognition · Convolutional neural network
Deep learning · Support vector machine · Boosting · Classification
Visibility loss · Video camera

1 Introduction

Determining the video camera visibility loss from the obtained images is an important subtask in the construction of any software and hardware video surveillance systems or vision systems in autonomous transport and production. This problem is the important factor in reducing the performance of tracking and object recognition.

Researchers distinguish several types of visibility loss of the video camera and use different approaches for their detection.

For example, in paper [1] the following categories are considered:

- Defocusing occurs as a result of imposing a lens or filter on the object, fog in the field of view of the camera, condensation on the lens. For control, the authors use the following algorithm: the image is divided into squares of 8×8 pixels and inside each such square the number of contrast pixels is counted. If their number is close to zero, then it is considered that the square is blurred. After that, the percentage of the blur is calculated, and if the threshold is exceeded, the camera is considered defocused.

A. Abraham et al. (Eds.): IITI 2018, AISC 874, pp. 434–443, 2019.
https://doi.org/10.1007/978-3-030-01818-4_43

- Unexpected camera rotation is expressed in a change in the orientation of the camera as a result of a breakdown of the camera's support or the predetermined lapel of the camera by the violator. To track the rotation, the authors suggest storing up to 5 frames of memory and working out if the threshold for changing the borders when turning is exceeded. The algorithm is based on comparing the previous and current frames in terms of brightness.
- Overlap is expressed in a strong darkening of the lens when it is closed, as well as in the event of failure of the illuminators or if the camera's automatic exposure system fails. Algorithm used by authors is the same as for checking the unexpected camera rotation.
- Backlight is expressed in strong clarification of the image, as a reaction to the "blinding" of the camera with an external light source, or if the automatic exposure system fails. The algorithm is based on the control of the image histogram. Based on the results obtained, the mean, median and standard deviation values are calculated. If any of the parameters are abruptly changed, a message is displayed about the exposure.

In the paper [2] authors use wavelet domain methods for image sequence analysis and detection of obscured camera view (obscuring by an object, full visibility loss) and reduced visibility (defocusing, backlight exposure).

Authors of research [3] describe detection of three camera tamper categories: covered, defocused and moved. They use signal activity as a measure of information in the image, its Kalman filtering, modeling likelihood probabilities as a mixture of Gaussian for tamper detection. To learn model parameters authors use synthetically generated image variants.

Paper [4] is devoted to detection of several camera artifacts – raindrops, dirt and scratches. For this task authors proposed algorithm based on normalized cross-correlation for two sequential images with delay.

Night fog detection for in-vehicle cameras is described in detail in the paper [5]. It based on detection and analysis of backscattered veil and halos around light sources.

Active visibility detection in foggy environment is proposed in [6]. It uses the system with laser and camera for fog recognition based on behavior of a laser beam under various conditions. However, experiments were carried out only in the laboratory.

Local Fog Detection in the image vanishing point considered in [7]. This approach based on saturation and RGB-correlation was tested only on datasets with the synthetic pictures.

Research [8] proposes more universal approach to classification of fog and fog-free scenes in day and night conditions. Authors generated image features using Fourier transform and Gabour filters and then classify them by linear classifier based on Fisher's Linear Discriminant Analysis.

Approach to mitigation of visibility loss in daylight fog conditions is considered in the research [9]. It uses road scene segmentation on the road, road-marking, sky areas and 3-D Model-Based Contrast Restoration.

Most of the existing methods use a image sequence to recognize the visibility loss while it may be necessary to analyze a single image. High quality results are obtained only for the case of fog detection on images and a case with total loss of visibility. Some of methods are successfully implemented for commercial cameras but have a closed algorithm description.

To detect various situations of partial visibility loss, for example, camera overlapping by obstacles, it is difficult to find the features for creating a reliable classification algorithm. At the same time, there is a sufficiently effective mathematical apparatus of deep convolutional neural networks [10] that allows solving similar problems for single images or image sequences.

This article is devoted to the investigation of various architectures of convolutional neural networks for the classification of images in order to detect the visibility loss of the video camera.

2 Task Formulation

Visibility loss detection for video camera is defined as the process of image classification based on image features automatically generated by deep convolutional network. Each of the classes represents a type of camera visibility loss:

- visibility loss,
- good visibility,
- partial visibility,
- blur or defocusing,
- overlapping by obstacle.

Formally: let P be a collection of N photographs, $P = \{p_1, p_2, \ldots, p_N\}$; the recognition purpose is to find photographs subsets E_i, where $i = 1, 2, \ldots, k$; $k = 5$ is the number of classes; each subset E_i is associated with one class.

For solving this task we have dataset which contains 6869 marked images with a size of 2704 × 1520 pixels, of which 5451 were assigned to the training sample, and 1418 to the test sample. The dataset has markup with five classes corresponding to different cases of camera visibility loss: “Visibility loss” (1840 images in training sample and 460 in test sample), “Partial visibility” (545 images in training sample and 136 in test sample), “Good visibility” (1616 images in training sample and 460 in test sample), “Blur” (684 images in training sample and 171 in test sample), “Obstacle” (766 images in training sample and 191 in test sample).

To measure the overall performance of the algorithm when recognizing each of the classes, we used the precision (P), recall (R), and accuracy (A) quality measures defined as follows:

$$P = \frac{TP}{TP + FP}, \; R = \frac{TP}{TP + FN}, \; A = \frac{TP + TN}{TP + FN + TN + FP}.$$

Measure R indicates that there are errors of the second kind – omissions, while P – about false positives. A shows the overall accuracy of the classification algorithm, equal to the ratio of correctly recognized images to the total number of images. Here, TP is true-positive, FP is false-positive, FN is false-negative and TN is true-negative classification results.

3 Classification of Image Features Using Support Vector Machine and Categorical Boosting

In the described cases of camera visibility loss, the entire image or part of it is defocused. Since only part of the image will be distorted in the camera partial visibility case, it is advisable to split the image into several regions of interest (ROI), as shown in Fig. 1. On each ROI various widely known features can be calculated [11], for example, the Brenner's focus measure. The calculation of this p_{ROI} characteristic for the image region I by the $N \times M$ size is carried out using a simple formula

$$p_{ROI} = \sum_{i=1}^{N} \sum_{j=1}^{M-2} (I(i,j) - I(i,j+2))^2.$$

In the task under consideration, the image is reduced to a resolution of 540 × 300 pixels and decomposed by 16 ROIs (Fig. 1). The Brenner's focus measures for each of the regions are then calculated and combined into a feature vector. It can be classified into one of the effective and fast methods of machine learning: support vector machine (SVM) [12] or categorical boosting (CatBoost) [13]. The latter method is actively gaining popularity after its publication in open access by the Russian IT-company Yandex.

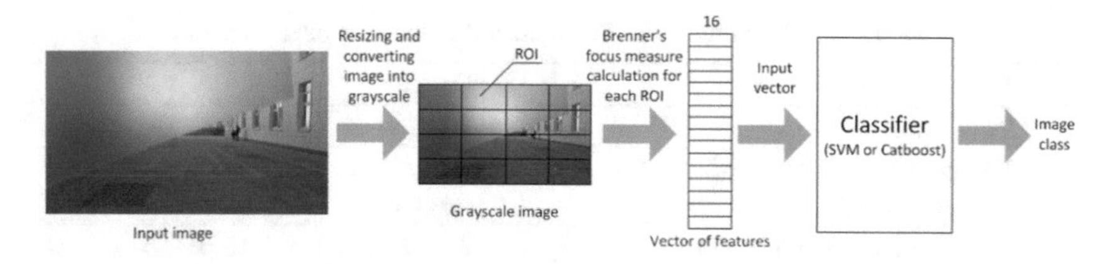

Fig. 1. Scheme of image classification using SVM or CatBoost

In this paper, these rapid and simple approaches will be considered, but the manual choice of characteristics is subjective, requires a lot of knowledge in the subject area and a search of various combinations of such features. The use of such features in the considered task is of the greatest difficulty in identifying the case of a partial visibility and camera overlapping by an obstacle.

The implementation of the automatic feature formation for image classification task is currently carried out using deep convolutional neural networks. In them input convolutional layers provide automatic formation of the image features of different levels of abstraction, and output fully connected layers (perceptron) are responsible for the classification of these features.

4 Neural Network Architectures for Camera Visibility Loss Detection

In this paper the main popular architectures of deep neural networks were used to classify images for the task of detection of different cases of camera visibility loss.

The VGG architecture, proposed in [14] in 2015, uses a 3 × 3 convolutional cascade and a fully connected layer at the output. It is more efficient in classification tasks than the earlier architecture AlexNet [15] of 2012. In this paper, a modification of the architecture [14] – VGGm (Table 1) is applied. It contains 12 convolutional layers and one fully connected. The authors used a similar method for complex objects classification in [16], where it showed its effectiveness.

Also in 2015, the ResNet architecture [17] was successfully presented and demonstrated high results. It consists of the so-called Residual (Identity) blocks, which allow the transmission of features from network layers close to the input, to layers close to the output of the network using summation operations. In this paper we apply a simplified modification of the architecture [17] – ResNetm, which contains 10 convolutional layers and one fully connected output layer (Table 2).

In 2014–2017, architectures like GoogLeNet [18] were developed. They are based on the Inception modules of several versions. These modules allow us to simultaneously generate features of images on various scales. This architecture, as a rule, gives more accurate results of classification, but is much more cumbersome and resource-intensive than VGG or ResNet. In this paper a modification of the modern architecture [19] – InceptionV3m (Table 3) is used. It contains Mixed-modules of several types (see Fig. 2).

Table 1. The architecture of the neural network model VGGm

No.	Type of layer	Details
1	Input layer	90 × 51 × 3 neurons
2	2D convolutional layer with ReLu activation function	7 × 7 core, 1 × 1 strides, 64 output maps with sizes 84 × 45
3	2D convolutional layer with ReLu activation function	3 × 3 core, 1 × 1 strides, 64 output maps with sizes 82 × 43
4	2D max pooling	Pool size 2 × 2, stride 2 × 2, 64 output maps with sizes 42 × 21
5	2D convolutional layer with ReLu activation function	3 × 3 core, 1 × 1 strides, 128 output maps with sizes 39 × 19
6	2D convolutional layer with ReLu activation function	3 × 3 core, 1 × 1 strides, 128 output maps with sizes 37 × 17
7	2D max pooling	Pool size 2 × 2, stride 2 × 2, 128 output maps with sizes 18 × 8
8	2D convolutional layer with ReLu activation function	3 × 3 core, 1 × 1 strides, 256 output maps with sizes 16 × 6
9	2D convolutional layer with ReLu activation function	3 × 3 core, 1 × 1 strides, 256 output maps with sizes 14 × 4
10	2D convolutional layer with ReLu activation function	3 × 3 core, 1 × 1 strides, 256 output maps with sizes 12 × 2
11	Dense (fully connected) layer with ReLu activation function and dropout	200 neurons, dropout with probability 0.5
12	Dense (fully connected) layer with softmax activation function	5 neurons

Table 2. The architecture of the neural network model ResNetm

No.	Type of layer	Details
1	Input layer	90 × 51 × 3 neurons
2	2D convolutional layer with batch normalization and ReLu activation function	7 × 7 core, 2 × 2 strides, 64 output maps with sizes 45 × 26
3	2D max pooling	Pool size 3 × 3, stride 2 × 2, 64 output maps with sizes 45 × 26
4	Convolution block	3 × 3 core, 1 × 1 strides, 256 output maps with sizes 22 × 12
5	Identity block	3 × 3 core, 2 × 2 strides, 256 output maps with sizes 22 × 12
6	Identity block	3 × 3 core, 2 × 2 strides, 256 output maps with sizes 22 × 12
7	Convolution block	3 × 3 core, 2 × 2 strides, 512 output maps with sizes 11 × 6
8	Identity block	3 × 3 core, 2 × 2 strides, 512 output maps with sizes 11 × 6
9	Identity block	3 × 3 core, 2 × 2 strides, 512 output maps with sizes 11 × 6
10	Identity block	3 × 3 core, 2 × 2 strides, 512 output maps with sizes 11 × 6
11	Average pooling	5 × 5 core, 512 output maps with sizes 2 × 1
12	Dense (fully connected) layer with softmax activation function	5 neurons

Table 3. The architecture of the neural network model InceptionV3m

No.	Type of layer	Details
1	Input layer	90 × 51 × 3 neurons
2	2D convolutional layer with ReLu activation function	3 × 3 core, 32 output maps with sizes 88 × 49
3	2D convolutional layer with ReLu activation function	3 × 3 core, 64 output maps with sizes 88 × 49
4	2D max pooling	Pool size 3 × 3, stride 2 × 2, 64 out-put maps with sizes 43 × 24
5	2D convolutional layer with ReLu activation function	1 × 1 core, 80 output maps with sizes 43 × 24
6	2D convolutional layer with ReLu activation function	3 × 3 core, 192 output maps with sizes 41 × 22
7	2D max pooling	Pool size 3 × 3, stride 2 × 2, 192 output maps with sizes 20 × 10
8	Mixed0 (Fig. 1a)	256 output maps with sizes 20 × 10
9	Mixed1 (Fig. 1a)	288 output maps with sizes 20 × 10
10	Mixed2 (Fig. 1a)	288 output maps with sizes 20 × 10
11	Mixed3 (Fig. 1b)	768 output maps with sizes 9 × 4
12	Mixed4 (Fig. 1c)	768 output maps with sizes 9 × 4
13	Global average pooling	768 maps
14	Dense (fully connected) layer with softmax activation function	5 neurons

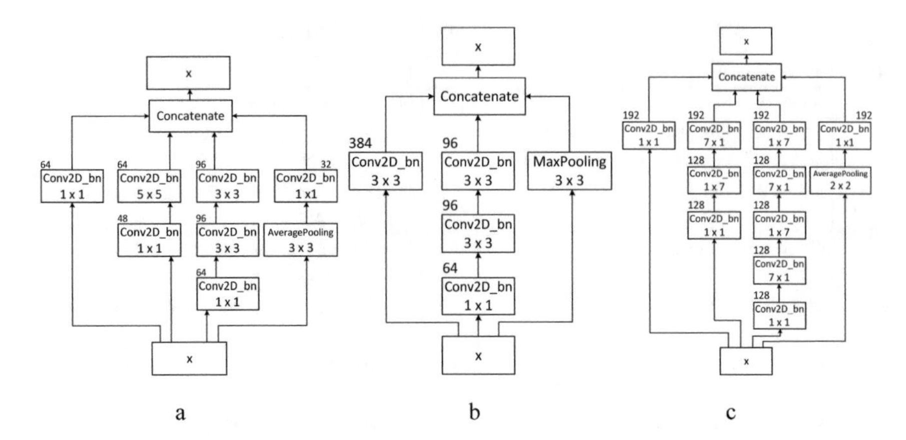

Fig. 2. Structures of various Mixed blocks for the InceptionV3 network.

The training process of described various architectures of neural networks is shown in Fig. 2. The highest learning efficiency (in terms of accuracy and speed) is provided by the ResNetm architecture for the task in question (Fig. 3).

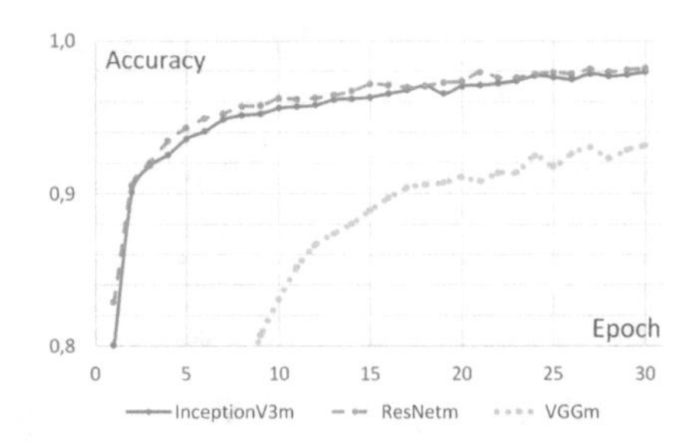

Fig. 3. The training process of various architectures of neural networks

5 Experimental Results

For experiments with classifiers described above we have used a personal computer with an Intel Core i5-2500 CPU, 3.30 GHz × 4, 8.00 GB RAM, NVIDIA GeForce GTX 1050 Ti (RAM: 4 GB) GPU, 64-bit operating system Windows 10. Visibility loss detection algorithms are implemented on Python 3.5.

SVM and CatBoost methods are tested using appropriate machine learning packages [12] and [13]. They are processed with CPU. Deep neural networks are implemented using deep classification software developed by authors [20] and are processed with GPU. This software is based on the Keras and TensorFlow libraries.

Comparative experimental results of classification quality for all used classifier models are presented in Tables 4 and 5.

Average classification time per frame in Table 5 is the time of processing of reduced input image: 540 × 300 pixels for the first two methods and 90 × 51 pixels for last three classifiers.

Average Classification Time Per Frame Significantly Varies for Different Image Sizes and Different Neural Network Architectures, as Shown in Fig. 4.

Table 4. Quality measures in the classification of the test sample images

Classifier	Loss of visibility		Good visibility		Partial visibility		Blur		Obstacle	
	P	*R*	*P*	*R*	*P*	*R*	*P*	*R*	*P*	*R*
SVM	0.83	0.96	0.91	0.45	0.81	0.09	0.37	0.99	0.55	0.44
CatBoost	0.96	0.98	0.95	0.98	0.82	0.68	0.97	0.94	0.87	0.90
ResNetm	**0.98**	0.99	**0.97**	0.99	0.91	0.75	**0.99**	**1.00**	0.91	**0.95**
VGGm	0.98	0.99	0.95	**1.00**	0.84	0.74	0.94	**1.00**	0.92	0.82
InceptionV3m	0.98	**1.00**	**0.97**	**1.00**	**0.95**	**0.87**	**0.99**	**1.00**	**0.96**	0.91

Table 5. Accuracy and speed of image processing

Classifier	Accuracy		Average classification time per frame, ms
	Training sample	Test sample	
SVM	0.6544	0.6671	**1.00**
CatBoost	0.9349	0.9030	1.99
ResNetm	**0.9823**	0.9668	8.36
VGGm	0.9484	0.9478	4.29
InceptionV3m	0.9814	**0.9753**	11.99

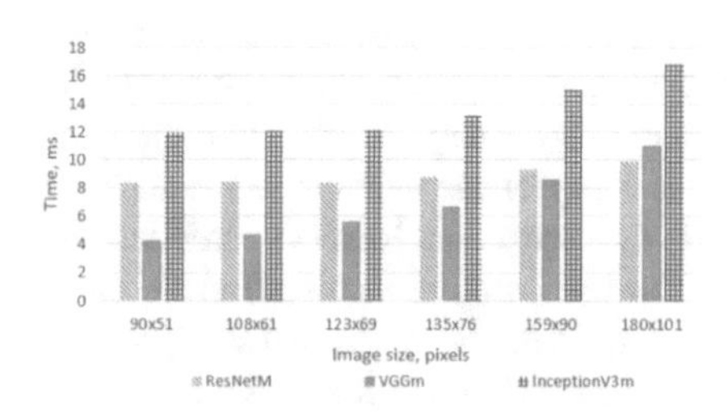

Fig. 4. Average classification time per frame by different neural networks

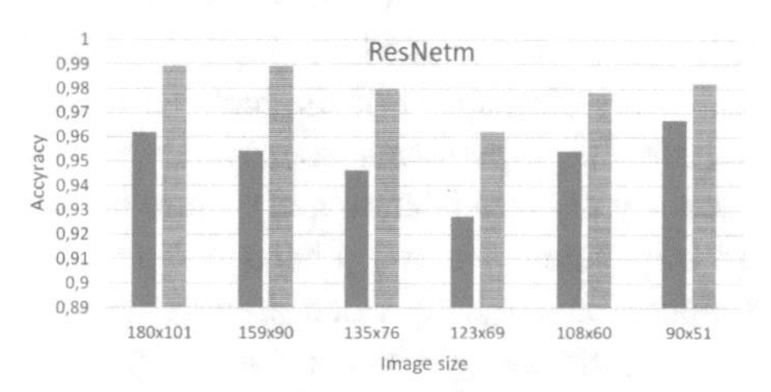

Fig. 5. Accuracy of convolutional neural network with ResNetm architecture

From the Tables 4 and 5 we can see that InceptionV3m net has the best classification quality on test sample but it also has the highest classification time. SVM method is the fastest but the most inaccurate. ResNetm is compromise classifier in the sense of accuracy on test sample (0.9668) and processing time per frame of 90 × 51 pixels size (8.36 s). It was found that the accuracy of training is practically independent of the size of the input images (Fig. 5). In Fig. 5 blue color shows the recognition accuracy on the test sample, a red color – on the training sample. In addition, thanks to its architecture, ResNet allows providing much more layers, but the number of parameters is smaller in comparison of other net architectures.

It should be noted than CatBoost classifier shows good quality results for 3 classes: loss of visibility, good visibility, blur. It also extremely fast because spend only 1.99 ms using CPU. While the described deep neural networks require the usage an energy-consuming GPU.

Examples of camera visibility loss detection using the developed software are shown in Fig. 6.

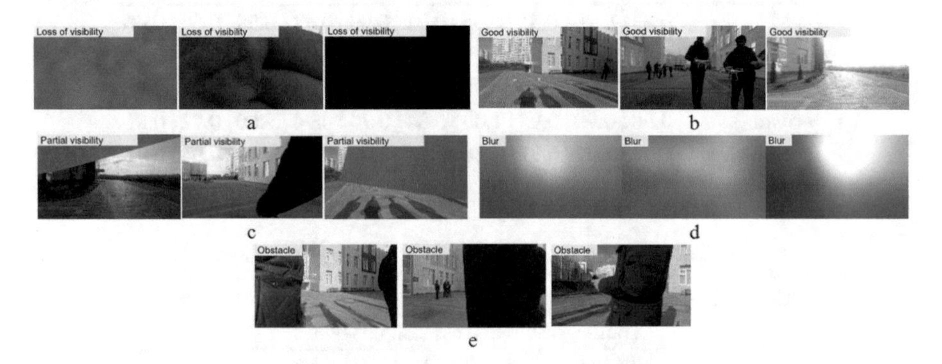

Fig. 6. Examples of camera visibility loss detection: a – visibility loss, b – good visibility, c – partial visibility, d – blur, e – obstacle

6 Conclusions

The methods for image classification considered in this paper allow solving the task of detecting five cases of camera visibility loss. Proposed deep convolutional neural networks have on the one hand high classification accuracy (0.94–0.97 on the test sample) and processing speed of the reduced original image (4–12 ms), and on the other hand require a high-performance graphics processor. A compromise network in the sense of accuracy of classification and computing speed is the considered ResNetm architecture. To implement this and other architectures, we can use both conventional video cards suitable for analyzing video streams of surveillance systems and embedded solutions, for example, an energy efficient and compact NVidia Jetson TX2 module that is widely used for embedded on-board vision systems for vehicles and driver assistance systems.

For cheaper implementation on the central processor it is more appropriate to use the method of image features classification based on categorical boosting for recognition cases of complete visibility loss, blur or good visibility. It demonstrates comparable classification accuracy for these classes with neural networks, but it has less computational complexity. CatBoost is much less demanding of the computing resources and will be able to operate in machine vision systems using budget single-board computers.

Acknowledgment. Research is carried out with the financial support of The Ministry of Education and Science of the Russian Federation within the Public contract project 2.1396.2017/4.6.

References

1. Tikhova, J.: Development sabotage detectors for surveillance systems Macroscop. Perm State National Research University (2013). https://www.scienceforum.ru/2013/pdf/7653.pdf. Accessed 11 May 2018
2. Aksay, A., Temizel, A., Cetin, A.E.: Camera tamper detection using wavelet analysis for video surveillance. In: Proceedings of the 2007 IEEE Conference on Advanced Video and Signal Based Surveillance, London, UK, pp. 558–562 (2007)
3. Mantini, P., Shah, S.K.: A signal detection theory approach for camera tamper detection. In Proceedings of the 14th IEEE International Conference on Advanced Video and Signal based Surveillance, Lecce, Italy (2017)
4. Einecke, N., Gandhi, H., Deigmöller, J.: Detection of camera artifacts from camera images. In: IEEE 17th International Conference on Intelligent Transportation Systems (ITSC), pp. 603–610 (2014)
5. Gallen, R., Cord, A., Hautière, N., Aubert, D.: Towards night fog detection through use of in-vehicle multipurpose cameras. Intelligent Vehicles Symposium (IV). IEEE (2011)
6. Miclea, R.-C., Silea, I.: Visibility detection in foggy environment. In: 20th International Conference on Control Systems and Computer Science (CSCS), pp. 959–964 (2015)
7. Alami, S., Ezzine, A., Elhassouni, F.: Local fog detection based on saturation and RGB-correlation. In: 13th International Conference on Computer Graphics Imaging and Visualization (CGiV) (2016)
8. Pavlic, M., Rigoll, G., Ilic, S.: Classification of images in fog and fog-free scenes for use in vehicles. In: Intelligent Vehicles Symposium (IV), pp. 481–486. IEEE (2013)
9. Hautiere, N., Tarel, J.-P., Aubert, D.: Mitigation of visibility loss for advanced camera-based driver assistance. IEEE Trans. Intell. Transp. Syst. **11**(2), 474–484 (2010)
10. LeCun, Y., Bengio, Y., Hinton, G.: Deep learning. Nature **521**, 436–444 (2015)
11. Pertuz, S., Puig, D., Garcia, M.A.: Analysis of focus measure operators for shape-from-focus. Pattern Recognit. **46**, 1415–1432 (2013)
12. Sklearn.svm.LinearSVC. Linear Support Vector Classification. http://scikit-learn.org/stable/modules/generated/sklearn.svm.LinearSVC.html. Accessed 11 May 2018
13. Prokhorenkova, L., Gusev, G., Vorobev, A., Dorogush, A.V., Gulin, A.: CatBoost: unbiased boosting with categorical features. arXiv:1706.09516v2 (2018)
14. Simonyan, K., Zisserman, A.: Very Deep Convolutions for Large-Scale Image Recognition. In: ICLR. arXiv:1409.1556 (2015)
15. Krizhevsky, A., Sutskever, I., Hinton, G.E.: ImageNet classification with deep convolutional neural networks. In: Proceedings of the 25th International Conference on Neural Information Processing Systems, NIPS 2012, vol. 1, pp. 1097–1105 (2012)
16. Yudin, D., Knysh, A.: Vehicle recognition and its trajectory registration on the image sequence using deep convolutional neural network. In: The International Conference on Information and Digital Technologies, pp. 435–441 (2017)
17. Kaiming, H., Xiangyu, Z., Shaoqing, R., Jian S.: Deep Residual Learning for Image Recognition. In: ECCV. arXiv:1512.03385 (2015)
18. Yudin, D., Zeno, B.: Event recognition on images by fine-tuning of deep neural networks. Adv. Intell. Syst. Comput. **679**, 479–487 (2018)
19. Szegedy, C., Vanhoucke, V., Ioffe, S., Shlens, J., Wojna, Z.: Rethinking the inception architecture for computer vision. In: ECCV. arXiv:1512.00567 (2016)
20. DeepClassificationTool. Deep image classification tool based on Keras. https://github.com/yuddim/deepClassificationTool. Accessed 11 May 2018

Neural Network Control Interface of the Speaker Dependent Computer System «Deep Interactive Voice Assistant DIVA» to Help People with Speech Impairments

Tatiana Khorosheva(✉), Marina Novoseltseva, Nazim Geidarov, Nikolay Krivosheev, and Sergey Chernenko

Kemerovo State University, Kemerovo, Russia
tkhorosheva@yandex.ru

Abstract. With the development of modern informational communication systems, voice control interface and speech recognition systems find application in various fields of activity. One application of such systems is for people with special needs who have speech impairments, and thus find using speech-dependent voice interfaces challenging. Our research team is developing a speaker dependent computer system «Deep Interactive Voice Assistant» (DIVA), which allows recognizing an arbitrary set of commands to control the computing system. The article presents the results of testing various artificial neural networks to train the machine to recognize vocal inputs. We examine such architectures as associative memory, multilayer perceptron and convolutional network. The research justifies the use of multilayer perceptron for the speaker dependent computer system DIVA as a training solution that demonstrated high results on a small selection. DIVA will be implemented in voice-user interface of such systems as «Smart House», mobile applications and IT-based assistive systems.

Keywords: Voice interface technology · Speech recognition technology
Assistive technologies · Neural network · Multilayer perceptron
Pattern recognition · Associative memory

1 Introduction

Voice user interface and speech recognition technologies have entered people's lives relatively recently and have since become indispensable. Development of these systems is a difficult task, as it requires the knowledge of the rules of a certain language, speech-to-text transformation technologies, as well as computational accuracy and efficiency while processing vocal information. A great amount of literature has dealt with creating such systems and increasing maximum accuracy and computational efficiency. One reason for the necessity to develop new systems, concepts and algorithms of voice processing is that voice interaction is becoming more convenient and, in some cases, the only available means to control intellectual information and technical systems. This

A. Abraham et al. (Eds.): IITI 2018, AISC 874, pp. 444–452, 2019.
https://doi.org/10.1007/978-3-030-01818-4_44

is extremely relevant to people with disabilities, whose integration into modern society is an issue that currently enjoys much attention [1]. Partial loss of voice or peculiarities thereof are main obstacles to speech recognition for people with disabilities. Modern speech recognition systems cannot be adapted to the individual features of said people, which prevents them from using these systems. A way to overcome this obstacle is to create trainable speaker-dependent systems based on artificial neural networks.

2 Overview of Available Voice User Interfaces for Speech Recognition and Control of Computer Systems

Consider the existing speech recognition systems and how people with speech impairments can use them. This overview allows determining a range of software solutions that enable speech recognition, and their key features [2, 3].

All speech recognition systems belong to two categories. The former is speaker-dependent systems that adjust themselves to the speaker's speech in the process of learning [4, 5]. These systems require complete readjustment in order to operate with a different speaker. The dependency on the speaker is contingent upon how the system uses data on the user's voice features. Speaker-dependent systems enable efficient voice recognition, albeit adjusting the system to each individual speaker is quite a laborious task. For instance, the Russian company VoiceLock has created automatic speech recognition software «Gorynych PROF» [6], which can be customized for individual users. At its core are achievements in the field of processing and recognizing Russian speech. Another example is the work of «Speech Technology Center» in Saint-Petersburg [7].

Speaker-independent systems are created for any speaker and require no training [4, 5]. Individual pronunciation features (such as speech tempo, timbre, etc.) complicates developing these systems. Nowadays, the developers are more focused on creating speaker-independent systems, which use vast amounts of data and do not take into account how people with speech impairments pronounce the commands to the computer.

The majority of modern software products have either proprietary or free license. It is impossible to adapt proprietary software for people with speech impairments since these applications are not open-source, disallowing changing the code. The world's largest international corporations create such systems. The examples are voice assistants, for instance, Apple's Siri, Google Assistant, Amazon's Alexa and Alisa, created by Yandex.

The code of free software open-source, which makes it possible to refine and adapt them for various situations. The most popular open-source systems are CMU Sphinx, Julius, Kaldi, RWTH ASR, iATROS. However, due to lack of comprehensive documentation, support for the Russian language in many systems and their limited commercial or scientific use [2, 3, 8–12], it is difficult to adapt these systems for people with disabilities.

As a result, this leads to the conclusion that modern voice user interface systems are speaker-independent and have proprietary license, which greatly limits their use in controlling computer technologies by people with disabilities.

3 Basic Approaches for Voice Recognition by Artificial Neural Networks

In most cases (using Google and Yandex as an example), voice recognition is carried out by algorithms based on numerous speakers who, albeit differently, pronounce the words «correctly». What interests us is creating a tool that will recognize a relatively limited set of commands, while it is irrelevant whether the speaker, i.e. user, might pronounce the words «incorrectly», since their speech might be incoherent. That is why the relevance utilizing existing pre-trained algorithms is rather dubious.

In order to recognize a limited set of commands, it is possible to use a classic multilayer perceptron [13], a number of feedforward layers or a convolutional network [14] with additional convolutional and pooling layers. Such networks have proven their worth in recognizing test samples, but they require a comparatively large number of training samples.

Another group of artificial neural networks is the recurrent neural networks. In these feedback structures, one neuron's output affects the others. Within the framework this research, of great interest is associative memory, designed to recognize images [15]: the network input receives a number of feature vectors, which form certain network states. After that, when a new feature vector is received, the network will go into state that resembles one of the memorized vectors, most similar to the new one.

The simplest to implement associative memory that compares with the training sample is the Hopfield Network [16], which in its original state has quite a limited memory capacity and is unstable. For speech recognition, the Hopfield Network is used in [17], while in [18] Hopfield Network is used to locate the audio signal, in [19] cost function is added into Hopfield Network by analogy to a mechanical system. Associative memory was further developed in the three-layer Hamming network [20] and bidirectional autoassociative Kosko network [21]. Willshaw network is also worth noting [22, 23]. Some studies [24, 25] suggest a dynamic associative memory, which remeasures weights while not only learning, but also operating as a response to input. This network is used to recognize the audio signal of voice commands.

Combined use of feedforward multilayer networks and recurrent networks is of high significance. [26] examines the use of associative memory based on multilayer perceptron and a connected recurrent network in noisy speech recognition, and [27] suggests a multilayer network with polynomial activation functions for recognizing the samples.

Originally, the developed networks were used to store rows of zeroes and ones. However, audio information is more complex, thus making it necessary to store floating-point values in the range of 0 to 1, which is why [28] suggests the Hopfield Network with an additional layer of neurons. In addition, [29] examines fuzzy Kosko networks, aimed at learning and recognizing the initially noisy input signal – here, the data might be floating-point values from 0 to 1.

In self-organizing networks [30], according to the rules of competition, activating a set of parameters defines the winning neuron, which corresponds to this set. It enables breaking the input data down into clusters and allocating each new vector to its own category. In [31], the authors suggest an artificial neural network based self-organizing incremental neural network. Such network is used to recognize both images and voice commands.

4 Testing Artificial Neural Networks to Evaluate the Quality of Voice Recognition

4.1 Developing Program Modules

The commands recognition module was realized on the Python 3 programming language. Python is a cross-platform language and can be launched on any machine with the Python interpreter installed and it contains a multitude of libraries for sound processing and machine learning. The following libraries were used during the research: the NumPy mathematical library, Wave sound library, Python speech features library, deep learning library Keras.

The user first trains the module, pronouncing an arbitrary command several times, and then moves on to pronounce the commands. Thus, the conduct of work of the module consists of the following steps:

1. Pre-processing the input audio signal. Here it is possible to perform Fourier transforms, allocate the mel-frequency cepstral coefficients (MFCCs), and obtain the image of the audio signal graph.
2. Learning/recognizing by the artificial neural network interface.

4.2 Preparing the Audio Signal

Training the neural network directly on the audio signal is inefficient. Due to the features of sound perception, the most suitable set of features is MFCCs, a logarithm of the spectrum capacity in a mel-frequency area that describes the power of the spectrum envelope, which characterizes the model of the vocal tract. To do this, the entire audio input is broken down into frames, with each undergoing the following:

1. The original voice input is recorded as:

$$x[n],\ 0 \le n \le N \tag{1}$$

2. The original signal's spectrum is found by means of the Fourier transform.

$$X_a[k] = \sum_{n=0}^{N-1} x[n] e^{\frac{-2\pi\pi i}{N}kn},\ 0 \le k \le N \tag{2}$$

3. The spectrum is mapped onto the mel scale.
4. The signal's power, which falls into each frame of the analysis, is obtained through the multiplication of the signal spectrum vectors and the window function.

$$S[m] = \ln\left(\sum_{k=0}^{N-1} |X_a[k]|^2 H_m[k]\right),\ 0 \le m \le N \tag{3}$$

5. The discrete cosine transform is applied to the obtained coefficients.

$$c[n] = \sum_{m=0}^{M-1} S[m]\cos(\pi n(m + 1/2)/M, 0 \le n \le N \quad (4)$$

The result is a set of MFCCs. We used a various number of frames (nframe – the number of frames: 10, 50, 100 and 200 frames) and coefficients in the window (ncoeff – the number of coefficients in one window: 10, 13, 15, 20 and 40 coefficients). In the end, we obtain a vector, which consists of nframe*nocoeff. This vector is sent to the ANN input at the next stage.

4.3 Testing the Artificial Neural Network Interface

The following types of artificial neural networks were tested for commands recognition

1. Multilayer perceptron with two hidden layers that contain 50 neurons with activation functions ReLu, tanh, Sigmoid. The output layer was given the activation function SoftMax.
2. Single-layer convolutional neural network.
3. Multilayer convolutional neural network, consisting of various layers: convolutional, subsampling layers and layers of fully connected of neural network.
4. Bidirectional autoassociative Kosko network (the Kosko network);
5. Self-organizing map (the Kohonen map, SOM).

These networks were tested on a test bench, which is an application in Python. The test bench carries out a series of actions corresponding to the modes of training and commands recognition (see Fig. 1). Work with the test bench begins with training. The user must record a training sample consisting of audio recordings. The recorded words are transformed into MFCCs, and each set of MFCCs is assigned a transcription of the recorded word. Afterwards, the MFCC feature vector is used to train the neural network. Finally, when training is complete, the program can work in the commands recognition mode.

We used 28 audio recordings for training, while testing included 12 audio recordings, broken down into four classes, each assigned a certain command, such as «open», «close», etc. Upon processing each audio recording, a vector consisting 390 MFCCs was created. Consider the test results of the chosen networks at various parameters.

Multilayer Perceptron. During our test, we used neural networks of various topology, differing in the number of layers, the number of neurons on a layer and activation functions: 390-4(softmax), 390-10(ReLU)-4(softmax), 390-100(ReLU)-100(ReLU)-4 (softmax), etc. Training time ranged from 10 to 16 s, while the accuracy of each network varied from 0.83 to 1. Simultaneously, an ensemble of neural networks was created. Apparently, changing the number of layers, neurons on a layer, the activation function and utilizing the network ensemble does not affect the accuracy.

Convolutional Network. During our test, we used convolutional networks of various topology: 390-1(1 × 3, ReLU)-4(softmax), 390-10(1 × 3, ReLU)-4(softmax), 390-10 (1 × 3)-50(ReLU)-4(softmax), 390-50(1 × 3)-50(ReLU)-4(softmax), etc. Recognition

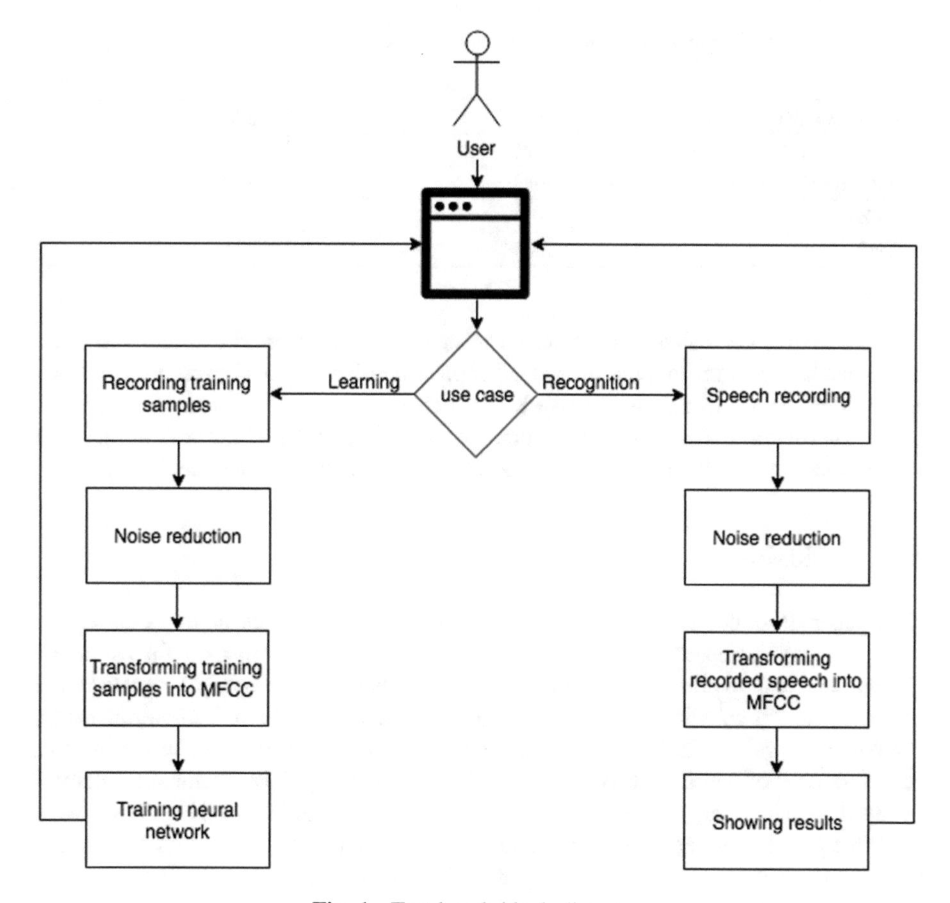

Fig. 1. Test bench block diagram.

accuracy is not higher than that of multilayer perceptron and varied from 0.75 to 1. However, this network requires less memory. Increasing the number of layers and neurons on a layer does not result in higher accuracy, while decreasing the number of neurons lowers the accuracy. Adding fully connected layers to the convolutional ones does not affect the accuracy.

Bidirectional Autoassociative Kosko Network. At first, MFCCs were used in training and testing the Kosko network. Each audio recording was transformed into 440 numbers within the range of 0 to 1. Binary features vectors were obtained from these numbers. Recognition accuracy peaked at 0.66. Training time: 1–2 s.

Self-organizing Maps (SOM). During our test, we used neural networks with the following topology: one-dimensional array of 4 neurons and a 10×10 two-dimensional array, a 50×50 two-dimensional array. Recognition accuracy varied from 0.6 to 0.85 (Table 1).

Table 1. A comparative analysis of the tested networks.

Network	Accuracy on the test sample	Average execution time
Multilayer perceptron	100%	~200 s
Single-layer convolutional network	100%	~200 s
Multilayer convolutional network	100%	~200 s
Kosko	66%	~1 s
SOM	85%	~3 min

As a result, we realized and used the following neural networks as the core of the command recognition module on test examples: multilayer perceptron, convolutional network, bidirectional autoassociative Kosko network and SOM. Tests have shown that activation function does affect the accuracy of the artificial neural network application. Perceptrons have demonstrated the highest accuracy on the test sample.

5 Conclusion

This paper describes the testing of neural networks with various input parameters in order to recognize speech commands, which was carried out by means of a test bench. During our test of the abovementioned neural networks, the optimal structure was chosen: a neural network with three convolutional layers of 50 size-3 one-dimensional convolutions and two hidden fully connected layers of 100 neurons. A further increase in the number of neuron layers does not affect the accuracy. The obtained results will be used in the speaker dependent computer system «Deep Interactive Voice Assistant» for people with speech impairments, which is being developed by the authors.

This work was supported by the grant «Computer appliance for verbal communication for people with speech impairments».

References

1. Convention on the Rights of Persons with Disabilities (CRPD): http://www.un.org/development/desa/disabilities/convention-on-the-rights-of-persons-with-disabilities.html. Accessed 01 May 2018
2. Gaida, C.: Comparing open-source speech recognition toolkits. http://suendermann.com/su/pdf/oasis2014.pdf. Accessed 01 May 2018
3. Gazetić, E.: Comparison Between Cloud-based and Offline Speech Recognition Systems. https://mediatum.ub.tum.de/doc/1399984/1399984.pdf. Accessed 01 May 2018
4. Rybka, J., Janicki, A.: Comparison of speaker dependent and speaker independent emotion recognition. Appl. Math. Comput. Sci. **4**(23), 797–808 (2013)
5. Lee, K., Huang, X.: On speaker-independent, speaker-dependent, and speaker-adaptive speech recognition. IEEE Trans. Speech Audio Process. **2**(1), 150–157 (1993)
6. Senkevich, G.: Computer for People with Disabilities. BHV-Petersburg, St. Petersburg (2014)
7. Center of Speech Technologies: https://www.speechpro.ru/. Accessed 01 May 2018

8. El Amrania, M., Hafizur Rahmanb, M., Wahiddinb, M., Shahb, A.: Building CMU Sphinx language model for the Holy Quran using simplified Arabic phonemes. Egypt. Inform. J. **3** (17), 305–314 (2016)
9. Tampel, I.: Automatic speech recognition - the main stages of 50 years. Sci. Tech. Her. Inf. Technol. Mech. Opt. **6**(15), 957–968 (2015)
10. Roebuck, K.: Speech Recognition: High-Impact Emerging Technology - What You Need To Know: Definitions, Adoptions, Impact, Benefits, Maturity, Vendors. Emereo Publishing, Australia (2012)
11. Povey, D.: The Kaldi speech recognition toolkit. In: IEEE 2011 Workshop on Automatic Speech Recognition and Understanding, pp. 1–4 (2011)
12. Lange, P., Suendermann-Oeft, D.: Tuning Sphinx to outperform Google's speech API. In: Proceedings of the ESSV 2014, Conference on Electronic Speech Signal Processing, Dresden, Germany (2014)
13. Simon, O.: Haykin Neural Networks and Learning Machines, 3rd edn. Pearson, Upper Saddle River (2009)
14. Zhang, Y., Pezeshki, M., Brakel, P., Zhang, S., Bengio, C.L.Y., Courville, A.: Towards end-to-end speech recognition with deep convolutional neural networks, CoRR, vol. abs/1701.02720. http://arxiv.org/abs/1701.02720 (2017)
15. Vazquez, R.A., Sossa, H.: Associative Memories Applied to Image Categorization. In: Martínez-Trinidad, J.F., Carrasco Ochoa, J.A., Kittler, J. (eds.) CIARP 2006. LNCS, vol. 4225, pp. 549–558. Springer, Heidelberg (2006)
16. Hopfield, J.J.: Neural networks and physical systems with emergent collective computational abilities. Proc. Natl. Acad. Sci. **79**, 2554–2558 (1982)
17. Vaishnavi, Y., Shreyas, R., Suhas, S., Surya, U.N., Ladwani V.M., Ramasubramanian, V.: Associative memory framework for speech recognition: adaptation of Hopfield network. In: IEEE Annual India Conference (INDICON), Bangalore, pp. 1–6 (2016)
18. Ladwani, V.M., Vaishnavi, Y., Shreyas, R., Vinay Kumar, B.R., Harisha, N., Yogesh, S., Shivaganga, P., Ramasubramanian, V.: Hopfield net framework for audio search. In: Communications (NCC), pp. 1–6. https://doi.org/10.1109/ncc.2017.8077074 (2017)
19. Barra, A., Beccaria, M., Fachechi, A.: A relativistic extension of Hopfield neural networks via the mechanical analogy. arXiv:1801.01743v1 (2018)
20. Hamming, R.: Coding and Information Theory. Prentice-Hall, Englewood Cliffs (1968)
21. Kosko, B.: Adaptive bidirectional associative memories. Appl. Opt. **26**(23), 4947–4960 (1987)
22. Willshaw, D.J., Buneman, O.P., Longuet-Higgins, H.C.: Non-holographic associative memory. Nature **222**, 960–962 (1969)
23. Stöckel, A.: Design Space Exploration of Associative Memories Using Spiking Neurons with Respect to Neuromorphic Hardware Implementations. Universität Bielefeld, Bielefeld (2016)
24. Vázquez, A.: New associative model with dynamical synapses. Neural Process. Lett. **28**(3), 189–207 (2008)
25. Vázquez, R. Sossa, H.: Voice translator based on associative memories. In: Advances in Neural Networks, pp. 341–350 (2008)
26. Minghu, J., Biqin, L., Baozong, Y.: Speech recognition by using the extended associative memory neural network (EAMNN). In: IEEE International Conference on Intelligent Processing Systems, vol. 2, pp. 1777–1780 (1997)
27. Krotov, D., Hopfield, J.: Dense associative memory for pattern recognition. In: Advances in Neural Information Processing Systems 29, pp. 1172–1180 (2016)

28. Giovanni, C.: Design of associative memory for gray-scale images by multilayer Hopfield neural networks. In: Proceedings of the 10th WSEAS International Conference on CIRCUITS, Vouliagmeni, Athens, Greece, pp. 376–379 (2006)
29. Sussner, P., Esmi, E., Villaverde, I., Graña, M.: The Kosko subsethood fuzzy associative memory (KS-FAM): mathematical background and applications in computer vision. J. Math. Imaging Vis. **42**, 134–149 (2012)
30. Kohonen, T.: Self-organizing Maps, 3rd Extended edn. Springer, New York/Heidelberg (2001)
31. Furao, S., Ouyang, Q., Kasai, W., Hasegawa, O.: A general associative memory based on self-organizing incremental neural network. Neurocomputing **104**, 57–71 (2013)

Hierarchical Reinforcement Learning with Options and United Neural Network Approximation

Vadim Kuzmin[1] and Aleksandr I. Panov[2,3](✉)

[1] National Research University Higher School of Economics, Moscow, Russia
[2] Moscow Institute of Physics and Technology, Moscow, Russia
panov.ai@mipt.ru
[3] Federal Research Center "Computer Science and Control" of the Russian Academy of Sciences, Moscow, Russia

Abstract. The "curse of dimensionality" and environments with sparse delayed rewards are one of the main challenges in reinforcement learning (RL). To tackle these problems we can use hierarchical reinforcement learning (HRL) that provides abstraction both on actions and states of the environment. This work proposes an algorithm that combines hierarchical approach for RL and the ability of neural networks to serve as universal function approximators. To perform the hierarchy of actions the options framework is used which main idea is to utilize macro-actions (the sequence of simpler actions). State of the environment is the input to a convolutional neural network that plays a role of Q-function estimating the utility of every possible action and skill in the given state. We learn each option separately using different neural networks and then combine result into one architecture with top-level approximator. We compare the performance of the proposed algorithm with the deep Q-network algorithm (DQN) in the environment where the aim of the magnet-arm robot is to build a tower from bricks.

Keywords: Hierarchical reinforcement learning · Options
Neural network · DQN · Deep neural network · Q-learning

1 Introduction

Most of the real life problems where the reinforcement learning (RL) can be used are prone to exponential growth of the size of stored data with the increase of the problem's scale. This is called the "curse of dimensionality". Also it can happen that the feedback from the environment is delayed and sparse which leads to the long sequence of actions the agent needs to perform in order to get it. This makes it harder for the agent to find and learn that long sequence. Hierarchical reinforcement learning (HRL) aims to contend with the described difficulties performing the abstraction on the set of available actions. Botvinick [3] defines HRL as computational techniques that allow RL to support temporally extend

A. Abraham et al. (Eds.): IITI 2018, AISC 874, pp. 453–462, 2019.
https://doi.org/10.1007/978-3-030-01818-4_45

actions. HRL also tries to decompose the given problem into smaller sub-tasks so that their sequential solution is more effective than the attempt to solve the initial task at once. Especially the decomposition of the problem is important in robotic tasks, in which the robot often has to apply certain sequences of actions. In our work we consider a problem from the Brick world, which is close to real robotic tasks.

Nowadays, artificial neural networks (ANNs) is the most effective tool for image recognition and representation. Their ability to extract useful features from the raw data can be used in robotics and RL in particular. Usually the data from robotic system's sensors is an image of the environment's state and based on it the robot is planning its actions. The capability of neural networks to serve as universal function approximators allows to use them as a value function or a Q-function and in this case the network is called the Q-network. The usage of neural network also dispenses with the necessity of storing all the data explicitly which helps with the dimensionality problem.

The aim of this work is to develop an algorithm of hierarchical reinforcement learning that exploits neural networks. To test the algorithm we simulate the work of a magnet arm robot which goal is to build a tower of required height from bricks. We compare its performance with the deep Q-network algorithm (DQN) [6].

2 Related Work

Hierarchical reinforcement learning has been a field of extensive research efforts for more than 20 years and, as with other disciplines, rapid development of ANNs has opened new perspectives and allowed to solve a bigger amount of tasks. The main trend of the last works in this field is building end-to-end solutions with automatic discovery of profitable hierarchy of actions. We will start with the description of fundamental works in HRL and then move to the latest advances in integration of neural networks into the HRL frameworks.

2.1 Hierarchical Reinforcement Learning

Markov decision process (MDP) is the base of a standard RL problem, while in the HRL setup semi Markov decision process (SMDP) is used because it takes into account that the action can take more than one time unit. More precisely, in SMDP the time of the process being in a particular state is a random variable. SMDP is defined as a tuple $<S, A, T, R, \gamma>$, where S and A are the set of states and actions respectively, $T(s', N|s, a)$ - the transition function (N is a number of time units the action a took, $R(s, a)$ - the reward function, γ - the discount factor needed for the process that can take infinite number of steps. To solve an SMDP means to find an optimal policy (π^*) that maps every environment's state $s \in S$ with the action $a \in A$ so as to maximize the cumulative reward.

Three basic frameworks of HRL can be noted, all of them use the notation of SMDP and initially implied that a programmer sets a hierarchy manually.

The first framework is called "options" [8], according to it the agent can choose between not only basic actions, but also macro-actions (skills) that are defined as a triple $<I, \pi, \beta>$, where $I \subseteq S$ - the set of states where the option can be chosen, $\pi : S \times A \rightarrow [0, 1]$ - the option's policy that for every possible state $s \in I$ and basic action a defines the probability of this action to be taken, $\beta : S \rightarrow [0, 1]$ - the option's termination condition. In its first version the framework operates with the preliminary given options with optimal policies. The policy over options can be learned with the standard RL algorithms but adjusted for SMDP. Further we provide the formalized Q-learning algorithm adjusted for options described in [8].

Algorithm 1. Q-learning on Options with ϵ-greedy exploration

```
Initialize Q(s, o_i) = 0 ∀o_i : s ∈ I_i, ∀s ∈ S ;
for number of episodes do
    reset the environment;
    observe the state s;
    while episode in not terminated do
        select an option o with ε-greedy from {o_i : s ∈ I_i};
        while o is not terminated do
            follow the option's policy;
            accumulate the reward, time and observe the state s';
            if s' terminates the episode then
                break
        observe the option's accumulated reward r, the next state s'
        (where the option terminated) and the time the option took k;
        Q(s, o) ← (1 − α)Q(s, o) + α[r + γ^k max_{o'∈{o_i:s'∈I_i}} Q(s', o')];
        if s' terminates the episode then
            break
        s ← s'
```

Hierarchies of Machines (HAMs) [2,7] is the second basic approach that constitutes of the set of non-deterministic finite-state machines. They are the programs that set restrictions on the actions the agent can perform in a given state. This is different from the options framework where the choice of action enlarges. Every machine is defined as the set of states, transition function and starting function that determines the initial state of the machine. There are 4 types of machines states: action performs a step in the environment, call is needed to lunch the other machine, choice non-deterministically picks the next state, stop terminates the execution of the current machine and returns back to the last call state. With the set of machines it is possible to reach the hierarchy of any depth by calling one machine through another with the call state. With the HAMs the set of possible policies is bounded and in this case the aim of the agent is to find the right sequence of transitions between the machines states. It is proven that the Cartesian product of the set of the environments states and the set of the machines states defines the SMDP solving which is equivalent to solving the initial MDP. If the Q-learning idea is used then for the HAMs the Q-function update rule is the following:

$$Q([s_c, m_c], a) \leftarrow (1 - \alpha)Q([s_c, m_c], a) + \alpha[r_c + \beta_c \max_{a'} Q([t, n], a')],$$

where s_c - state of the environment on a previous step, m_c - state of the machine on a previous step, t and n - current state of the environment and machine respectively, a - action chosen at the previous step, r_c and β_c - cumulative reward and discount from the last choice state respectively.

The third approach, MAXQ [4], performs the decomposition of the value function for the initial MDP into the set of value functions for subtasks, this set has a tree structure where the solution of the root subtask solves the whole MDP. It is implied that a programmer manually chooses the subtasks the agent has to solve. When the initial value function is decomposed the standard algorithms of RL for SMDP can be used.

In the current work we use the options framework to perform the hierarchy as they are easily interpretable, well studied and represent the most natural way of building the hierarchy.

2.2 Approximation with Neural Networks

It is known that an ANN can approximate complex function with the high precision, therefore, it can be used to replace tabular format of value function and Q-function storage. Furthermore, there are several successful applications of neural networks to HRL.

The work [1] operates exactly with the options framework and relies on the policy gradient theorems. A new option-critic (the critic is a part of the algorithm that evaluated the policy) architecture is proposed and it aims to find options automatically learning them in parallel with the policy over options. Under the assumption that functions representing the options policies and their termination functions are differentiable, the stochastic gradient descent (SGD) is used to perform the optimization. In the Arcade Learning Environment (ALE) a deep convolutional neural network is used to approximate intra-option policies, termination functions and policy over options. Presented approach is an end-to-end solution that requires only the number of options to be learned. Shown results confirm the ability to find interpretive options profitable for playing the testing games.

A combination of options and deep Q-networks is presented in [5]. Two-level hierarchy is proposed where meta-controller chooses sub-goals for options and sends it to the controller which tries to reach that goal operating only basic actions, the controller gets the reward from the internal critic. Meta-controller and controller are deep convolutional neural networks that receive image as an input (plus the sub-goal in case of internal controller) and output the Q-function values. They use SGD to train the networks and the Q-learning update rule extended on options. In the beginning of the process only controller is trained, after some time it is followed by the mutual training process. There is an attempt to automatically discover options, however, in the experiment with the Atari game "Montezumas Revenge" the problem was not solved to the full extent.

The novel recurrent network architecture is introduced in [9]. It allows to learn macro-actions automatically only interacting with the environment. The architecture consists of two modules: the "action-plan" determines the sequence of elementary actions, the "commitment-plan" decides at what moment the macro-action terminates and the "action-plan" is updated. The difference from the previous models is that instead of planning only one next step this network outputs a matrix where every column is a step starting from the current one and every row is a basic action. With such matrix the network builds a plan for some horizon of steps, follows this plan and updates the matrix to change the macro-action. The experiments are conducted on Atari games where the convolutional network is used to decode the state of the environment. The results showed that using the found macro-actions helped to achieve higher game scores. In the task of text generation LSTM network is used instead of convolutional layers, the experiment resulted in frequent n-grams that also proves the correctness of the model.

Thus, in all the reviewed works there is a tendency for using convolutional and recurrent networks and so-called "managers" controlling the hierarchy.

3 Methodology

As it was mentioned before, the options framework is used to implement the hierarchy. The set of options must be determined by the programmer in advance where every option is a separate Q-network pretrained on a training environment with the standard DQN algorithm [6].

The main challenge of using the non-linear approximator for the Q-function is that convergence can not be guaranteed. But there are techniques able to enhance the stability, they are "experience replay" and usage of "target network". The experience replay implies using a special buffer that stores the transition made by the agent (the current state, the action, the next state, reward and the flag of termination) then every time we sample a batch of transitions that is used to perform the optimization with the ADAM optimizer. Such technique ensures the data used for training is not correlated. We need a separate "target network" that has the same architecture as the main one and copies its weight periodically. It is used to produce the target value for computing the error and also provides the stability of the learning process.

Hereafter we provide the algorithm that uses the framework of options and neural networks as approximators for intra policies and the policy over options. Let us introduce the following notation: s - the environment's state, o - the option, r - the reward, Q_ϕ - the Q-function for the main network, $Q_{\phi'}$ - the Q-function for the target network, ϕ, ϕ' - the weights of the main and target network respectively, y - the target value for the $Q_\phi(s, a)$. See the Algorithm 2.

Algorithm 2. DQN with options and ϵ-greedy exploration

Data: environment, Q_ϕ - network for the Q-function, α - learning rate, γ - discount factor, replay buffer size, batch size, N - frequency of updating the target network, $O = \{o_i\}_{i=1}^n =< I_i, \pi_i, \beta_i >_{i=1}^n$ - set of options;
reset the environment and observe the state s_i;
last observation := s_i;
for *number of steps* **do**
  save to the buffer the last observation;
  select an option o_i with ϵ-greedy from $\{o_i : s \in I_i\}$;
  while o_i *is not terminated* **do**
    follow the option's policy;
    accumulate the reward, time and update the state s';
    if s' *terminates the episode* **then**
      $done_i$:= True ;
      break
  observe the option's accumulated reward (r_i), the next state (s_i') (where the option terminated) and the time the option took (k_i);
  save the effect ($o_i, r_i, r_i, done_i, k_i$) in the state s_i to the buffer;
  last observation := s_i';
  Perform the training:;
  sample batch ($s_j, o_j, s_j', r_j, done_j, k_j$);
  compute the target value: $y_j = r_j + (1 - done_j)\gamma^{k_j} \max_{o_j'} Q_{\phi'}(s_j', o_j')$;
  update the wights of the main network using the following gradient:
  $\phi \leftarrow \phi - \alpha \sum_j (\frac{dQ_\phi}{d\phi}(s_j, o_j)(Q_\phi(s_j, o_j) - y_j))$;
  update the weight of the target network every N steps: $\phi' \leftarrow \phi$

4 Experiments

The goal of the experiments is to verify the convergence of the algorithm, check how successfully it overcomes the "curse of dimensionality" and the problem of sparse delay reward.

4.1 Experimental Environment

As a testing environment we have chosen a model of a magnet arm robot that can grab and move bricks. The base step is to mathematically describe the environment, the agent and their interplay, which can be done by defining an MDP. There is a two dimensional space where the agent can perform 6 basic actions: one move for each direction and turning the magnet on or off. The state is represented by the two dimensional matrix where brick is coded by one, turned on magnet - by two, turned off magnet - by three, empty cell - by zero. Such matrix can be used to produce an RGB image representing the state (Fig. 1). The goal of the agent is to build a tower from bricks, the required height is specified. At the beginning of the episode the agent is placed at the top left cell of the grid and the bricks are places from the right to the left. The reward function is set the way that the agents gets a small negative reward for every action and a large positive reward only when the tower of required height is built and the magnet is off. The built MDP is deterministic as all the transitions between states are executed with the probability equal to one.

Figure 1 represents three different states of the environment with the size 4×3 pixels and 2 bricks. The left image is the starting position, the middle one shows that the magnet is on and the brick is being moved, the right one is the finish state when the tower is built and the magnet is off.

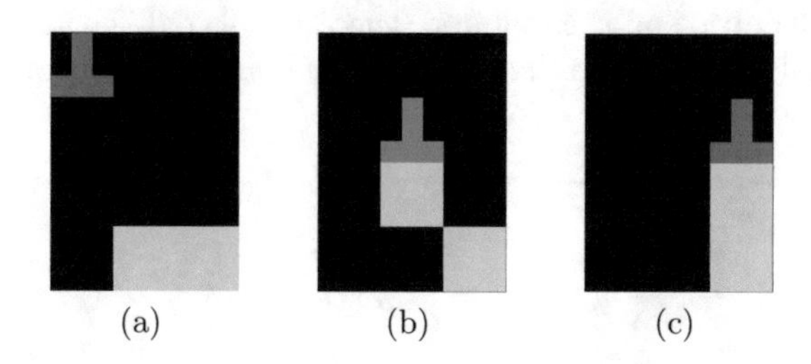

Fig. 1. Magnet arm environment example. Yellow squares are blocks, a red inverted T is an arm with working magnet, a blue inverted T is an arm with idle magnet.

The chosen environment is easily scalable as it is possible to change the size of the grid, the amount of bricks and the required height of the tower. This will allow us to understand how the learning time and convergence changes with the growth of the problem's size. As the reward function is set to give positive feedback only when the tower is built the problem has delayed feedback. This task is also relevant because it can be modelled in 3D to test a real robot.

4.2 Experimental Details

With the described method any amount of options can be utilized and basic actions are treated as options that take exactly one time unit. We have implemented two options, the first one makes the robot to go down and reach the brick or the bound and the second one makes the robot to go down, grab the brick and lift it to the top. Both of this options can be used for different manipulations with bricks.

To train the options' Q-networks we use separate training environments. The first difference of the training environment from the experimental one is their termination condition that tells that the episode is over when the condition of the option is fulfilled. The second difference is that in every new episode the position of the bricks is arbitrary and the agent can be anywhere in the top raw of the matrix. This will allow options to be flexible enough and be initiated from any environment's state.

In the experiments we use the buffer with the size of 1 million transitions, the learning starts after 5500 steps and the target network is updated every 200 steps. The baseline is the DQN algorithm that does not include any hierarchy. The neural network architecture is similar to what have been used in [6] for playing Atari games. The input to the network is an RGB image of varying size depending on a problem. The network has three convolutional layers followed by two fully connected layers where the last one has as many neurons as there are actions in the environment including options. The convolutional layers have 32, 64 and 64 neurons respectively with the kernel size 8, 4 and 3 respectively, and the stride 2, 4 and 1 respectively.

We are testing out algorithm on different sizes of the problem starting with the 4×3 grid, 3 cubes and the required height equal to 3. Then move to a larger 6×4 grid with 4 bricks where the robot has to build a tower from all the bricks.

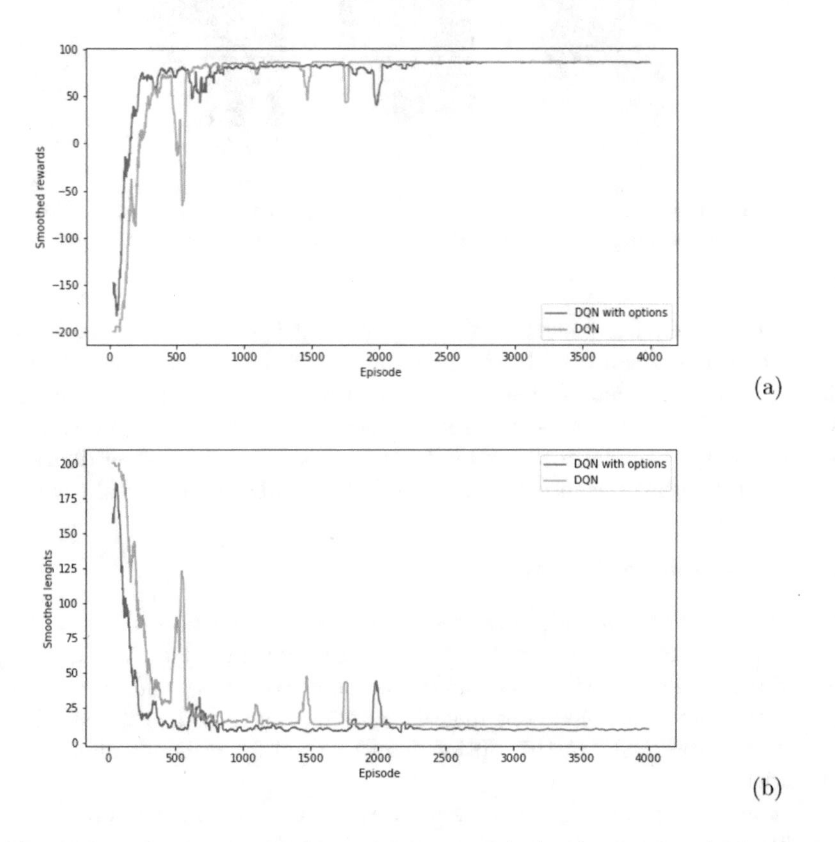

Fig. 2. Learning process in the environment with the size 4×3 and 3 bricks

4.3 Results

The graphs represent the smoothed curves of the learning process. The curves show how the episode's reward or length was changing. The Fig. 2 is the case for the first experimental setup. It can be seen that the algorithm exploiting

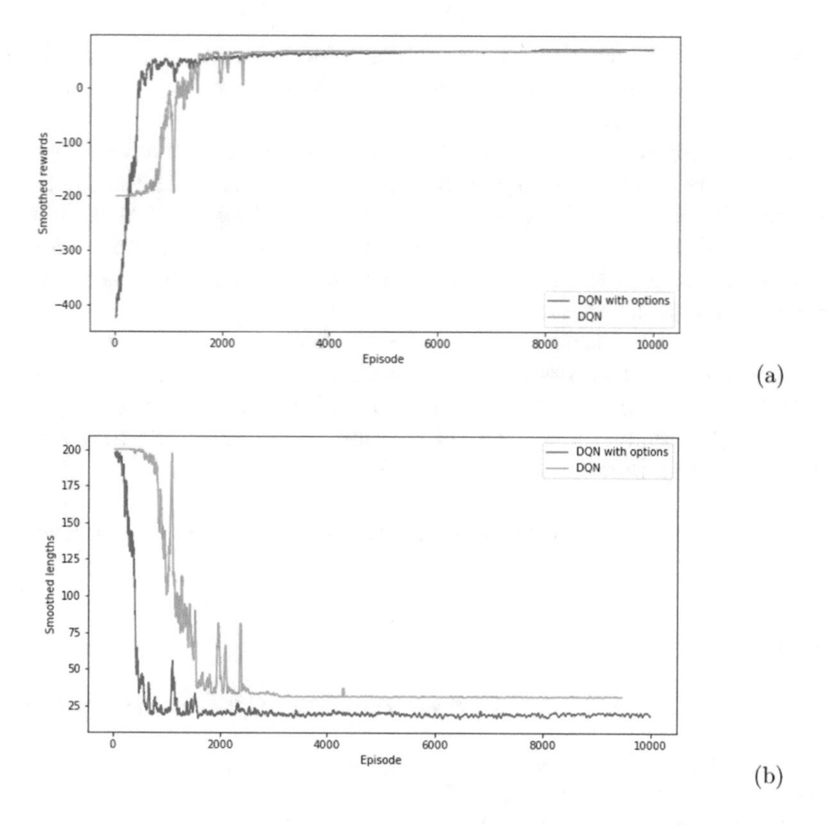

Fig. 3. Learning process in the environment with the size 6×4 and 4 bricks

options finds the right solution faster, gets the positive reward earlier and needs less actions (the option counts as one action). However, the 4×3 environment is not large enough to see the profit of the options. In this case the agent can handle the process of building a tower without options and the problem of the delayed reward is not that challenging.

The larger problem helps to show that the options can give noticeable acceleration of the learning process. The Fig. 3 show the results for the 6×4 environment with 4 bricks. Here the usage of options allows to outperform the standard DQN significantly. The agent with the options starts to reach the positive episode reward in double-quick time and takes less actions to finish the episode (almost half as much actions as the agent without options).

5 Conclusion and Future Work

Taking into account the derived results it can be stated that using options the agent explores the environment more efficiently and can find the profitable plan of actions faster. The proposed algorithm exploiting the combination of neural networks and the HRL can be used to tackle the "curse of dimensionality" and delayed rewards. There is also a strong possibility of transfer learning, because the learned options are not specific to any problem in the environment with the magnet arm and bricks.

There is a minor drawback connected to the experimental environment that can lead to the agent ending up without using the options. As the agent gets a small negative reward, taking an option results in a larger negative cumulative reward which leads to the penalty for taking the option. To overcome this issue, we can either change the reward function or eliminate the basic actions being sure that the options are enough to build the tower.

It is evident that because of the same architecture of all the networks lower layers can be shared between them. Therefore, the next step is to develop an architecture that will combine intra policies with the policy over option, joining the network's architectures. Another possible direction of the future work is finding a way of automatic discovery of sub-goals for options.

Acknowledgements. This work was supported by the Russian Science Foundation (Project No. 18-71-00143).

References

1. Bacon, P.-L., Harb, J., Precup, D.: The Option-Critic Architecture. arXiv:1609.05140v2 (2016)
2. Bai, A., Russell, S.: Efficient reinforcement learning with hierarchies of machines by leveraging internal transitions. In: Proceedings of the Twenty-Sixth International Joint Conference on Artificial Intelligence. Main track, pp. 1418–1424 (2017)
3. Botvinick, M.M.: Hierarchical reinforcement learning and decision making. Curr. Opin. Neurobiol. **22**, 956–962 (2012)
4. Dietterich, T.G.: Hierarchical reinforcement learning with the MAXQ value function decomposition. arXiv:cs/9905014 (1999)
5. Kulkarni, T.D., Narasimhan, K.R., Saeedi, A., Tenenbaum, J.B.: Hierarchical deep reinforcement learning: integrating temporal abstraction and intrinsic motivation. arXiv:1604.06057 (2016)
6. Mnih, V., et al.: Human-level control through deep reinforcement learning. Nature **518**(7540), 529533 (2015)
7. Parr, R., Russell, S.: Reinforcement learning with hierarchies of machines. In: Advances in Neural Information Processing Systems: Proceedings of the 1997 Conference. MIT Press, Cambridge (1998)
8. Sutton, R.S., Precup, D., Singh, S.: Between MDPs and semi-MDPs: a framework for temporal abstraction in reinforcement learning. In: Artificial Intelligence (1999)
9. (Sasha) Vezhnevets, A., et al.: Strategic attentive writer for learning macro-actions. In: Proceedings of NIPS (2016)

New Approaches to Discrete Modeling of Natural Neural Networks

Oleg Kuznetsov, Ludmila Zhilyakova, Nikolay Bazenkov, Boris Boldyshev, and Sergey Kulivets(✉)

Trapeznikov Institute of Control Sciences of RAS, Moscow 117997, Russia
{olpkuz, skulivec}@yandex.ru

Abstract. A discrete model of multitransmitter interactions between neurons in a common extracellular space (ECS) is proposed. Neurons in the model are heterogeneous in three different senses. They differ in (i) the type of endogenous change in the membrane potential, (ii) the type of secreted neurotransmitter, and (iii) the set of receptors, wherein each receptor is sensitive to a particular neurotransmitter. The model is characterized by the broadcast nature transmission of the signals: the neurotransmitter appeared in the ECS is treated as an input signal for all neurons with receptors sensitive to it. It is shown that the extrasynaptic interaction of neurons combined with multitransmitter environment enables to reproduce the rhythms generated by simple natural neural networks.

Keywords: Discrete model of natural neural network · Chemical interactions Neurotransmitters · Endogenous activity

1 Introduction

In recent years, fundamentally new network models of nervous systems have appeared and begun to develop intensively. Their emergence and explosive growth are caused by two factors: the development of new equipment and devices that allow obtaining large data sets of high resolution, and the emergence of powerful computer technologies capable of processing this data. It turned out that in many biological and social systems the structure of connections between their elements is described by complex networks with similar properties. And the brain networks were are not an exception. Two mutually complementary directions of approaches to the graph-theoretic studies of brain networks were called “structural and functional connectomics” [1, 2].

Artificial neural networks and structural connectomics are based on the idea of a “wire brain”, where the brain is represented as an electrical network with a rigidly defined topology formed by “wires” (axons) connecting identical neurons. However, many properties inherent to living neural networks cannot be described in such terms. Modern research shows that neurons are not identical: they can generate different patterns of endogenous activity, respond to different neurotransmitters, as well as

This work was partially supported by the RFBR, project No. 17-29-07029.

A. Abraham et al. (Eds.): IITI 2018, AISC 874, pp. 463–472, 2019.
https://doi.org/10.1007/978-3-030-01818-4_46

release (or secret) different neurotransmitters, i.e. be transmitter-specific. Thus, the reorganization of the network topology and the change in the modes of neuron activity can occur ad hoc under the action of neurotransmitters [3–6].

The main goal of this work is to show, by the example of small networks, the informational significance of the chemical composition of extracellular space, the important role of extrasynaptic chemical interactions in the formation of steady rhythms in neural ensembles and different neuromodulating effects. This role has been repeatedly confirmed at the experimental level [3–11], however, the corresponding mathematical models do not exist. To reveal this significance explicitly, we propose a model of the extrasynaptic mechanism of communication between neurons, which uses only multitransmitter interactions via the extracellular space (ECS). Such communication is broadcast: the neurotransmitter appeared in the ECS is an input signal for all neurons with receptors sensitive to this neurotransmitter. This reduction is our conscious model idealization, designed to emphasize the informational role of multitransmitter interactions. It does not deny the presence of synaptic contacts, but treats them as a special case of local communication between two neurons.

The realization of this goal is connected with the choice of an appropriate mathematical apparatus. The model we proposed is discrete, but it takes into account much more parameters than the formal McCulloch-Pitts neuron model. Approaches to modeling heterochemical interactions are described in [12, 13].

2 Definitions of the Basic Entities of the Model

Let us define the terms used in the model. Basically, these terms coincide with corresponding biological definitions; however some concepts represent generalized biological characteristics.

Neurotransmitters (transmitters, chemical messengers) are biologically active chemical agents used for neuronal interactions. They transmit signals among neurons. Neurotransmitters are released by active neurons. In the model, all neurons with receptors to the given neurotransmitter react to its presence in one of two ways: excite or inhibit the neuron. In different neurons, the receptors to the same transmitter may differ in sign: excite one neuron and inhibit the other.

Receptors are protein molecules built into the neuron membrane and sensitive to a certain neurotransmitter. As noted above, the variety of receptors is reduced to two main types according to the principle of the effect to the electrical properties of a neuron: excitatory and inhibitory.

Membrane potential (MP) is a generalized characteristic of the change in all currents flowing through channels in a neuron membrane. The membrane potential of the neuron can change due to both external influences, and internal processes occurring in the cell. In the second case, the change in MP will be called *endogenous*. And the activity of the neuron, resulting from endogenous growth of MP, will be called *endogenous activity*.

The activation threshold of a neuron completely coincides with the concept of activation threshold in artificial neural networks. If the value of the membrane potential exceeds the threshold, the neuron, while being activated, generates spikes and releases its specific neurotransmitter in the extracellular space.

Extracellular space (ECS) is a set of extracellular structures providing mechanical support of cells and transport of chemicals. The model treats the ECS as the common chemical surrounding of a small group of neurons. It is believed that there is no concentration gradient: the transmitters are uniformly distributed throughout the volume of the ECS.

Heterochemical interactions are interactions among neurons through the access to the common extracellular space.

Extrasynaptic transmission is a way of chemical interaction of neurons without synapses. In the proposed model, we have completely eliminated synapses to investigate the possibility of constructing non-synaptic stable neural ensembles.

Activity pattern is the stable rhythmic activity of a group of neurons unfolding over time.

3 Object of Modeling

We consider a set of neurons in the common extracellular space. ECS contains various neurotransmitters that can both be released by active neurons, and come from the outside.

Neurons have receptors, and each receptor is able to react to a certain type of neurotransmitter. The receptor's effect on the membrane potential of a neuron can have different signs (excitatory or inhibitory) and different strength of influence (weight).

Each neuron is characterized by the following set of indicators.

1. Type of released neurotransmitter. It is believed that each neuron releases one specific transmitter.
2. A set of receptors. Receptors sensitive to the same transmitter are collected in slots. Each generalized receptor slot is characterized by type and weight. The type coincides with the type of transmitter which it reacts to, and the weight is determined by the total sensitivity of all receptors. The weight can be positive, then the transmitter is excitatory for a given neuron, and negative, then this transmitter has an inhibitory effect.
3. The type of endogenous change of membrane potential (MP). The model presents three types of neurons:
   a. *oscillating* (*bursting*) *neuron* is a neuron that, in the absence of external influences, produces bursts of spikes in regular intervals;
   b. *tonic neuron* is a neuron, constantly active in the absence of inhibition;
   c. *reactive neuron* is a passive neuron that is activated only with external suprathreshold excitation.
4. A set of parameters characterizing the change in the membrane potential: the lower and upper boundaries that define the range of these changes, the resting potential which denotes the equilibrium state of the membrane, and the excitation threshold.

A neuron is activated if its MP has exceeded the threshold value specific for each neuron. Activation occurs as a result of either endogenous processes or external influences, when the sum of receptor reactions (taking into account their weights) exceeds the threshold value.

The qualitative chemical characteristics of an individual neuron are shown in Fig. 1.

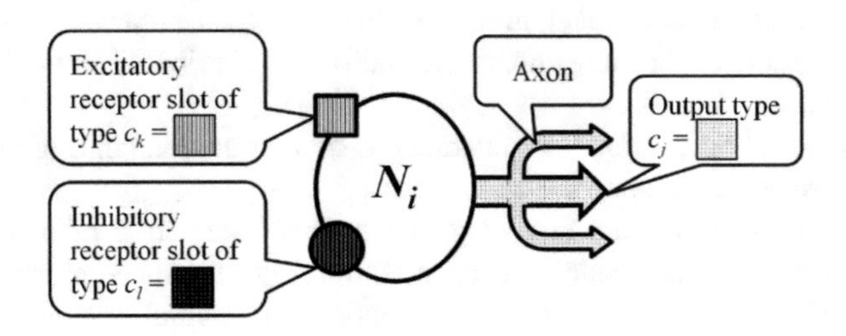

Fig. 1. Qualitative chemical characteristics of a neuron N_i; c_j, c_k, c_l denote the types of transmitters

Quantitative chemical characteristics include the transmitter emission value per time unit and the weight of the receptor slots. They are represented in Fig. 2.

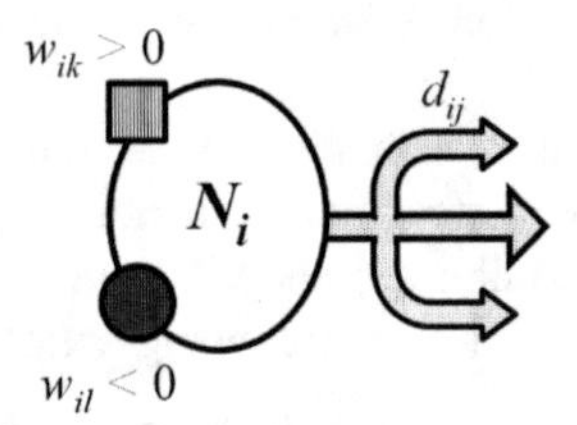

Fig. 2. Quantitative chemical characteristics of a neuron; d_{ij} denotes the value of releasing transmitter, w_{ik}, w_{il} denote the weights of receptor slots

The electrical characteristics of a neuron, describing the range of its activity and the overall changes in the membrane potential, are shown in Fig. 3. This figure shows a piecewise-linear approximation of the change in the membrane potential for an endogenous oscillator in the absence of external influences. The linear sections of the function of the membrane potentials changing correspond to different ion currents at each activity interval, which are assumed to be constant in the model.

Let us describe the chemical interactions of a system of three neurons N_1, N_2, and N_3 which have an access to a common extracellular space. Let these three neurons release two different transmitters: neuron N_1 releases transmitter c_j, neurons N_2 and N_3 release transmitter c_l (see Fig. 1). Each neuron has a unique set of receptor slots (Fig. 4).

Let these three neurons be placed in the common extracellular space (Fig. 5).

If neurons have a certain endogenous activity, due to structure of chemical bonds they can organize a stable pattern of three-phase rhythmic activity, successively releasing their specific neurotransmitters into the ECS. In the example in Fig. 5, with certain properties of endogenous changes in MP of neurons, this activity will form a

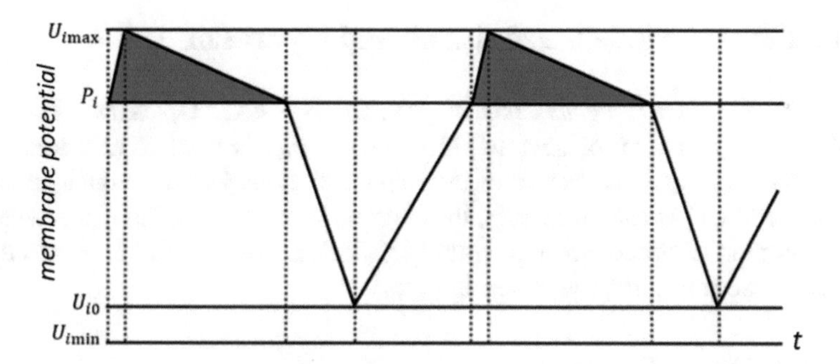

Fig. 3. Electrical characteristics of a neuron and dynamics of its membrane potential. U_{imin} and U_{imax} determine the range of possible MP values; P_i is activation threshold value; grey triangles denote the bursts of spikes; U_{i0} denotes the balanced value of MP that turns the direction of MP change from descending to ascending.

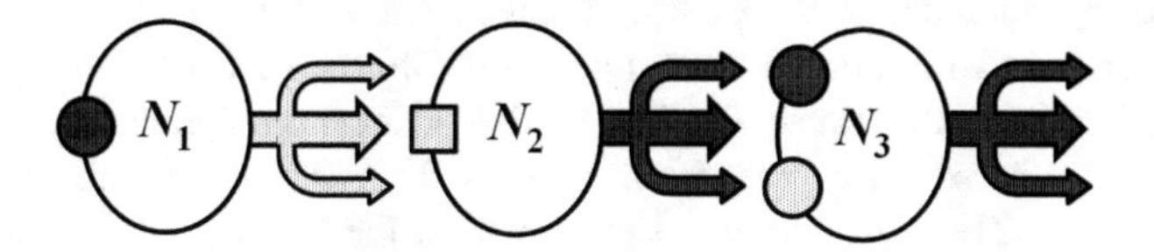

Fig. 4. Three neurons with chemical outputs and receptor slots

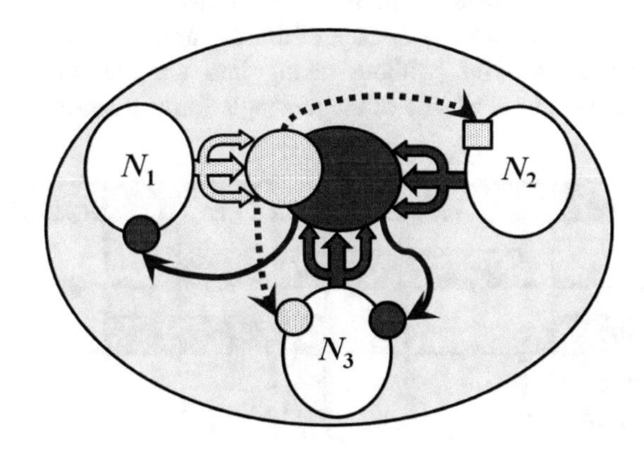

Fig. 5. Extrasynaptic interaction of neurons through the common ECS

simple cycle $N_1 - N_2 - N_3 -$…. Such stable patterns of rhythmic activity in small ensembles and the mechanisms of their building and self-organization are the objects of our research.

Graphs of changes in membrane potentials of neurons for the heterochemical ensemble presented in Fig. 5 will be given in Sect. 5.

4 The Formal Model: Definitions and Operation Rules

We define *a heterogeneous neural system* **S** = <**N**, **C**, **X**(*t*), **T**>, where **N** = $\{N_1, \ldots, N_n\}$ is a set of neurons, **C** = $\{c_1, \ldots, c_m\}$ is a set of neurotransmitters, **X**(*t*) = $(x_1(t), \ldots, x_m(t))$ is a vector of their concentrations in extracellular space, **T** is time of the system. At each time step, the neurons interact, releasing neurotransmitters into the common ECS and reacting with them. Each neuron has full access to the ECS and reacts to all transmitters it is sensitive to.

4.1 Event Time of System

Let us assume that the system time **T** is continuous, and inside this time the *events* occur.

Events include: changing the state of any neuron – from active to passive and vice versa; appearance of a new neurotransmitter in the ECS; change in concentration (including disappearance) of the existing neurotransmitter in the ECS.

Events are points on a continuous time scale. This scale is divided by events into time steps of different durations. The boundaries of the steps (points on this scale) are successively numbered by nonnegative integers 0, 1, 2, … and are called discrete moments of time. The step number coincides with the number of the moment of its beginning: the step t is a half-interval $[t, t + 1)$: its left boundary is included in the half-interval, the right boundary belongs to the next step. No events occur inside the step.

Figure 6 shows a simplified diagram of the formation of the time moments: the shaded rectangles show the activity of each of three neurons N_1, N_2, and N_3; events are formed by alternating activity (without taking into account the concentrations of neurotransmitters); each vertical side of the rectangle forms a *moment* on the time axis.

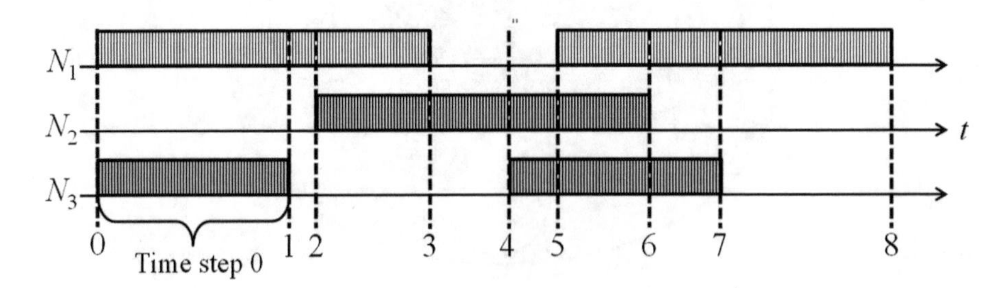

Fig. 6. Time steps and moments formed by events

4.2 Characteristics of Neurons

Each neuron of a system is characterized by a set of parameters. They can be divided into two categories: electrical and chemical.

Electrical Characteristics

1. The type of endogenous electrical activity (type of change in MP). The description of this type includes all characteristics of the MP change function, in particular, for

example in Fig. 3, four slope angles of linear segments corresponding to four different generalized currents must be specified. Two of them depolarize the membrane (below and above the threshold) and two hyperpolarize the membrane (also on both sides of the threshold). As it mentioned above, the neurons of the model are divided into three large classes, according to the type of activity: oscillating (bursting) neurons, neurons with tonic activity, and reactive neurons. However, each individual neuron can have unique values for the MP change parameters.
2. The range of MP changes, activation threshold, and rest potential.

Chemical Characteristics

1. Type and amount of the specific transmitter d_{ik}, where the index i corresponds to neuron N_i, and the index k corresponds to neurotransmitter type c_k.
2. A set of receptors and their weights w_{ik}.

Thus, the chemical properties of the neurons can be represented in the form of Table 1. Each row describes a single neuron

Table 1. Chemical characteristics of neurons

Neuron	Output D	Weights of slots W		
		c_1	…	c_m
N_1	d_{1k}	w_{11}	…	w_{1m}
…	…	…	…	…
N_n	d_{nl}	w_{n1}	…	w_{nm}

4.3 Characteristics of ECS

The state of the ECS at time t is represented by a vector $X(t) = (x_1(t), \ldots, x_m(t))$, where $x_j(t), j = 1, \ldots, n$ is total amount of neurotransmitter c_j presenting in the ECS during the time step t. The following assumptions are made:

1. the neurotransmitter is distributed uniformly throughout the extracellular space, the receptors react to it equally (taking into account the weights);
2. amount d_{ij} of the neurotransmitter c_j exists in the ECS during the activity of N_i plus a certain duration time τ_{c_j}, independent of i.

The concentration of the transmitter c_j for the step t is calculated by the formula:

$$x_j(t) = \sum\nolimits_{i=1}^{n} d_{ij} \cdot (y_i(t) + I_{d_{ij}}(t)), \tag{1}$$

where $y_i(t) \in \{0, 1\}$ is a state of a neuron at step t; $y_i(t) = 1$ means that the neuron is active at step t; $y_i(t) = 0$ means that it is passive; $I_{d_{ij}}(t) \in \{0, 1\}$ is the indicator of the presence of neurotransmitter c_j in the ECS, equal to 1 for time τ_{c_j} after neuron N_i stops spiking.

4.4 Interactions of Neurons

Concentrations of transmitters in the ECS are recalculated according to the formula (1) at each time moment (after the onset of the nearest event). Then, new currents of all neurons in the system are calculated. They characterize changes in the membrane potentials:

$$v_i(t) = s_i(t) + v^{\alpha}_{ien},$$

where $s_i(t)$ is an *exogenous component* of currents, equal to the strength of external influences:

$$s_i(t) = \sum\nolimits_{j=1}^{m} w_{ij}x_j(t);$$

v^{α}_{ien} is an endogenous current of neuron N_i in this phase of activity determined by parameter α.

Further, the nearest event in the system is to be determined according to the new dynamics of change in the MPs of neurons and the lifetime of the transmitters. And then all states (the values of the neuron' MPs, their activities and new concentrations of transmitters in the ECS) are recalculated again for this new time moment.

In general case, a system of several neurons and several transmitters for arbitrary parameters can generate complicated and non-periodic behavior. Note that the presented model of system's behavior is not simulative – every system state is unambiguously calculated for a given initial state and parameters of neurons and transmitters listed in this section.

Let's specify the tasks that can be solved with the help of this computational model.

1. Selection of neuron parameters (types and weights of receptors, type and volume of released neurotransmitter, currents that determine the dynamics of the membrane potential) and neurotransmitter parameters (lifetime), which enable the system to generate stable patterns of periodic activity.
2. Identification of biologically adequate parameters for building models of known motor central pattern generators in simple organisms.
3. Realization the effects of rhythm restructuring (neuromodulation) with the help of neurotransmitters.
4. Modeling the learning of individual cells and of a network as a whole by changing the weights of the receptors.

5 Example of Neural Interactions

Consider an ensemble of three neurons (Fig. 5), where N_1 and N_3 are oscillators, N_2 is a reactive neuron. N_1 releases the neurotransmitter c_1, and N_2 and N_3 release the transmitter c_2.

For a given set of parameters, this group of neurons generates a three-phase rhythmic activity. The graph in Fig. 7 is constructed using a discrete mathematical model that calculates neuron' MPs at each time moment.

It can be seen from Fig. 7 that in a few first steps the neurons generate a quasiperiodic activity: the phases of activity of the three neurons alternate periodically, but the dynamics of their membrane potentials is not reproduced exactly. Search and selection of parameters for generation of a stable cycle from arbitrary initial conditions is a separate problem that will be solved in subsequent works.

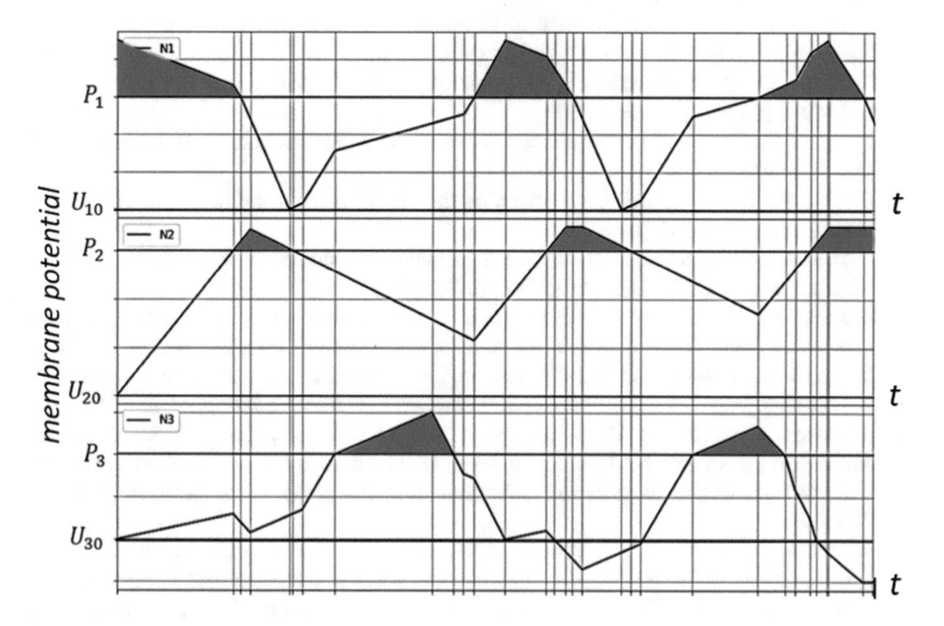

Fig. 7. Change in the membrane potentials of interacting neurons, which forms a three-phase rhythm

6 Conclusion

A discrete asynchronous model of chemical interactions between neurons with endogenous electrical activity is proposed. All the parameters of neurons and the ways of their interactions are formalized. The model has more expressive power than the "wire" models based on local synaptic connections between homogeneous formal neurons. On the other hand, its computational complexity is relatively low, which will allow building networks of a sufficiently large dimensions and investigating their properties, such as the formation of activity patterns, regions of stability of parameters, neuromodulation as a reorganization of rhythms and other problems.

References

1. Baronchelli, A., Ferrer-i-Cancho, R., Pastor-Satorras, R., Chater, N., Christiansen, M.H.: Networks in cognitive science. Trends Cogn. Sci. **17**(7), 348–360 (2013)
2. Bullmore, E., Sporns, O.: Complex brain networks: graph theoretical analysis of structural and functional systems. Nat. Rev. Neurosci. **10**, 186–198 (2009)
3. Bargmann, C.I.: Beyond the connectome: How neuromodulators shape neural circuits. BioEssays **34**(6), 458–465 (2012)
4. Dyakonova, V.E.: Neurotransmitter mechanisms of context-dependent behavior. Neurosci. Behav. Physiol. **44**, 256–267 (2014)
5. Dyakonova, V.: Neurotransmitternye mechanismy context-zavisimogo povedenija. Zh. Vyssh. Nerv. Deyat. **62**(6), 1–17 (2012). (in Russian)
6. Sakharov, D.: Biologicheskij substrat generatsii povedencheskih actov. Zh. Obsch. Biol. **73** (5), 334–348 (2012). (in Russian)
7. Bloom, F.E.: The functional significance of neurotransmitter diversity. Am. J. Physiol. **246**, 184–194 (1984)
8. Getting, P.: Emerging principles governing the operation of neural networks. Annu. Rev. Neurosci. **12**, 185–204 (1989)
9. Sakharov, D.: The multiplicity of neurotransmitters: the functional significance. Zh. Evol. Biokhim. Fiziol. **26**(5), 733–741 (1990). (in Russian)
10. Brezina, V.: Beyond the wiring diagram: signalling through complex neuromodulator networks. Philos. Trans. R. Soc. Lond. B Biol. Sci. **12; 365**(1551), 2363–2374 (2010)
11. Moroz, L.L., Kohn, A.B.: Independent origins of neurons and synapses: insights from ctenophores. Phil. Trans. R. Soc. Lond. B. Biol. Sci. **371**(1685), 20150041 (2016)
12. Bazenkov, N., Vorontsov, D., Dyakonova, V., Zhilyakova, L., Zakharov, I., Kuznetsov, O., Kulivets, S., Sakharov, D.: Discrete modeling of inter-neuron interactions in multitransmitters networks. Iskusstvenny Intellekt i Prinyatie Reshenii [Artif. Intell. Decis. Mak.] **2**, 55–73 (2017)
13. Bazenkov, N., Dyakonova, V., Kuznetsov, O., Sakharov, D., Vorontsov, D., Zhilyakova, L.: Discrete modeling of multi-transmitter neural networks with neuronal competition. In: Biologically Inspired Cognitive Architectures (BICA) for Young Scientists. Advances in Intelligent Systems and Computing, vol. 636, pp. 10–16. Springer International Publishing AG (2018)

Meta-Optimization of Mind Evolutionary Computation Algorithm Using Design of Experiments

Maxim Sakharov(✉) and Anatoly Karpenko

Bauman MSTU, Moscow, Russia
max.sfn90@gmail.com, apkarpenko@mail.ru

Abstract. This paper presents a new technique for solving a meta-optimization problem for Mind Evolutionary Computation (*MEC*) algorithm using a full factorial designed experiment. This approach can be also generalized for other global optimization population-based algorithms. In general, design of experiments allows one to determine the influence of input factors and their interaction on the output of a process. It's proposed to use such an approach to identify the most important free parameters as well as to estimate their interaction and determine the optimal values of those parameters for specific classes of objective functions. The paper contains the description of proposed method and software implementation along with the results of numerical experiments conducted to determine optimal values of the *MEC* algorithm's free parameters.

Keywords: Mind Evolutionary Computation · Global optimization
Meta-optimization · Design of experiments

1 Introduction

In [1–3] the original version of the Mind Evolutionary Computation (*MEC*) algorithm along with several of its modifications were utilized for solving various real world global optimization problems. The original version, Simple *MEC* or *SMEC*, is a population-based algorithm; this class of methods is very popular nowadays due to their simplicity of implementation and flexibility of application. In addition, these algorithms can localize so called suboptimal solutions, in other words, solutions which are close to the global optimum, with a high probability. In real world optimization problems, such solutions are often sufficient.

On the other hand, the main disadvantage of such algorithms is their slow convergence to the neighborhood of global optimum because these methods don't utilize any local information about the objective function's landscape [4, 5]. This fact often restricts their usage in real world problems, where computation time is a crucial factor.

Today researchers, working in the field of population-based optimization methods, are trying to increase their efficiency by means of either hybridization or meta-optimization. A development of hybrid algorithms implies a combination of various methods in such a way that advantages of one method would overcome disadvantages of another one [4]. Meta-optimization, on the other hand, implies the adjustment of free

A. Abraham et al. (Eds.): IITI 2018, AISC 874, pp. 473–482, 2019.
https://doi.org/10.1007/978-3-030-01818-4_47

parameters' values that would provide the maximum efficiency of an algorithm being used [4].

In [6, 7] a wide experimental study was carried out by the authors to identify the most important free parameters for *SMEC* and their influence on the objective function's value. While those experiments helped to reveal optimal values of the free parameters for different classes of objective functions, they were costly in terms of computational resources and didn't take into account possible interaction between the parameters.

In this paper we propose a new technique for adjusting values of the free parameters utilizing the design of experiments. Proposed method considers not only influence of every single parameter but also their interaction. In addition, this method is less costly in terms of computation and, thus, can be used as a part of Landscape Analysis (*LA*) procedure [8] before solving any optimization problem.

The paper also contains the description of the *SMEC* software implementation along with the results of numerical experiments with a use of multi-dimensional benchmark functions of different classes.

2 MEC Algorithm and Its Free Parameters

In this paper we consider a deterministic global unconstrained minimization problem

$$\min_{X \in R^{|X|}} \Phi(X) = \Phi(X^*) = \Phi^*. \tag{1}$$

Here $\Phi(X)$ – the scalar objective function, $\Phi(X^*) = \Phi^*$ – its required minimal value, $X = (x_1, x_2, \ldots, x_{|X|})$ – $|X|$-dimensional vector of variables, $R^{|X|}$ – $|X|$-dimensional arithmetical space.

Initial values of vector X are generated within a domain D, which is defined as follows

$$D = \{X | x_i^{min} \leq x_i \leq x_i^{max}, i \in [1 : |X|]\} \subset R^{|X|}. \tag{2}$$

The Mind Evolutionary Computation algorithm belongs to a class of algorithms inspired by human society. Algorithms of that class simulate some aspects of human's behavior. An individual s is considered as an intelligent agent, which operates in a group S made of analogous individuals. During the evolution process each individual is affected by other individuals within a group.

Canonical *MEC* was introduced in [9]. This algorithm is composed of three main stages: initialization of groups, similar-taxis and dissimilation. Operations of similar-taxis and dissimilation are repeated iteratively while the best obtained value of an objective function $\Phi(X)$ is changing. When the best obtained value stops changing, the winner of the best group from a set of leading ones is selected as a solution to the optimization problem.

A population in the *SMEC* algorithm is made of leading groups $S^b = (S_1^b, S_2^b, \ldots, S_{|S^b|}^b)$ and lagging groups $S^w = (S_1^w, S_2^w, \ldots, S_{|S^w|}^w)$, which include $|S^b|$ and $|S^w|$ groups correspondingly; the number of individuals within each group is set to be the same and equals $|S|$. The ratio between a number of leading and lagging groups η is a free parameter of the algorithm.

Each group has its own communication environment named a local blackboard and denoted as C_i^b, C_j^w correspondingly. In addition to this, the whole population has a general global blackboard C^g.

The initialization stage creates groups S^b, S^w and puts them in the search domain. We illustrate the initialization stage by an example of the group S_i.

1. Generate a random vector $X_{i,1}$ whose components are distributed uniformly within the corresponding search subdomain. Identify this vector with the individual $s_{i,1}$ of the group S_i.
2. Determine the initial coordinates of the rest of the individuals in the group using the formula

$$X_{i,j} = X_{i,1} + N_{|X|}(0, \sigma), j \in [2 : |S|], \tag{3}$$

where $N_{|X|}(0, \sigma)$ – $(|X| \times 1)$-dimensional vector of independent random real numbers, distributed normally with math expectation and standard deviation equaling 0 and σ respectively.

The similar-taxis stage implements a local search inside every group S^b, S^w and can be described as follows.

1. Take information on the current best individual s_{i,j_b}, $j_b \in [1 : |S|]$ of the group S_i from the blackboard C_i.
2. Determine new coordinates of the rest individuals $s'_{i,j}$, $j \in [1 : |S|]$, $j \neq j_b$ in this group using formula (3).
3. Calculate the objective function's values for all individuals in the group $\Phi'_{i,j} = \Phi(X'_{i,j})$, $j \in [1 : |S|]$.
4. Determine a new winner of the group s'_{i,k_b}, $k_b \in [1 : |S|]$ as an individual with the lowest value of the objective function $\Phi(X)$.
5. Put information on the new winner on the blackboards C_i, C^g.

The dissimilation stage implements a global search between all groups and uses the following steps.

1. Take information on the best individuals of all groups S^b, S^w from a global blackboard C^g.
2. Compare their scores and rank them. If a score of any leading group S_i^b is less than a score of any lagging group S_j^w, then the latter becomes a leading group and the leading group becomes a lagging one. If a score of any lagging group S_j^w is lower than scores of all leading groups for ω consecutive iterations, then it's removed from the population.

3. Replace each removed group with a new one using the initialization procedure.

In [6] we demonstrated that the following free parameters have the most significant effect on the optimization process: σ – standard deviation at the initialization stage, ω – removing frequency, η – ratio between a number of leading and lagging groups. These parameters are used in this paper as input factors for the designed experiment.

3 Design of Experiments

In general, designed experiments consist of series of runs where input variables of any process are simultaneously changed and responses are observed [10, 11]. Such an approach often is more efficient than studying one factor at a time. With designed experiments, in far fewer tests, many factors can be studied simultaneously, while providing both independent estimates of factor effects and valuable information about their interactions. Without properly designed experiments, the effects of these interactions are often overlooked.

We propose to use this technique for studying the importance of *MEC* algorithm's free parameters and their interaction. In [6] we performed a wide performance investigation on how the specified free parameters affect the optimization process. For example, if one increases standard deviation σ from 0.5 to 0.75, the optimized value for some classes of objective functions will increase by 100%. However, that experimental study required a large number of trials and computational resources.

The new approach, with a use of designed experiments will allow one to study not only parameters but also their interaction at a smaller number of trials. As a result, it's possible to identify the optimal values of those parameters for different classes of objective functions.

While there are many different experimental designs, in this work a full factorial design was used in order to analyze all combinations of the experimental factor values. As specified above, for *SMEC* algorithm we consider three different free parameters. For each factor the lower and upper limits were set as follows: $\sigma_{min} = 0.05$; $\sigma_{max} = 1.0$; $\omega_{min} = 5$; $\omega_{max} = 40$; $\eta_{min} = 0.1$; $\eta_{max} = 0.9$. These limits are referred as levels in the context of design of experiments (DoE). Subsequently, a three-factor design with two levels for each factor is used in this study (Fig. 1). Each corner point represents a unique combination of factor levels.

In order to take into account possible non-linear dependencies between the input values of free parameters and the output values of the objective function four center points were used to compose four combinations with mid-level values.

As it's usually the case with the population-based algorithm, their efficiency also depends on the problem dimension. In order to reveal possible patterns in the optimal values of the free parameters across the objective functions of the same class but different dimensions two blocks of experiments were conducted simultaneously – the first block is for 4D functions, the second one – for 8D function.

The plan of the full factorial experiment was created automatically using *Minitab Statistical Software* which is widely used in industry to perform DoE analysis for manufacturing processes. The experiment plan consists of 48 runs. Each run represents a combination of three parameters either in the corner or in the center point. As we use blocks to deal with different dimensions, there were 24 runs for 4D function and 24 runs for 8D function. The combinations are the same for two blocks.

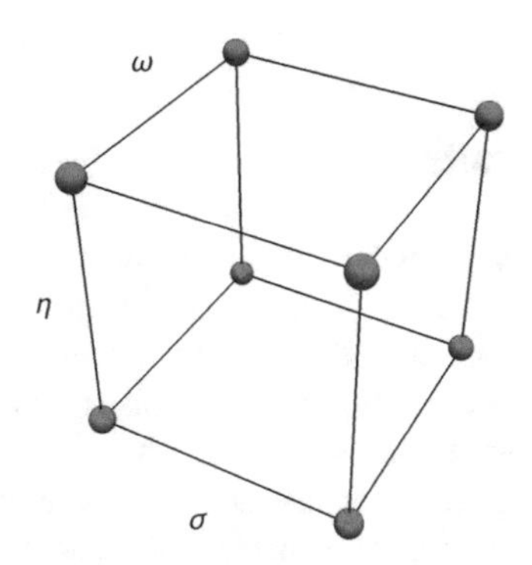

Fig. 1. A three-factor design with two levels of each factor

4 Performance Investigation

Simple *MEC* algorithm was implemented by authors in *Wolfram Mathematica*. Software implementation has a modular structure, which allows one to modify algorithms easily and also include some additional assisting methods.

Results of the optimization process heavily depend on the initial vector X. In order to decrease this influence all numeric experiments for each combination of the free parameters according to the design were carried out using the multi-start method with $K = 10$ launches. The average value of an objective function $\bar{\Phi}$ based on the results of all launches was utilized in the analysis.

Multi-dimensional benchmark optimization functions [12] are considered in this paper (Table 1). An original search domain for every function equals

$$D = \{X| - 10 \leq x_i \leq 10, i \in [1 : |X|]\}. \tag{4}$$

Table 1. Definitions of the benchmark functions.

Function	Definition	Global minimum
Rastrigin	$\Phi(X) = \sum_{i=1}^{\lvert X\rvert} (10 + x_i^2 - 10\cos(2\pi x_i))$	$\Phi(X^*) = 0$; $X^* = (0, 0, \ldots, 0)$
Rosenbrock	$\Phi(X) = \sum_{i=1}^{\lvert X\rvert-1} \left(100(x_{i+1} - x_i^2)^2 + (1 - x_i)^2\right)$	$\Phi(X^*) = 0$; $X^* = (1, 1, \ldots, 1)$
Zakharov	$\Phi(X) = \sum_{i=1}^{\lvert X\rvert} x_i^2 + \left(\sum_{i=1}^{\lvert X\rvert} 0,5ix_i\right)^2 + \left(\sum_{i=1}^{\lvert X\rvert} 0,5ix_i\right)^4$	$\Phi(X^*) = 0$; $X^* = (0, 0, \ldots, 0)$

During the experiments the following values of free parameters were used for the *SMEC* algorithm: total number of groups $\gamma = |S^b| + |S^w| = 40$; number of individuals in each group $|S| = 20$. The number of stagnation iterations $\lambda_{stop} = 50$ was used as a termination criterion for the algorithms. Tolerance used for identifying stagnation was equal to $\varepsilon = 10^{-5}$.

5 Analysis of the Results

Each benchmark functions was studied independently using *Minitab Statistical Software*. Figure 2 displays the Pareto Chart of the standardized effects for Rastrigin function. This chart helps to identify the most important factors and their interactions based on the results of statistical modelling [11]. Rastrigin function has a large number of local minima unlike Rosenbrock and Zakharov functions. Obtained results demonstrate that all three parameters are not important on their own, however all interactions are statistically significant, especially the effect from simultaneous changes in standard deviation σ and ratio η.

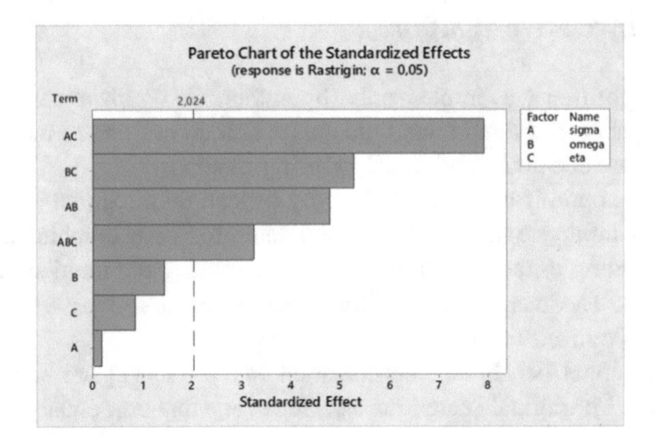

Fig. 2. Pareto chart of the standardized effects for Rastrigin function

In the Fig. 3 main trends for three parameters are presented. Red points are associated with the center points and their positions reveal non-linear dependencies of the objective function's values on the free parameters' values. For Rastrigin function the minimal value can be achieved using the highest value of remove frequency ω, the lowest value of ratio η and the middle value of standard deviation σ. These results agree with ones obtained in the previous work [6], where all factors were studied independently.

Figure 4 demonstrates the presence of strong interaction between factors. Recalling Pareto chart where the most important interaction was one between standard deviation σ and ratio η one can conclude that in order to achieve the minimal value of Rastrigin function when $\sigma = 0.05$, the value of ratio between number of groups should be equal to $\eta = 0.1$. Alternatively, if $\sigma = 1.0$, the value of ratio should equal $\eta = 0.9$.

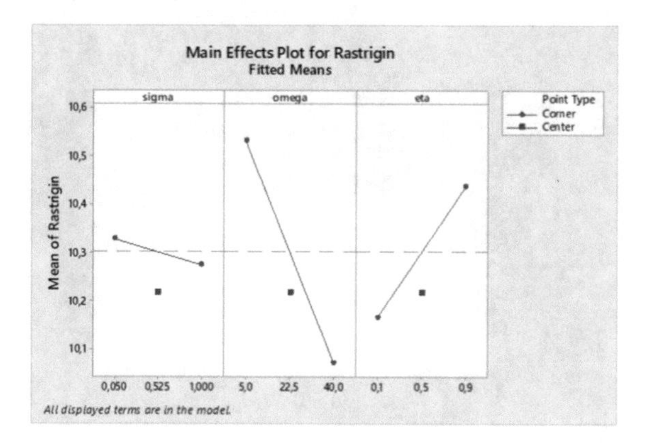

Fig. 3. Main effects for Rastrigin function

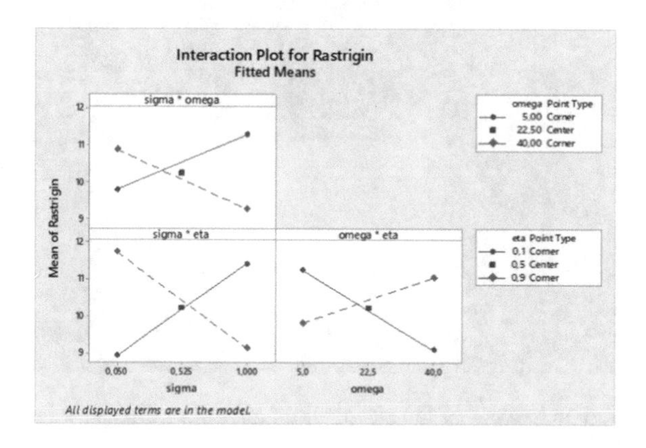

Fig. 4. Interaction plot for Rastrigin function

Unlike Rastrigin function, both Rosenbrock and Zakharov functions don't have many local minima. This similarity is clear when one looks at the Pareto charts for two functions (Figs. 5 and 6). While all factors and interactions are above the importance threshold, standard deviation σ and interaction of η and ω are the most crucial for both functions.

Figures 7 and 8 reveal certain patterns for the dependencies between the objective function's values and free parameters' values. Center points demonstrate that all dependencies are non-linear, like ones for Rastrigin function, and can be further investigated. For both functions the most important interaction is between ratio η and removing frequency ω. Interaction plots in the Figs. 7 and 8 show that minimal values of both functions can be achieved with the lowest value of ω and the highest values of η or vice versa. All other interactions are weaker; the curves follow the same trend but with different angles.

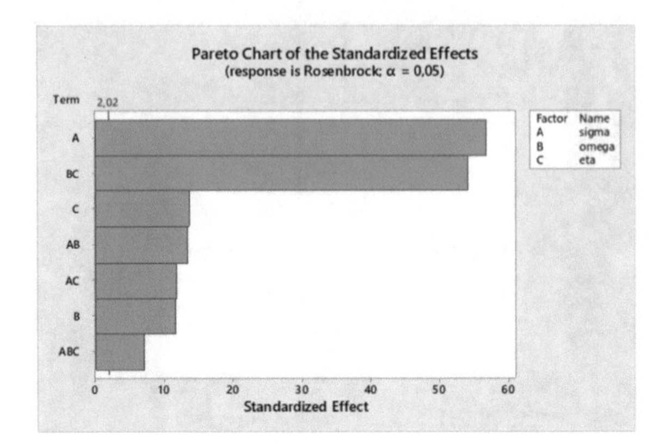

Fig. 5. Pareto chart of the standardized effects for Rosenbrock function

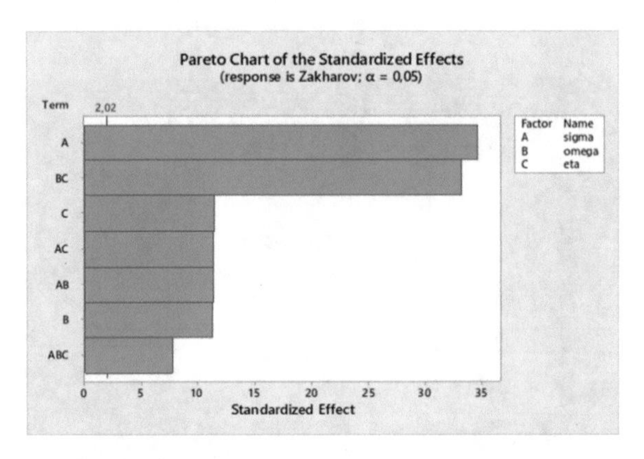

Fig. 6. Pareto chart of the standardized effects for Zakharov function

It's worth mentioning, that these conclusions were obtained with a relatively small number of evaluations and also are scalable across dimensions. This allows one to apply such a technique for adjusting the values of the *MEC* algorithm's free parameters prior to solving any real world optimization problem, subsequently obtaining a higher quality solution.

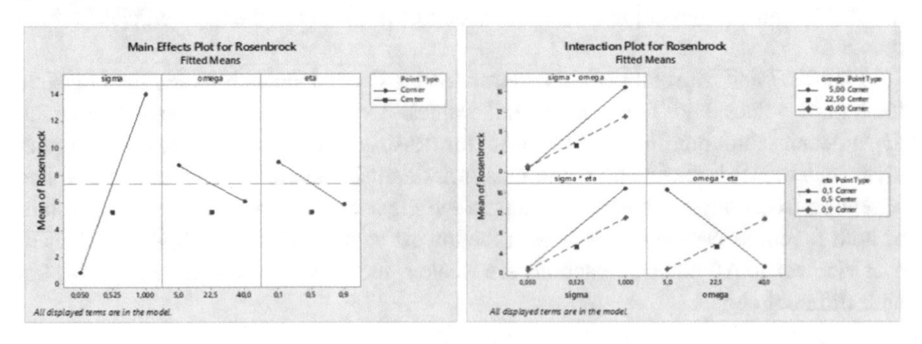

Fig. 7. Main effects and interaction plot for Rosenbrock function

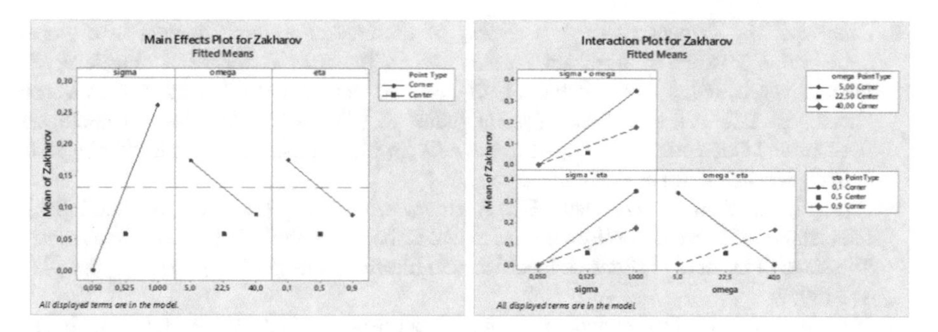

Fig. 8. Main effects and interaction plot for Zakharov function

6 Conclusions

This paper presents a new technique for solving a meta-optimization problem for the canonical *MEC* algorithm. With a use of designed experiments, namely a full factorial experiment, dependencies between the objective function's value and algorithm's free parameters were studied.

Revealed patterns help to select optimal values of the free parameters (σ – standard deviation at the initialization stage, ω – removing frequency, η – ratio between a number of leading and lagging groups) for different classes of objective functions before launching an optimization process. Tuned *SMEC* algorithm, thus, will be capable of finding not only a better solution but achieving it at less iterations.

Further research will be devoted to investigating other types of designed experiments which utilize less runs as well as studying other classes of objective functions.

References

1. Karpenko, A.P., Sakharov, M.K.: Multi-memes global optimization based on the algorithm of mind evolutionary computation. Inf. Technol. **7**, 23–30 (2014). (in Russian)
2. Jie, J., Zeng, J.: Improved mind evolutionary computation for optimizations. In: Proceedings of 5th World Congress on Intelligent Control and Automation, Hang Zhou, China, pp. 2200–2204 (2004)
3. Jie, J., Han, C., Zeng, J.: An extended mind evolutionary computation model for optimizations. Appl. Math. Comput. **185**, 1038–1049 (2007)
4. Weise, T.: Global Optimization Algorithms - Theory and Application. University of Kassel, 758 p. (2008)
5. Karpenko, A.P.: Modern algorithms of search engine optimization. In: Nature-Inspired Optimization Algorithms, 446 p. Bauman MSTU Publ., Moscow (2014). (in Russian)
6. Sakharov, M., Karpenko, A.: Performance investigation of mind evolutionary computation algorithm and some of its modifications. In: Proceedings of the First International Scientific Conference "Intelligent Information Technologies for Industry" (IITI 2016), pp. 475–486. Springer (2016). https://doi.org/10.1007/978-3-319-33609-1_43
7. Sakharov, M.K.: Study on mind evolutionary computation. In: Technologies and Systems 2014, pp. 75–78. Bauman MSTU Publ., Moscow (2014)

8. Sakharov, M., Karpenko, A.: A new way of decomposing search domain in a global optimization problem. In: Abraham, A., Kovalev, S., Tarassov, V., Snasel, V., Vasileva, M., Sukhanov, A. (eds.) Proceedings of the Second International Scientific Conference "Intelligent Information Technologies for Industry" (IITI 2017). Advances in Intelligent Systems and Computing, pp. 398–407, vol. 679. Springer, Cham (2017). https://doi.org/10.1007/978-3-319-68321-8_4
9. Chengyi, S., Yan, S., Wanzhen, W.: A survey of MEC: 1998–2001. In: 2002 IEEE International Conference on Systems, Man and Cybernetics IEEE SMC 2002, Hammamet, Tunisia. 6–9 October. Institute of Electrical and Electronics Engineers Inc., vol. 6, pp. 445–453 (2002)
10. Montgomery, D.C.: Design and Analysis of Experiments, p. 752. Wiley, Hoboken (2012)
11. Hardwick, C.: Practical Design of Experiments: DoE Made Easy, p. 50. Liberation Books Ltd., United Kingdom (2013)
12. Floudas, A.A., Pardalos, P.M., Adjiman, C., Esposito, W.R., Gümüs, Z.H., Harding, S.T., Klepeis, J.L., Meyer, C.A., Schweiger, C.A.: Handbook of Test Problems in Local and Global Optimization, 441 p. Kluwer, Dordrecht (1999)

An Interpreter of a Human Emotional State Based on a Neural-Like Hierarchical Structure

Konstantin V. Sidorov(✉), Natalya N. Filatova, and Pavel D. Shemaev

Department of Information Technologies, Tver State Technical University, Afanasy Nikitin Quay 22, Tver 170026, Russia
{bmisidorov,nfilatova99}@mail.ru, pshemaev@rambler.ru

Abstract. The paper considers possibility of hybrid system utilization that integrates the strategies of neural network models and methods of fuzzy data processing, for production rules construction and interpretation of biomedical signals by using the developed interpreter. The signals interpreter is based on the idea of growing pyramidal networks, which was adapted to work with fuzzy descriptions of objects. Objects fuzzy descriptions are generated by using informative attributes extracted from the time series attractors reconstructions. During training process, the class models are forming in interpreter's hierarchical structure and then transforming into fuzzy statements serving as a set of production rules for fuzzy logic inference system. Fuzzy statements reflect the main characteristics of all objects from training set and presented in verbal terms understandable to experts. This paper provides detailed description of interpreter's software realization and demonstrates construction algorithm of the neural-like hierarchical structure (NLHS) with production rules forming procedure. Article also contains the results of interpreter's program implementation on biomedical signals represented by electroencephalography (EEG) and speech signals. All collected biomedical signals (from training and test sets) are characterize changes in emotional reactions of subjects undergoing audiovisual stimulation. Mathematical apparatus of attractor reconstruction for EEG and speech signals allows to monitor changes in the individual properties of this structures on perception stage and after stimulation. As a result of monitoring a number of regularities determined, which are reveal themselves in attractor's characteristics and variations of emotional reactions correlated with them, and also with experts estimations as well as subjects self-estimations.

Keywords: Interpreter of emotions · Human emotions · Software tool
EEG · Speech signal · Attractor · Neural-like hierarchical structure
Fuzzy set · Fuzzy sign · Production rule · Training set · Test set

The work has been done within the framework of the grant of the President of the Russian Federation for state support of young Russian PhD scientists (MK-1898.2018.9).

A. Abraham et al. (Eds.): IITI 2018, AISC 874, pp. 483–492, 2019.
https://doi.org/10.1007/978-3-030-01818-4_48

1 Introduction

For at least last two decades, the problem of analysis and interpretation of various biomedical signals reflecting psychophysiological processes in human body is in the field of interests for developers of robotic complexes and automated systems [1]. This is caused by increased requirements in supporting tools for effective dialogue interaction between computer and user. A mechanism model for perception and transmission of emotions can become one of the ways for solving this problem.

Emotions, being primarily the psychological science prerogative, nevertheless, as an object of research, took an important place in artificial intelligence and robotics long time ago. The methods proposed by developers for human emotional state interpretation allows generating of regularities descriptions and formulating rules characterizing specific classes of emotional states through analyzing speech [2–4], bioelectrical signals (EEG, ECG, EMG, EOG, GSR, etc.) [5–7], and also subject's face videos reflecting mimic and gestures [8, 9]. The following fact should addressed that in most cases the received results are based on experimental-statistical methods of expert's evaluation and differentiation of signal processing methods depending on channels type for different responses.

In this paper, we consider the possibility of the hybrid system [10] utilization, which incorporate neural networks models strategies and fuzzy data processing methods, for classification rules construction and interpretation of biomedical signals by means of developed interpreter. The interpreter is filled with the real signals data sets forming secondary characteristics representation. These secondary characteristics are transformed into fuzzy form, which allows to create classification rules in understandable for experts verbal terms. The rules are formed through interpretation of regularities containing in neural-like hierarchical structure (NLHS), constructed during training process. Each rule represents itself as a fuzzy model for one of signal classes. Expert testifies the obtained production rules.

2 Neural-Like Hierarchical Structure Generation Algorithm

NLHS construction algorithm partially presented in the works [11, 12]. NLHS is the advancement of ideas of growing pyramidal networks [13] and represents oriented acyclic graph in a tiered-parallel form without peaks and with one incoming arc, all arcs are oriented from the lower to the upper levels.

All peaks of network are divided into two types: receptors and associative elements. Receptors are stand on zero level of hierarchy and act as network's inputs. The top of network considered excited if adjacent peaks of previous level are excited.

The initial network consists of receptors connected by outgoing arcs to the final peak of the top level Y, which plays a secondary role in the process of network generation. Each term of attribute's linguistic variable matched to input receptor; as a result, their number is equal to the total number of terms from all attributes.

Each associative element characterized by following parameters: network hierarchy level; state (excited/not excited); parenting class for this element; the number of receptors l with outputs to this element; the values of excitation counters $mcl1$–$mclk$

determining the response of the peak to input objects of classes *cl*1–*clk*, respectively, where *k* is the number of classes represented in the training set.

There are two stages of NLHS construction. The first stage provides conjunctive dependencies formation for attribute's values. The new peaks creation process realized through sequential execution of rules A and B after submission of the next object on network's input.

2.1 Rule A

An array of unexcited associative elements *Vd*, sorted by increasing peak's level in the network's hierarchy, is selected from the graph. Next, we take another peak *vt* from *Vd* array, selecting adjacent excited peaks located at the previous level of the hierarchy and put them into *Va* array. If the power of *Va* array is less than two, we go to the next observed peak *vi* + 1, otherwise a new associative element *vn* introduced into the network. The arcs from *Va* peaks to *vi* are eliminated and new arcs connecting *Va* peaks with new peak *vn* are introduced instead of them. The output of new element *vn* is connected by outgoing arc with *vi* input.

The levels of all peaks with existed paths from the new associative element *vn* are raised recursively by 1 (the graph is reduced to a tiered-parallel form) and the rule's execution is interrupts. The rule moves to the next peak only if the rule's execution for previous peak didn't changed the network structure.

2.2 Rule B

An array of excited peaks *Ve* with lower, comparing to penultimate, hierarchy level (associative elements and receptors) is selected from the graph. Then each peak *vi* from *Ve* array is explored. If *vi* does not have adjacent excited peaks located at the next level of hierarchy, *vi* is added to the array *S* and we go to the next peak *vi* + 1 of *Ve* array. At the end of *Ve* array scanning, a new associative element *vn* is added to the network. Arcs from the peaks of *S* array are placed at the input *vn*, besides the outgoing arcs to *Y* peak are eliminated from elements of *S* array. *Vn* is connecting with the final peak of *Y*'s network upper level by outgoing arc.

In result of construction first stage the associative elements adjacent to the *Y* peak are form *Vc* array, where each element is associated with the training set's object characterized by unique combinations of fuzzy attributes values.

2.3 NLHS Introduction

The *Vc* array (see Fig. 1) fully describes all objects from training set. At the second stage, the algorithm selects minimum possible number of control elements from the pyramids with peaks located in *Vc* array. The control elements – associative elements with the most common set of attributes values typical to the objects from particular class. The control elements separation method is described below in this paper.

Objects from the training set are step by step fed to the network's input. After object submission, the excitation spread among all peaks, correcting the parameters values *mcl*1–*mclk* and *l* of the peaks. Then, a set of control elements is selected from all network's associative elements according to the following conditions (in order of priority): (1) only one of the counters *mcl*1–*mclk* is filled in the associative element; (2) the maximum value of *mclj* counter among other associative elements from network, where *j* is the class represented by this peak; (3) the maximum value of *l* counter. All found peaks are recorded in the network's control elements set.

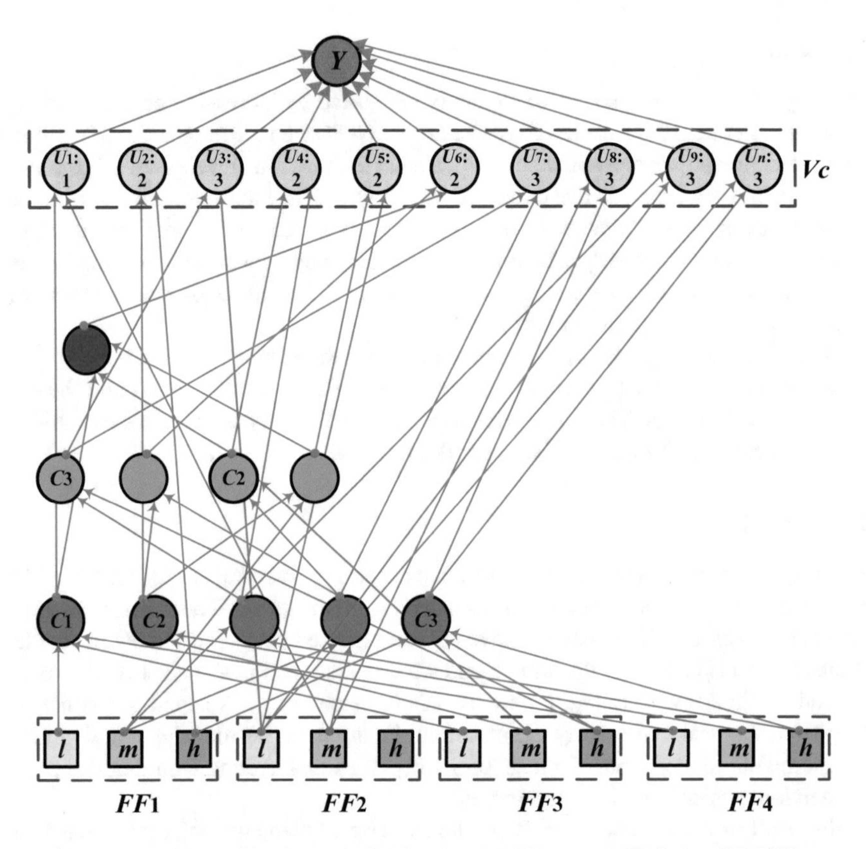

Fig. 1. Graphical representation NLHS (*l* – "LOW", *m* – "MID", *h* – "HIGH").

The network's peaks from penultimate level, adjacent to the terminal peak *Y* (U_1–*Un*), are form the set *Vc*, and each of them characterizes a group of close objects with same combination of the attribute's values *FFp*. The associative elements C_1, C_2, C_3 – control peaks of classes 1, 2 and 3, respectively. FF_1, FF_2, …, *FFp* – fuzzy attributes ($p = 1, 2, \ldots, P$ is the serial number of the fuzzy attribute, where *P* is the total number of fuzzy attributes).

The production rules for each class are selected based on his control elements analysis. Receptors with outgoing paths to control elements are joined up by conjunctive connection RS_n^F, where *F* is the class number, *n* is the class control element's number.

The production rule for each class is constructed from set of RS_n^F, connected by disjunctive connection. For example, for class number 3, 2 control peaks of C_3 are selected and the following rule will be generated: IF (FF_4 IS B AND FF_3 IS B AND FF_1 IS C) OR (FF_4 IS B AND FF_3 IS B AND FF_1 IS B) THEN CLASS IS 3.

3 Interpreter Software Implementation

The interpreter is implemented in C# 3.0 for the .NET Framework 3.5 and later. This paper presents a general structure of the software (see Fig. 2).

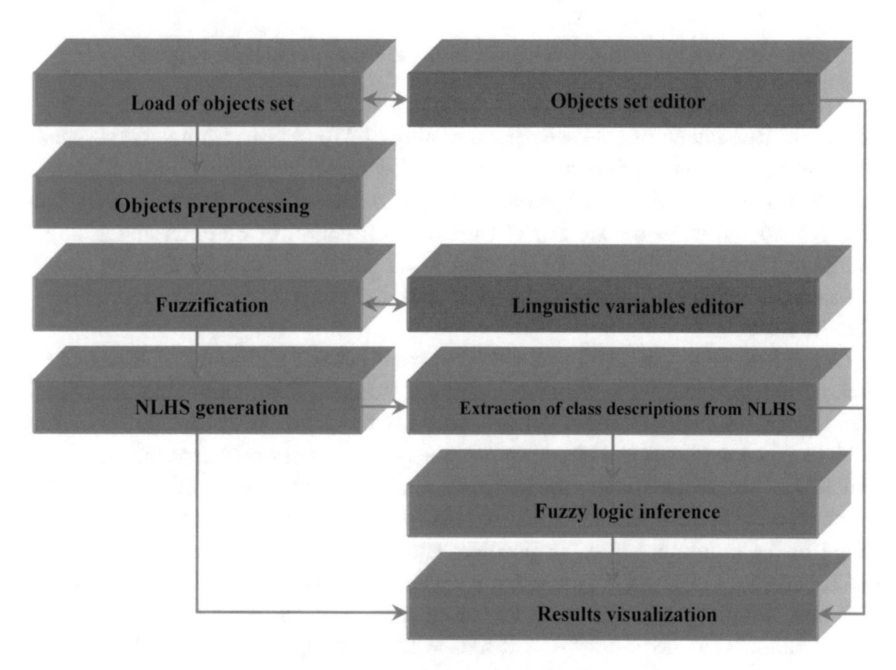

Fig. 2. The interpreter's structure.

The first version of the interpreter is briefly described in [7, 14]. The main blocks of improved interpreter include following modules: preprocessing, fuzzification, generation of NLHS and class descriptions, fuzzy logic inference.

The signal preprocessing module is designed for initial data processing: filtering; normalization; automatic optimal value determination for parameters of attractor reconstructed upon investigated acoustic signal; calculation of attractor and power spectral density; generation of secondary attributes vector basing on attractor. The result of the module is the X vector, which represents the signal with a set of discrete attributes.

Fuzzification module performs transformation of the X vector into linguistic scale and transition to fuzzy attributes.

The NLHS generation module is designed to construct a graph model (G) of classes from the training set. G structure reflects relationships between attributes of objects from training set.

The NLHS class descriptions extraction module is intended to select the sets of production rules from constructed NLHS.

Fuzzy logic inference module is solving classification task by using constructed production rules and calculates the function's value of object's correspondence to the class description.

4 Human Emotional State Monitoring

4.1 Emotion Valence Interpretation from EEG and Speech Signals

To create the interpreter of emotions manifesting in natural speech we created the database of Russian emotional speech, consisting of 270 phrases from various speakers. For database creation, 16 people (11 men and 5 women), aged from 18 to 60, acted as test subjects. For emotion activation (positive and negative), we used video stimuli accompanied by audio tracks. For objective confirmation of emotion changes, a 19-channel electroencephalogram (EEG) was recorded for each subject, allowing real-time monitoring of changes in brain electrical activity while perceiving of stimuli with different valence.

At the first stage, the EEG was recorded from subject during presentation of three series of video stimuli: 1 – positive emotions (scenes of humorous nature involving people and animals); 2 – neutral state (scenes with natural landscapes); 3 – negative emotions (scenes of surgical operations, scenes of the cruel treatment with peoples and animals). At the second stage, the subject's speech signal was recorded. Each subject pronounced the control phrase "And my voice sounds like this" (at times when he felt a change in the emotional state, or according to the operator's conventional sign). All speech samples (270 test phrases) with 5 s duration were saved in *.wav files with 22050 Hz sampling rate and a 16 bits resolution. EEG patterns (270 patterns) with 12 s. duration were saved in *.ASCII files with sampling frequency of 250 Hz.

For specification of samples from registered signals, a homogeneous as well as a heterogeneous set of characteristics can be used, however, in both cases it is necessary to consider the amplitude-frequency composition of signals [15]. This can be performed by using attractor's attributes from EEG and speech signals, whose performance was confirmed in [16–18].

The paper proposes an algorithm for rough estimation of attractors' m-dimensional projections sizes. For all m-projections with four quadrants we calculate a maximum length vector. Based on such estimates with all quadrants we determine an average vector for one (i-th) attractor's projection $\bar{A}^i_{\max} = 0,25\sum_{j=1}^{4} A^{i,j}_{\max}$, $A^{i,j}_{\max} = \max\{\sqrt{x_h^2 + x_{h+\tau}^2}\}$, where x_h, $x_{h+\tau}$ are time series values for h-th and $h+\tau$-th moments of time; i is a projection number; $i = \overline{1,3}$ ($i = 1$ when x_n–$x_{n+\tau}$; $i = 2$ when x_n–$x_{n+2\tau}$; $i = 3$ when $x_{n+\tau}$–$x_{n+2\tau}$); j is a projection quadrant number; $j = \overline{1,4}$. Each description of EEG pattern and speech sample is represented as follows [18]:

$$A(z) = \bigcup_{i=1}^{3} \bar{A}^{i}_{\max} \bigcup_{j=1}^{4} A^{i,j}_{\max}, \tag{1}$$

$$B(k) = |b_w|_{p \times 1}, b_w = \bigcup_{i=1}^{3} \bar{B}^{i}_{\max} \bigcup_{j=1}^{4} B^{i,j}_{\max}(w), w = \overline{1,P}, P \leq 8, \tag{2}$$

where *A(z)*, *B(k)* are descriptions of attractor's characteristics that are reconstructed according to speech signals and EEG, respectively; $A^{i,j}_{\max}$, $B^{i,j}_{\max}$ is a maximal vector length of *j*-th quadrant in *i*-th projection; $\bar{A}^{i}_{\max}$, $\bar{B}^{i}_{\max}$ is a length of an aggregate vector of *i*-th projection; *z* is a number of speech signal object; $z = \overline{1,270}$; *k* is a number of EEG object; $k = \overline{1,270}$; *w* is a number of EEG lead (the EEG patterns analysis has been carried out in 8 most informative leads of the right hemisphere [18]).

Therefore, according to the descriptions (1) and (2) every speech signal object in *A (z)* form is described by 15 characteristics (see Fig. 3), every EEG object in *B(k)* form is described by 120 characteristics (see Fig. 4).

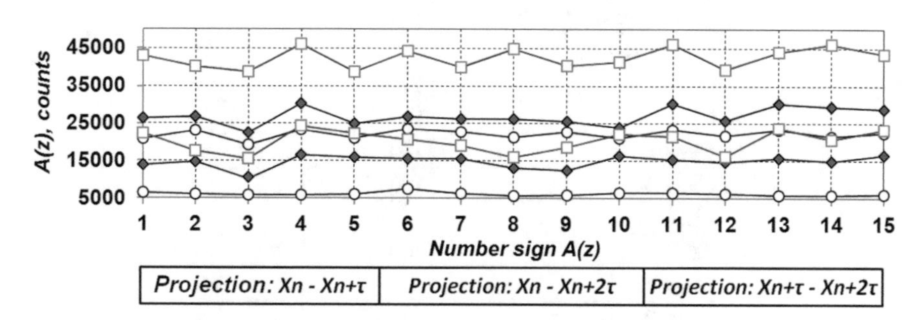

Fig. 3. A range of *A*(*z*) (1) characteristics according to speech signals (○ – negative emotions (Class 1), ♦ – neutral state (Class 2), □ – positive emotions (Class 3)).

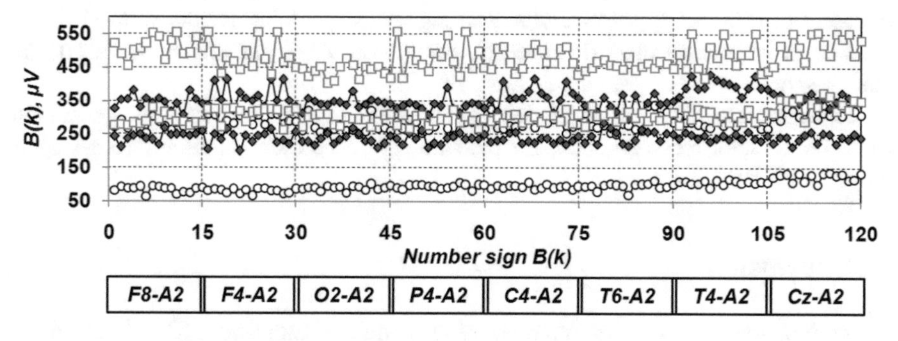

Fig. 4. A range of B(*k*) (2) characteristics according to 8 EEG leads in the right hemisphere ("The 10–20 System of Electrode Placement").

The obtained characteristic curves (see Figs. 3 and 4) show the lack of linear split classes (Class 2 and Class 3), as well as significant classes intersections (Class 1 and Class 2) according to speech signals and all EEG leads. Descriptions of characteristics (1 and 2) is transferred into a linguistic scale and are applied to monitor human emotions valence.

Emotion interpreter testing was performed through non-crossing training and test sets (see Table 1). Based on training sets analysis a several variants of NLHS was created and classification rules for emotion classes were retrieved. The following results (see Table 2) demonstrate the classification rules application to signals test sets.

Table 1. The training and test sets structure (Speech signals/EEG).

Set	All	Class 1	Class 2	Class 3
Training set	120	40	40	40
Test set	150	50	50	50

Table 2. The results of interpretation for emotions valence (Speech Signals/EEG).

Successful of classification%	Training set	Test set
Speech signals		
All	100	95
Class 1	100	92
Class 2	100	96
Class 3	100	98
EEG		
All	100	94
Class 1	100	91
Class 2	100	94
Class 3	100	96

The obtained results (see Table 2) demonstrated an acceptable accuracy for classification of different types of biomedical signals. Combining results showed that in most cases incorrectly classified speech samples and EEG patterns are belong to the same persons.

Thus, results of the work proves effectiveness of discrete attributes generation, based on attractor reconstruction, for emotion's valence differentiation through application of NLHS.

5 Conclusion

The core of software complex (preprocessing module, fuzzification, NLHS generation, extraction of class descriptions from NLHS, fuzzy logic inference) is invariant to subject field thus can be considered as universal feature that can be used for construction of human's emotions interpreters from different biomedical signals.

The conducted research demonstrates the NLHS application possibility for biomedical signals classification recorded in process of subject's stimulation with audiovisual stimuli, which cause emotions of different valence.

The set of generated classification rules, reflecting patterns in NLHS structure, is in most cases corresponds by form with logical conclusions of experts analyzing the same samples.

Further research can be conducted with interpreter detalization by new attributes (such as estimation of human emotional state via electromyography and electrocardiography).

References

1. Rangayyan, R.M.: Biomedical Signal Analysis, 2nd edn. Wiley-IEEE Press, New York (2015)
2. Farhoudi, Z., Setayeshi, S., Rabiee, A.: Using learning automata in brain emotional learning for speech emotion recognition. Int. J. Speech Technol. **20**(3), 553–562 (2017)
3. Abhang, P.A., Gawali, B.W.: Correlation of EEG images and speech signals for emotion analysis. Br. J. Appl. Sci. Technol. **10**(5), 1–13 (2015). https://doi.org/10.9734/BJAST/2015/19000
4. Filatova, N.N., Sidorov, K.V., Iliasov, L.V.: Automated system for analyzing and interpreting nonverbal information. Int. J. Appl. Eng. Res. **10**(24), 45741–45749 (2015)
5. Fortin, P.E., Cooperstock, J.R.: Laughter and tickles: toward novel approaches for emotion and behavior elicitation. IEEE TAC **8**(4), 508–521 (2017)
6. Lan, Z., Sourina, O., Wang, L., Liu, Y.: Real-time EEG-based emotion monitoring using stable features. Vis. Comput. **32**(3), 347–358 (2016). https://doi.org/10.1007/s00371-015-1183-y
7. Filatova, N.N., Sidorov, K.V., Terekhin, S.A.: A software package for interpretation of nonverbal information by analyzing speech patterns or electroencephalogram. Softw. Syst. **111**(3), 22–27 (2015). https://doi.org/10.15827/0236-235x.111.022-027. (in Russia, Programmnye Produkty i Sistemy)
8. Okada, G., Yonezawa, T., Kurita, K., Tsumura, N.: Monitoring emotion by remote measurement of physiological signals using an RGB camera. ITE Trans. MTA **6**(1), 131–137 (2018)
9. Bobkov, A.S., Dmitriev, A.S., Zaboleeva-Zotova, A.V., Orlova, Y., Rozaliev, V.L.: Detailed analysis of postures and gestures for the identification of human emotional reactions. WASJ **24**, 151–158 (2013)
10. Kolesnikov, A.V.: Gibrid intelligent systems: a theory and development technology. St. Petersburg State Polytech. Univ. Publ., St. Petersburg (2001). (in Russia, Gibridnye Intellektualnye Sistemy: Teoriya i Tekhnologiya Razrabotki)
11. Filatova, N.N., Khaneev, D.M.: Use of neurolike structures for automatic generation of hypotheses for classification rules. Fuzzy Syst. Soft Comput. **8**(1), 27–44 (2013). (in Russia, Nechetkie Sistemy i Mjagkie Vychislenija)
12. Khaneev D.M., Filatova N.N.: The pyramidal network for classification of objects, presented by fuzzy features. Eng. Sci. **134**(9), 45–49 (2012). Izvestiya SFedU. (in Russia, Izvestiya SFedU. Tekhnicheskie Nauki)
13. Gladun, V.P.: Growing pyramidal networks. News Artif. Intell. **2**(1), 30–40 (2004). (in Russia, Novosti Iskusstvennogo Intellekta)

14. Filatova, N.N., Khaneev, D.M., Sidorov, K.V.: Signals interpreter based on neural-like hierarchical structure. Softw. Syst. **105**(1), 92–97 (2014). (in Russia, Programmnye Produkty i Sistemy)
15. Rabinovich, M.I., Muezzinoglu, M.K.: Nonlinear dynamics of the brain: emotion and cognition. Adv. Phys. Sci. **180**(4), 371–387 (2010). https://doi.org/10.3367/ufnr.0180.201004b.0371. (in Russia, Uspekhi Fizicheskikh Nauk)
16. Filatova N.N., Sidorov K.V., Shemaev P.D.: Prediction properties of attractors based on their fuzzy trend. In: Abraham, A., et al. (eds.) Proceedings of the Second International Scientific Conference "Intelligent Information Technologies for Industry" (IITI 2017) Advances in Intelligent Systems and Computing, vol. 679, pp. 244–253. Springer, Switzerland (2018). https://doi.org/10.1007/978-3-319-68321-8_25
17. Filatova, N.N., Sidorov, K.V., Terekhin, S.A., Vinogradov, G.P.: The system for the study of the dynamics of human emotional response using fuzzy trends. In: Abraham, A., et al. (eds.) Proceedings of the First International Scientific Conference "Intelligent Information Technologies for Industry" (IITI 2016), Advances in Intelligent Systems and Computing, vol. 451, part III, pp. 175–184. Springer, Switzerland (2016). https://doi.org/10.1007/978-3-319-33816-3_18
18. Filatova, N.N., Sidorov, K.V.: Interpretation of the emotion characteristics through the analysis of attractors reconstructed on EEG signals. Fuzzy Syst. Soft Comput. **11**(1), 57–76 (2016). (in Russia, Nechetkie Sistemy i Mjagkie Vychislenija)

A Vessel's Dead Reckoning Position Estimation by Using of Neural Networks

Victor V. Deryabin(✉) and Anatoly E. Sazonov

Admiral Makarov State University of Maritime and Inland Shipping, Dvinskaya str. 5/7, 198035 Saint-Petersburg, Russian Federation
gmavitder@mail.ru, sazonst@yandex.ru

Abstract. The article aims to prove the feasibility of implementation of a neural network based approach for a vessel's dead reckoning positioning. For this purpose, four dead reckoning algorithms have been developed on the basis of neural networks. Each of these algorithms is characterized by a certain set of navigational equipment used. Training samples are prepared with the use of differential and/or kinematic equations, depending on navigational equipment being used. Testing of the algorithms has been conducted with a vessel's motion modelling, based on solving corresponding differential equations. The parameters of five vessels of significantly different types were used. A vessel's sailing during four hours under wind and wave influence is named a navigational situation. It has been considered 300 such navigational situations. Each situation belongs to one of three classes, characterized by certain time behaviour of control actions and external influences. The results of the testing have shown that neural network based dead reckoning algorithms may be used for calculation of a vessel's trajectory.

Keywords: Vessel's trajectory · Dead reckoning · Neural network

1 Introduction

Traditional methods of dead reckoning may be based on solving differential equations (DE). Forces calculation techniques may be inadequate to the reality. Furthermore, there are errors of numerical integration methods used for solving DE. Inherent inadequacies of the above traditional methods of dead reckoning lead us to find more precise solutions. Probably, such solutions may be found with the use of neural networks, as they are known as universal approximation tools [1–3].

There is a vast amount of literature about neural network technologies implementation for prediction of a ship's motion parameters [4–7]. The general feature of the existing approaches is that there is no such problem formulation that a neural network is intended for a vessel's motion prediction in various navigational situations, so that dead reckoning can be really carried out with the use of the network. In this regard, it is reasonable to find solutions which allow using neural networks in a large variety of sailing conditions.

A. Abraham et al. (Eds.): IITI 2018, AISC 874, pp. 493–502, 2019.
https://doi.org/10.1007/978-3-030-01818-4_49

2 Neural Network Architecture

Let examine dead reckoning positioning in more details. Let assume that Earth's surface area around initial position of a vessel at a point O may be approximated by a tangent plane. The axis Ox is directed off the point O along a meridian to the North, while the axis Oy is directed along a corresponding parallel to the East. Thus, x, y are coordinates of a vessel. They are generally calculated on the basis of a log, gyrocompass, INS, with the use of information about wind, current, wave and other. In the simplest case, gyro heading and a longitudinal component of speed through the water are used. The most suitable neural network architecture seems to be NARX (nonlinear autoregressive with exogenous inputs), as it is described in [8] (See Fig. 1).

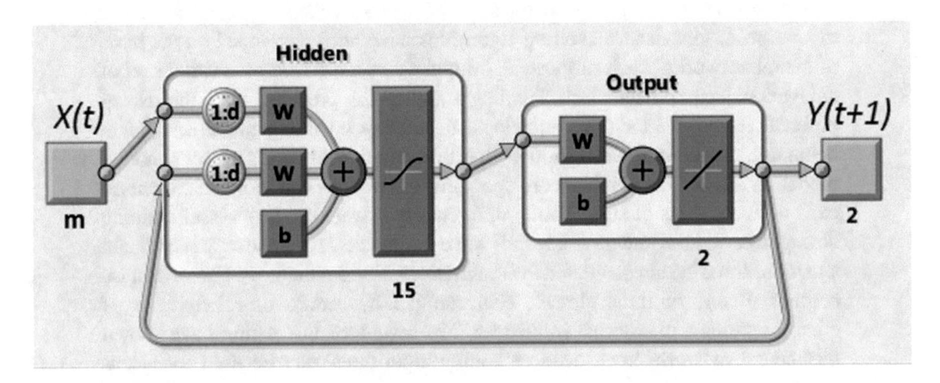

Fig. 1. The general view of the neural network architecture (W is a matrix of weights, b is a bias vector, $X(t)$ is an input signal value at a moment of time t, $Y(t + 1)$ is an output signal value at a moment of time $t + 1$, d is a number of time delays, m is a size of the input vector X).

3 Input/Output Signals, a Number of Delays

3.1 Gyrocompass and Log, Measuring a Longitudinal Component of Speed Through the Water Only (GC + RLOG_1)

The input signal is defined as follows:

$$X = \left(\sin K, \cos K, V_{ox1}, \omega V_{ox1}, \delta, n, V_R^2 \sin \alpha_R, F_{Wy1}\right)^T \tag{1}$$

where K is a vessel's true heading, V_{ox1} is a longitudinal component of speed through the water, ω is a vessel's rate of turn, δ is rudder angle, n is propeller speed of rotation, V_R is speed of relative wind, α_R is direction (as to central plane) of relative wind, F_{Wy1} is force, acting on a vessel's hull from sea regular waves. The last is calculated by integration of pressure over the wet surface of a vessel's hull. In its turn, the pressure is determined on the basis of Bernoulli's equation [9].

The output signal is determined as:

$$Y = \begin{pmatrix} x(t_{k+1}) - x(t_k) - [V_{ox1}(t_{k+1}) \cos K(t_{k+1}) - V_{ox1}(t_k) \cos K(t_k)] \cdot dt \\ y(t_{k+1}) - y(t_k) - [V_{ox1}(t_{k+1}) \sin K(t_{k+1}) - V_{ox1}(t_k) \sin K(t_k)] \cdot dt \end{pmatrix} \quad (2)$$

where k is a current moment of time, dt is a time step of numerical integration that is equal to 1 s.

The number of delays d is chosen to be equal to 2.

3.2 Gyrocompass and Log, Measuring Both Components of Speed Through the Water (GC + RLOG_2)

The input signal is defined as follows:

$$X = \left(\sin(K), \cos(K), V_{ox1}, V_{oy1}\right)^T \quad (3)$$

where V_{oy1} is a transverse component of a vessel's speed through the water.

The output vector is formed with a vessel's coordinates, i.e.

$$Y = (x, y)^T \quad (4)$$

The number of delays d is set to 15.

3.3 Gyrocompass and Log, Measuring Both Components of Speed Over the Ground (GC + TLOG_2)

The input signal is

$$X = \left(V_x, V_y\right)^T \quad (5)$$

where V_x is the northern component of speed over the ground, and V_y is the eastern component.

The output of the network is defined by formula (4). The number of delays d is 5.

3.4 INS with a Stable Platform (INS)

The input signal is

$$X = \left(a_x, a_y\right)^T \quad (6)$$

where a_x and a_y are the northern and the eastern components of a vessel's (her centre of gravity) acceleration vector, respectively.

The output signal is calculated by using the formula (4). This network is a combination of two networks with the structure defined in the Subsect. 3.3. The first neural network predicts the vector (5), taking the input vector (6). The input signal of the second one is (5) while the output is defined with (4).

4 Training Set Forming

This significantly depends on which dead reckoning scheme is used. Let consider all cases like it has been done in the previous section.

4.1 GC + RLOG_1

Term "model situation" means a case of a vessel's motion during a defined period of time, which can be characterized by certain control actions (CA) and external influences (EI). The control actions are formed by a navigator with setting propeller speed of rotation n and rudder angle δ values. For different vessels, limits of δ and, especially, n may differ significantly. Let consider a standard steering gear with rudder angle limits $\pm 35°$. Let assume that propeller is of a fixed pitch. The speed of rotation of the propeller varies between 0 to $n_{\max}$ values. The upper limit $n_{\max}$ may vary significantly for different vessels. Therefore, let speak about N_n, i.e. the number of the values of n, uniformly distributed over the interval $[0;\, n_{\max}]$.

The training set consists of two subsets. The first one is a purely training subset. Samples of this subset are directly used to correct the net weights and biases during a learning process. Samples of the second, testing (validating) subset are used to control a generalization capability of the network during training. The purely training subset is formed in the following way. The interval of values of rudder angle is substituted for a set of these values uniformly distributed with a step of 5° at the interval. That is −35°, −30°… 0, +30°, +35° rudder angle values are used. In its turn, there are values of propeller speed of rotation: $0,\ n_{\max}/(N_n - 1),\ 2 \cdot n_{\max}/(N_n - 1), \ldots, n_{\max}$. Finally, there are "mean" hydro-meteorological conditions within a sailing area. These conditions are defined as follows. True wind speed varies from 15 to 20 m/s, height of waves varies from 5 to 6 m. Wavelength is chosen from 140 to 160 m; wave steepness is assumed to be not more than 0.2. All values of hydro-meteorological parameters are chosen randomly for each case of one-hour sailing, but each case is characterized by a combination of time-constant values of n and δ. All combinations of these values are considered. Variable N_n equals 10. The testing subset is formed in an analogical way. The only difference is that step of variation of rudder angle is set to 7°, and value N_n equals 7. Thus, the number of samples from purely training set is 540150, while the number of testing (validating) samples equals 277277. All input-output samples are formed in accordance with the formulas (1), (2) and are calculated as a result of solving the system of differential equations for a vessel's motion [10].

4.2 GC + RLOG_2

Forming training data set is carried out in two stages. At the first stage, components of the input vector (3) are performed as finite sets of corresponding values, uniformly distributed at their intervals with appropriate steps. Table 1 contains all necessary information for calculation of components of the input vector. At the second stage, the values of elements of created sets are randomly chosen, and the corresponding components of the output vector (4) are calculated by using the following formulas:

$$x(t_{k+1}) = x(t_k) + (V_{ox1}(t_k) \cdot \cos[K(t_k)] - V_{oy1}(t_k) \cdot \sin[K(t_k)]) \cdot dt,$$
$$y(t_{k+1}) = y(t_k) + (V_{ox1}(t_k) \cdot \sin[K(t_k)] + V_{oy1}(t_k) \cdot \cos[K(t_k)]) \cdot dt, \quad (7)$$

In this case, the "etalon model" for a generation of output samples is Euler's numerical integration algorithm (7). There are no samples for testing (validating).

Table 1. Intervals and steps for components of input/output vector of the network.

Name	Symbol	Measurement units	Interval	Step
True heading	K	°	0…360	0.1
Speed through the water (longitudinal)	V_{ox1}	knot	−10…30	0.1
Speed through the water (transversal)	V_{oy1}	knot	−10…10	0.1
Abscissa	x	m	−120.7… +104.9	-
Ordinate	y	m	−112.2… +125.0	-

4.3 GC + TLOG_2

Similarly to the previous case, the method of training samples forming for this network includes two stages. At the first stage, both components of the vector (5) are considered to be elements of the following set: $-15, -15+0.1, -15+0.2, \ldots, +15$ m/s. Then, a time sequence of the input samples is formed as follows. Primarily, a sufficient number of training samples is defined as [11]:

$$N \geq W_b/e \quad (8)$$

where W_b is a total number of weights and biases, e is a percent number of incorrect responses of a network after testing. In the case, W_b is equal to 1397. So, if 1% error level is selected, number of training samples is equal to 139700 according to (8). Thus, each of two sequences has 139700 elements. The time step is equal to 1 s. These elements are randomly chosen from the sequence mentioned above. At the second stage, randomly selected components of the input vector are integrated with Euler's method, and output samples are formed. But there is one particularity. It is extremely desirable that the output samples vary in restricted limits (like abscissa and ordinate in Table 1). This may be achieved by repeating random selection procedure until the output vector components lie in designated limits ($\pm 100 - \pm 150$ m for the case). Two rules are implemented to minimize the number of the repetitions:

- the corresponding components of the input signal are to be a different sign ("plus" and "minus") at any two serial moments of time

– for these moments an absolute value of the difference between absolute values of corresponding components must be not more than a designated value (3 m/s for the case)

Briefly, the methodology of training set forming for the considered network looks as follows: sampling (quantization) of the input signal, its numerical integration, and the process repetition until the output signal lies within designated limits.

4.4 INS

As it was mentioned above, the structure of the network for this case is a combination of two NARX networks with a similar architecture. The training set forming for one network is described in details in the previous subsection. For the other network, training samples are prepared in a similar way. At first, two time sequences of the input vector (6) components with 139700 elements are formed. Then, each element of each sequence is randomly selected from the following set $-5, -5 + 0.05, -5 + 2 \cdot 0.05, \ldots, +5\,\text{m/s}^2$. As noted in the Subsect. 4.3, there are two rules intended for keeping components of the output signal within desired limits ($\pm 15 - \pm 20$ m/s for the case). The maximum absolute value of the difference between absolute values of corresponding components is set to $0.35\,\text{m/s}^2$.

5 Training Process

Firstly, let us consider training procedure for the network mentioned in Subsect. 3.1. Levenberg–Marquardt's method [12, 13, 15] with Bayesian regularization [14] is used as a learning algorithm. Training is carried out in a batch mode and interrupted over every five epochs (iterations). After each interruption, generalization capability of the net is controlled by using samples from the testing (validating) set. The maximum value of the distance between corresponding points of a vessel's trajectories (modelled and predicted) is used as a measure of the network prediction precision. Then, training continues. When it became evident that further learning failed to achieve better results, the training process had been stopped. Time of training varied from 61 to 126 h for ships of different types. The MATLAB software was used as a calculation tool.

For other networks (see Subsects. 3.2, 3.3 and 3.4), Levenberg–Marquardt's training method with Bayesian regularization was implemented as well. The particularity of training is that there is no generalization capability control. Training of the networks is carried out when they are performed in a serial-parallel form, i.e. without a feedback loop. When it became evident that further training would not cause significant performance function decreasing, the training procedure had been stopped. The performance function is the mean squared error.

6 Testing of Neural Networks

As the networks are considered to be used as a vessel's trajectory prediction instruments, it is necessary that their generalization capabilities are to be checked in a large variety of model (navigational) situations. Hydrometeorological situation (HS) is a complex of parameters, describing external influences irrespective of a certain vessel. True wind or wave direction are the examples of the above parameters. Navigational situation (NS) is a certain combination of CA and EI, acting on a vessel's hull. Relative (as to a certain vessel) wind direction and speed, direction and period of waves are the examples of NS parameters. Testing for all existing navigational situations is impossible; therefore, it is reasonable to consider only a limited number of such situations. All navigational situations may be classified on the basis of the behaviour of control and external influences in time. From this point of view, four combinations of the mentioned behaviour are possible: (1) CA and EI are time-independent; (2) CA are time-independent, but EI not; (3) CA are time-dependent, but EI not; (4) CA and EI are time-dependent. The above types of NS may be approximately grouped into three classes, as it is shown in Table 2.

Table 2. The classes of navigational situations

Class number	Characteristics	Rules of navigational situation modelling
I	CA are approximately constant, and EI may change or not (depending on an autopilot tuning)	HS is constant. A vessel is heading stabilized with constant propeller rate of revolution
II	CA are time-independent in mean, and EI change	HS is constant or time-dependent. Propeller revolution rate and rudder angle remain unchanged during sailing
III	CA and EI are time-dependent	HS is constant or time-dependent. Propeller revolution rate and rudder angle vary during sailing

Let us consider HS forming in more details. If HS is time-independent, a random number generator with uniform probability density function will be used to choose HS parameters from their intervals (see Table 3).

Table 3. Parameters of hydrometeorological situation

Name of parameter	Symbol	Measurement unit	Interval
True wind speed	V_{tr}	m/s	[0; 30]
True wind direction	K_{tr}	°	[0; 360)
Height of wave	h	m	[0; 10]
Regular waves direction (from which waves come)	K_w	°	[0; 360)
Length of wave	λ	m	[10; 250]

Regular waves are modelled on the basis of the linear theory [16].

Time-depended wind and waves parameters change as follows:

$$x_m(t) = x_{m0} + A_1 \sin(2\pi t/\tau_1) + A_2 \sin(2\pi t/\tau_2) + A_3 \sin(2\pi t/\tau_3) \quad (9)$$

where $x_m(t)$ is a value of a modelled parameter for a moment of time t, x_{m0} is an initial value of this parameter (at zero time moment), A_1, A_2, A_3 are amplitudes and τ_1, τ_2, τ_3 are periods of the modelled parameter. The amplitudes and periods values are performed in Table 4. If the variable $x_m(t)$ exceeds its limits (see Table 3), it will be equal to the nearest boundary value of the corresponding interval.

Table 4. Amplitudes and periods for harmonics

Parameter name	Symbol	Measurement unit	First harmonic		Second harmonic		Third harmonic	
			A	τ	A	τ	A	τ
True wind speed	V_{tr}	m/s	[0; 30]	2–10 h	[0; 5]	10 min–1 h	[0; 2]	10 s–1 min
True wind direction	K_{tr}	°	[0; 360]	(*)	[0; 30]	(*)	[0; 10]	(*)
Height of waves	h	m	[0; 5]	(**)	-	-	-	-
Regular waves direction	K_w	°	[0; 90]	(***)	-	-	-	-
Length of wave	λ	m	[30; 100]	1–10 h	-	-	-	-

(*) Periods of true wind speed and direction may differ not more than 1 h, 1000 and 30 s for corresponding harmonics.

(**) Waves height period of oscillation may differ from true wind speed first harmonic period, not more than 1 h.

(***) Waves direction period of oscillation may differ from true wind direction first harmonic period, not more than 1 h.

While the third class navigation situation modelling, it is considered that variables, characterizing control actions, changes as follows:

$$\begin{aligned} \delta(t) &= 35^\circ \sin(2\pi t/\tau_\delta), \\ n(t) &= (n_{\max}/2) \cdot (1 + sin(2\pi t/\tau_n)), \end{aligned} \quad (10)$$

where τ_δ is a period for rudder angle oscillation, τ_n is a period for the oscillation of propeller speed of rotation.

Minimum values of τ_δ, τ_n are set in accordance with limitations of the steering gear and propulsive complex, while their maximum values are equal to 8 h.

Neural network functionality has been checked by a vessel's motion modelling in a horizontal plane on the basis of the differential equations [10]. Parameters of five types of ships were used for solving DE. The main characteristics of these vessels are performed in [10].

300 model situations of three classes (see Table 2) have been considered. 60 of situations belong to the first class, 120 belong to the second class, and 120 represent the third class. The time duration of NS was set to four hours. Any navigational situation is modelled with a certain set of parameters which define HS (Tables 3 and 4). Meanwhile, data from Table 4 are only used for NS modelling when HS is time-dependent, i.e. for the second and the third classes only. In each situation, the maximum distance between trajectories, based on DE solving and neural network solutions, was fixed. The results of the tests are presented in Table 5. In this table the maximums of the above-mentioned maximum distances are printed in bold, while their mean values are placed in brackets.

Table 5. The results of the neural networks testing

Dead reckoning scheme	Vessel type				
	Project «232»	Project «B-352»	Project «584E»	«Sevmorput'»	Project «1511»
GC + RLOG_2	**14.2** (2.7) m	**13.9** (3.1) m	**17.1** (3.3) m	**15.8** (3.5) m	**12.8** (2.9) m
GC + TLOG_2	**2.1** (1.3) mm	**2.2** (1.4) mm	**2.4** (1.3) mm	**2.3** (1.3) mm	**2.3** (1.3) mm
INS	**6.9** (2.0) m	**3.7** (1.6) m	**4.0** (1.4) m	**4.5** (1.2) m	**3.7** (1.3) m
GC + RLOG_1	**10.2** (1.6) nautical miles	**9.5** (0.8) nautical miles	**5.3** (0.7) nautical miles	**6.9** (0.9) nautical miles	**5.1** (0.5) nautical miles

7 Conclusions

The main result of this research is that neural network based algorithms have been developed for a vessel's dead reckoning. These have been tested with computer modelling. Samples for training (learning) and testing were obtained with the use of differential and/or kinematic equations. The system of differential equations may be considered as an approximation (model) of the reality and their implementation – as a preparation stage before natural experiments. Herewith, particular attention has been paid to the development of the testing method that lets to consider networks of being working (or not) algorithms for the purposes of a vessel's dead reckoning. At less, it is possible to state that there are dead reckoning algorithms based on the neural networks, working ability of which has been checked in 300 model situations with four-hour sailing at the computer experiment stage. The accuracy of the networks depends significantly on a type of navigational equipment used. For the case with a gyrocompass and log, measuring longitudinal speed only, the network accuracy is low enough, but for the other cases, the accuracy is very good. Results of the additional test for these cases have also shown that the constructed neural network systems are able to determine a vessel's trajectory with high accuracy. Besides one neural network was trained with the samples, obtained with Euler's method (for INS case), the network itself let to determine a ship's trajectory about 286 m (5%) accurately than Euler's method. In other terms, this neural network has found itself as a more accurate algorithm than its "teacher".

As to the further investigations, some directions may be named. First of all, it is necessary to improve the accuracy of a vessel's neural network based dead reckoning systems, which weights and biases depend on physical characteristics of a certain vessel (as it is true for the system in the Subsect. 3.1). For an ideal case, such systems are to be trained on the basis of samples obtained as a result of natural experiments. The key problem is how to conduct those experiments. Secondly, it is reasonable to improve neural net testing methods to minimize the number of navigational situations considered at a test stage.

References

1. Cybenko, G.: Approximation by superpositions of a sigmoidal function. Math. Control Sig. Syst. **2**, 303–314 (1989)
2. Funahashi, K.: On the approximate realization of continuous mappings by neural networks. Neural Netw. **2**(3), 183–192 (1989)
3. Hornik, K., Stinchcombe, M., White, K.: Multilayer feedforward networks are universal approximators. Neural Netw. **2**(5), 359–366 (1989)
4. Ebada, A.: Intelligent techniques-based approach for ship maneuvering simulations and analysis (Artificial Neural Networks Application). Dr.-Ing. genehmigte Dissertation, Institute of Ship Technology und Transport Systems, University Duisburg-Essen, Germany (2007)
5. Waclawek, P.: A neural network to identify ship hydrodynamics coefficients. In: Chislett, M. S. (eds.) Marine Simulation and Ship Maneuverability: Proceedings of the International Conference, MARSIM 1996, Balkema, Rotterdam, pp. 509–514 (1996)
6. Xu, T., Liu, X., Yang, X.: A novel approach for ship trajectory online prediction using bp neural network algorithm. Adv. Inf. Sci. Serv. Sci. (AISS) **4**(11), 271–277 (2012)
7. Deryabin, V.: Adaptive filtering algorithms in vessel's position prediction problem (in Russian). Vestnik Gosudarstvennogo universiteta morskogo i rechnogo flota imeni admirala S.O. Makarova, vol. 1, pp. 12–19 (2014)
8. Haykin, S.: Neural Networks and Learning Machines, 3rd edn. Prentice Hall, New York (2009)
9. Faltinsen, O.: Sea Loads on Ships and Offshore Structures. Cambridge University Press, Cambridge (1999)
10. Deryabin, V.: Model ship traffic above the horizontal plane. Transp. Bus. Russia **6**, 60–67 (2013). (in Russian)
11. Callan, R.: The Essence of Neural Networks. Prentice Hall Europe, London (1999)
12. Levenberg, K.: A method for the solution of certain problems in least squares. Q. Appl. Math. **2**, 164–168 (1944)
13. Marquardt, D.: An algorithm for least-squares estimation of nonlinear parameters. J. Soc. Ind. Appl. Math. **11**(2), 431–441 (1963)
14. Foresee, F., Hagan, M.: Gauss-Newton approximation to Bayesian learning. In: Proceedings of the 1997 IEEE International Conference on Neural Networks, vol. 3, pp. 1930–1935. IEEE, New Jersey (1997)
15. Yu, H.: Advanced Learning Algorithms of Neural Networks. Ph.D. dissertation, Auburn, USA (2011)
16. Newman, J.: Marine Hydrodynamics. The MIT Press, Cambridge (2017)

Evaluation of Neural Network Output Results Reliability in Pattern Recognition

Daniil V. Marshakov(✉), Vasily V. Galushka, Vladimir A. Fathi, and Denis V. Fathi

Don State Technical University, Rostov on Don 344000, Russian Federation
daniil_marshakov@mail.ru

Abstract. Artificial neural networks (ANN) are well-known effective parallel systems which have successfully approved themselves in solving of complicated artificial intelligence problems. The practice of the widespread ANN application due to their high efficiency in solving non-formalized or hard-formalized problems associated with the need for ANN training on experimental material particularly in pattern recognition. In solving problems of pattern recognition the feedforward neural network is usually used due to the simplicity of algorithmic implementation, the availability of advanced training methods, the possibilities of multi-parallel computations. When neural network classifier implementing within decision support systems, it is necessary to assess the ANN results reliability based upon interpretation of the output signals to establish trust between users and the neural network algorithm. In this paper the ANN output results reliability evaluation method in terms of the degree of belonging of the recognized patterns to the originally specified classes is considered. The proposed method is based on the computation of Euclidean distance between the actual ANN output vector characterizing the class of the object recognition and a set of sample vectors defining a priori known classes at the training stage followed by its evaluation by the curve construction of the normal probability distribution law coinciding with the Gaussian function. A feature of this method is the construction of an individual probability distribution curve computed for each ANN output vector. An experimental research of the proposed method in MATLAB is presented on the example of solving the known Fisher's Iris Database classification for input data without noise and with noise. The obtained results confirm the adequacy of the proposed method which can be used both in independent neural network pattern recognition systems and within decision support complexes.

Keywords: Artificial neural network · Pattern recognition · Classification Reliability

1 Introduction

Artificial neural networks (ANN) are well-known effective parallel systems which have successfully approved themselves in solving of complicated artificial intelligence problems. The practice of the widespread ANN application due to their high efficiency

A. Abraham et al. (Eds.): IITI 2018, AISC 874, pp. 503–510, 2019.
https://doi.org/10.1007/978-3-030-01818-4_50

in solving non-formalized or hard to formalize problems associated with the need for ANN training on experimental material particularly in pattern recognition [1, 2].

The basis of the ANN functioning is the possibility of knowledge application given in the learning by the pre-formed training patterns sample to form a neural network of certain decisions. At the same time, in the learning process, the generalized features of the input patterns are distributed in inter-neural connections and the ANN produces a certain internal algorithm for solving the problem. The result achievement process is hidden and "nontransparent" for a user because of the impossibility of justification of generated responses to an input signals value by the neural network. When using ANN as a part of decision support complexes among others the necessary condition for trust and "transparency" between users and a neural network algorithm [3] is an ANN results certainty value based on the interpretation of the output signals.

To bring to the validity of the output solutions of neural networks, methods [4, 5] of bringing neural network algorithms to the form of production rules of the IF-THEN type are known, on the basis of which neuro-fuzzy models are built as well as hybrid models using ANN and neuro-expert system technologies. Neuro-fuzzy models are represented as adaptive functional equivalents of fuzzy inference model. The basis of neuro-expert systems is a knowledge base organized in the form of a neural network the knowledge of which is presented in the form of an adaptive distributed information field.

For this purpose, the following facts should be taken into account. In case of neuro-fuzzy models the system becomes more "transparent" for the analysis of the structure of relations however, the redundancy of information field of the neural network system is reduced, which can negatively affect its functional resistance to destabilizing effects. For neuro-expert systems activation of neural network knowledge base is similar to extraction of rules from the neural system information field of the neural network, which usually requires high computational complexity [6].

In practice, there are cases when the neural network knowledge base is used in its "pure" form without its adaption to the system of rules when the neural network output values are considered as the initial assumptions for the inference engine and the finished rule base of the expert system. However, in this case additional mechanisms for assessing the ANN output states are required.

In this paper we consider estimation of ANN output results reliability to pattern recognition neural network systems regardless of the internal algorithm of their work.

2 Problem of Pattern Recognition by Artificial Neural Network

The pattern in pattern recognition systems is the set of data at the input of the system. In general terms, the problem of pattern recognition is identified with the classification problem, which consists in determining the belonging of the input pattern, represented by the feature vector to one or more deliberately designed classes. The solution of such a problem by a neural network involves the construction of a function that takes the value equal to one in the area of space where the objects of a given class and zero outside this area are located. However, in practice the exact value of the ANN output signal is not of fundamental importance. It is important only to localize the neural network response which serves as an interpretation for the input signal being processed. Herewith, it is predicted that only one of m neural network output signals gives "an active" response and the rest elements of an output vector are passive.

The disadvantage of this ANN output results representation is the lack of neural output activity level evaluation i.e. the degree of pattern belonging to each of the classes. Meanwhile, in actual practice (limited accuracy of objects parameters measurement, strong noise in the input signal, objects movement and others) object recognition signs are changed in random and the input pattern may be close to the boundary values of two or more classes. This situation significantly reduces the recognition quality of the neural network up to possible erroneous classification.

A reasonable way out of this situation is a probabilistic evaluation of the ANN output results [7]. Among the solutions available for this purpose it is possible to emphasize the use of Radial Basis Function Network (RBF) [8] or Probabilistic Neural Network (PNN) [9]. However, in the first case, the evaluation of the reliability of the ANN output results showing the degree of probability of belonging to a certain class, is achieved for neural networks with a well-known function of output data values distribution, while for other neural network solutions the output data statistics will be described by their previously unknown laws of distribution values. In the case of PNN, the Parzen window is used for approximation of unknown probability densities for training samples, which allow solving the problem of probability density estimation according to available data but for the implementation of such algorithms, it is required to use a large enough volume of memory to store training samples.

The most often applicable in practice for solving the problem of neural network pattern recognition is a multilayer perceptron [10, 11] due to its relatively simple algorithmic implementation, the availability of advanced training methods, parallel computing capabilities. ANN type perceptron is also called Multilayer Feed-forward neural network because there are no feedback signals propagate from ANN input to its output through the unification of the neural layers with usually sigmoidal activation functions.

3 Methodology for ANN Output Results Reliability Evaluation

Methods based on the concept of a distance between classes are more often used for multilayer feed-forward neural network distribution for supervised learning. These methods can be classified as comparison methods, K-Nearest-Neighbors method, classification of patterns by the Angels and others.

The peculiarity of these methods is the decisive rules construction on the basis of a training sample providing for the reference classes parameters by which the similarity degree between a priori known classes represented by the set of desired vectors $\{T\}$ and the actual output vector Y ANN can be evaluated.

All observed vectors are represented in a space with some metric [12]. A metric is a property of any set of elements characterized by a function $d(Y, T)$, called distance specified for any pair elements.

One of the most natural distance functions possessing specific geometrical properties and reproducing intuitive distance properties between vectors is Euclidean metric. According to Euclidean metric a distance $d_E(Y, T)$ between an actual output vector $Y = (y_1, y_2, \ldots, y_m)$ and a sampling one $T = (t_1, t_2, \ldots, t_m)$ is determined in correlation:

$$d_E(Y,T) = \sqrt{(y_1 - t_1)^2 + (y_2 - t_2)^2 + \ldots + (y_m - t_m)^2} = \sqrt{\sum_{i=1}^{m} (y_i - t_i)^2}, \quad (1)$$

where m – elements in a vector is equal to the number of recognition classes.

Using the computed distances between actual output vector Y ANN characterizing an object recognition class and a set of sampling vectors $\{T\}$, defining a priori known classes at the stage of training, the degree of similarity between resulting classes can be evaluated. Thus, for ANN output vector having a minimum vector distance defining a priori class, the output result of neural network with most probability can be interpreted as belonging to the appropriate class while the most remote vector characterizes the least probability of this affiliation.

To evaluate the results reliability in this problem the curve of the normal distribution is applied. The normal distribution gives a good model for real phenomena in which there is a strong tendency of data to cluster around the center with a rapid decrease in the frequency of deviations with increasing deviation from the center.

Hence, formation of the assessment of the degree of recognized object belonging to a certain class is possible by constructing a probability density function of the normal distribution which falls with the Gaussian function:

$$f(d) = \frac{1}{\sigma\sqrt{2\pi}} \cdot e^{-\frac{(d-\mu)^2}{2\sigma^2}}, \quad (2)$$

where μ – mathematical expectation, recognizing the probability distribution center; σ – standard deviation; σ^2 – variance, describing the dispersion of data around its mathematical expectation.

The classical distribution curve according to the normal law has got a symmetric bell curve, the maximum ordinate of which is equal to $\frac{1}{\sigma\sqrt{2\pi}}$, correspond to $d = \mu$. With the distance from the point μ the distribution density falls and with $d \rightarrow d_{\max}$ the curve asymptotically tends to X-axis (absciss). $d_{\max}$ is a maximum possible value of Euclidean distance for the definite ANN. Since the area of the distribution of probability for the considered distances is a final value and is in the interval $d \in [0, d_{\max}]$, $\mu = 0$.

The maximum normal distribution density is determined by a numerical value $\frac{1}{\sigma\sqrt{2\pi}}$, however, in the formation of the probabilistic value of the recognized object belonging degree to a particular class, the probability should be distributed in the range of [0, 1].

Because of these observations, the probability density function for this problem-solving will be defined by the formula:

$$f(d) = e^{-\frac{d^2}{2\sigma^2}} \tag{3}$$

Traditionally, the standard deviation is determined through the square root of its variance expressed through the difference of random variable d and its mathematical expectation μ. However, with limited arrays of distance samplings the standard arithmetic mean d_{mean} of the distance set can be used instead of the mathematical expectation in the standard deviation parameter. In this case the formula for calculating the standard deviation for the distribution is as follows:

$$\sigma = \sqrt{\frac{1}{m}\sum_{i=1}^{m}(d_i - d_{mean})^2} \tag{4}$$

For the most complete coverage of the distribution and subsequent formation of an adequate assessment of the ANN output result reliability it is also necessary to include in the sample of m distances the final value of the distribution area, which is the maximum permissible value of the distance between classes defined as:

$$d_{\max} = \sqrt{m} \tag{5}$$

Thus, an arithmetic mean of the samples will be:

$$d_{mean} = \frac{\sum_{i=1}^{m} d_i + d_{\max}}{m+1} \tag{6}$$

Hence, the final formula for calculating a measure of dispersion is as follows

$$\sigma = \sqrt{\frac{1}{m+1}\left[\sum_{i=1}^{m}(d_i - d_{mean})^2 + (d_{\max} - d_{mean})^2\right]} \tag{7}$$

Based on (3) and (7) a Gaussian curve is constructed, where the ordinate axis corresponds to the probability in the interval [0, 1] and the abscissa axis corresponds to the Euclidean distance between ANN output class and the allowable classes.

4 Experimental Part

Consider the example of evaluation of the ANN recognition output results in the solution of the classical problem of "Fisher's Iris Database" classification [11] for vectors without noise and with noise. The original problem includes iris class recognition according to the results of four parameters measures: sepal length and width, petal length and width. Set for training the ANN contains 150 samples belonging to three different classes (Iris setosa, Iris virginica, Iris versicolor) with 50 samples in each class.

Modeling of ANN learning and recognition procedure is performed in MATLAB. The object of the research will adopt multi-layer feedforward neural network with configuration 4-20-3. Sigmoidal activation functions are used as the activation function of each neuron. For ANN training the standard method of backward propagation of error with optimization of weights by means of Gradient descent with momentum and adaptive learning rate backpropagation (traingdx) is used. To estimate the quality of ANN training the sum of error squares with the permissible error 0, 01 is used.

We now show the applicability of the proposed evaluation approach to the reliability of the ANN output result on the example of a specific input vector $X = \{5.0,\ 3.6,\ 1.4,\ 0.2\}$ from the training sample corresponding to the first variety of iris. ANN output result is $Y = \{0.9972,\ 0.0007,\ 0.0000\}$. For illustration purposes the result is given as a value histogram of ANN output signals (Fig. 1a). According to the formula (1) Euclidean distances between an output vector Y and permissible classes which makes up $D = \{0.0028,\ 1.4118,\ 1.4122\}$ are determined. On the basis of formulas (3), (5), (6), (7) the Gaussian curve is constructed and the conformity assessment (Fig. 1b) of ANN output classes to permissible classes is carried out: $f(D) = \{1.0000,\ 0.1081,\ 0.1079\}$.

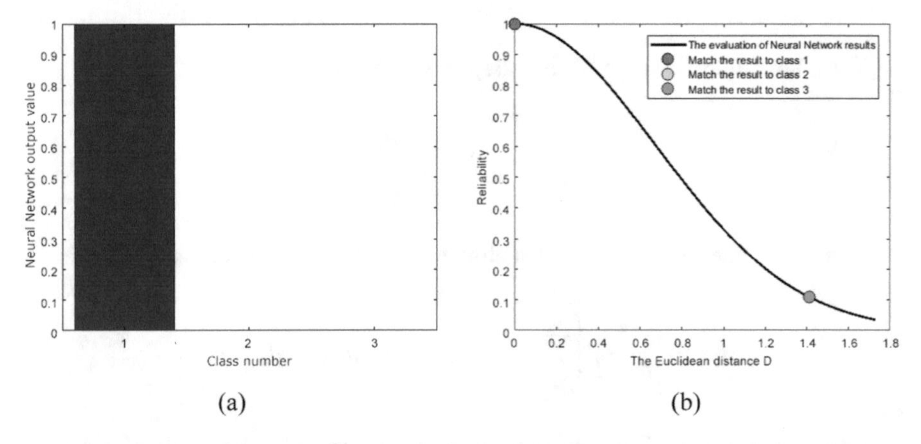

Fig. 1. Analysis of ANN recognition procedure of input vector without noise: (a) ANN output signals value histogram; (b) ANN output class reliability evaluation.

The obtained results reflect the correspondence of the output vector to the first class of irises.

We carry out a similar research for the ANN input vector which is not included in the training sample. For this purpose, use MATLAB to noise the original input vector to reduce the quality of its recognition. After that, ANN output vector takes the form $\hat{Y} = \{0.8586,\ 0.1507,\ 0.0621\}$ and histogram takes the form shown in Fig. 2a. According to the formula (1) Euclidean distances between an output vector $\hat{Y}$ and permissible classes makes up $\hat{D} = \{0.2158,\ 1.2093,\ 1.2805\}$. By performing similar to the first case calculations according to the formulas (3), (5), (6), (7) the conformity evaluation of ANN output classes to permissible classes makes up $f(\hat{D}) = \{0.9268,\ 0.0919,\ 0.0689\}$ (Fig. 2b).

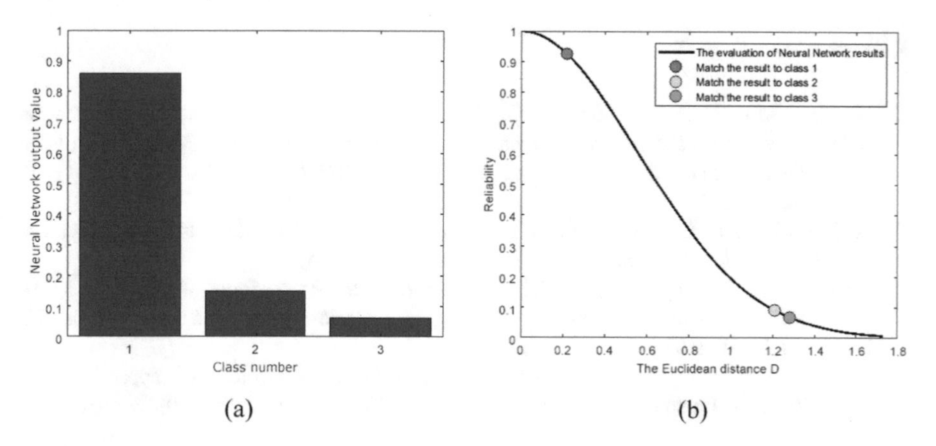

Fig. 2. Analysis of ANN recognition procedure of output vector with noise: (a) ANN output signals value histogram; (b) ANN output class reliability evaluation.

From the obtained results of modeling it is shown that the proposed method of ANN output results estimation provides a sufficient degree of interpretation of the output result of recognition, doesn't make contradictory assessments in the recognition quality evaluation, and therefore is adequate.

5 Conclusions and Further Work

The proposed method of ANN output results estimation provides a sufficient degree of interpretation of the output result of recognition and doesn't make contradictory assessments in the recognition quality evaluation. The peculiarity of this method is construction of an individual probability distribution curve computed for each ANN output vector depending on the deviation degree from the a priori known class. The advantage of this method is that the output result is evaluated depending on the "activity" rate of each of the neural network outputs. However, the method has got some disadvantages including ANN low-level output signals amplification as can be seen in Fig. 1b.

The ANN output results reliability evaluation considered in this paper can be used both in the independent neural network systems of pattern recognition and in complex systems of decision support.

The analysis and comparison along with Euclidean distance taken as a basis in this publication, of other measures for measuring similarity between classes seems promising for the study of the recognition quality assessment. So, among the measures of similarity and distance measurement in vector spaces are also known [12]: patterns comparison on the basis of their correlation, measurement in terms of the guides of the sinuses, weighted similarity measures, Measures of Similarity in the Minkowski Metric and Tanimoto Similarity Measure.

References

1. Khong, L.M.D., Gale, T.J., Jiang, D., Olivier, J.C., Ortiz-Catalan, M.: Multi-layer perceptron training algorithms for pattern recognition of myoelectric signals. In: Proceedings of 6th Biomedical Engineering International Conference (BMEiCON-2013), pp. 1–5. IEEE, Piscataway (2013)
2. Al-Fatlawi, A.H., Ling, S.H., Lam, H.K.: A comparison of neural classifiers for graffiti recognition. J. Intell. Learn. Syst. Appl. **6**(2), 94–112 (2014)
3. Lipton, Z.C.: The mythos of model interpretability. In: Proceedings of the 2016 ICML Workshop on Human Interpretability in Machine Learning (WHI-2016), New York, NY (2016). arXiv:1606.03490
4. Kotenko, I.V., Shorov, A.V., Nesteruk, P.G.: Analysis of bio-inspired approaches for protection of computer systems and networks. SPIIRAS Proc. **3**(18), 19–73 (2011). (in Russian)
5. Benitez, J.M., Castro, J.L., Requena, I.: Are artificial neural networks black boxes? IEEE Trans. Neural Netw. **8**(5), 1156–1164 (1997)
6. Chorowski, J., Zurada, J.M.: Extracting rules from neural networks as decision diagrams. IEEE Trans. Neural Netw. **22**(12), 2435–2446 (2011)
7. Marshakov, D.V.: About the method of probabilistic interpretation for the results of the artificial neural network. In: Proceedings of the VII International Workshop "Sistemnyy analiz, upravleniye i obrabotka informatsii", pp. 102–106. Don State Technical University, Rostov-on-Don, Russian Federation (2016). (in Russian)
8. Kachaykin, E.I., Ivanov, A.I., Bezyaev, A.V., Perfilov, K.A.: Reliability estimation of the automated neural network expertise of authorship hand-written handwriting. Voprosy kiberbezopasnosti **2**(10), 43–48 (2015). (in Russian)
9. Savchenko, A.V.: Statistical pattern recognition based on probabilistic neural network with homogeneity testing. Knowl. Acquis. Autom. Reason. **4**, 45–56 (2013). (in Russian)
10. Chauhan, S., Goel, V., Dhingra, S.: Pattern recognition system using MLP neural networks. IOSR J. Eng. **2**(5), 990–993 (2012)
11. Sharma, L., Sharma, U.: Neural network based classifier (pattern recognition) for classification of iris data set. Int. J. Recent Dev. Eng. Technol. **3**(2), 64–66 (2014)
12. Kohonen, T.: Self-Organizing Maps, 3rd edn. Springer, Heidelberg (2001)

A Neural-Like Hierarchical Structure in the Problem of Automatic Generation of Hypotheses of Rules for Classifying the Objects Specified by Sets of Fuzzy Features

Konstantin V. Sidorov(✉), Natalya N. Filatova(✉), and Pavel D. Shemaev(✉)

Department of Information Technologies, Tver State Technical University, Afanasy Nikitin Quay 22, Tver 170026, Russia
{bmisidorov,nfilatova99}@mail.ru, pshemaev@rambler.ru

Abstract. The article describes a toolkit from the class of hybrid neural network models, which integrates a strategy of generalization and methods of fuzzy data processing. Evolution of properties of these models through the tasks of rules hypotheses construction for the purpose of classification of objects specified by fuzzy features can improve an efficiency of interpreters, designed for signals of bioengineering systems and for other data recorded from medical equipment. Considered the possibility of neural-like hierarchical structure (NLHS) utilization for the task of automatic generation of rules hypotheses for classification of objects determined by discrete features. We propose a hybrid algorithm to solve this problem, which acts as generalization generator, creating descriptions close in form to expert's constructions. The algorithm includes in a composition of classification rules only the most significant features. The apparatus of fuzzy sets can be used for description of feature spaces. This article describes the software which implements the presented algorithm and also demonstrates results of its work with artificial data sets. The Fisher's irises classification task was used to assess program capabilities. The variants of NLHS-based classification rules, created for a training set, considers as hypotheses, which verification is carried out using a test set and an expert analysis of their structure. Options of rules correction are checked with expert's support besides the accuracy and completeness of the corrected rules are also evaluated. The constructed NLHS and corresponding training set is saved as a basic archive explaining formalisms contained in interpreter rules.

Keywords: Algorithm · Classification · Neural-like hierarchical structure Fuzzy set · Fuzzy sign · Production rule · Training set · Test set

The reported study was funded by RFBR according to the research project № 18-37-00225.

A. Abraham et al. (Eds.): IITI 2018, AISC 874, pp. 511–523, 2019.
https://doi.org/10.1007/978-3-030-01818-4_51

1 Introduction

The development of instrumental methods for exploration of a human state as well as creation of bioengineering medical systems is associated with formation problem of rules describing individual classes of patient states, or signals recorded by instruments during a research.

Methods of automatic detection and generating of descriptions of regularities, appearing in sufficiently large sample sets, usually create a structures (decision trees, neural networks, associative rules, etc.) difficult for human perception. An expert, solving the generalization problem, performs 4 operations [1]:

(1) selects common intervals of feature values for specific classes of objects;
(2) transforms this intervals into qualitative scale, forming for them the appropriate qualitative values of features;
(3) forms conjunctions from this qualitative features which are determine the appropriate classes of objects;
(4) if some of the objects lie at the intersection of classes, specifies descriptions of this objects trying to find distinctions between them, and corrects, according to this, the found conjunctions.

Data Mining systems (WizWhy, See 5, etc.) while solving construction task for classification rules, similarly to the expert, perform only the first operation, using the found feature intervals for creation of classification rules, or coefficients correction of neural network-based classifiers. We should address a fact that in this case decent classifiers are created only for tasks with sufficiently representative training sets [2, 3]. However, in medical practice we should to address not only stereotyped situations but also situations characterizing initial stages of disease, when part of signs (symptoms) may not reveal themselves yet.

A feature of Data Mining programs is the use of different probabilistic characteristics for a membership degree assessment of an object lying on classes (C_1 and C_2) intersection. However, this method poorly conform with accepted in medicine formation algorithms of diagnostic conclusion. Specialist of a subject field should to know which signs (symptoms) are separate object from C_1 or C_2, and also to assess tendency of features changing, i.e. to assess a "move direction" of object in feature space: there is a transition from C_1 to C_2 or vice versa.

Obviously, that existed methods of automatic generation of classification (diagnosis) rules are viable only as an instrument for formation of hypothesis [2] that are then should be verified by an experts of subject field and might undergo correction procedure. However, for realization of this strategy, the production rules should be easily interpretable for expert and also be extended by an additional information about properties of the training and test sets.

As far as all results of object analysis and classification should be easily interpretable, to solve such task we need a toolkit from class of hybrid neural network-based models, which integrates strategy of generalization and methods of fuzzy data processing.

2 Features of Data Interpreted by Bioengineering Systems

The objects undergoing analysis in bioengineering systems are characterizing with detailed description formed through both homogeneous and heterogeneous feature systems. However, since the most of initial data's obtained from measuring equipment, we can assume that the initial descriptions of classification objects are created by using numerical characteristics [4, 5]. Moreover, as this might be time series or other functional dependencies 100 and more features can be utilized for their feature representation. In case of utilization of such descriptions for automatic construction of classification/diagnostic rules, the rules will be basing on an interval list of feature values that will complicate their interpretation by an expert.

The expediency of transition to fuzzy variables is also related to an inaccuracy of physical data characterizing biological objects. This might be related to inaccuracy in localization of a signal registration points for different patients, to individual features of symptom's expression, to nonstationarity of characteristics, to subjectivity of feature evaluation methods etc. [1].

In [6, 7], algorithms and a model of a growing pyramidal network are proposed to automatically form concepts (an effective classification rules), determining a functions describing core of interpreted classes. A feature of suggested model is the possibility of each constructed concept to reflect not only structure of the class core but also properties of objects lying on class intersection. The model demonstrated good results in practice with objects with specified quantitative or attributive (categorical) characteristics.

The development of properties of this model for a task of concepts constructon for classes of objects, specified with fuzzy features, can potentially improve an efficiency of interpreters utilizing for signals of bioengineering systems and for other data registered with a medical equipment.

3 The Algorithm of Automatic Generation of Hypotheses of Rules for Classifying the Objects Specified with Fuzzy Features

The algorithm of automatic generation of hypotheses of rules is based on integration of a fuzzy inference mechanism with a neural-like hierarchical structure (NLHS), which reflects relationships between informative features in the descriptions of objects from corresponding classes. The NLHS is realized as oriented acyclic graph in a tiered-parallel form. The main states of growing pyramidal networks model [7] are used for NLHS construction, which provides creation of a well interpretable class descriptions.

The main stages of rules generation algorithm work are:

(1) fuzzification of training set's objects descriptions;
(2) NLHS construction;
(3) parametric adjustment of NLHS on the structure of training set;
(4) formation of descriptions of hypotheses of classification rules;
(5) verification of hypotheses of rules;
(6) correction of hypotheses, the formation of working classification rules.

The fuzzification process can be conducted through utilization of membership functions specified by the expert, or through the formal construction procedure. In the last case the boarders of basic set for an each i value from chosen linguistic variable (LS_k) are determining through the minimum and maximum values of i feature on a set of objects from training set [8, 9]:

$d_{i\max} = \underset{j \in K}{\max}(x_{ji})$; $d_{i\min} = \underset{j \in K}{\min}(x_{ji})$; where K – number of objects in training set; i – feature sequence number; j – training set object's sequence number.

Parameters of membership functions $\{\mu_{T1}(S(f_i)), \mu_{T2}(S(f_i)), \mu_{T3}(S(f_i))\}$ are determining through relations:

$$b = d_{i\min} + 0.5 * (d_{i\max} - d_{i\min}); \Delta = 0.1 * (d_{i\max} - d_{i\min}); d_{in} = d_{i\min} + \Delta; d_{ix} = d_{i\max} - \Delta.$$

Then, (by using 3 terms for all linguistic variables) the description of each object, defined by a vector $X = \{x_1, x_2, \ldots, x_i, \ldots, x_m\}$, from the training set will be represented by array X_t of the following form (Fig. 1).

X_t	T_1	T_2	T_3
x_1	μ_{11}	μ_{12}	μ_{13}
x_2	μ_{21}	μ_{22}	μ_{23}
...	...	...	...
x_m	μ_{m1}	μ_{m2}	μ_{m3}

Fig. 1. X_t array description.

In the first step, the NLHS is represented by a graph of the following form:

$$G^0 = <R, S, Y>, \quad (1)$$

where R is a set of receptors for entering objects into the network; Y is a final peak of top level, performing a service role; S is a set of arcs directed from receptors to the Y peak.

The NLHS construction algorithm works with the training set (V_1), all objects of which are described through linguistic variables and corresponding term-sets (T_i). Each term is associated with one receptor, i.e. if $\forall j \| T_j \| = n$, m – number of linguistic variables, the number of receptors in the NLHS will be $nr = n * m$. For each term-set the fuzzification algorithm based on calculation of t-norm determines a receptor which induces an output signal corresponding to the value of the membership function.

For a new object $e_1 \in V_1$ in form $e_1 = <xt_1 \backslash \mu_{11}, xt_2 \backslash \mu_{22}>$ a subset of excited elements is singled out from R. From S a subset of arcs (KS) is singled out to connect this elements with Y: $(S(r_{11}, Y), S(r_{22}, Y)) \in KS$.

From the initial set of arcs, links with the excited receptors are eliminated and a new, so-called. associative peak v_1 is introduced in NLHS. It is connected by incoming arcs to *KS* receptors, and its outgoing arc is connected to the *Y* input. Thus, a set of arcs and peaks of the NLHS is corrected: $S_0 = S \backslash KS$, $S_1 = S_0 \cup S(t_{11}, v_1) \cup S(t_{22}, v_1) \cup S(v_1, Y)$.

The constructed graph NLHS G^0 (1) further takes the following form:

$$G^1 = <R, S, Y, A> , \quad (2)$$

where *A* is a set of associative elements $(v_1 \in A)$.

Following introduction of all other objects from the training set, the set *A* expands, which automatically implies a change in *S*. Each associative element $(v_j \in A)$ corresponds to conjunctive dependence of the features values determined by receptors, which has outgoing path to v_j. The rules of introduction of associative peaks into the network for various situations are described in detail in [6, 7], and a formation algorithm of the NLHS structure is described in [10, 11].

A distinctive feature of the graph constructed by NLHS is its hierarchical structure. If in G^1 (2), we take the service peak *Y* as a zero level, then peaks of a first level form a subset $VS \subseteq A$, which determines the values of features extracted for description of classes represented in the training set. Each element *VS* is a peak of some fragment from G^1 graph: $G^1 = \cup GS_i$, $i = 1, 2, \ldots, p$, $||VS|| = p$, $(\forall i) GS_i = <AS_i, SS_i, RS_i>$, $(\exists j) GS_i \cap GS_j \neq 0$.

The analysis of the constructed graph is performed within the framework of the third stage of rules generation algorithm. The parametric adjustment of the NLHS includes allocation of a subset of so-called control elements in the *A* set, and evaluation of the *A* parameters. The control element is an associative element with the most common set of feature values from one of the classes. The parameters of $v_j \in A$ are: (1) a network hierarchy level; (2) state (excited/not excited); (3) a parenting object's class from the training set $(e_1 \in V_1)$ for this associative element; (4) a number of receptors (*l*) in $cl1 - clk$ classes pyramid with outputs to associative element (v_j); (5) values of excitation counters (counters $m_{cl1} - m_{clk}$ determine the response of peak to input objects of classes $cl1 - clk$, respectively, where *k* is a number of classes represented in the training set).

The NLHS configuration algorithm is based on reexamination of objects from training set. When an object e_j from scope of the notion of class Class 1 is entered, the receptors (R_i) from object description become excited as well as associative elements (A_i) with incoming paths from peaks of R_i. For each peak in the sets R_i and A_i, the parameters m_{cl-1} are defined as follows:

$$m_{cl-1}(i) = m_{cl-1}(i) + \mu(i), \quad (3)$$

where $m_{cl-1}(i)$ is a coefficient characterizing the number of excitations of *i-th* peak when the objects from Class 1 are entered. The estimation of $m_{cl\text{-}1}$ (3) is based on the values of membership functions of the corresponding features for training set's objects.

Loaded with parameters, the NLHS is a complete classifier and can serve as the basis to form production rules. The algorithm for hypotheses generation of classification rules is based on examination of NLHS fragments, each of which starts at one of the control peaks.

Let us consider the features of the algorithm in a following example. Suppose that there is a network fragment with a peak KV^C characterizing class C (Fig. 2). A set of receptors, with outgoing path to KV^C is selected. Each receptor corresponds to the value of a linguistic variable.

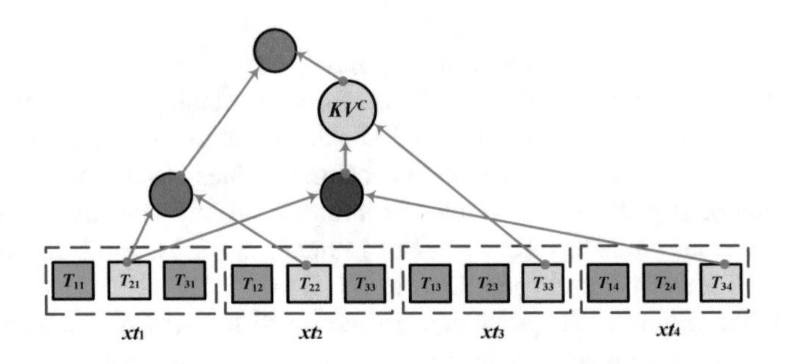

Fig. 2. NLHS fragment with control peak KV^C characterizing class C.

Each feature in example is presented by linguistic variable receiving values on a term set $xt_i = \{T_{1i}, T_{2i}, T_{3i}\}$ (see Fig. 2). The NLHS fragment exploration allows to generate a logical expression of the form:

$$RS(KV^C) = \{xt_1 \text{ is } T_{21} \wedge xt_2 \text{ is } T_{22} \wedge xt_3 \text{ is } T_{33} \wedge xt_4 \text{ is } T_{34}\}. \quad (4)$$

A conjunctive utterance defines a stable combination of the feature values of class C objects set. Examination of all class C control elements allows to select the set of conjunctive utterance $\{RS(KV^C)\}$ of the form (4). Basing on them the rule that allows to separate class C objects from other classes is generating:

$$(RS(KV_1^C)) \vee RS(KV_2^C)) \vee \ldots RS(KV_n^C)) \rightarrow \text{Class is } C. \quad (5)$$

The constructed NLHS-based variants of classification (diagnosis) rules are considered as hypotheses which verification is performed through test set and expert analysis of their structure. With the help of the expert, we can check the options for correcting the rules (5) and evaluate an accuracy and completeness of corrected rules. The working variants of classification rules are utilized during a program development for interpretation of results of instrumental research. The constructed NLHS and corresponding training set are stored as a basic archive explaining the formalisms laid down in the rules of interpreter or expert system.

4 An Exploration of Algorithm for Automatic Generation of Classification Rules for Objects Specified by Discrete Features

To analyze the reviewed algorithm for generating of classification rules we used artificial data sets. Since the main task was to construct a generator of generalizations that creates descriptions close in form to expert's logical constructions, the well-known control task of irises classification (Fisher's Irises task [12, 13]) was used to evaluate a program's capabilities. The Fisher's Irises task is distinguished by small dimension of the feature space and their homogeneous nature. Fisher's irises are consist of data on 150 copies of iris, 50 copies from 3 varieties (classes) – Iris Setosa (Class 1), Iris Virginica (Class 2), Iris Versicolor (Class 3). Each object is represented by features of S_i form:

- 4 features characterizing morphological features of objects (in cm: S_1 – sepal length; S_2 – sepal width; S_3 – petal length; S_4 – petal width);
- 1 separation feature (S_5 – variety of iris).

A training and test sets including 25 object for each class are generated.

The analysis of training set using basic scales of features showed that one of the classes (Class 1) could be linearly divide with another two, and two classes Class 2 and Class 3 are having intersection in all features S_1–S_4 (Fig. 3).

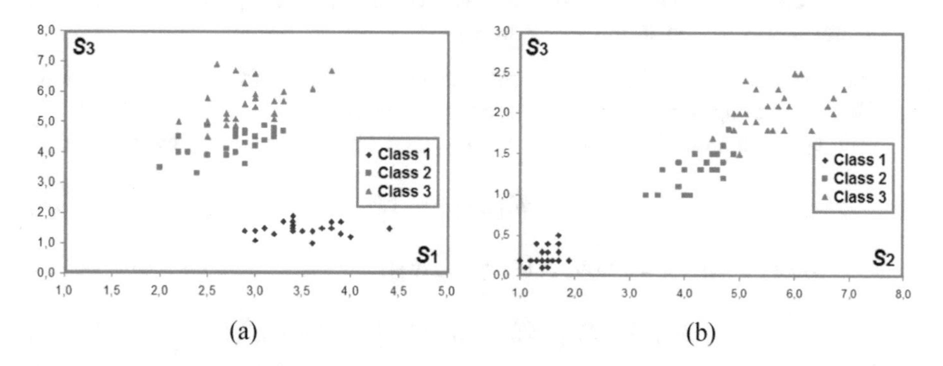

Fig. 3. Distribution of objects from training set in irises classification task: a – basic scales of features S_1 and S_3; b – basic scales of features S_2 and S_3.

The descriptions fuzzification of the training set's objects is conducted with one term-set for all features: LOW (l) – low feature value, MID (m) – middle feature value, HIGH (h) – high feature value. The transition to linguistic variables (Table 1) clearly reveals features of the training set:

- Class 1 objects are distinguished by the S_1 and S_2 features, besides the estimations of membership functions μ_i of corresponding fuzzy variables are showing different expression power of these properties and vary significantly (0.55–0.99). All objects

Table 1. Fuzzification of the training set.

Class	Fuzzy feature	Percentage of classes in the training set, %			
		S_1	S_2	S_3	S_4
Class 1	l	88	–	100	100
	m	12	56	–	–
	h	–	44	–	–
Class 2	l	12	56	–	–
	m	72	44	100	96
	h	16	–	–	4
Class 3	l	4	44	–	–
	m	44	48	12	4
	h	52	8	88	96

from this class in the training set have low ("LOW") feature S_3 and S_4 values with equal value of membership functions $\mu = 0.992$;

- most of objects from Class 2 (96%) have medium feature S_3 and S_4 values, however those features are showing themselves in varying rate (the membership function μ_i in the range of 0.5–0.9);
- about 70% of objects from Class 3 have "HIGH" S_3 and S_4 feature values with membership function μ_i not less than 0.7.

The application of NLHS-based rule generator (Fig. 4) for discussed training set allowed creating logic functions describing three classes of irises.

With consideration of template (5) we acquired implications of following form:

$$\text{Rule 1} :: RS\left(KV_1^1\right) \rightarrow \text{Class 1}, \tag{6}$$

in case $RS\left(KV_1^1\right) = ((S_3 = \text{LOW}) \wedge (S_4 = \text{LOW}))$;

$$\text{Rule 2} :: RS\left(KV_1^2\right) \vee RS\left(KV_2^2\right) \rightarrow \text{Class 2}, \tag{7}$$

in case $RS\left(KV_1^2\right) = ((S_3 = \text{MID}) \wedge (S_4 = \text{MID}))$, $RS\left(KV_2^2\right) = ((S_1 = \text{MID}) \wedge (S_2 = \text{MID}) \wedge (S_3 = \text{MID}))$;

$$\text{Rule 3} :: RS\left(KV_1^3\right) \vee RS\left(KV_2^3\right) \rightarrow \text{Class 3}, \tag{8}$$

in case $RS\left(KV_1^3\right) = ((S_3 = \text{HIGH}) \wedge (S_4 = \text{HIGH}))$, $RS\left(KV_2^3\right) = ((S_2 = \text{LOW}) \wedge (S_4 = \text{HIGH}))$.

Estimation of the rules accuracy demonstrated zero classification error for all three classes on the training set. It should be noted that such result can be achieved only by using a fuzzy logic inference for calculation conorm disjunction of following form $(\mu_{a \cup b} = \mu_a + \mu_b - \mu_a * \mu_b)$.

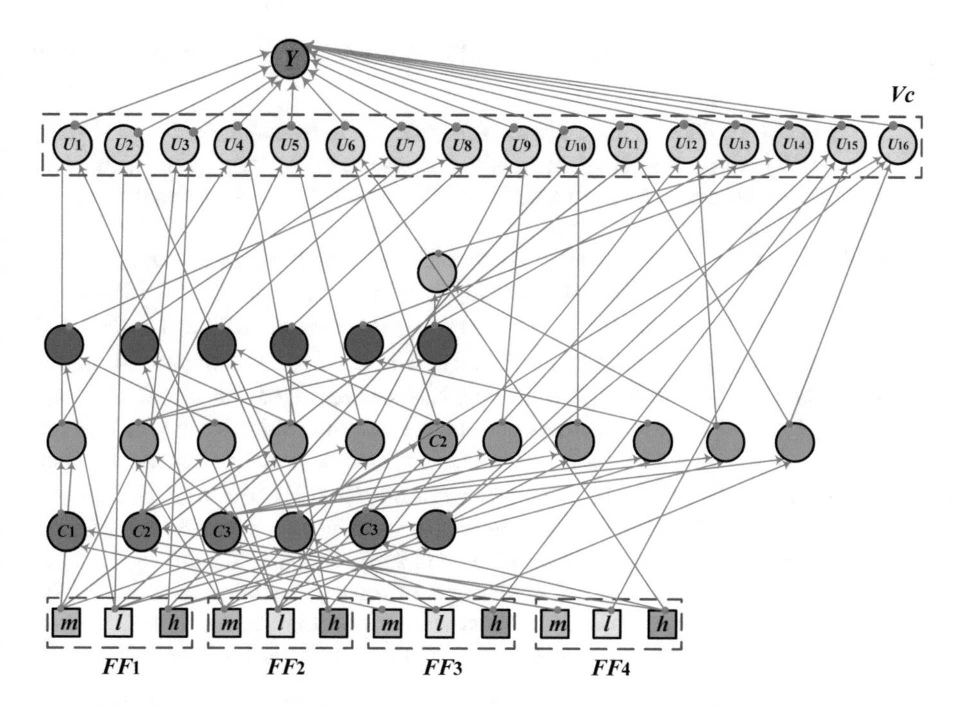

Fig. 4. NLHS structure for Fisher's Iris classification task (FF_1–FF_4 – fuzzy attributes).

All objects from Class 1 according to rule (6) are classified with $\mu_i \approx 1$ (Fig. 5), while belonging of 44% of objects from the training set, to Class 3 the rule (8) evaluates with $\mu_i \geq 0.9$. The significant lower values of μ_i, obtained by the rule (7) during classification of objects from Class 2, are explained not so much by the composition of the set as by triangular form of the membership function for terms of the "MID" type.

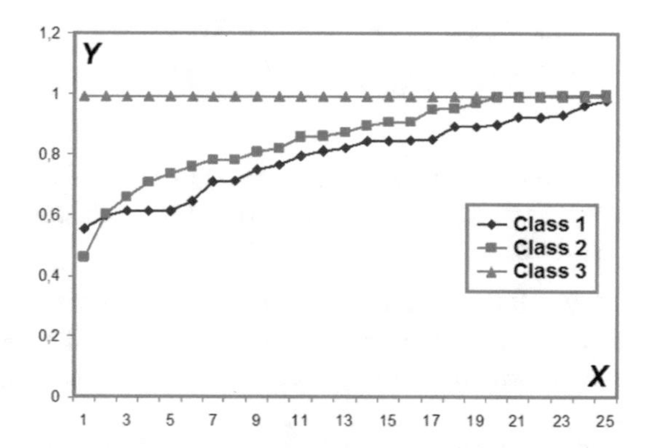

Fig. 5. The membership functions for training set's objects (Rules 1–3 (6–8)): X – the object's number; Y – the membership function μ_i of object to class.

The expert analysis of the rules (6–8) structure allows to make following conclusions:

- The rule (6) for Class 1 determination includes a single condition ($S_3 = \text{LOW} \wedge S_4 = \text{LOW}$) which is well conform with the results of the training set's structure analysis executed by the expert (see Table 1). The completeness of the rule, on the training set, is 100%;
- The rule (7) for Class 2 determination includes two alternative conditions and according to recommendations of developers of expert systems [2, 14, 15], can be converted into 2 implications with the same right-hand side: Rule 2_1 :: $RS(KV_1^2) \rightarrow$ Class 2; Rule 2_2 :: $RS(KV_2^2) \rightarrow$ Class 2. The Rule 2_1 determines 24 objects from Class 2 while Rule 2_2–8 objects. Intersection of the rules area of action reveals 7 objects (*SL* subset) that could be correctly classified by any of this two rules. If we use Rule 2_1 for *SL* classification, then only one object will be identified by using $RS(KV_2^2)$. The completeness of rule Rule 2_1 is 96%, Rule 2_2–4%;
- The rule (8) for Class 3 determination is also includes two alternative conditions. Similar reasoning allows us to assess the completeness of the Rule 3_1–84%. Since there is also an intersection of the rules area of action, then after exception of *SL*-type objects from the rule's Rule 3_2 area of action its completeness is 16%. In fact, the condition $RS(KV_2^3)$ identifies 4 objects lying on the class boundary.

The estimation of the rules accuracy with the test set revealed a classification error of 6.6% (for rule (7) – 1 mismatch; for rule (8) – 4 mismatches). By using logical function "AND" of triangular norm like a max $(0, \mu_a + \mu_{b-1})$ for calculation of membership function the classification error is 5%. In general, with the test set, the generated hypotheses of rules are also providing high accuracy and completeness (Table 2).

Table 2. The classification parameters for test set.

Rule/Class	Accuracy	Completeness
Rule 1/Class 1	1 (100%)	1 (100%)
Rule 2/Class 2	0.857 (85.7%)	0.96 (96%)
Rule 3/Class 3	0.955 (95.5%)	0.84 (84%)

The analysis of hypotheses of the rules (6–8) and of results of their verification, performed by an expert, showed the necessity in correction of classification rules which is justified by several facts.

(1) Rules Rule 2_1 and Rule 3_1 are highlighting the concentration areas of objects from corresponding classes. Objects from the test set, classified with an error, are located on the boundaries (or abroad) of action areas of this rules and form the *Err* set (Table 3).
(2) The NLHS-classifier selected S_3 and S_4 features as most significant.

(3) The comparison of linguistic descriptions of the objects from *Err* set showed that all features are determining with a fairly high correspondence degree $(\mu_i > 0.6)$.

Table 3. The test set's objects classified with an error.

Rule	S_1	S_2	S_3	S_4
Class 2_34	MID/0.987	LOW/0.649	HIGH/0.601	MID/0.601
Class 3_28	MID/0.899	MID/0.737	MID/0.987	HIGH/0.642
Class 3_34	MID/0.721	LOW/0.520	HIGH/0.601	MID/0.736
Class 3_35	MID/0.899	LOW/0.778	HIGH/0.857	MID/0.862
Class 3_39	MID/0.987	MID/0.737	MID/0.987	HIGH/0.642

An attempt to interpret them as objects from one of the classes leads to small values of μ_i. So, if the object $\{\text{Class 3\_28}\} \in \text{Class 3}$, then $\mu_i = 0.141$. As result, we can assume that they really take an intermediate position between Class 2 and Class 3. If we proceed from analogy with diagnostic rules, the *Err* set can be considered as a description of initial stage of disease corresponding to Class 3. In this case, the NLHS can be supplemented with pyramids corresponding to descriptions of objects from *Err* set:

Table 4. The comparisons of NLHS-classifier work results and expert conclusions.

Rule's name	Formalized representation	Verbal interpretation
Rule 1	$(S_3 = \text{LOW} \wedge S_4 = \text{LOW})$ → **Class 1**	IF (Petal length = "Low") AND (Petal width = "Low") THEN **Class 1**
Rule 2	$((S_3 = \text{MID} \wedge S_4 = \text{MID}) \vee (S_1 = \text{MID} \wedge S_2 = \text{MID} \wedge S_3 = \text{MID}))$ → **Class 2**	IF [(Petal length = "Middle") AND (Petal width = "Middle")] OR [(Sepal length = "Middle") AND (Sepal width = "Middle") AND (Petal length = "Middle")] THEN **Class 2**
Rule 3	$((S_3 = \text{HIGH} \wedge S_4 = \text{HIGH}) \vee (S_2 = \text{LOW} \wedge S_4 = \text{HIGH}))$ → **Class 3**	IF [(Petal length = "High") AND (Petal width = "High")] OR [(Sepal width = "Low") AND (Petal width = "High")] THEN **Class 3**
Rule 4	$((S_1 = \text{MID} \wedge S_2 = \text{MID} \wedge S_3 = \text{MID} \wedge S_4 = \text{HIGH}) \vee (S_1 = \text{MID} \wedge S_2 = \text{LOW} \wedge S_3 = \text{HIGH} \wedge S_4 = \text{MID}))$ → **Class 4**	IF [(Sepal length = "Middle") AND (Sepal width = "Middle") AND (Petal length = "Middle") AND (Petal width = "High")] OR [(Sepal length = "Middle") AND (Sepal width = "Low") AND (Petal length = "High") AND (Petal width = "Middle")] THEN **Class 4**

$$\text{Rule 4} :: RS\left(KV_1^4\right) \vee RS\left(KV_2^4\right) \rightarrow \text{Class 4}, \tag{9}$$

in case $RS\left(KV_1^4\right) = ((S_1 = \text{MID}) \wedge (S_2 = \text{MID}) \wedge (S_3 = \text{MID}) \wedge (S_4 = \text{HIGH}))$, $RS\left(KV_2^4\right) = ((S_1 = \text{MID}) \wedge (S_2 = \text{LOW}) \wedge (S_3 = \text{HIGH}) \wedge (S_4 = \text{MID}))$.

The conducted researches experimentally confirmed a decent capability of the NLHS-classifier as productions generator for the task of objects classification specified by independent discrete features. Unlike models of growing pyramidal networks the new algorithm incorporates (with the help of membership functions), during a parametric configuration of network, the expression power of features in the descriptions of the training set's objects. The created rules do not require additional interpretation; in fact, they are immediately obtained in a form close to the verbal representation (Table 4). The structure of the rules (6–9), constructed by NLHS for the training set, fully conform with logical conclusions of the expert which have been made during analysis of the same set.

5 Conclusion

The NLHS-classifier algorithm is implemented in C# 3.0 for the .NET Framework 3.5 and later execution environment. The program is serve as a tool for construction of production models determining classes of objects specified by sets of discrete features.

Further experimental studies with the NLHS will be oriented on tasks of a graphical dependencies classification representing examples of the time series from biomedical signals (electroencephalogram, electromyogram, rheoencephalogram) with fuzzy estimates of ordinates characterizing changes in valence and human emotional reactions level during audiovisual stimulation.

References

1. Filatova, N.N., Dmitriev, G.A., Grigorieva, O.M.: Methods and algorithms for classifying graphic objects in problems of medical diagnostics. Tver State Technical University Publisher, Tver (2011). (in Russian, Metody i Algoritmy Klassifikatcii Graficheskikh Obektov v Zadachakh Meditcinskoi Diagnostiki)
2. Vagin, V.N., Golovina, E.Ju., Zagorjanskaja, A.A., Fomina, M.V.: Reliable and Plausible Conclusion in Intellectual Systems, 2nd edn. Fizmatlit Publisher, Moscow (2008). (in Russian, Dostovernyj i Pravdopodobnyj Vyvod v Intellektual'nyh Sistemah)
3. Yarushkina, N.G., Afanaseva, T.V., Perfileva, I.G.: Intelligent Time Series Analysis: Textbook. Ulyanovsk State Technical University Publisher, Ulyanovsk (2010). (in Russian, Intellektual'nyy Analiz Vremennykh Ryadov: Uchebnoe Posobie)
4. Loskutov, A.Yu.: Time Series Analysis: Lectures. Moscow State University Publisher, Moscow (2006). (in Russian, Analiz Vremennykh Ryadov: Kurs Lektsiy)
5. Ifeachor, E.C., Jervis, B.W.: Digital Signal Processing: A Practical Approach, 2nd edn. Pearson Education, London (2002)

6. Gladun, V.P.: Partnership with Computers. Man-Machine Purposeful Systems. Port-Royal Publisher, Kiev (2000). (in Russian, Partnerstvo s Kompiuterom. Cheloveko-Mashinnye Tceleustremlennye Sistemy)
7. Gladun, V.P.: The growing pyramidal networks. Artif. Intell. News **1**, 30–40 (2004). (in Russian, Novosti Iskusstvennogo Intellekta)
8. Khaneev, D.M., Filatova, N.N.: The pyramidal network for classification of objects, presented by fuzzy features. Izvestiya SFedU. Eng. Sci. **134**(9), 45–49 (2012). (in Russian, Izvestiya SFedU. Tekhnicheskie Nauki)
9. Filatova, N.N., Khaneev, D.M.: Use of neuro-like structures for automatic generation of hypotheses for classification rules. Fuzzy Syst. Soft Comput. **8**(1), 27–44 (2013). (in Russian, Nechetkie Sistemy i Mjagkie Vychislenija)
10. Filatova, N.N., Khaneev, D.M., Sidorov, K.V.: Signals interpreter based on neural-like hierarchical structure. Softw. Syst. **105**(1), 92–97 (2014). (in Russian, Programmnye Produkty i Sistemy)
11. Filatova, N.N., Khaneev, D.M., Sidorov, K.V.: Diagrams classification algorithm with consequent enlarging of features. Softw. Syst. **107**(3), 78–86 (2014). (in Russian, Programmnye Produkty i Sistemy)
12. Fisher, R.A.: The use of multiple measurements in taxonomic problems. Ann. Eugen. **7**, 179–188 (1936)
13. UCI Machine Learning Repository: Iris Data Set. http://archive.ics.uci.edu/ml/datasets/Iris. Accessed 05 May 2018
14. Marcellus, D.H.: Expert Systems Programming in Turbo Prolog. Prentice Hall Inc., Upper Saddle River (1989)
15. Kolesnikov, A.V.: Gibrid Intelligent Systems: A Theory and Development Technology. St. Petersburg State Polytechnic University Publisher, St. Petersburg (2001). (in Russian, Gibridnye Intellektualnye Sistemy: Teoriya i Tekhnologiya Razrabotki)

Author Index

A. Abraham et al. (Eds.): IITI 2018, AISC 874, pp. 525–527, 2019.
https://doi.org/10.1007/978-3-030-01818-4

Printed in the United States
By Bookmasters